Geophysical Monograph Series
American Geophysical Union

Geophysical Monograph Series
A. F. Spilhaus, Jr., Managing Editor

geophysical monograph 24

Mechanical Behavior of Crustal Rocks

The Handin Volume

N. L. Carter
M. Friedman
J. M. Logan
D. W. Stearns
editors

American Geophysical Union
Washington, D.C.
1981

American Geophysical Union

Published under the aegis of the AGU
Geophysical Monograph Board:
Thomas E. Graedel, Chairman;
Donald H. Eckhardt, William I. Rose, Jr.,
Rob Van der Voo, Ray F. Weiss, William R. Winkler

Library of Congress Cataloging in Publication Data

Main entry under title:

Mechanical behavior of crustal rocks.

(Geophysical monograph; no. 24)
A collection of papers in honor of John Handin.
Includes bibliographies.
1. Rock deformation—Addresses, essays, lectures. 2.
Earth—Crust—Addresses, essays, lectures. 3. Geodynamics—Addresses, essays, lectures. 4. Handin, John. I.
Handin, John. II. Carter, N. L. (Neville Louis), 1934- .
III. American Geophysical Union. IV. Series.
QE604.M42   551.8   81-4626
ISBN 0-87590-024-0   AACR2

Published by

American Geophysical Union
2000 Florida Avenue, N.W.
Washington, D.C. 20009

Printed in the United States of America

CONTENTS

Mechanical Behavior of Crustal Rocks

The Handin Volume

John Walter Handin

John Walter Handin was born June 27, 1919, in Salt Lake City, but his family moved soon thereafter to west Los Angeles, then a lovely and exciting place in which to grow up. The Santa Monica Mountains were a wilderness to be explored, and the Pacific surf could be dared for a dime's carfare.

Handin's paternal grandparents came from Sweden on that vast crest of immigration during the latter half of the nineteenth century. His maternal grandparents descended from English pilgrims who settled New England early in the seventeenth century. His parents had no formal education beyond high school; however, despite much hardship, they encouraged him to go on to college during the great depression.

Handin wanted very much to study linguistics because instruction in his high school had been superior in English and foreign languages, but mediocre in mathematics, chemistry, and physics, and utterly ignored in earth science. However, the pragmatism of the time led him to civil engineering at UCLA. As his professor in the required course in geology he luckily drew William C. Putnam, an inspiring teacher and remarkable human being, who taught his students to believe as did Charles Darwin that 'geology carries the day.' But the world was in turmoil and devotion to scholarship all but impossible. Handin was awarded the bachelor's degree in geology without distinction and promptly went to war in 1941 as a 2nd lieutenant in the Coast Artillery Corps. Later converted to field artillery, his battalion fought for the liberation of the Philippine Islands and in the battle for Okinawa; he is now a retired Lt. Colonel in the U. S. Army Reserve. His interest in geology having been sustained by regular correspondence with Putnam during the war, he returned to UCLA five years later. Its geology department had suddenly become outstanding by the addition of Milton Bramlette, James Gilluly, and George Tunell to its already good faculty. Handin had intended to train for a career in exploration geophysics in the oil industry, finishing with the master's degree in geology and the minor in physics. However, in 1947 he married Frances Robertson who persuaded him to continue toward the doctorate and worked to provide their livelihood as well, for in those days the generous financial support, taken for granted by today's graduate students, was unknown.

Handin became infected by U. S. Grant's enthusiasm for sedimentological research and earned his doctorate in 1949 for work on the source, transportation, and deposition of beach sands. During the course of his study he worked part-time for Wally Pentegoff of the U. S. Army Corps of Engineers and acquired a lifelong interest in engineering geology. Meanwhile, David Griggs had joined the Institute of Geophysics, and knowing of him from readings in Gilluly's courses, Handin took the first seminar that Griggs offered at UCLA—an experience at once humbling, provoking, and loving. After years of groping, Handin abruptly discovered what to him was the new world of tectonophysics. In Grigg's words, 'John made this field his own, and started on his remarkable career.' He remained at the Institute as a postdoctoral fellow for a year, working with Griggs in the high-pressure laboratory and Frank Turner of the University of California at Berkeley on dynamic petrofabric analysis of experimentally deformed marble.

In 1950, Handin accompanied Griggs to the annual meeting of the American Geophysical Union, where he presented a scientific paper for the first time and met Francis Birch, Walter Bucher, Maurice Ewing, Harry Hess, King Hubbert, and William Rubey, a most exhilarating experience for a young researcher. He learned of Hubbert's plan for a research program in structural geology at Shell's Exploration and Production Research Laboratory in Houston. There were only two very high-pressure (ca. 1 GPa) rock-deformation facilities in the country: at UCLA (Griggs) and Harvard (F. Birch, P. W. Bridgman, and later, Gene Robertson). Because federal funding of academic research was very small (except in nuclear physics), the opportunity to build a relatively expensive high-pressure laboratory was unique. Handin eagerly grasped it and he joined Shell in late 1950. Hubbert and Rubey became his lifelong friends along with Gilluly and Griggs and, by example, these four men largely shaped Handin's professional life.

At Shell, Handin created a laboratory for experimental rock deformation that was to become world famous, and he adopted and perfected the theoretical, experimental, and observational investigative approach that has become the 'new' structural geology. He was the major figure in the development of the research group in rock mechanics and structural geology then unmatched in accomplishment, breadth, or talent in a single organization. In 1967 Handin moved to Texas A&M University, where he became Distinguished Professor of Geology and Geophysics and established and was the first Director of the Center for Tectonophysics, also known internationally for its achievements, this time including graduate education. The Center has sponsored 22 doctoral dissertations and 34 master's theses. Handin has served as either chairman or member of advisory committees for most of these as well as having supported them through research grants and contracts. He is now Director of the Earth Resources Institute, which encompasses several Centers and programs (including Tectonophysics) and is simultaneously Associate Dean for the College of Geosciences at Texas A&M University.

Only truly outstanding individuals are able to change the course of a science. During his remarkable career as a scientist, educator, and research administrator, John Handin has achieved this distinction. He has contributed much to our understanding of, and interest in, the mechanical properties of rocks and their application to geological, geophysical, and engineering problems. Handin has become one of the world's foremost authorities in rock mechanics and tectonophysics. His creative achievements are recorded in some 75 scientific papers, including 'landmark' publications on the effects of confining pressure, pore pressure, and temperature on the strength and ductility of a wide variety of rocks. Special studies include those on strain rate, intermediate principal stress, gamma irradiation, hydrous phase transformations, fracture surface energy, thermal cracking, and testing machine stiffness; on slip systems in calcite, dolomite, mica, and plagioclase; on dynamic petrofabric analyses of calcite marble and dolomite rock; on the independence of tensile strength on mean stress; and on the flow and fracturing of folded-rock beams.

In addition, Handin has himself pioneered in the development of many different kinds of high-pressure and high-temperature techniques to make possible his work and that of his and other laboratories. He has clarified the concepts of internal friction in the Coulomb-Mohr failure criterion and of brittle versus ductile behavior. He has initiated studies of the frictional sliding of rocks, of deformation mechanisms and their use in dynamic petrofabric interpretations, and of residual strains (stresses) in rocks and rock masses. He has applied basic information on the mechanical properties of rocks to formulate a hypothesis for earthquakes and a model for earthquake control. Outside the field of rock mechanics, Handin's early work on beach accretion and erosion and its application to coastal engineering is still regarded as classic.

In November 1970, Handin received the award of the American Institute of Mining, Metallurgical, and Petroleum Engineers for distinguished achievement in rock mechanics. This award recognizes his outstanding contributions 'as a research scientist and educator,' and was given for 'his profound interest and significant contributions to the field of rock mechanics over the past two decades.' Handin's total commitment to his science and his primacy in the field are reflected in an impressive number of professional activities, including his election as Vice President of the International Society of Rock Mechanics and by the fact that he is continually asked to help organize professional meetings or chair work sessions, notably the two Penrose Conferences on Fracture Mechanics and Earthquake Source Mechanisms, the Third International Congress on Rock Mechanics, and the Chapman Conference on the State of Stress in the Lithosphere.

The breadth of his interests and influence is evident from current extramural activities. He serves on the U. S. Geological Survey's Advisory Panel on Earthquake Studies, the Policy Board of the National Geotechnical Centrifuge Facility, and the JOIDES Panel on Sedimentary Petrology and Physical Properties. He is co-chairman with William R. Judd of the Panel on Research Requirements of the U. S. National Committee for Rock Mechanics and is a past chairman of the parent committee. He has consulted for the U. S. Air Force, U. S. Army of Corps of Engineers, and Defense Nuclear Agency (formerly the Defense Atomic Support Agency) on underground defense facilities; for the Advanced Research Projects Agency on seismic decoupling of nuclear expolsions; for the Los Alamos Scientific Laboratory and Sandia Laboratories on exploitation of geothermal energy; for the Lawrence Berkeley Laboratory, Lawrence Livermore Laboratory, Oak Ridge National Laboratory, Environmental Protection Agency, Department of Energy, Intera Environmental Consultants, and the states of California and Louisiana on underground isolation of nuclear waste; for the U. S. Geological Survey on earthquake prediction, especially with C. Barry Raleigh on the Rangely experiment; for Terra Tek on in situ stress and field tests of fractured rock-masses with William F. Brace, Howard R. Pratt and Henri S. Swolfs; and for the Lake Powell Research Project on hydrological problems with Orson L. Anderson.

On a more personal level, John with his booming voice and appearance of the 'big bad bear,' has tended to terrify young scientists initially but, after getting to know him, they have found that he is a very gentle, kind, sensitive, and understanding man. Through his low-key approach to scientific problems, he fosters an environment in which both students and colleagues thrive and may develop their capabilities to their potential. He brings to the technical discussion a sound understanding of fundamentals, nearly a photographic knowledge of the literature, and an awareness of current research around the world. Although patient, he attacks the core of research problems in a direct, no-nonsense fashion, having grasped the essence of an apparent enigma immediately. He has the ability to recognize talent at once and to blend bright young research scientists with diverse backgrounds and specialities into cohesive and effective research teams. In addition, he is a true scholar, in the classical sense, as well as a superb scientific writer and editor. Those of us who have been so fortunate to have been associated with John are keenly aware of these outstanding personal and professional qualities and have been impressed, guided and inspired by them. Thus, we owe John a great deal of love and gratitude and this volume in his honor is one way of expressing our respect for him and for his remarkable impact and accomplishments in his chosen field.

We salute this volume's contributors, whose gifts of time and talent have built a fitting monument to his achievements. On Handin's behalf as well as our own, we also wish to express appreciation to his co-workers, co-authors, fellow committeemen and many other friends for their affection, loyalty and support throughout his career.

The Editors

Mechanics of Deformation of Crustal Rocks: Historical Development

M. KING HUBBERT

On July 1, 1945, Shell Oil Company and its overseas associates, the Royal Dutch/Shell Group, authorized the establishment in Houston, Texas, of the Shell Exploration and Production Research Laboratory. I was appointed an associate director of that laboratory and had the challenging experience of helping to design the laboratory, to employ its staff, and to formulate its research program.

One of the projects proposed at the outset was for fundamental research in the mechanics of rock deformation and associated problems in tectonophysics and structural geology. This was to involve the development and integration of three separate aspects of structural geology: (1) the appropriate mechanical theory of rock behavior under various stress conditions, (2) the experimental and petrofabric investigations of rocks under stress and temperature conditions occurring within the earth to depths of 10 km or more, and (3) the geological field work integrating the theoretical and experimental results into the interpretation of geologic structures as observed in nature.

This program was not to begin until a cadre of the best qualified people available should be assembled. In late 1946, M. A. Biot, one of the world's leaders in applied mechanics, joined the staff on a part-time basis as an outside consultant. For the next 20 years, Biot wrote a succession of papers on a wide variety of problems, such as elastic-wave transmission through fluid-filled porous sedimentary rocks, the mechanics of folding and of the instability of bedded salt in the initiation of salt domes, and, finally, the extension of the theory of the thermodynamics of irreversible processes to the deformations of rocks.

However, the program did not begin officially until late in 1950 when John Handin, who had been a research associate of David Griggs at the Institute of Geophysics at the University of California, Los Angeles, joined the staff in charge of the experimental work. At the same time, Helmer Odé, a Dutch geologist who had come shortly before from the University of Leiden with advanced education in mathematics and physics, was assigned to the project for theoretical support. Work began immediately on the design and construction of high-pressure apparatus for deformation of rocks at high confining pressure and temperature.

Once this work got under way, the recruitment of other staff members followed spontaneously: Hugh C. Heard from Griggs' laboratory for experimental work; Donald V. Higgs from the University of California, Los Angeles, Melvin Friedman from Rutgers University, and Iris Borg from the University of California, Berkeley (as a part-time consultant) in structural petrology; John J. Prucha, formerly senior geologist with the New York State Geological Survey, and David W. Stearns from South Dakota School of Mines for field studies; and George M. Sowers, experienced in Shell's exploration programs, for numerical and physical modeling. Later, after Higgs became a research administrator, Neville L. Carter came from Griggs' laboratory primarily as a structural petrologist. Subsequently, Handin and Stearns moved to the company's office in Ventura, California, to apply the laboratory's research results to tectonic problems, and John M. Logan from the University of Oklahoma joined the staff as an experimentalist.

By the time I retired from Shell at the end of 1963, this group had already achieved international renown. Indeed, for some 15 years under the able leadership first of Harold Gershinowitz and later of Noyes Smith, the laboratory was surely the best in the earth sciences in industry and quite probably the best anywhere in the world. Unfortunately, it was inevitable that a state of financial austerity would develop within the petroleum industry, restricting research activities to more immediately utilitarian objectives. This situation developed in Shell during the second half of the 1960's when it was decided to discontinue the basic-research programs. There was imminent danger that the group would disband and disperse, principally to several different universities. At this stage, Texas A&M University saw the opportunity to take over most of the team, and in 1967 the Center for Tectonophysics was established under the direction of John Handin, who was also appointed Distinguished Professor of Geology and Geophysics. Accompanying him were five colleagues from the Shell staff: Melvin Friedman, David W. Stearns, George M. Sowers, Jack N. Magouirk, and John M. Logan. Prucha had already left for Syracuse University; Carter went to Yale University, then to the State University of New York at Stony Brook, and rejoined the group at Texas A&M as Head of its Department of Geophysics and Research Associate in the center in 1978; and Heard joined the Lawrence Livermore Laboratory.

Preservation of this team, which had collaborated productively for many years, and the rebuilding of the high-pressure, petrofabrics, and photomechanical laboratories at Texas A&M were made possible by the strong support of the then President Earl Rudder and Dean Horace R. Byers and Associate Dean Earl F. Cook of the College of Geosciences and by the enlightened policy of the State of

Texas to grant its institutions of higher learning seed money to initiate new research programs.

George Sowers died in an automobile accident while he was leading his students on a field trip to Central America in 1974. In 1975, Brann Johnson, who had studied with Barry Voight at Pennsylvania State University, and David K. Parrish, who had been a student of Clark Burchfiel at Rice University and post-doctoral fellow under Neville Carter at Stony Brook, joined the center. James E. Russell, geophysicist and mining engineer, came from South Dakota School of Mines in 1978 to strengthen the group in engineering rock mechanics. In 1980 Stearns moved to the University of Oklahoma, and Parrish took a job in industry, while John H. Spang and Raymond C. Fletcher, both former students of William M. Chapple at Brown University, were added to the staff. Handin became Associate Dean of the College of Geosciences in 1973 and Director of the newly established Earth Resources Institute in 1979 at which time Friedman became Director of the Center for Tectonophysics. The center's staff and students have benefited from the active support of Anthony F. Gangi, a former student of Leon Knopoff at UCLA; Earl R. Hoskins, who had studied with John C. Jaeger at the Australian National University; and Richard A. Schapery, from the California Institute of Technology, who was a leading authority on the mechanical properties of composite materials.

No better evidence need be cited for the esteem in which the work of Handin and his associates is regarded by his professional contemporaries and students than the fact that they have prepared this commemorative volume in his honor. Hence, in this introduction, rather than to review work which is already well known, it may be more appropriate to offer some account of the background of events which led to the establishment of this group and the initiation of its program of research. This involves principally the state of knowledge of structural geology during the preceding two decades.

By the 1920's, after more than a century of field mapping over most of the land areas of the earth, the observational phenomena of structural geology were reasonably well known. These consisted principally of the geometrical configurations of deformed and metamorphosed rocks and of kinematic interpretations of the associated deformations. The principal weakness in the understanding of geologic structures resulted from the fact that, while these structures involve the complex mechanics of deformable bodies, a knowledge of this kind of mechanics was largely lacking among structural geologists. Rocks were presumed to be deformed by poorly defined, intuitive one-sided thrusts. The distinction between the dimensionally different quantities, force, and stress (or force per unit area) was largely unknown, and in textbooks of structural geology, when attempts at stress analysis were made at all, they were usually done erroneously by parallelogram resolutions applicable to forces but not to stresses.

A common justification for this state of affairs was that while geological phenomena admittedly must satisfy the rules of mechanics, they are so complex in detail as to be unamenable to mathematical analysis. Hence, structural geology was constrained to the natural-history methods of interpretation current at that time. Consequently, geological education rarely included even elementary mechanics or mathematics through differential and integral calculus.

Actually, the fundamental theories of the mechanics of deformable bodies—the theory of elasticity, of hydrodynamics, and of both infinitesimal and finite strain—already had been developed during the eighteenth and nineteenth centuries. As early as 1773, the study of the failure of solids stressed to rupture was initiated by the French scientist, C. A. Coulomb, in a paper presented in that year (but not published until 1776) [*Coulomb*, 1776]. Coulomb, who had also established the laws of sliding friction, concluded that the shear stress required to produce rupture and slippage in a solid consists of two parts, one, which is the intrinsic shear strength τ_0 of the solid, and a second, which is proportional to the normal component of stress acting across the surface of potential failure. By using modern notation, Coulomb's law of failure can be expressed by the equation,

$$\tau_{\text{crit}} = \tau_0 + \mu\sigma \tag{1}$$

where τ_{crit} is the shear stress required to produce rupture; τ_0, the shear strength of the material; σ, the normal stress component; and μ, a dimensionless constant of the material. For rocks and loose soils, μ subsequently has been found to have a value of about 0.6.

Principally during the decade 1820-1830, the theory of elasticity was worked out by the French mathematicians, Navier, Cauchy, and Poisson. One of the fundamental results established was that the state of stress at any given point inside a continuous body may be defined by three normal and six tangential stress components acting upon any three mutually perpendicular planes passing through the point and parallel to their respective lines of intersection. From those nine stress components, only six of which are mutually independent, the stress across any other plane whose normal makes arbitrary angles with the normals to the reference planes can be determined. In particular, it was shown that through each point there exists one set of mutually perpendicular planes, across each of which all shear stresses vanish, and the stresses reduce to the three principal normal stresses, σ_1, σ_2, and σ_3.

One of the leading structural geologists of the late nineteenth century was Charles R. Van Hise of the University of Wisconsin and the U.S. Geological Survey, who for years had worked with the tightly folded Precambrian sediments of the northern Great Lakes region. In 1896, Van Hise published a major monograph [*Van Hise*, 1896] in which the structural geological phenomena of this region were described and ably discussed.

For concurrent publication with this paper, Van Hise had an engineering colleague, L. M. Hoskins, write a companion paper [*Hoskins*, 1896]. Here Hoskins outlined the principal properties of the state of stress in a solid and also the main outlines of the theory of finite strain. He also stated the Coulomb principle, that the angle of rupture in a solid is influenced by both the strength of the material and the magnitude of the normal stress across the plane of rupture. He showed that for an isotropic solid under compressive stress, shear fracture should occur along sets of intersecting planes whose acute angle of intersection is bisected by the axis of greatest principal compressive stress. He then cited 31 experiments on the rupture of cylinders of cast iron in which the average angle between the plane of rupture and the axis of greatest compressive stress was 35°. With regard to rocks, Hoskins added,

> If these principles hold in the case of rocks under pressure, rupture by shearing would be expected to occur along planes oblique to the axis of greatest and least intensity of compressive stress but (if the material is isotropic) inclined at angles of less than 45° to the axis of greatest stress. In such a case, two intersecting sets of planes of rupture may develop, cutting each other at an oblique angle, the greatest pressure bisecting the acute angle [*Hoskins*, 1896, p. 875].

With this lucid beginning, what happened next is particularly ironic. C. K. Leith, a younger colleague of Van Hise, became one of the most able field geologists in the mapping and interpretation of the geologic structures of the northern Great Lakes region during the first quarter of the present century. He also taught structural geology at the University of Wisconsin and published a textbook, *Structural Geology* [*Leith*, 1913]. In this book he attempted to apply the principles stated by Hoskins to the interpretation of the observed systems of fracture and slaty cleavage in the tightly folded and indurated Precambrian sediments.

For this purpose, Leith constructed an equilateral-parallelogram, wooden frame, hinged at each corner. On this frame, with the wire strands parallel to its edges, was stretched a wire screen. At the center was riveted a circular cardboard disc, and on the screen, with the frame in its unstrained rectangular position, was painted a circumferential circle and two intersecting lines, each parallel to the wire mesh.

By pressing the frame along one diagonal, shortening would occur along this diagonal and elongation along the other. The circular cardboard would remain as a reference, but the painted circle on the screen would be deformed into a strain ellipse; the painted diameters parallel to the wire strands would rotate from mutual perpendicularity to an obtuse and acute angle of intersection. The obtuse angle would be bisected by the axis of shortening and the acute angle by the axis of elongation. These lines were supposed to represent the planes on which fracture in rocks would occur. Hence, according to this model, the obtuse angle of the set of conjugate fractures should be bisected by the axis of shortening and of greatest compres-

sive stress, contrary to what was plainly stated by Hoskins and demonstrated experimentally.

Another use of the wire model was to demonstrate rotational strain and stress. With the rectangular model set on a table in a vertical plane, the top of the model could be translated parallel to the table top. In this case, the strain ellipse on the wire screen, the long axis of which is parallel to the elongation diagonal of the frame, undergoes rotation with respect to the table top. This is rotational strain. However, it was presumed to be caused by rotational stress, an unbalanced-force couple acting about an axis perpendicular to the plane of the frame. It was not recognized that in response to such a torque, the frame would have to spin in angular acceleration. Since the spin did not occur, the total torque must have been zero and the supposed rotational stress a fiction.

This wire model and its associated theory was widely used in courses in structural geology in north-central universities. It was deceptive in that it had a qualitative resemblance to the relation of fractures and slaty cleavage in folded sediments, yet it was fundamentally fallacious mechanically. Its use impeded any real understanding of the mechanics of rock deformation for a quarter of a century.

With regard to faulting, four classes of faults had long been recognized by the 1920's: normal faults, reverse faults, wrench faults with horizontal shear along a vertical plane, and overthrust faults. Reverse faults, overthrusts, and wrench faults were clearly associated with compressive stresses, although overthrusts with 15 km or more of displacement were a mechanical enigma. Normal faults, which involved an elongation in the horizontal direction perpendicular to the strike of the fault plane, were also commonly interpreted as being caused by tensile stresses and hence were frequently referred to as 'tension faults.'

An interesting corollary to this latter view was a contemporary interpretation of faulting in the thick section of Tertiary sediments along the Gulf Coast of Texas and Louisiana. During the early 1920's, the known structures of these sediments were salt domes, associated with which some very large oil fields were being discovered. In 1926, a former Shell (then Roxana) geologist, as a visiting professor, gave a course on oil geology at the University of Chicago. He was asked by a student whether there was also faulting in the Gulf Coast sediments. His reply was, since reverse faults were due to compressive stresses which also produce folding, and since folding had not occurred in these sediments, then reverse or thrust faults would not be expected. On the other hand, normal faults were due to tension, but the Tertiary sediments consist of loose sands and clays which have no tensile strength. Hence, normal faults could not be formed. Therefore, except around salt domes, faulting in the Gulf Coast Tertiary sediments should not be expected.

It is now known that these sediments are cut by an abundance of normal faults, striking principally parallel to the coastline and to the strike of the sediments. Yet, I have been informed by a senior Shell geologist that the foregoing dogma persisted until well into the 1930's, when electric logs finally provided indisputable evidence that such faults do occur. In fact, one geologist, who had the temerity to show a fault on his map, was transferred out of the area, and another was assigned to make a proper interpretation.

One type of study of geologic structures that had been widely pursued by the 1920's was the formation of laboratory models in which stratified plastic materials had been deformed principally by layer-parallel compression. In this manner, folds and thrust faults could be produced that more or less resembled the behavior of rocks in the field. The most famous of such experiments were those performed by Bailey Willis of the United States Geological Survey, described in a major paper [*Willis*, 1893]. For his strata, Willis used such materials as plaster of Paris for the strong layers and beeswax for the weaker. Although he did achieve folding that resembled Appalachian structures, he had to load his models with an overburden of about $\frac{1}{2}$ m of lead shot to hold them down. Since the Appalachians required no such enormous overburden, this was evidence that Willis' materials were much too strong. The same had been true of most other model experiments and led to doubt as to the significance of such experiments.

A new development came about in the work of Hans Cloos of the University of Bonn in Germany [*Cloos*, 1929, 1930]. At the International Geologic Congress held in Washington in the summer of 1933, Cloos showed an exhibit of a dozen or so models of geologic structures which closely resembled real systems reduced in size to about 30 cm square. These models had become hardened, but upon seeing them, I asked Cloos what materials he had used. He said that they had been very soft, almost liquid clays.

My first thought was that this was even further from reality than had been the previous experiments. However, in my hotel that evening I made a rough analysis which indicated that the materials used by Cloos were in fact very close to those of the required strength for dimensional similarity. Indeed, Cloos had chosen his materials on the basis of a simple but valid similarity criterion.

By the 1920's, not many data existed on the mechanical properties of rocks. Some information was available on the crushing strengths of building stones, and qualitative experiments had been performed on the plastic deformation of marble and other rocks and minerals by F. D. Adams and J. T. Nicolson and Adams, but not under conditions in which triaxial stresses could be measured accurately [*Adams and Nicholson*, 1901; *Adams*, 1910]. Quantitative experiments on the deformation of marble under confining pressure had been conducted by *Th. von Kármán* at the University of Göttingen [*von Kármán*, 1911]. During the 1930's, some very important work on the mechanical properties of rocks and minerals was done at Harvard through a collaboration between Reginald A. Daly in geol-

ogy and P. W. Bridgman in physics and their students and research associates, Francis Birch and David T. Griggs.

In 1931, I joined the staff of the geology department at Columbia University, and one of my first assignments was to develop a course in structural geology. It was obvious by that time that one of the greatest needs was to incorporate the already existing knowledge of the mechanics of deformable bodies into the theoretical understanding and teaching of structural geology. Shortly after my arrival, I received a visit from a modest little gentleman who introduced himself as A. Nádai, formerly a professor of applied mechanics at Göttingen but recently arrived in the United States to take charge of the research on plasticity at the Westinghouse Research Laboratory in East Pittsburgh, Pennsylvania. Nádai brought with him a book by himself, *Plasticity* [*Nádai*, 1931a], which had been just translated and enlarged from the original German edition, *Der Bildsame Zustand der Werkstoffe* [*Nádai*, 1927] and published by McGraw-Hill as the first volume of the distinguished series of the Engineering Societies Monographs. Part 2 of this book was devoted to 'Some Applications of the Mechanics of the Plastic State of Matter to Geology and Geophysics.' Nádai also gave me a reprint of his recent Edgar Marburg Lecture given before the Society of Testing Materials [*Nádai*, 1931b]. He said that he had long been interested in rock deformation in geology as an example of the type of mechanics in which he was working, and he was looking for a geologist who might have similar interests.

To me, this was almost the answer to a prayer. I read the lecture and promptly bought and studied the book and also visited Nádai in his laboratory in East Pittsburgh. One of the experiments described in the Marburg lecture was on the phenomenon of slip in loose sand. If, in a box of loose sand, a thin blade, such as a broad putty knife, is inserted vertically and then translated horizontally in the direction normal to the blade, two slip surfaces will be produced, extending from the bottom edge of the blade to the surface of the sand. The slip surface in front of the blade will be a reverse fault with a dip of about 30° and to its rear there will be a normal fault with a dip of about 60°. This behavior was also shown to be predictable by the method of stress analysis developed by the German engineer, Otto Mohr [*Mohr*, 1914].

This behavior of sand proved to be of fundamental importance in the understanding of the mechanics of faulting in rocks. Here, too, it had long been established that for normal faults in unfolded strata, the average angle of dip of the fault surfaces is about 60° and for reverse faults, about 30°. With only a slight modification of the Mohr theory applicable to loose sand, to account for the shear strength of the rock, one could show that the same theory also accounted for the observed behavior of rocks. It also exposed the error of regarding normal faults as resulting from tensile stress. Loose sand has zero tensile strength, yet it responds to horizontal extension with well-defined normal faults with 60° dips. In both normal and reverse faults in loose sand, the slip failure is in response to compressive principal stresses of unequal magnitude. In each case the slip surface occurs parallel to the axis of intermediate principal stress and makes an angle of about 30° to the axis of greatest compressive stress.

An apparatus demonstrating this phenomenon was constructed, and the associated Mohr theory was developed for use in the course in structural geology at Columbia during the remainder of the 1930 decade.

Another important development during the 1930's was the establishment in 1935 of the Interdivisional Borderlands Committee of the National Research Council, between the divisions of Geology, Physics, and Chemistry. The chairman of this committee was the geologist, Thomas S. Lovering, then of the University of Michigan. I can no longer remember the complete membership of this committee initially, but it included, besides Lovering, Reginald A. Daly from the geology department and Francis Birch from the physics department at Harvard, George W. Morey, physical chemist with the Carnegie Institution Geophysical Laboratory in Washington, I. S. Bowen, an astronomer with the Mount Wilson Observatory in Pasadena, California, and another geologist.

Although the assignment of this committee was the 'borderlands' between geology, physics, and chemistry, the problems actually considered were almost exclusively the physical and chemical aspects of geological phenomena. At the end of the first year, the anonymous geological member had to withdraw from the committee, and I was appointed to fill out the term. The record showed that during its first year, this committee had made a comprehensive review of geological problems of a physical and chemical nature upon which research needed to be done. One of these was the determination of physical properties of rocks; another was the need for development of the theory of scale models applicable to problems of structural geology.

At the conclusion of the committee's work, Francis Birch agreed to undertake the assembly of a handbook on the physical properties of rocks. I agreed to prepare a paper on the theory of scale models. I wrote the paper, 'Theory of Scale Models as Applied to the Study of Geologic Structures' [Hubbert, 1937]. To assemble the physical properties of rocks, a new committee was appointed consisting of Birch (chairman), J. F. Schairer of the Geophysical Laboratory, and H. Cecil Spicer of the U.S. Geological Survey. This resulted in the monograph, *Handbook of Physical Constants* [Birch et al., 1942]. For the revised edition of this handbook, published in 1966 as Memoir 97 of the Geological Society of America, John Handin wrote the chapter on 'Strength and Ductility.'

The study on the theory of scale models largely resolved the paradox of an earth composed of strong rocks behaving in isostatic equilibrium, postglacial uplift, and mountain making as if composed of very weak materials. The

theory showed (as Cloos had demonstrated earlier) that a laboratory model for an earth feature with a length reduction of, say, a million fold, but with materials of the same density as the real ones, would have to be constructed of materials with a strength reduction of exactly a million fold if it were to behave like the original feature. Comparable reductions in viscosity must be made when both the length and the time scales are reduced.

Utilizing these results, David Griggs published a paper [Griggs, 1939] based upon a viscous scale model of large-scale convection currents, whose significance is more apparent in present-day plate tectonics than it was in 1939.

In recognition of the need for research, for academic facilities, and for publication outlets for the physical aspects of structural geology, the American Geophysical Union in about 1938 or 1939 appointed a committee consisting of M. King Hubbert (chairman), David T. Griggs, and A. Nádai to study the problem and recommend what should be done about it. This committee recommended the creation of a new section of the American Geophysical Union to be known as the Section of Tectonophysics, a name suggested by Norman L. Bowen. Its establishment was authorized by the council on April 9, 1940.

In October 1943, I joined Shell Oil Company in Houston, Texas, as a research geophysicist. As was noted earlier, in July 1945, Shell established the Exploration and Production Research Laboratory. In the Gulf Coast it was by that time well established that the Tertiary sediments were intersected by numerous normal faults, but those faults were still being regarded by many geologists as 'tension faults.' At about this time, I discovered and obtained a copy of a new book by the British geologist, E. M. Anderson [Anderson, 1942]. Actually, this book was essentially an elaboration and extension of an analysis published much earlier by Anderson [Anderson, 1905]. Unfortunately, that paper had been overlooked by most American structural geologists. Even a cursory inspection of this book clearly revealed that it was a valid and able mechanical analysis. I called it to the attention of one of my colleagues, Willy Hafner, a geologist-geophysicist from the Swiss Federal Technical University at Zürich. Hafner borrowed the book and studied it thoroughly. We both arrived at about the same idea: that there was need for in-house education of Shell geologists and geophysicists on the mechanics of geologic structures. Accordingly, at the following annual Shell Geophysical Conference, we presented two complementary papers. Hafner's was essentially an exposition of the Anderson book; mine involved the construction of an apparatus for demonstrating faulting in loose sand and the development of the associated theoretical analysis, based upon the aforementioned works of Mohr and Nádai.

These two papers, bound in a single volume, were widely used internally in Shell during the next several years as a textbook on certain aspects of modern structural geology. We realized finally that this information was not generally known in geology and that it should be published. The result was two companion papers [*Hubbert*, 1951; *Hafner*, 1951]. At about the same time, we learned that Jean Goguel, professor in the School of Mines in Paris and Director of the Geological Map Service of France, was also making significant physical analyses of rock deformation in geologic structures [*Goguel*, 1948, 1952].

About 1946, soon after the establishment of the Shell laboratory, but before the initiation of the program on rock mechanics and structural geology, an engineering problem arose with respect to drilling in the Gulf Coast Tertiary sediments: what to do about the troublesome problem of lost circulation, which was being discussed by some of my engineering colleagues. When I was called into this conference, I had been entirely unfamiliar with the nature of this problem. What I learned was that in drilling into these sediments, the water and other fluid pressures to depths of about 1,500–2,000 m were essentially 'normal' or hydrostatic. Then, rather abruptly, the well would encounter abnormal pressures about 1.5 times higher than normal. To avoid a blowout and possible loss of the well, the mud density had to be increased to a specific gravity high enough to produce a well pressure higher than the fluid pressures in the rocks. If the mud density were too small, a blowout was risked; if it were too large, lost circulation would occur, and the drilling mud being pumped down the drill pipe would fail to return. In that case, the pressure would also drop, and again a blowout might occur. The lost circulation was caused by the disappearance of the mud into a pressure-induced fracture in the rock. The question was, How is it mechanically possible for mud with a pressure of only about 0.7 of that owing to the weight of the sedimentary overburden to open a fracture?

Once the nature of the problem was understood, the answer was obvious. It was being assumed that the stress state in the rocks was one of hydrostatics, with all three principal stresses equal to the vertical pressure of the overburden. By means of the sandbox experiment, it could be shown that when normal faulting in the sand occurred, the least compressive stress was horizontal and perpendicular to the strike of the fault and of a magnitude only about half the maximum stress owing to the weight of the overburden. In the Gulf Coast sediments, in which normal faulting had occurred repetitively throughout the Tertiary period, the stress state most of the time must have been very near the rupture condition, with the least stress much less than that of the overburden. Hence, the mud pressure would only have to exceed slightly this minimum stress to form a pressure-induced fracture. Besides, these fractures should be vertical and stress oriented; that is, as Anderson had already shown for igneous dikes, they should be perpendicular to the least principal stress and parallel to the strike of the regional system of normal faults.

About 2 years later, engineers with the Stanolind Company (now Amoco) in Tulsa announced an important new development in petroleum production [*Clark*, 1949; *Howard and Fast*, 1950]. They showed that by means of hydraulic pressure applied in oil or gas wells, the reservoir rocks could be fractured and the fractures propped open by the injection of a sand-laden gel. In describing their previous work, they showed charts of the measured fluid pressures that had been required to produce hydraulic fractures at various depths. For several hundred such cases in the midcontinent region, the average value of the fracturing pressures was only about 0.8 of the pressure owing to the weight of the overburden. Nevertheless, these authors insisted that the induced fractures were horizontal bedding-plane fractures. For the next 6 or 7 years, this interpretation was accepted within the industry with rarely a dissent.

About 1947 or 1948, while planning the Shell program on rock mechanics, I visited the laboratory of the U.S. Bureau of Reclamation in Denver to obtain some information concerning a machine they had developed for making triaxial tests on large cores of rock and concrete. There I met Douglas McHenry, who gave me a preprint copy of a paper [*McHenry*, 1948] which he was to present at the forthcoming Troisième Congrès des Grands Barrages in Stockholm. His theory was entirely new to me. Its purpose was to determine experimentally with triaxial tests on both jacketed and unjacketed cylinders of concrete the magnitude of the surface porosity in accordance with a theory of Karl Terzaghi. [*Terzaghi*, 1932]. According to Terzaghi, the uplift pressure across a surface inside the concrete should be the total pore pressure reduced by the porosity along the surface. If, for example, the porosity along the surface should be 0.3, then the force exerted by the fluid pressure on the two opposite blocks across the given surface should be only 0.3 of that across an open fracture with a porosity of 1.0.

McHenry described tests on 337 concrete cylinders, both jacketed and unjacketed. Then, using the Mohr theory for the normal stress across the surface of fracture, he solved for the magnitude of the surface porosity in accordance with the Terzaghi hypothesis. A least squares solution gave a mean value of 1.02 ± 0.019. This to me seemed incredible, because in a stress state of equal principal stresses on a jacketed specimen, the pore pressure surely would be propagated through fluid and solid alike and hence would be exerted upon the total surface of the jacketed exterior boundary and not just over the pore area.

The question of the mechanical behavior of stressed porous rocks with internal fluid pore pressure became of critical importance in two subsequent investigations: the mechanics of hydraulic fracturing of rocks in oil wells [*Hubbert and Willis*, 1957] and the role of fluid pressure in the mechanics of overthrust faulting [*Hubbert and Rubey*, 1959; *Rubey and Hubbert*, 1959].

In 1951 the Shell Exploration and Production Research Laboratory was transferred from Shell Oil Company to its research subsidiary, Shell Development Company, while I was retained temporarily by Shell Oil Company as a consultant in general geology. In 1954 I was joined by David G. Willis, a geologist-geophysicist who had just completed his graduate work at Stanford University. One of our assignments was to try to resolve the problem of what actually was happening underground when rocks were fractured by applied hydraulic pressure. By this time the use of this technique for stimulation of production had become widespread, and thousands of wells had been treated. Yet the original Stanolind interpretation, that this fracturing occurred along horizontal bedding planes, was still accepted almost universally, despite the fact that the pressures required were consistently less than that owing to the total weight of the overburden. This anomaly was obscurely accounted for by the Stanolind engineers by an assumed 'effective weight' of the overburden which was less than the actual weight. The applied pressure had only to exceed this so-called 'effective pressure.'

In our study, Willis and I made an analysis of the stresses about the well bore under various conditions of initial undisturbed tectonic stresses, along the lines I had suggested to my colleagues in 1946-1947 with respect to the problem of lost circulation. We obtained the same results, that in tectonically relaxed regions such as the Gulf Coast and the midcontinent, characterized by normal faulting, the fractures almost had to be vertical.

A new problem was encountered, however, because instead of being dry, these rocks were porous and were normally filled with water with a pore pressure of its own. What effect would this have upon the mechanical behavior of the rocks? In particular, could the total overburden be lifted in a horizontal fracture by a fluid pressure less than the total weight of the overburden in case the fluid were flowing upward through the porous rocks instead of being applied statically against an impervious barrier at the base?

We investigated this problem both theoretically and experimentally by supporting a column of loose sand filled with water both statically and hydrodynamically. With an impermeable membrane at the base of the sand, the water pressure required to support the water-filled column was equal to the pressure owing to the total weight of the column. With the membrane removed and the water flowing upward through the sand, the buoyancy force was distributed hydrodynamically throughout the volume of the sand column. The pressure at the base, however, when lifting occurred, was the same as in the static case. Our conclusion was that it is impossible to produce a bedding-plane fracture with a hydraulic pressure less than that owing to the total weight of the overburden. Hence, the fractures actually being produced in tectonically relaxed regions by lesser pressures must be vertical. By now this has been amply confirmed. Unequivocal evidence indicates that except at shallow depths of less than a few hundred meters, nearly all hydraulic fractures are vertical, except in regions of active compressional tectonism as in parts of California.

Extending these ideas to the mechanics of rocks under-

ground, Willis and I found that the McHenry data on the fracturing of concrete were consistent with the later *Terzaghi* [1943] resolution of the total stress into a neutral component equal to the fluid pore pressure, and an effective component in the solid, but not with the earlier Terzaghi hypothesis regarding the reduction of the uplifting force by the surface porosity. Another result was the realization of the weakening effect on rocks by the increase of the pore pressure. Consider a loose sand at a depth of 2 or 3 km. With zero pore pressure, this sand in response to the weight of the overburden would act like a solid rock. If the pore pressure were increased until it supported the total weight of the overburden, the sand would be reduced in strength to that of loose sand. Hence, the pore pressures in underground rocks should have a strong influence in weakening these rocks during tectonic deformation.

A companion study of verification of the Anderson thesis that igneous dikes are emplaced on surfaces normal to the trajectories of the lines of least principal compressive stress was assigned to Helmer Odé. Using the dike pattern associated with the Spanish Peaks in Colorado, Odé made an analysis of the regional stresses associated with the central Spanish Peaks intrusion and the nearby boundary of the Rocky Mountain Front Range [*Odé*, 1957]. The observed dike pattern was found to be in excellent agreement with Anderson's hypothesis.

The potential effect of pore pressures in tectonic deformation came into sharper focus in 1955 when I was given a guided field trip across the Swiss Alps and was shown the famous Glarus overthrust in which Permian sediments had been thrust over Tertiary sediments on a nearly horizontal surface for a distance of some 30 km. The question arose, How thick must the overburden have been when this faulting occurred? And what must have been the pore pressure of the associated water? Almost certainly the overthrust block must have been a few kilometers thick, and in such an orogeny the pore pressure must have been increased to the maximum possible, enough to float the overburden. This would greatly reduce the force required to move such a block and would also resolve the century-old enigma of the apparent mechanical impossibility of such an event.

A paper on this subject was about to be written when by chance it was learned that William W. Rubey of the U.S. Geological Survey was approaching the same conclusion as a result of his 30 years of field work on the overthrust belt of southwestern Wyoming. This led to a collaboration and to the publication of the two companion papers [*Hubbert and Rubey*, 1959; *Rubey and Hubbert*, 1959].

About 1957, when part 1 of these papers was being written, I was approached by John Handin who was troubled because McHenry's results appeared to be inconsistent with those *Griggs* [1936] had obtained earlier from triaxial-compression tests on unjacketed specimens of Solenhofen limestone, the strengths of which did indeed increase with confining pressure. He suspected that this paradox had to do with the greatly different permeabilities of this rock and concrete, and he then repeated the experiments on highly permeable Berea sandstone and returned to report the new results in which he was sure I should be interested. He showed me a Mohr diagram of his extensive series of experiments with a range of effective stresses up to 600 MPa and of pore pressures from 0 to 200 MPa. The data plotted as Mohr circles of effective stress gave a rectilinear fracture envelope with the equation

$$\tau(\text{MPa}) = 15.4 + \sigma \tan 29°$$

This was a wider range of stresses and pore pressures than McHenry had used, and the measurements were of higher precision, yet the results agreed essentially with those of McHenry. This was exactly the information needed in the overthrust-fault paper. So, with Handin's permission, his diagram was reproduced as Figure 18 in the *Hubbert and Rubey* [1959] paper, and, together with the McHenry data, provided an essential experimental basis for much of the analysis. This was the beginning, and apparently also the first publication, except for an abstract by Handin in 1958, of a program of research by Handin and his associates on the relation between rock strength and pore pressure. In their definitive paper [*Handin et al.*, 1963], publication of which had been delayed several years for proprietary reasons, they have shown unequivocally from tests on a wide variety of sedimentary rocks that the law of effective stress is indeed applicable provided that (1) the interstitial fluid is inert relative to the mineral constituents of the rock so that the pore-pressure effects are purely mechanical, (2) the permeability is sufficient relative to the rate of deformation to allow pervasion of the fluid and continuous equilibration of the pore pressure, and (3) the rock is an aggregate with connected pore space such that the interstitial fluid pressure is fully transmitted throughout the solid phase. Important too is their demonstration that like concrete, soils, and other materials, rocks may compact or dilate during inelastic deformation, depending on their composition, fabric, and porosity and upon mean effective pressure. Experimental confirmation of Terzaghi's principle has been crucial to the correct solutions of most rock-mechanics problems.

Experimental tectonophysics may be said to have come of age in 1956 when a Symposium on Rock Deformation was cosponsored by UCLA's Institute of Geophysics and by the Shell Development Company. Most of the active workers in experimental and theoretical tectonophysics, then a small band numbering not more than 50, was assembled probably for the first time, including, besides David Griggs' associates and students and Shell's researchers, P. W. Bridgman, W. F. Brace, J. C. Maxwell, G. J. F. MacDonald, E. Orowan, M. S. Paterson, E. C. Robertson, and D. J. Varnes. The papers presented on that occasion were eventually published [*Griggs and Handin*, 1960] after a delay of nearly 3 years, because there was some rethinking of the critical problems, especially in the light of evolving data on pore-pressure effects. The paper by Griggs and Handin, 'Observations on Fracture and a Hypothesis of Earthquakes,' is still one of the most often cited in the literature of tectonophysics.

Since the early 1960's, the field of structural geology and associated subjects has grown exponentially. By now there must be more than 20 research centers throughout the world devoted to these subjects, and the membership of the Section of Tectonophysics of the American Geophysical Union has grown to some 1100. These developments and achievements during the last 30 years not only have justified but have surpassed the hopes and expectations of 1945 that a research program combining physical theory, proper experimentation, and related field work should be able within a decade or two to establish structural geology on a firm foundation of mechanical science.

On this happy occasion, my compliments and well-wishes are extended not only to John Handin and his present colleagues, but also to his former associates, David Griggs, M. A. Biot, Helmer Odé, John J. Prucha, David G. Willis, Kenneth Deffeyes, Peter Weyl, Donald V. Higgs, Rex V. Hager, Jr., H. W. Fairbairn, Iris Y. Borg, Hugh C. Heard, John H. Howard, and others who contributed in various ways to the success of this pioneering program. Thanks are also due to the management of the Shell companies for its foresight in the initiation and support of this work during its formative stages and to Texas A&M University for its wisdom in providing the facilities and favorable administrative environment for its continuing prosperity.

References

Adams, F. D., and J. T. Nicolson, An experimental investigation into the flow of marble, *Philos. Trans. R. Soc. London, Ser. A, 195*, 363-401, 1901.

Adams, F. D., An experimental investigation into the action of differential pressure on certain minerals and rocks, employing the process suggested by Professor Kick, *J. Geol., 18*, 489-525, 1910.

Anderson, E. M., The dynamics of faulting, *Edinburgh Geol. Soc. Trans., 8*, 387-402, 1905.

Anderson, E. M., *The Dynamics of Faulting and Dyke Formation with Applications to Britain*, Oliver and Boyd, London, 191 pp., 1942. (Revised ed., 206 pp., Oliver and Boyd, Edinburgh, 1951; reprint ed., 206 pp., Hafner, New York, 1972.)

Birch, F., J. F. Schairer, and H. C. Spicer (Eds.), Handbook of Physical Constants, *Spec. Pap. Geol. Soc. Am., 36*, 325 pp., 1942.

Bridgman, P. W., *The Physics of High Pressure*, 398 pp., MacMillan, New York, 1931.

Bridgman, P. W., Shearing phenomena at high pressure of possible importance to geology, *J. Geol., 44*, 653-669, 1936.

Clark, J. B., A hydraulic process for increasing the productivity of wells, *Trans. Am. Inst. Min. Metall. Pet. Eng., 186*, 1-8, 1949.

Clark, S. P., Jr. (Ed.), *Handbook of Physical Constants*, *Mem. Geol. Soc. Am.*, *97*, 586 pp., 1966.

Cloos, H., and K. Gebirge, Kunstliche Gebirge, 1, *Natur Museum*, *59*, 225-243, 1929.

Cloos, H., and K. Gebirge, Kunstliche Gebirge, 2, *Natur Museum*, *60*, 258-269, 1930.

Coulomb, C. A., Essai sur une application des règles des maximis et minimis a quelques problèmes de statique, relatifs a l'architecture, *Acad. R. Sci. Paris, Mem. Math. Phys.*, *7*, 343-382, 1776.

Daly, R. A., *Our Mobile Earth*, 342 pp., Charles Scribner's Sons, New York, 1926.

Goguel, J., Introduction à l'étude méchanique des déformations de l'écorce terrestre, in *Mém. Carte Géologique de France*, 2nd ed., 530 pp., Imprimerie National, Paris, 1948.

Goguel, J., *Traité de Tectonique*, translated from French by Hans E. Thalmann, 384 pp., W. H. Freeman, San Francisco, 1952.

Griggs, D. T., Deformation of rocks under high confining pressures, *J. Geol.*, *44*, 541-577, 1936.

Griggs, D. T., A theory of mountain-building, *Am. J. Sci.*, *237*, 611-650, 1939.

Griggs, D. T., and J. Handin (Eds.), Rock Deformation, *Mem. Geol. Soc. Am.*, *79*, 382 pp., 1960.

Hafner, W., Stress distributions and faulting, *Geol. Soc. Am. Bull.*, *62*, 373-398, 1951.

Handin, J., Effects of pore pressure on the experimental deformation of some sedimentary rocks, *Geol. Soc. Am. Bull.*, *69*, 1576-1577, 1958.

Handin, J., R. V. Hager, Jr., M. Friedman, and J. N. Feather, Experimental deformation of sedimentary rocks under confining pressure: Pore-pressure tests, *Am. Assoc. Petroleum Geol. Bull.*, *47*, 717-755, 1963.

Hoskins, L. M., Flow and fracture of rocks as related to structure, *Annu. Rep. 16*, pp. 845-874, U.S. Geol. Surv., Washington, D. C., 1896.

Howard, G. C., and C. R. Fast, Squeeze cementing operations, *Trans. Am. Inst. Min. Metall. Pet. Eng.*, *189*, 53-60, 1950.

Hubbert, M. K., Theory of scale models as applied to the study of geologic structures, *Geol. Soc. Am. Bull.*, *48*, 1459-1520, 1937.

Hubbert, M. K., Mechanical basis for certain familiar geologic structures, *Geol. Soc. Am. Bull.*, *62*, 355-372, 1951.

Hubbert, M. K., and W. W. Rubey, Role of fluid pressure in mechanics of overthrust faulting, 1, Mechanics of fluid-filled porous solids and its application to overthrust faulting, *Geol. Soc. Am. Bull.*, *70*, 115-166, 1959.

Hubbert, M. K., and D. G. Willis, Mechanics of hydraulic fracturing, *Trans. Am. Inst. Min. Metall. Pet. Eng.*, *210*, 153-168, 1957.

Leith, C. K., *Structural Geology*, 169 pp., Henry Holt, New York, 1913. (Revised ed., 390 pp., 1923.)

McHenry, D., The effect of uplift pressure on the shearing strength of concrete, in *Troisième Congres des Grands Barrages*, vol. 1, pp. 1-24, Stockholm, 1948.

Mohr, O., *Abhandlungen aus dem Gebiete der Technischen Mechanik*, 2nd ed., 567 pp., W. Ernst und Sohn, Berlin, 1914.

Nádai, A., *Der Bildsame Zustand der Werkstoffe*, 171 pp., Springer, Berlin, 1927.

Nádai, A., *Plasticity*, Eng. Soc. Monogr., 349 pp., McGraw-Hill, New York, 1931*a*.

Nádai, A., Phenomenon of slip in plastic materials, *Am. Soc. Test. Mat. Proc.*, *31*, 11-46, 1931*b*.

Odé, H., Mechanical analysis of the dike pattern of the Spanish Peaks area, Colorado, *Geol. Soc. Am. Bull.*, *68*, 567-575, 1957.

Rubey, W. W., and M. K. Hubbert, Role of fluid pressure in mechanics of overthrust faulting, 2, Overthrust belt in geosynclinal area of western Wyoming in light of fluid-pressure hypothesis, *Geol. Soc. Am. Bull.*, *70*, 167-205, 1959.

Terzaghi, K., Tragfáhigkeit der Flachgründungen, in *First Congress of the International Association of Bridge and Structural Engineering*, pp. 659-683, Paris, 1932.

Terzaghi, K., *Theoretical Soil Mechanics*, 510 pp., John Wiley, New York, 1943.

Van Hise, C. R., Principles of North American pre-Cambrian geology, *Annu. Rep. 16*, pp. 571-843, U.S. Geol. Surv., Washington, D. C., 1896.

von Kármán, T., Festigkeitsversuche unter allseitige Druck, *Zeits. Ver. deutsch. Ingeniere*, *55*, 1749-1757, 1911.

Willis, B., The mechanics of Appalachian structure, *Annu. Rep. 13*, pp. 211-281, U.S. Geol. Surv., Washington, D. C., 1893.

Calcite Fabrics in Experimental Shear Zones

M. Friedman and N. G. Higgs

Center for Tectonophysics, Texas A&M University, College Station, Texas 77843

Cylindrical specimens of Tennessee sandstone, with dry crushed calcite along 35° precut surfaces, are deformed at 200-MPa confining pressure, from 25° to 910°C, and at a shear strain rate of $10^{-2}\,s^{-1}$. Under these conditions the inelastic deformations are contained within the calcite layer. Shear displacements range between 1.5 and 3.0 mm, with engineering shear strain $\gamma < 5.7$. The compacted gouge (about 0.5 mm thick) is deformed in simple shear with non-coaxial incremental stress/strain and finite strain axes. The shear strength of specimens decreases from 300 MPa at 25°C to 30 MPa at 910°C, and the sliding mode changes from stick slip to stable sliding between 250° and 400°C. At <250°C the calcite deforms primarily by cataclasis, with Riedel shears that transect matrix and porphyroclasts, and conspicuous microfractures and twin lamellae in the porphyroclasts. The sliding mode is unstable (stick slip), but stress drops decrease between 25° and 250°C by a factor of 6. Between 250° and 650°C, slip on r $\{10\bar{1}1\}$ and twin gliding on e $\{01\bar{1}2\}$ predominate and produce highly elongated porphyroclasts with strong dimensional and crystallographic orientations. Average axial ratios change from about 2 at ≤250°C to as much as 16 at 650°C ($\gamma = 4.96$); average apparent long axes increase two- to three-fold, but average nominal grain areas remain constant. It is shown that the grain shape fabrics fit very closely predicted fabrics based upon homogeneous simple shear. The c axes of the porphyroclasts are strongly oriented within 20° of the axis of finite maximum shortening and geometrically track the axis of shortening with increasing shear strain. The sliding mode is stable. At temperatures of 550°-600°C, recrystallization occurs, after shearing by slip. Mosaics of very fine grained neoblasts are produced, with axial ratios <2.0. At 805°-910°C, two populations of neoblasts occur in each of four specimens. The first consists of larger, conspicuously twinned neoblasts with a ghost structure that tracks the finite strains and a strong crystallographic fabric that geometrically tracks neither the incremental stress nor the finite strain. The second group of neoblasts are smaller, strain free, and more equant and exhibit a very scattered crystallographic orientation. Neglecting differences in shear strain and temperature, the size of the larger neoblasts D is inversely proportional to the differential stress σ, with $D = 41\sigma^{-1.42}$. These larger neoblasts are well developed at a shear strain of only 0.61 (910°C) with an axial ratio of 1.7. This ratio and the crystallographic fabric remain constant for shear strains of 3.25, 4.71, and 5.69. It would appear that these neoblasts are syntectonic and are continually reformed during the shearing. The smaller strain-free neoblasts are probably due to annealing late in each experiment.

Introduction

The first major contributions that emerged from the programs of experimental rock deformation initiated by D. Griggs (Institute of Geophysics, University of California at Los Angeles) were studies of calcite single crystals and marbles [*Griggs and Miller*, 1951; *Handin and Griggs*, 1951; *Turner and Ch'ih*, 1951; *Griggs et al.*, 1951, 1953; *Borg and Turner*, 1953; *Turner et al.*, 1954, 1956; *Griggs et al.*, 1960a, b; *Heard*, 1963; *Heard et al.*, 1965]. Indeed, J. Handin's first publication in this field is his collaboration with Griggs on predicted fabric changes in Yule marble. These investigations firmly established the necessity to complement experimental efforts with fabric studies in order to explain and extrapolate the experimental results through a knowledge of the corresponding deformation mechanisms. Significant results of the observational work on Yule marble and calcite single crystals include understanding of twin and translation gliding systems in calcite, internal and external rotation phenomena, development of preferred crystallographic and dimensional fabrics from slip in constrained crystals, and fabrics and textures that develop as a result of annealing and syntectonic recrystallization. These results have led to the development of dynamic petrofabric tools for structural analyses of both experimentally and naturally deformed rocks [*Friedman*, 1978]. One major limitation of this work is the coaxial nature of the stress and strain fields in cylindrical specimens deformed by compression ($\sigma_1 > \sigma_2 = \sigma_3$) or extension ($\sigma_3 < \sigma_1 = \sigma_2$). In such tests it is not possible to determine whether particular fabric changes relate directionally to the stresses or the permanent strains. Accordingly, with the exception of strain analyses [e.g., *Groshong*, 1972, 1974; *Friedman et al.*, 1976; *Teufel*, 1980], most interpretations of calcite fabrics traditionally relate data to the state of stress [e.g., *Turner*, 1953; *Friedman*, 1963, 1964; *Carter and Raleigh*, 1969]. The primary purpose of the present work is to describe the deformation of calcite in shear zones where the axes of incremental stress and strain are not parallel to those of the finite strain.

Specific results from the previous coaxial experiments pertinent to our work include:

1. τ_c for translation gliding on r {$10\bar{1}1$} in calcite decreases with increasing temperature, so at 400°–500°C it is about equal to that for twin gliding on e {$01\bar{1}2$}, i.e., about 10 MPa (see figure 2 by H. C. Heard in *Friedman* [1967]).

2. In specimens shortened >25% at conditions where gliding flow on r {$10\bar{1}1$} is favored (e.g., 300°–500°C), pronounced grain elongation occurs normal to the axis of shortening and σ_1, the normals to e {$01\bar{1}2$} twin lamellae form a point maximum parallel to this axis, and the corresponding c axes form a similarly oriented maximum or are distributed in a small circle girdle of 10°–26° radius about this principal deformation axis [*Turner et al.*, 1956; *Rutter and Rusbridge*, 1977].

3. The fabric of point 2 is fully consistent with rotations of constrained crystals in homogeneously deformed aggregates, following closely *Handin and Griggs'* [1951] modification of *Taylor*'s [1938] hypothesis, and with slip on r {$10\bar{1}1$} [*Wenk et al.*, 1973] and twin gliding on e {$01\bar{1}2$} as the active gliding systems [*Turner et al.*, 1956].

4. Annealing recrystallization of sheared aggregates produces a de-orientation of any strong crystallographic orientation existent prior to annealing. In single crystals the c axes of neoblasts tend to be inclined at 25°–30° to their host [*Griggs et al.*, 1960a].

5. In syntectonic recrystallization the orientation of neoblasts also is controlled by the orientation of the host, with the c axis of the neoblast at 30° to that of the host. Neoblasts developed at grain boundaries show two influences: their c axes tend to lie at 20°–30° to that of the primary grain that they replace, and they also tend to be parallel to the axis of shortening (also σ_1) [*Griggs et al.*, 1960b].

In the programs of experimental deformation of calcite single crystals and aggregates there are several accounts

of the odd specimen for which the axes of stress and strain inadvertently are not coaxial [e.g., *Turner et al.*, 1956, figures 9E–F], but specific attempts to achieve this condition are limited: *Rutter and Rusbridge* [1977] for calcite marble, *Tullis* [1977] for quartzite, *Lister* [1977], *Lister et al.* [1978], *Lister and Price* [1978], and *Lister and Paterson* [1979] with computer simulations. Rutter and Rusbridge obtained this condition in direct shear tests (but only in zones a few grain diameters wide) and in two-stage deformations in which a pronounced foliation (dimensional alignment of grains) produced during stage 1 is oriented at 25° to σ_1 across the boundaries of the specimen in stage 2. Using the two-stage technique, experiments are made at 400°C, 300-MPa confining pressure, and 10^{-5} s^{-1}, conditions at which slip on r {$10\bar{1}1$} is the dominant orienting mechanism and for which resulting c axis orientations are predicted assuming slip on r [*Wenk et al.*, 1973]. *Rutter and Rusbridge* [1977, p. 83] find the grain shape orientation reflects the finite strain, and the crystallographic fabric rotates faster than the grain shape fabric so that the c axes and normals to the e twin planes are inclined about midway between the normal to the foliation and the greatest principal compressive stress across the boundaries of the specimen. They point out that similar c-axis fabrics oblique to the grain shape fabric are not uncommon in nature [*Wenk et al.*, 1973; *Wenk and Shore*, 1975; *Eisbacher*, 1970].

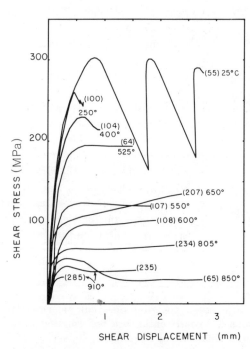

Fig. 1. Typical shear stress displacement curves for dry calcite gouge specimens deformed at 200-MPa confining pressure, a shear strain rate of 10^{-2} s^{-1}, and at temperatures from 25° to 910°C. Each curve is labeled with a specimen number (in parentheses) and temperature in degrees centigrade.

TABLE 1. Experimental Conditions, Results, and Textural Data for Porphyroclasts and Neoblasts in Calcite Gouge Deformed Dry, at Confining Pressure of 200 MPa and a Shear Strain Rate of $10^{-2} \, s^{-1}$

Specimen	Temperature, °C	Displacement, mm	Shear Strain, γ	Ultimate Shear Strength, MPa	Length of Long Axes L_A, Average/ Standard Deviation, mm	Axial Ratios, Average/ Standard Deviation, L_A/S_A	Average Nominal Area, $L_A \cdot S_A$, mm²	Comments
250	25	0.0	0	⋯	0.05/0.04	2.0/0.77	0.0013	Cataclasis predominates.
251	600	0.0	0	⋯	0.06/0.03	1.8/0.56	0.0020	Cataclasis predominates.
270[a]	25	0.0	0	⋯	0.12/0.08	1.7/0.52	0.0085[b]	Cataclasis predominates; larger porphyroclasts have dense twin lamellae.
55	25	3.71	⋯	300	0.07/0.04	2.0/0.73	0.0025	Cataclasis predominates.
100	250	0.66	⋯	265	0.07/0.04	2.4/0.94	0.0020	Cataclasis and slip; some clast elongation.
109	250	1.85	⋯	170	0.08/0.05	2.4/0.85	0.0027	Cataclasis and slip; some clast elongation.
104	400	0.92	1.65	230	0.08/0.05	4.0/1.8	0.0016	Slip predominates; clasts markedly elongated.
110	400	1.10	1.99	245	0.08/0.06	3.7/1.7	0.0017	Slip predominates; clasts markedly elongated.
64	525	1.50	2.46	195	0.17/0.14	7.7/4.6	0.0038	Slip; marked elongation of clasts.
107	550	1.80	2.98	120[d]	0.11/0.07	9.1/5.6	0.0013	Incipient recrystallization among matrix and elongated clasts.
108	600	1.92	3.63	100	0.01/0.01	1.4/0.36	0.0001	Very fine-grained mosaic of neoblasts.
207[a]	650	2.59	4.96	135	0.38/0.26	16.4/9.5	0.0088[b]	Slip, maximum elongation of clasts; no neoblasts.
234L[c]	805	2.25	4.71	70	0.07/0.06	1.7/0.78	0.0029	Mosaic of coarser, twinned neoblasts followed by smaller, strain-free neoblasts.
65L[c]	850	2.75	5.69	30[d]	0.09/0.06	1.7/0.74	0.0048	Mosaic of coarser, twinned neoblasts followed by smaller, strain-free neoblasts.
235L[c]	910	1.57	3.25	40[d]	0.17/0.08	1.8/0.52	0.0161	Mosaic of coarser, twinned neoblasts followed by smaller, strain-free neoblasts.
285L[c]	910	0.27	0.61	35	0.11/0.05	1.7/0.50	0.0071	Mosaic of coarser, twinned neoblasts followed by smaller, strain-free neoblasts.
285S[c]	910	0.27	0.61	35	0.03/0.01	1.4/0.39	0.0006	Mosaic of coarser, twinned neoblasts followed by smaller, strain-free neoblasts.

[a]Initial crushed calcite obtained from ball-mill only. In all other specimens, calcite crushed first in ball-mill and then with mortar and pestle.

[b]Note larger areas of porphyroclasts arise from use of calcite crushed only in ball-mill.

[c]Specimen contains two distinct sizes of neoblasts, these are separated in large and small, hence specimen designation 234L, 285L, or 285S (see text).

[d]Steady state, i.e., residual stress.

The primary purpose of the present work is to describe and discuss the evolution of calcite fabrics with increasing temperature (25° to 910°C) in experimental shear zones in which the axes of incremental stress/strain and those of finite strain are not coincident. Results are relevant to the mechanisms of deformation in ductile mylonites [e.g., *Ramsay and Graham*, 1970; *Lister and Price*, 1978; *Etheridge and Wilkie*, 1979; *Weathers et al.*, 1979], and to the dynamic interpretation of their fabric. Changes in strength and ductility, sliding mode, fabric and texture of the calcite gouge are related to changes in deformation mechanism from cataclasis, through gliding to recrystallization flow.

Stress–Strain States in the Shear Zone

It is stated in the previous section that the experiments described here involve deformations in which the axes of

stress/incremental strain and those of finite strain are not coincident. This condition is a direct consequence of the geometry of the test specimen in which displacement boundary conditions closely approximating those of progressive simple shear are imposed on a thin, deformable layer bounded by parallel, rigid members. The premise is now examined further.

It is commonly assumed that the principal compressive stress σ_1 in 35°-precut experiments is parallel to the cylinder axis throughout the specimen. During the initial stages of a test, this assumption probably holds. However, during post-yield deformation of a shear zone, three lines of evidence suggest that σ_1 across the boundaries of and in the shear zones, rotates to an angle of 45° to the shear zone boundary:

1. Petrofabric data of *Conrad and Friedman* [1976] demonstrate that extension microfractures adjacent to 35° precuts and faults in intact specimens reflect a local

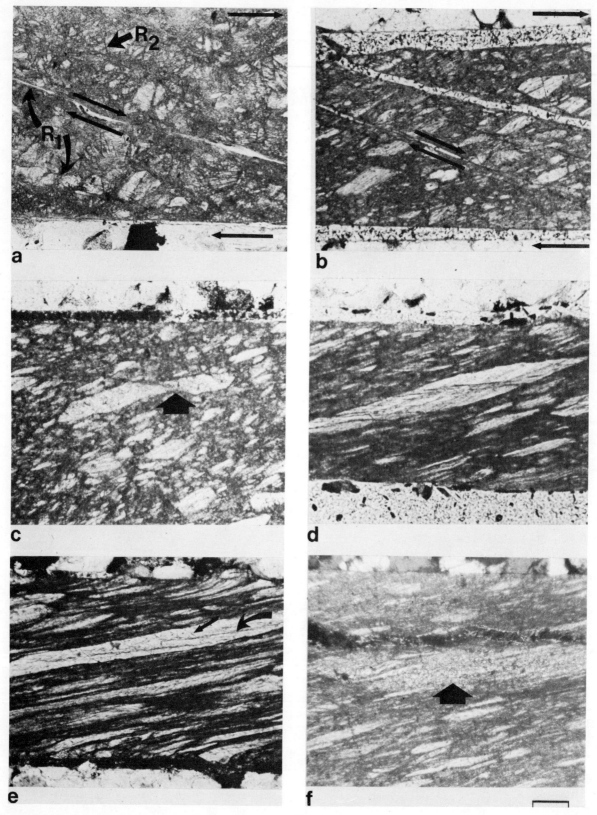

Fig. 2

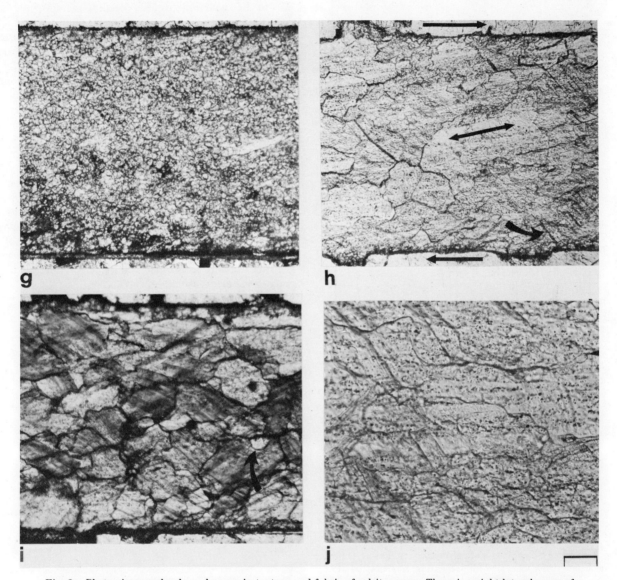

Fig. 2.   Photomicrographs show changes in texture and fabric of calcite gouge. There is a right lateral sense of shear along the shear zone in each, and all views are taken in plane-polarized light. Scale line located beneath (*f* and *j*) is 0.1 mm for views *a–i* and is 0.025 mm for *j*. (*a*) Specimen 55, 25°C. Cataclasis predominates with conspicuous R_1 and R_2 Riedel shears. (*b*) Specimen 100, 250°C. Riedel shears still conspicuous; porphyroclasts show alignment at about 25° to shear zone boundary. (*c*) Specimen 104, 400°C, $\gamma = 1.65$. R_1 shear is inclined at smaller angle to shear zone and offsets an elongated porphyroclast (bold arrow). (*d*) Specimen 64, 525°C, $\gamma = 2.46$. Elongated and aligned porphyroclasts are developed. Note planar *e* twin lamellae subparallel to shear zone boundary. (*e*) Specimen 207, 650°C, $\gamma = 4.96$. Highly elongated, completely twinned porphyroclasts occur with irregular *e* twin lamellae (curved arrow) and late planar twin lamellae (straight arrow). (*f*) Specimen 107, 550°C, $\gamma = 2.98$. Bold arrow points to a zone of incipient recrystallization. (*g*) Specimen 108, 600°C, $\gamma = 3.63$. Pervasive fine-grained neoblasts occur with remnant elongated porphyroclasts and R_1 shear. (*h*) Specimen 65, 850°C, $\gamma = 5.69$. Mosaic of large neoblasts shows ghost structure (double-ended arrow) and late, rational twin lamellae (curved arrow). (*i*) Specimen 235, 910°C, $\gamma = 3.25$. View shows larger, twinned neoblasts and smaller, strain-free ones (curved arrow) within the shear zone and even smaller neoblasts at the calcite host rock interface. (*j*) Specimen 65, 850°C, $\gamma = 5.69$. Detail of ghost structure (*h*) with alignment of impurities set E-W.

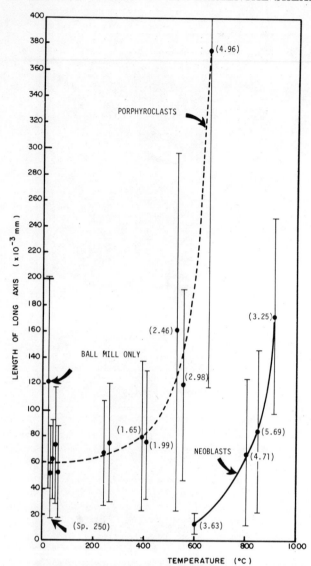

Fig. 3. Changes in long axes of porphyroclasts (dashed curve) and neoblasts (solid curve) with increasing temperature. For each specimen, a closed circle represents average length from 100 measurements, and vertical bars are ±1 standard deviation (Table 1). Corresponding shear strains (Table 1) are given in parentheses near each average length.

in the experiments is taken to lie at >35°, and probably at 45°, to the shear zone walls. Indeed, if the deformation is identified as progressive simple shear (and we will show that it is), the stress and incremental strain axes are required to lie at 45° to the shear zone boundaries [*Chapple*, 1968].

In contrast, the principal finite strains during progressive simple shear rotate with continued deformation. At engineering shear strains γ of 6 ($\epsilon_{12} = 3$, the maximum attained in our tests) the principal finite shortening ϵ_1 is oriented at 81° to the shear zone boundaries. This establishes a separation of the stress/incremental strain and finite strain axes of 36°. Note the angular separation would be larger if σ_1 was only 35° to the boundary. We will show, however, that at the higher temperatures (>400°C) where intracrystalline glide and recrystallization processes are dominant, grain shape fabrics are developed which very closely fit predicted fabrics based on an assumption of homogeneous simple shear. Accordingly, the orientations of the finite strain tensor and the principal stresses in the gouge zone are reasonably well known.

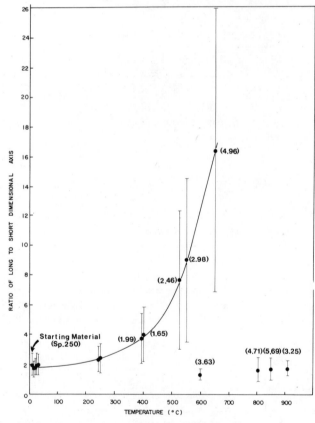

Fig. 4. Change in aspect ratio of porphyroclasts (solid curve) and neoblasts with increasing temperature (Table 1). Neoblast ratios show no change between 600° and 910°C. A closed circle shows an average from 100 measurements, and vertical bars are ±1 standard deviation. Corresponding shear strain (Table 1) is given in parentheses near each average ratio.

reorientation of σ_1 to 40°-50° to the shear surfaces. Simple photoelastic fault models support their interpretation.

2. Photoelastic stress meters within the experimental shear zones of *Mandl et al.* [1977] provide direct evidence for the orientation of σ_1 at 45° to the shear zone walls during shearing and show that the shear zone is the plane of maximum shear stress.

3. Finite element models (calculated by N. G. Higgs) of the specimen configuration used herein, indicate reorientation of σ_1 in the shear zone to an angle 10°-15° higher than the boundary σ_1. From these observations, σ_1

Methods

Cylindrical specimens of Tennessee sandstone (4 cm long by 1.75-cm diameter) with thin layers (1 mm) of dry crushed calcite along 35° precut surfaces are deformed at 200-MPa confining pressure from 25° to 910°C, and at a shear strain rate of 10^{-2} s^{-1} (Table 1, Figure 1). Final gouge thickness is 0.5-0.6 mm. Tests are conducted in a triaxial apparatus with internal furnace and argon confining medium, as described in *Friedman et al.* [1979]. Shear stresses are corrected for area changes along precut and the influence of the copper jacket. Chronologically, each specimen is confined at 200 MPa, heated to desired temperature at 10°-20°C/min, deformed, axial load removed, and quenched by switching off the furnace while confined. The cooling rate is about 80°-100°C/min.

Temperatures are accurate to ±2°C for most specimens. Exceptions are specimen 107, for which the temperature is most likely 550°C but might be as high as 600°C, and specimens 64 and 65, for which early furnace design problems permit only a best estimate of temperatures (probably ±50°C) based on similar textures and fabrics in specimens with reliable temperatures. Gradients exist along the length of the specimens, and at 900°C they reach a maximum of 30°C. There are no systematic variations in texture or fabric along the length of the shear zones, however, so these gradients do not seem to be important.

Calcite powder is prepared by crushing Yule marble in a ball mill and with mortar and pestle; the resulting grain size distribution ranges from 1 to 120 μm (Figure 2a), with the bulk of the material in the 1-10-μm range. The larger grains (average lengths about 60 μm and 120 μm, depending upon initial treatment, Figure 3) are con-

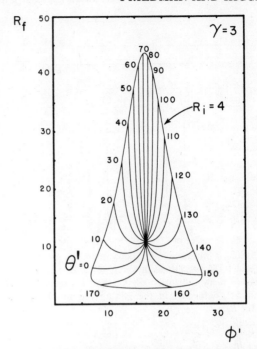

Fig. 6. R_f/ϕ' diagram showing a bounding $R_i = 4$ curve and internal angular zones (θ') in which grains with long axes initially inclined at corresponding angles to the shear zone would lie after deformation by simple shear to $\gamma = 3$.

spicuous as porphyroclasts within the finer matrix. Grain size estimates are based on light microscope examination alone, hence particles in the submicron range are not accounted for. This initial material simulates the grain size range in natural gouges [*Higgins*, 1971; *Engelder*, 1974], and the larger clasts provide useful displacement and strain markers in thin section. Average initial aspect ratios of the porphyroclasts are 2 and less (Figure 4).

The deformed samples are vacuum-impregnated with epoxy, and thin sections are cut parallel to the cylinder axis and perpendicular to the shear zone. Textural data are for 100 porphyroclasts or neoblasts per specimen that have been sampled randomly on traverse lines along the complete length of the shear zone. These data include the azimuth of long dimensional axes and the lengths of their long and short axes. In addition to plots of grain size, dimensional axial ratios, and nominal grain area against temperature, the grain shape fabrics are analyzed according to the techniques of *Ramsay* [1967] and *Dunnet* [1969]. The apparent long to short axial ratio R_f is plotted against angle of long axis to shear zone ϕ'. When deformed, initially circular grains plot as a single point in R_f/ϕ' space, while initially randomly oriented elliptical grains plot along a series of hyperbolic curves whose shapes depend upon the initial axial ratio R_i and the principal strain ratio R_s (Figure 5). The precise point that a deformed grain will occupy on any particular R_f/ϕ' curve depends upon its initial orientation, here defined with re-

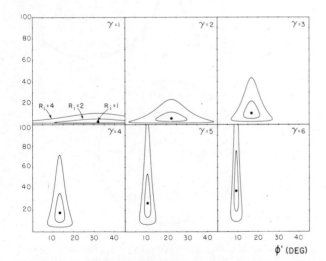

Fig. 5. Idealized R_f/ϕ' diagrams for shear strain from 1 to 6. R_f refers to the ratio of apparent long to short dimensions of grains, and ϕ' is the inclination of the long axis to the shear zone. See text for explanation of closed circle, $R_i = 1$, and curves $R_i = 2$ and 4, where R_i is the initial axial ratio.

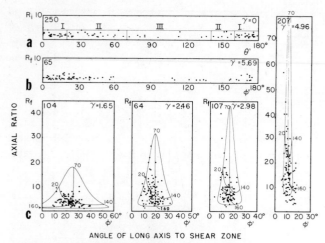

ANGLE OF LONG AXIS TO SHEAR ZONE

Fig. 7. R_f/θ' diagrams for (a) specimen 250, $\gamma = 0$, with angular regions I–III (explained in text), (b) specimen 65, $\gamma = 5.69$, deformed in recrystallization regime, and (c) specimens 104, 64, 107, and 207, deformed in the gliding regime, with bounding curves for $R_i = 4$.

spect to the shear zone as θ' (Figure 6). The curves are oriented symmetrically about the direction of principal finite elongation; thus if a suitable R_f/ϕ' locus can be fitted to the data, the orientation of the finite strain tensor is determined.

The technique has been applied to our experiments by combining the equations for R_f and ϕ [Dunnet, 1969, pp. 118–124] with the equations for progressive, homogeneous simple shear [Ramsay, 1967, pp. 83–88], and by calculating R_f/ϕ' curves based on a knowledge of the imposed shear strain and the ranges of R_i and θ' in the initial material. If the predicted curves are in accord with the data measured from thin section, one can conclude that progressive simple shear is a good description of the deformation and that the orientation of the finite strain tensor is then known with precision. Assumptions made in the analysis include homogeneity of deformation and elliptical grain shape. It will be shown below that for tests conducted at >400°C, the homogeneity assumption appears valid, and while the ellipticity assumption is not strictly correct, statistically, the data generate the expected scatter diagrams.

Results

Stress-Strain Data

The ultimate strength of the calcite gouge decreases with increasing temperature (Figure 1). Shear strength decreases from about 300 MPa at 25°C to about 30 MPa at 910°C. The transition from stick slip to stable sliding takes place between 250° and 400°C. Stress drops diminish by a factor of 6 between 25° and 250°C. From 400° to 800°C, most specimens exhibit steady state flow; work softening occurs at temperatures of 850° and 910°C.

Starting Texture and Fabric

Specimen 250, placed only under 200-MPa confining pressure at 25°C (i.e., $\gamma = 0$), is used to characterize the initial texture and fabric of the gouge. The gouge consists of twinned porphyroclasts, with long axes averaging 0.054 mm in length and with an average axial ratio of about 2.0 (Figures 3 and 4), embedded in a very fine grained, massive matrix (similar to Figure 2a). The porphyroclasts exhibit a modest dimensional orientation with their long axes tending to lie parallel to the shear zone boundaries (Figure 7a). This dimensional fabric is reflected crystallographically by a tendency for the c axes and normals to e {01$\bar{1}$2} twin lamellae to cluster about the normal to the shear zone (Figures 8a–b). The corresponding normals to r {10$\bar{1}$1} are widely dispersed (Figure 8c). This fabric arises because of a slight grain elongation to the cleavage-bounded porphyroclasts created during crushing. Then compaction causes the elongated porphyroclasts to align parallel to the precut.

STARTING MATERIAL

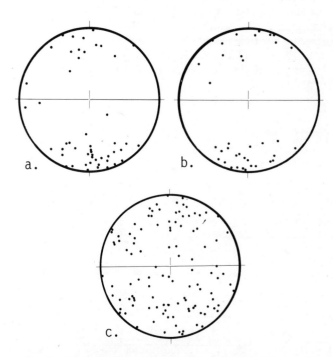

Fig. 8. Diagrams illustrate crystallographic fabric in starting material (specimen 250). The plane of each diagram is oriented perpendicular to the shear zone, with the latter placed E-W (see horizontal great circle), lower hemisphere, equal-area projection. Plotted are (a) c axes in 50 porphyroclasts, (b) normals to 37 sets of e_1 {01$\bar{1}$2} twin lamellae in 50 porphyroclasts, and (c) normals to 111 r {10$\bar{1}$1} planes.

Overview of Textural and Fabric Changes

Changes produced with increasing temperature and constant confining pressure (200 MPa), shear strain rate $10^{-2}\,s^{-1}$, and variable displacement are summarized briefly in Table 1 and are readily appreciated in photomicrographs (Figure 2). At room temperature, conspicuous conjugate Riedel shears (R_1 and R_2) are developed through the fine-grained matrix and porphyroclasts (Figure 2a). The acute angle between R_1 shears and the gouge-rock interface ranges from 15° to 20°. These shears, a conspicuous feature of shear zones [e.g., *Tchalenko and Ambraseys*, 1970], occur in all stages of development from incipient zones to zones of intense grain-size reduction, to discrete fractures with obvious loss of cohesion. Comparatively short R_2 fractures are developed at about 70°-80° to the gouge-host interface. Porphyroclasts are twinned and fractured. In general, there is a cataclastic texture of angular clasts arranged within a structureless, finer matrix, similar to that of a typical cataclasite [*Higgins*, 1971, pp. 5-6].

At 250°C the gouge layer looks essentially the same as in room temperature specimens, except a moderate preferred orientation of the long axes of elongate porphyroclasts (Figure 2b) is developed. Such clasts are oriented with their long axes oblique to the gouge-rock interface and at high angles to the cylinder axis. R_1 and R_2 shears are developed, but R_1 shears exhibit a wider range of orientation than at room temperature.

At 400°C, specimens 104 and 110 (stable sliding) are characterized by a well-developed preferred dimensional orientation of the porphyroclasts (Figure 2c). In both specimens, R_1 shears are conspicuous and are inclined at only 8°-10° to the gouge-host rock interface. Right lateral displacement along these shears is marked by offsets of porphyroclasts (Figure 2c). Measurements of these and other offsets indicate that the displacements average about 50 μm for tests where the total displacement parallel to the sliding surface is 2 mm.

At temperatures between 400° and 650°C the long axes and the axial ratios of the porphyroclasts are dramatically increased (Figures 2d-e). The grain shape fabric defines a foliation plane within the gouge that at 650°C (specimen 207, Figure 2e), for example, is inclined at 11° to the shear zone boundary.

Incipient recrystallization of both matrix and elongated porphyroclasts starts at about 550°C in specimen 107 (Figure 2f) and becomes pervasive at 600°C in specimen 108 (Figure 2g), where a few surviving highly elongated porphyroclasts attest to a shearing history prior to recrystallization. Most of the fine-grained neoblasts are strain free. At 805°-910°C, all porphyroclasts (and matrix) have been recrystallized into mosaics of coarse-grained, twinned neoblasts and a secondary population of smaller, strain-free neoblasts (Figures 2h-i). In specimens taken to shear strains of 3.25-5.69, a ghost structure of aligned impurities (Figure 2j) makes a progressively smaller

angle to the shear zone boundary with increasing shear strains. That is, for specimens 235, 234, and 65 (shear strains of 3.25, 4.71, and 5.69, respectively), the average angle between the ghost structure and the shear zone is 17°, 16°, and 9°, respectively.

Quantitative Textural Analyses

Length of long axes. Plots of long-axis length against temperature show two distinct trends, although it should be realized that the elongation is also a function of the shear strain (Figure 3, Table 1). From an initial average length of about 60 μm (120 μm for one specimen prepared only with a ball mill) the lengths increase about two- to threefold between 25° and 650°C. This is the temperature regime where the porphyroclasts become elongated and are not recrystallized. The second trend is for the larger population of neoblasts in the recrystallized specimens. Porphyroclast long axes also increase with increasing shear strain, but neoblast size is not clearly correlated with shear strain (Table 1).

Dimensional ratios. These data also show two distinct trends (Figure 4). Average ratios of porphyroclasts increase from an initial value of about 2 to as much as 16 in specimen 207 (650°C). Axial ratios at 600°, 805°, 850°, and 910°C are <2 for the larger neoblasts in the recrystallized material. These ratios are independent of temperature and of shear strain (Figure 4).

Grain shape fabrics. Changes in the dimensional alignment of the porphyroclasts and neoblasts are treated with the aid of R_i/θ' and R_f/ϕ' diagrams (Figure 7). The undeformed specimen (specimen 250) is characterized by porphyroclasts with $R_i \leq 4$ (average 2) and a weak preferred orientation of long axes subparallel to the shear zone (Figure 7a). This plot is divided into regions of angular density as follows: region I, >10 clasts per 10° interval; region II, <10 and >3 clasts per 10° interval; and region III, <3 clasts per 10° interval. For specimen 250, region I occurs at 0°-20° and 160°-180° (subparallel to the shear zone); region II occurs at 20°-70° and 140°-160°; and region III occurs in the interval 70°-140°. These initial axial ratios and accompanying pattern of preferred dimensional orientations are the prototype used for analysis of all other R_f/ϕ' data. The shear strain γ for each sample, calculated from the imposed shear displacement and final shear zone width, is used to generate the R_f/ϕ' curve ($R_i = 4$) in specimens 104, 64, 107, and 207, with angular boundaries at $\theta' = 20°$, 70°, 140°, and 160° (Figure 7c). These shear strains determine the position along the abscissa of the $R_i = 4$ curve on each diagram, assuming progressive, homogeneous, simple shear.

The theoretical R_f/ϕ' curves (Figure 7c) fit the R_f/ϕ' data for the four specimens extremely well and demonstrate that the homogeneous, simple shear model is reasonable for the calcite shear zone. For example, in specimen 64, most of the data points are on or within the $R_i = 4$ curve, the interval $\theta' = 70°-140°$ (region III) has rela-

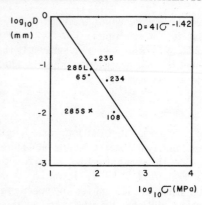

Fig. 9. Plot of neoblast grain size D against differential stress for the population of larger, strained neoblasts in five specimens (solid circles) and for the smaller strain-free neoblasts in specimen 285 (cross). Here σ is the differential stress at the end of each test (Figure 1).

tively few data points, and the interval $\theta' = 160°$-$180°$ and $0°$-$20°$ (region I) contains the bulk of the data, all as predicted. The same applies to specimens 104, 107, and 207, with the single exception that a significant proportion of the data in 207 fall outside the $R_i = 4$ curve and are better fit with $R_i = 6$.

The effect of recrystallization on the grain shape fabric is seen in the R_f/ϕ' diagram for specimen 65 (Figure 7b), which is representative of all the recrystallized shear zones. Recrystallization eliminates the large axial ratios characteristic of the porphyroclasts and tends to randomize the grain shape fabric, although not completely. It is evident that a weak preferred orientation of long axes occurs at about $\phi' = 20°$. This direction does not coincide, however, with the ghost structure in this specimen ($\phi = 9°$) nor the orientation of the principal finite elongation (9.5°) calculated from progressive homogeneous simple shear.

Neoblast grain size. Previously, it was noted that the long axes of the neoblasts increase with increasing temperature (Figure 3). Since the yield and steady state stress of the specimen decrease with increasing temperature (Figure 1), the neoblast size must also increase with decreasing stress if they are syntectonic, as recognized for other minerals by a number of workers [*Post*, 1973, 1977; *Kohlstedt et al.*, 1976; *Twiss*, 1977; *Mercier et al.*, 1977; *Ross et al.*, 1981; *Weathers et al.*, 1979]. The relation has been generalized in the form $D = A\sigma^{-n}$, where D is grain size in millimeters, A and n are constants, and σ is the differential stress in megapascals. Accordingly, we plot average neoblast size $[(\text{long axis} + \text{short axis})/2]$ against stress for five specimens (Figure 9). Specimen 108 contains only one population of neoblasts (Figure 2g), but specimens 65, 234, 235, and 285 exhibit two populations: larger, strained neoblasts and smaller, strain-free, more equant neoblasts (Table 1; Figures 2h-i). Only the average sizes of the larger, strained neoblasts are used to cal-

culate the least squares line. Note the data point for the smaller grains in specimen 285 (285S) lies considerably off the main trend. The least squares line defines the relation $D = 41\sigma^{-1.42}$ for the calcite neoblasts.

Crystallographic Fabric

Gliding regime. Although some gliding flow takes place between 25° and 400°C (Figures 2a-b), marked elongation of the porphyroclasts takes place between 400° and

PORPHYROCLASTS

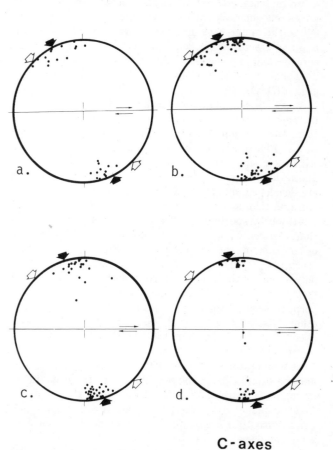

C-axes

Fig. 10. Diagrams illustrate orientation of c axes in porphyroclasts for specimens deformed at 400°-650°C, in which gliding is the dominant deformation mechanism. Plane of each diagram is perpendicular to the right lateral shear zone, as indicated. Open, broad arrow indicates the greatest principal compressive stress σ_1 oriented at 45° to the shear zone. The closed, broad arrow is the axis of principal finite shortening ϵ_1 defined by the elongated porphyroclasts and the R_f/ϕ' diagrams (Figure 7c). Data are plotted in lower-hemisphere, equal-area projection: (a) 27 c axes in specimen 104, 400°C, $\gamma = 1.65$, (b) 71 c axes in specimen 64, 525°C, $\gamma = 2.46$, (c) 50 c axes in specimen 107, 550°C, $\gamma = 2.98$, and (d) 39 c axes in specimen 207, 650°C, $\gamma = 4.96$.

650°C (Figures 3 and 4). These changes in shape occur without significant change in the nominal areas (Table 1), thus eliminating grain growth through recrystallization as playing any role. As reviewed previously, this is the temperature regime in which τ_c (10 MPa) for slip on r $\{10\bar{1}1\}$ approaches that for twin gliding on e $\{01\bar{1}2\}$ (see figure 2 by H. C. Heard in *Friedman* [1967]). It is the regime where it is established that fabric changes are produced by twinning and rotations due to slip [e.g., *Turner et al.*, 1956].

Fabric changes in this regime are documented for specimens 104, 64, 107, and 207, γ's equal to 1.65, 2.46, 2.98, and 4.96, respectively, and temperatures of 400°, 525°, 550°, and 650°C, respectively (Table 1). The c axis fabrics (Figure 10) change abruptly from those of the starting material (Figure 8a) and rotate toward the normal to the shear zone. Clearly, this pattern geometrically tracks the axis of greatest finite shortening as the latter rotates away from the stress axis with increasing shear strain. This strain axis is positioned from the symmetry axis in the R_f/ϕ diagram (Figure 7c). Corresponding e_1 $\{01\bar{1}2\}$ diagrams show point maxima of normals to e_1 lamellae coincident with the axis of shortening to γ of 2.98 (Figures 11a-c) and then a separation into two maxima in specimen 207 ($\gamma = 4.96$) (Figure 11d). This change in pattern reflects the transition from incomplete to complete twinning. That is, at a shear strain of 2.98, the twin lamellae normal to the axis of shortening are broad but still measurable, while at $\gamma = 4.96$ the porphyroclasts become completely twinned, and relict twin boundaries become indistinct (Figure 2e). Accordingly, the lamellae measurable in each porphyroclast are late rational features giving rise to the separate maxima (Figure 11d). Throughout this sequence of specimens the normals to the three sets of r $\{10\bar{1}1\}$ planes, determined from the c axis and normal to e_1 lamellae in each porphyroclast, are distributed in small circle girdles about the corresponding c axes, as, for example, in specimens 104 and 207 (Figure 11e-f).

Recrystallization regime. Recrystallization starts in specimen 107 at 550°C (Figure 2) and develops a pervasive, fine-grained mosaic of neoblasts at 600°C. This mosaic becomes coarser grained at higher temperatures (Figures 2h-i). On the other hand, specimen 207 shows no evidence of recrystallization at 650°C, perhaps a function of its larger initial grain size (Table 1). The crystallographic fabric resulting from the recrystallization is documented for the larger, twinned neoblasts in specimens 285, 235, 234, and 65 for which γ equals 0.61, 3.25, 4.71, and 5.69, respectively, and temperatures are 910°, 910°, 805°, and 850°C, respectively (Figure 12). In addition, the fabric of the population of small, strain-free, neoblasts is given for specimen 285. The c axis fabric from the larger neoblasts in specimen 285 (Figure 12a) is not significantly different from that of the starting material (Figure 8a). This perhaps reflects the small shear strain, 0.61, for this specimen. In the other three specimens, however, the c axis pattern from the larger neoblasts is strengthened,

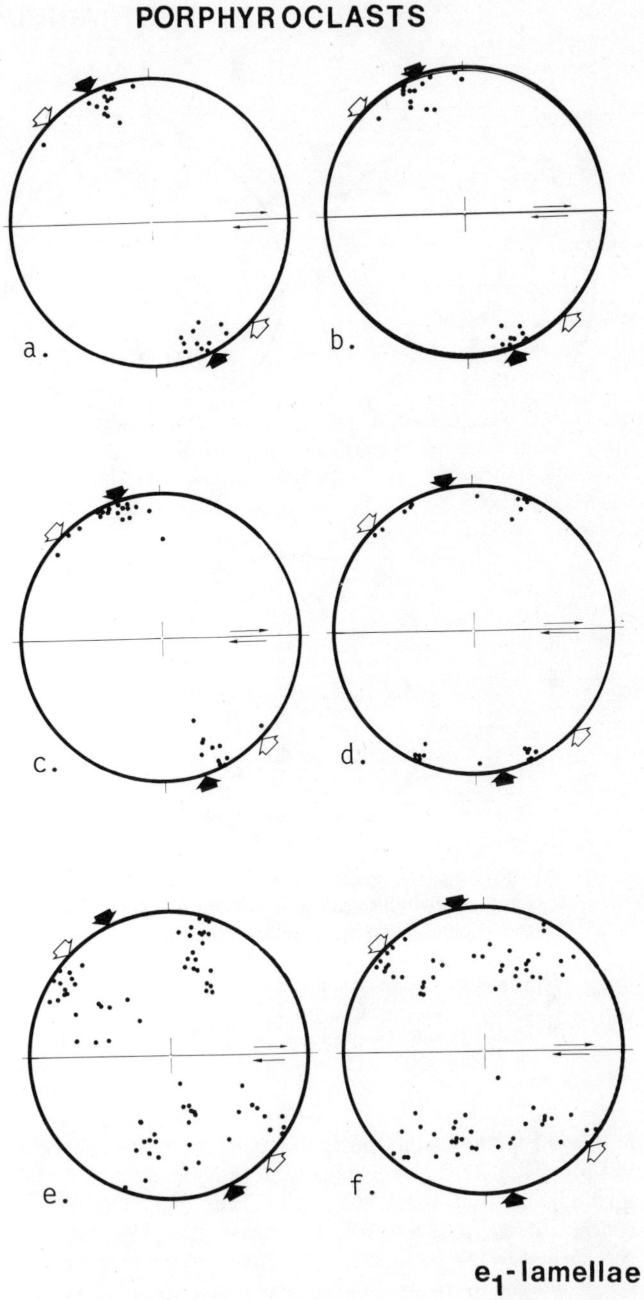

PORPHYROCLASTS

e_1-lamellae

Fig. 11.   Diagrams show orientation of the best developed set of $\{01\bar{1}2\}$ twin lamellae (e_1 lamellae) and r $\{10\bar{1}1\}$ planes in porphyroclasts for specimens in which gliding flow predominates. Plane and type of projection and nature of bold arrows are same as in Figure 10. Plotted are: (a) normals to 25 sets of e_1 lamellae, specimen 104, 400°C, $\gamma = 1.65$, (b) normals to 23 sets of e_1 lamellae, specimen 64, 525°C, $\gamma = 2.46$, (c) normals to 29 sets of e_1 lamellae, specimen 107, 550°C, $\gamma = 2.98$, (d) normals to 25 sets of e_1 lamellae, specimen 207, 650°C, $\gamma = 4.96$, (e) normals to 75 sets of r planes, specimen 104, 400°C, $\gamma = 1.65$, (f) normals to 72 sets of r planes, specimen 207, 650°C, $\gamma = 4.96$.

STRAINED NEOBLASTS

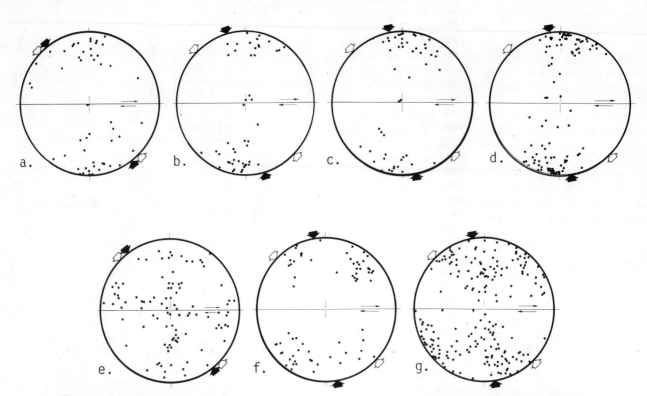

Fig. 12. Diagrams show crystallographic orientation of neoblasts in specimens deformed at 805°-910°C, in which recrystallization occurs. Plane and type of projection and nature of bold arrows same as in Figure 10, except the orientation of ϵ_1 (solid arrow) is determined from the known shear strain and the theory of progressive, homogeneous, simple shear. Plotted are: (a) 50 c axes in syntectonic neoblasts, specimen 285L, 910°C, $\gamma = 0.61$, (b) 50 c axes in syntectonic neoblasts, specimen 235L, 910°C, $\gamma = 3.25$, (c) 50 c axes in syntectonic neoblasts, specimen 234L, 805°C, $\gamma = 4.71$, (d) 100 c axes in syntectonic neoblasts, specimen 65L, 850°C, $\gamma = 5.69$, (e) 100 c axes in annealed neoblasts, specimen 285S, 910°C, $\gamma = 0.61$, (f) normals to 68 sets of e_1 lamellae in 100 syntectonic neoblasts, specimen 65L, (g) normals to 204 sets of r planes in 68 syntectonic neoblasts, specimen 65L.

and it is constant, apparently independent of shear strain (Figures 12b-d). These c axes tend to lie in a great circle girdle inclined at about 80° to the shear zone. The major concentration in this girdle is rotated clockwise beyond the normal to the shear zone. It coincides with neither the stress nor the finite strain axes. The c axes from the small, strain-free grains in specimen 285 are widely dispersed (Figure 12e). The normals to e and r planes in the larger neoblasts are typified by specimen 65 (Figure 12f-g, respectively). Those in the small grains are essentially random (not shown).

Discussion and Conclusions

Results of this study that warrant emphasis are as follows:

1. For dry calcite gouge at 200 MPa and with $\dot{\gamma}$ at 10^{-2} s^{-1}, the transition from stick slip to stable sliding occurs at 250°-400°C and coincides with a change in the dominant mechanism of deformation from cataclasis to gliding flow.

2. Quantitative textural analyses verify that deformation within the calcite gouge at 400°-650°C (i.e., in the gliding regime) can be approximated by progressive homogeneous simple shear. This is amply demonstrated by the good agreement between the distribution of R_f/ϕ' data points and the bounding envelope at $R_i = 4$ with internal angular divisions predicted from the theory of simple shear and the initial nonrandom distribution of porphyroclasts (Figure 7). It follows that within the shear zone the axes of stress and finite strain are not coaxial and diverge by 34° at a shear strain of 5.7. Further, the decreasing angle between the ghost structure and the shear zone in recrystallized specimens 235, 234, and 65 for increasing shear strains of 3.25, 4.71, and 5.69, respectively, is consistent with simple shear (measured angles of 17°, 16°, and 9°, respectively, compared to calculated angles of 15.5°, 11.5°, and 9.5°, respectively). This indicates that the ghost structure marks the finite strains in the recrystallized material.

3. Development of strong dimensional and c axis orientations within the 400°-650° temperature range is probably due to slip on r {10$\overline{1}$1} and twin gliding on e {01$\overline{1}$2} [*Turner et al.*, 1956]. The latter is conspicuous in thin section. Profuse thin twin lamellae are developed in the porphyroclasts at lower temperatures, and they are oriented as an array within the same maximum as the corresponding c axes (Figures 10 and 11). As the temperature increases, the lamellae broaden, and in specimen 207 (γ = 4.96, 650°C) the twinning is nearly complete. At this stage, the lamellae boundaries are irregular, parallel to the long axis of the porphyroclast, and are difficult to measure accurately with the U stage (Figure 2e). As a result, few of these lamellae are sampled, and only the generation of late twin lamellae is mapped (Figure 11d).

The predominant mechanism for affecting strong dimensional and crystallographic orientations is probably slip on r {10$\overline{1}$1}. In fact, at these temperatures, the critical resolved shear stress for r translation is about equal to that for e twin gliding (see figure 2 by H. C. Heard in *Friedman* [1967]). The external rotations associated with slip are discussed by *Handin and Griggs* [1951], *Turner et al.* [1954], and more recently by *Lister et al.* [1978] and are predicted by the Taylor-Bishop-Hill model [*Taylor*, 1938; *Bishop and Hill*, 1951; *Bishop*, 1953, 1954]. In principle, once a slip plane begins to operate at some high resolved shear stress coefficient S_0, which varies from a maximum of 0.5-0.0 [*Handin and Griggs*, 1951], it will rotate externally until the resolved shear stress vanishes, $S_0 = 0$. This rotation is not readily apparent in plots of normals to the three r planes in each calcite crystal constructed from measurements of the c axis and normal to e twin plane (Figures 11e-f). From an initial random distribution (Figure 8c), the normals to r tend to form small circle girdles about the axis of shortening (ϵ_1) with some concentration at low-S_0 values, i.e., normals parallel and perpendicular to σ_1 (Figures 11e-f). This pattern is maintained throughout the slip regime. These patterns are suggestive of slip on r, but they are not convincing. A more direct approach would be the identification of r as the slip plane affecting internal rotation [*Turner et al.*, 1954]; however, most e twin planes measured are late-formed rational planes. Special selective traverses would be required to define possible internal rotations, and these were not attempted.

Instead we test the hypothesis that as the shear strain increases, the S_0 value on the active r plane should decrease. Accordingly, the S_0 value is determined for that r plane and corresponding slip direction in each porphyroclast that is closest to being cozonal with the major shear zone and shearing direction. The data show that average S_0 values do decrease with increasing shear strain and that the corresponding standard deviations also decrease (Figure 13). Both results are consistent with the hypothesis and indicate the important role of slip on r.

Accordingly, within the temperature regime dominated by gliding flow, there is no reason to doubt the applicability of the Taylor-Bishop-Hill model for producing the strong dimensional and crystallographic orientations of the porphyroclasts. Primarily, slip on r {10$\overline{1}$1} and, secondarily, complete twinning on e {01$\overline{1}$2} are probably the active gliding systems. Constant average nominal areas of porphyroclasts as observed in thin section preclude recrystallization during this phase of the deformation.

4. Within this slip regime the crystallographic orientations (c axes, and e lamellae) rotate with increasing shear strain and geometrically track the axis of shortening defined statistically from the short axes of the porphyroclasts, i.e., the normal to the foliation produced by the elongated porphyroclasts. This fact is important for the dynamic interpretation of fabric data from mylonites and other shear zones. *Turner et al.* [1956, figures 9e-f] obtained exactly the same result as ours in a specimen shortened overall by 37%. In contrast, *Rutter and Rusbridge* [1977] found only a partial rotation of the c axis toward the normal to the final foliation (their oblique fabric), perhaps because their shear strain was too small. Our results suggest that in study of natural, calcite shear zones where gliding flow predominates, we can expect the c axes to track geometrically the axis of finite shortening. Thus with the orientation of the shear zone and this strain axis, one can use the theory of progressive homogeneous simple shear (point 2, above) to calculate a minimum shear strain (as well as the magnitudes of the principal finite strains) independent of knowledge of the actual shear displacement.

5. Incipient recrystallization begins at 550°C (specimen 107), becomes pervasive at 600°C (specimen 108), and at 805°-910°, it creates a mosaic of two sets of neoblasts. Most abundant are neoblasts with rational twin lamellae, a strong crystallographic orientation, an average axial ratio of about 1.7, and average long axes up to 0.172 mm. The second population consists of smaller, strain-free, more equant neoblasts with a nearly random crystallographic orientation that clearly have formed later in the deformation than their larger, strained hosts. It is perhaps significant that specimen 207, with a larger set of porphyroclasts than the other specimens (Table 1, footnotes a and b), is still in the gliding regime at 650°C, whereas the finer-grained material begins to recrystallize at 550°C. The latter would have a larger grain surface area and a higher surface energy per unit volume to drive the recrystallization process at a lower temperature.

6. The nature of the recrystallization processes that produced the larger, strained neoblasts and the smaller, strain-free neoblasts is evident from consideration of their size and fabric. The relationship between the average size of the larger neoblasts and differential stress ($D = 41\sigma^{-1.42}$) agrees very well with *Twiss'* [1977] prediction ($D = 19\sigma^{-1.47}$) based on equilibrated strain energies of grain boundaries and intragranular dislocations. In that this relationship holds solely for steady state flow and for syntectonic (dynamic) versus annealing recrystallization, we conclude the population of larger, strained neoblasts are the product of syntectonic recrystallization. In contrast, the population of smaller neoblasts are strain free,

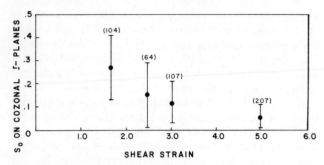

Fig. 13. Plot of the average resolved shear stress coefficient S_0 for the r {10$\bar{1}$1} plane in each crystal that is closest to being cozonal with the shear zone and shearing direction against shear strain for each of four specimens deformed in the gliding regime. Specimen numbers are in parentheses. Closed circle represents the average, and the vertical bars are ±1 standard deviation for S_0 calculated from 25, 22, 29, and 23 porphyroclasts in specimens 104, 64, 107, and 207, respectively.

have nearly random crystallographic orientation (Figure 12d), and form texturally at the expense of the larger, strained hosts. The smaller neoblasts most likely are the product of an annealing, de-orienting recrystallization that occurs very late in the loading or unloading history.

7. Deformation within the shear zone appears to be reasonably homogeneous and pervasive, as indicated by the internal homogeneity of the distorted porphyroclasts, the spatial development of the neoblasts, and the repeated nature of the dimensional and crystallographic fabrics among comparable specimens. At the very boundary of the shear zone, however, there is evidence of some heterogeneity. This is conspicuous in our recrystallized specimens where neoblasts at the boundary are smaller than those of the two internal populations, yet coarser than the groundmass in the non-recrystallized material (Figures 2h–i). This grain size reduction at the interfaces is not apparent in our specimens deformed in the regimes of cataclasis or gliding flow primarily, because the groundmass itself is so fine grained. In other gouge types, however, the development of very fine grained material at the interface is common and is thought to signify a shift in the locus of shearing from within the gouge to the interface [Logan et al., 1979]. The significance of the finer neoblasts at the shear zone boundary in our recrystallized materials is not known.

8. The axial ratios, dimensional and crystallographic fabrics, and ghost structure of the syntectonically recrystallized neoblasts shed light on the concept of 'recrystallization flow.' This is a topic reviewed recently in some depth by McQueen and Jonas [1975], Sandström and Lagneborg [1975], and Lister and Price [1978]. We recall that (1) axial ratios of the larger, strained neoblasts (about 1.7) are independent of shear strain from 0.61 to 5.69 (Table 1); (2) although these neoblasts are not highly elongated, their long axes are inclined to the shear zone boundary at 10°–30° (Figure 7b), reflecting some shear

strain; (3) their crystallographic fabric also is independent of shear strain at least from 3.25 to 5.69; and (4) the ghost structure within the mosaic of these neoblasts coincides with the axis of greatest finite elongation calculated from the known shear displacement and the theory of progressive homogeneous simple shear. Compatible with these facts is the hypothesis that the neoblasts form early in the history of shearing and continually reform during the deformation at each stage, maintaining their crystallographic fabric, low axial ratio, and average size, depending on differential stress. The ghost structure of impurities is the only record of the total shearing event, i.e., it persists and continually reflects the total finite strain. This idea is expressed by McQueen and Jonas [1975, pp. 438–439] as 'waves of repeated dynamic recrystallization.' The phenomenon is manifest as oscillations of the stress-strain curve after the initial yielding, as demonstrated for nickel by Luton and Sellars [1969], and in plain 0.25%-C steel [McQueen and Jonas, 1975, figure 30a]. In our specimen, there is only the post-yielding work-softening. The underlying mechanisms that continually reconstitute and reshape the neoblasts most likely are dislocation glide and dislocation creep [McQueen and Jonas, 1975]. This is also suggested by the fact that our stress versus temperature data plot along the correct strain rate contour in the fields of dislocation glide and creep in the calcite deformation map of Rutter [1976, figure 6], by the ghost structure, and by the modest dimensional alignment of the neoblasts that reflects some shearing late in the history of the deformation.

9. The crystallographic orientations of the syntectonic neoblasts also contribute to the above. At low shear strains (specimen 285, $\gamma = 0.61$), the neoblast c axis fabric appears to be directly inherited from that of the starting material (compare Figures 8a and 12a), but once γ reaches 3.25, the neoblast c axes not only do not change with shear strain (Figure 12b–d), but their orientation relative to stress/strain axes and to the shear zone boundary is significantly different from the fabric produced solely by gliding at a comparable shear strain (compare Figures 10 and 12). Thus, although dislocation glide may play a role in the recrystallization flow process, this mechanism alone does not determine directly the crystallographic orientation of the neoblasts. Lister and Price [1978, p. 53] review the oriented nucleation hypothesis and the growth selection hypotheses from the metals literature as ways to explain how recrystallization can modify an existing fabric. The first deals with the growth of neoblasts from nuclei that are oriented with respect to the host crystal. The second hypothesis indicates there is selection of nuclei that grow at the expense of less favorably oriented ones. Lister and Price [1978, p. 54] criticize both of these classical hypotheses because they do not consider effects of the deformation itself. They offer an alternative hypothesis that postulates (1) 'recovery and recrystallization processes are not dramatically important in affecting or modifying fabric development during deformation,' and (2) 'the

crystallographic fabrics have developed because of the mechanism of the deformation process.' Points (3 and 4) of their work pertain specifically to the quartz mylonite zone they studied and are not of interest here. Their main point seems to be that although recovery and recrystallization involve both dislocation glide and climb, grain boundary and point-defect migration, the mechanism that most rapidly produces reorientation is dislocation glide. They state [*Lister and Price*, 1978, p. 57] 'deformation fabrics formed ... during dynamic recovery with the formation of dislocation cell walls, are likely to reflect mainly the influence of the conservative component of dislocation motion within the subgrain interiors, and to behave in an approximately similar fashion to deformation fabrics formed at lower homologous temperatures.' For the calcite gouge study herein, we think it is very clear that the crystallographic fabrics resulting from gliding (at 'lower homologous temperatures') are not the same as those for syntectonic recrystallized neoblasts at higher temperatures, but at comparable shear strains. This fact would appear to question seriously the applicability of the Lister-Price hypothesis to the calcite gouge. Granted the underlying mechanisms during syntectonic recrystallization involve dislocation glide and climb, but the resulting crystallographic orientations clearly are not those expected from glide alone. Rather the neoblast fabric appears to be inclined at about 30° to that likely in the host, suggesting a host control similar to that recognized by *Griggs et al.* [1960b] and discussed by *Hobbs* [1968]. That is, early in each wave of recrystallization c axes may be rotated by dislocation glide toward the axis of shortening, and subsequent neoblasts are nucleated and grow inclined to this host fabric.

10. Regardless of its origin, the c axis fabric for the syntectonically recrystallized grain geometrically tracks neither the incremental stress nor the finite strain axes. This is important for those using the c axis fabric for dynamic analysis. For example, an interpretation based upon results from coaxial deformations holds that recrystallized calcite c axes group parallel to the greatest principal compressive stress. Our data show this clearly is wrong for simple shear, and that if used, one would in fact predict a sense of shear for the shear zone that also would be incorrect.

11. Finally, it is worth emphasizing that recrystallization under the ambient conditions of these experiments takes place in less than 60 s (specimen 285), and several generations of neoblasts may have evolved in specimens like specimen 65 ($\gamma = 5.69$) in a matter of 10^2-10^3 s. Similar natural fabrics may reflect essentially an instantaneous record of the geological event.

12. Much remains to be investigated to fully understand recrystallization flow. The crystallographic fabric of the very fine neoblasts developed at 550°C and 600°C where gliding flow obviously has preceded recrystallization needs to be studied by X ray diffraction. In these neoblast fabrics we may or may not find documenta-

tion for *Lister and Price*'s [1978] hypothesis. It would also be helpful to deform a specimen at 900°C to shear strains between 0.61 and 3.25 to see how the crystallographic fabric changes from that in Figures 12a-b. In addition, the dislocation arrays in elongated porphyroclasts and the two sets of neoblasts need to be investigated with TEM to test the conclusions reached here solely from light microscopy.

Acknowledgments. This work is part of our investigation of the behavior of simulated gouge materials at elevated temperatures and pressure supported by the U.S. Geological Survey as part of the Earthquake Hazards Reduction Program, grant 14-08-0001-G-460. J. M. Logan, Center for Tectonophysics, initially suggested the experiments on calcite gouge at elevated temperatures in order to investigate possible correlations between sliding mode and deformation mechanism [*Logan et al.*, 1979]. We are very grateful for his insight, encouragement, and helpful comments. We also wish to thank N. L. Carter and J. M. Christie for their helpful review of this paper.

References

Bishop, J. F. W., A theoretical examination of the plastic deformation of crystals by glide, *Philos. Mag., 44*, 51-64, 1953.

Bishop, J. F. W., A theory of the tensile and compressive textures of face-centered cubic metals, *J. Mech. Phys. Solids, 3*, 130-142, 1954.

Bishop, J. F. W., and R. Hill, A theory of the plastic distortion of a polycrystalline aggregate under combined stresses, *Philos. Mag., 42*, 414-427, 1951.

Borg, I., and F. J. Turner, Deformation of Yule Marble: Part VI-Identity and significance of deformation lamellae and partings in calcite grains, *Geol. Soc. Am. Bull., 64*, 1343-1352, 1953.

Carter, N. L., and C. B. Raleigh, Principal stress directions from plastic flow in crystals, *Geol. Soc. Am. Bull., 80*, 1231-1264, 1969.

Chapple, W. M., Finite homogeneous strain and related topics, in *Rock Mechanics Seminar*, edited by R. E. Riecker, Spec. Rep., pp. 317-354, Terrestrial Sci. Lab., Air Force Cambridge Res. Lab., Bedford, Mass., 1968.

Conrad, R. E., and M. Friedman, Microscopic feather fractures in the faulting process, *Tectonophysics, 33*, 187-198, 1976.

Dunnet, D., A technique of finite strain analysis using elliptical particles, *Tectonophysics, 7*, 117-136, 1969.

Eisbacher, G. H., Deformation mechanism of mylonite rocks and fractured granites in the Cobequid Mountains, Nova Scotia, Canada, *Geol. Soc. Am. Bull., 81*, 2009-2020, 1970.

Engelder, J. T., Cataclasis and the generation of fault gouge, *Geol. Soc. Am. Bull., 85*, 1515-1522, 1974.

Etheridge, M. A., and J. C. Wilkie, Grainsize reduction, grain boundary sliding, and the flow strength of mylonites, *Tectonophysics, 58*, 159-178, 1979.

Friedman, M., Petrofabric analysis of experimentally deformed calcite-cemented sandstones, *J. Geol.*, *71*, 12–37, 1963.

Friedman, M., Petrofabric techniques for the determination of the principal stress directions in rocks, in *State of Stress in the Earth's Crust*, edited by W. R. Judd, pp. 451–552, Elsevier, New York, 1964.

Friedman, M., Description of rocks and rock masses with a view towards their mechanical behavior, in *Proceedings of the First International Congress on Rock Mechanics, Lisbon, Portugal*, vol. 3, pp. 182–197, Lab. Nac. de Eng. Civil, Lisbon, Portugal, 1967.

Friedman, M., Structural uses of calcite twin lamellae, *Geol. Soc. Am. Abstr. Progr.*, *10*, 5, 1978.

Friedman, M., L. W. Teufel, and J. D. Morse, Strain and stress analyses from calcite twin lamellae in experimental buckles and faulted drape-folds, *Phil. Trans. R. Soc. London Ser. A*, *283*, 87–107, 1976.

Friedman, M., J. Handin, N. G. Higgs, and J. R. Lantz, Strength and ductility of four dry igneous rocks at low pressures and temperatures to partial melting, in *Proceedings of the 20th U.S. Symposium on Rock Mechanics*, pp. 35–50, Austin, Texas, 1979.

Griggs, D., and W. B. Miller, Deformation of Yule Marble: Part I–Compression and extension experiments on dry Yule Marble at 10,000 atmospheres confining pressure, room temperature, *Geol. Soc. Am. Bull.*, *62*, 853–862, 1951.

Griggs, D., F. J. Turner, I. Borg, and J. Sosoka, Deformation of Yule Marble: Part IV–Effects at 150°C, *Geol. Soc. Am. Bull.*, *62*, 1385–1406, 1951.

Griggs, D., F. J. Turner, I. Borg, and J. Sosoka, Deformation of Yule Marble: Part V–Effects at 300°C, *Geol. Soc. Am. Bull.*, *64*, 1327–1342, 1953.

Griggs, D. T., M. S. Paterson, H. C. Heard, and F. J. Turner, Annealing recrystallization in calcite crystals and aggregates, *Geol. Soc. Am. Mem.*, *79*, 21–37, 1960*a*.

Griggs, D. T., F. J. Turner, and H. C. Heard, Deformation of rocks at 500° to 800°C, *Geol. Soc. Am. Mem.*, *79*, 39–104, 1960*b*.

Groshong, R. H., Jr., Strain calculated from twinning in calcite, *Geol. Soc. Am. Bull.*, *83*, 2025–2038, 1972.

Groshong, R. H., Jr., Experimental test of least-square strain gage calculation using twinned calcite, *Geol. Soc. Am. Bull.*, *85*, 1855–1864, 1974.

Handin, J. W., and D. Griggs, Deformation of Yule Marble: Part II–Predicted fabric changes, *Geol. Soc. Am. Bull.*, *62*, 863–886, 1951.

Heard, H. C., Effect of large changes in strain rate in the experimental deformation of Yule Marble, *J. Geol.*, *71*, 162–195, 1963.

Heard, H. C., F. J. Turner, and L. E. Weiss, Studies of heterogeneous strain in experimentally deformed calcite, marble, and phyllite, *Univ. Calif. Publ. Geol. Sci.*, *46*, 81–152, 1965.

Higgins, M. W., Cataclastic rocks, *U.S. Geol. Surv. Prof. Pap.*, *687*, 1971.

Hobbs, B. E., Recrystallization of single crystals of quartz, *Tectonophysics*, *6*, 353–401, 1968.

Kohlstedt, D. L., W. B. Durham, and C. Goetze, High temperature creep in olivine, in *Proceedings of the Second International Conference on Mechanical Behavior of Materials*, pp. 383–387, American Society of Metals, New York, 1976.

Lister, G. S., Discussion: Cross-girdle c-axis fabrics in quartzites plastically deformed by plane strain and progressive simple shear, *Tectonophysics*, *39*, 51–54, 1977.

Lister, G. S., and M. S. Paterson, The simulation of fabric development during plastic deformation and its application to quartzite: Fabric transitions, *J. Struct. Geol.*, *1*, 99–115, 1979.

Lister, G. S., and G. P. Price, Fabric development in a quartz-feldspar mylonite, *Tectonophysics*, *49*, 37–78, 1978.

Lister, G. S., M. S. Paterson, and B. E. Hobbs, The simulation of fabric development and its application to quartzite: The model, *Tectonophysics*, *45*, 107–158, 1978.

Logan, J. M., M. Friedman, N. G. Higgs, C. Dengo, and T. Shimamoto, Experimental studies of simulated gouge and their application to studies of natural fault zones, in *Proceedings of Conference VIII: Analysis of Actual Fault Zones in Bedrock*, pp. 305–343, National Earthquake Hazards Reduction Program, Menlo Park, Calif. 1979.

Luton, M. J., and C. M. Sellars, Dynamic recrystallization in nickel and nickel-iron alloys during high temperature deformation, *Acta Metall.*, *17*, 1033–1043, 1969.

Mandl, G., L. N. J. de Jong, and A. Maltha, Shear zones in granular material, *Rock Mech.*, *9*, 95–144, 1977.

McQueen, H. J., and J. J. Jonas, Recovery and recrystallization during high temperature deformation, in *Treatise on Material Science and Technology*, edited by R. J. Arsenault, vol. 6, pp. 393–493, Academic, New York, 1975.

Mercier, J.-C. C., D. A. Anderson, and N. L. Carter, Stress in the lithosphere: Inferences from steady state flow in rocks, *Pure Appl. Geophys.*, *115*, 199–226, 1977.

Post, R. L., Jr., The flow laws of Mt. Burnet dunite, Ph.D. thesis, Dep. of Geol., Univ. of Calif., Los Angeles, 1973.

Post, R. L., Jr., High-temperature creep of Mt. Burnet dunite, *Tectonophysics*, *42*, 75–102, 1977.

Ramsay, J. G., *Folding and Fracturing of Rock*, McGraw-Hill, New York, 1967.

Ramsay, J. G., and R. H. Graham, Strain variation in shear belts, *Can. J. Earth Sci.*, *7*, 786–813, 1970.

Ross, J. V., H. G. Avé Lallemant, and N. L. Carter, Stress dependence of recrystallized grains and subgrain size in olivine, *J. Geophys. Res.*, in press, 1981.

Rutter, E. H., The kinetics of rock deformation by pressure solution, *Phil. Trans. R. Soc. London, Ser. A*, *283*, 203–219, 1976.

Rutter, E. H., and M. Rusbridge, The effect of non-coaxial strain paths on crystallographic preferred orientation development in the experimental deformation of a marble, *Tectonophysics*, *39*, 73–86, 1977.

Sandström, R., and R. Lagneborg, A model for hot working occurring by recrystallization, *Acta Metall.*, *23*, 387–398, 1975.

Taylor, G. I., Plastic strain in metals, *J. Inst. Met.*, *62*, 307-324, 1938.

Tchalenko, J. S., and N. N. Ambraseys, Structural analysis of the Dasht-e Baȳaz (Iran) earthquake fractures, *Geol. Soc. Am. Bull.*, *81*, 41-60, 1970.

Teufel, L. W., Strain analysis of experimentally superposed deformations using calcite twin lamellae, *Tectonophysics*, *65*, 291-310, 1980.

Tullis, J., Preferred orientation of quartz produced by slip during plane strain, *Tectonophysics*, *39*, 87-102, 1977.

Turner, F. J., Nature and dynamic interpretation of deformation lamellae in calcite of three marbles, *Am. J. Sci.*, *251*, 276-298, 1953.

Turner, F. J., and C. S. Ch'ih, Deformation of Yule Marble: Part III-Observed fabric changes due to deformation at 10,000 atmospheres confining pressure, room temperature, dry, *Geol. Soc. Am. Bull.*, *62*, 887-906, 1951.

Turner, F. J., D. T. Griggs, and H. C. Heard, Experimental deformation of calcite crystals, *Geol. Soc. Am. Bull.*, *65*, 883-934, 1954.

Turner, F. J., D. T. Griggs, R. H. Clark, and R. H. Dixon, Deformation of Yule Marble: Part VII-Development of oriented fabrics at 300°-500°C, *Geol. Soc. Am. Bull.*, *67*, 1259-1294, 1956.

Twiss, R. J., Theory and applicability of recrystallized grain size paleopiezometer, *Pure Appl. Geophys.*, *115*, 227-244, 1977.

Weathers, M. S., J. M. Bird, R. F. Cooper, and D. L. Kohlstedt, Differential stress determined from deformation-induced microstructures of the Moine Thrust zone, *J. Geophys. Res.*, *84*, 7496-7509, 1979.

Wenk, H. R., and J. Shore, Preferred orientation in experimentally deformed dolomite, *Contrib. Mineral. Petrol.*, *50*, 115-126, 1975.

Wenk, H. R., C. S. Venkitsubramanyan, and D. W. Baker, Preferred orientation in experimentally deformed limestone, *Contrib. Mineral. Petrol.*, *38*, 81-114, 1973.

Anisotropy in the Rheology of Hydrolytically Weakened Synthetic Quartz Crystals

Mark F. Linker and Stephen H. Kirby

U.S. Geological Survey, 345 Middlefield Road, Menlo Park, California 94025

Samples cut from a hydrothermally grown synthetic quartz crystal with 370 ppm hydroxyl were tested at constant compressive load at atmospheric pressure over a range of temperature $T = 400$-$800°C$, axial stress $\sigma = 800$-2000 bar, and duration of load $t = 27$ min to 2 months. Two orientations of compression were chosen: O^+ ($45°$ to $[2\bar{1}\bar{1}0]$ and to $[0001]$) and $\perp m$ (perpendicular to $(01\bar{1}0)$). Plastic strains up to 10% shortening were achieved. Microcracking was minor or absent. We report the following contrasts in behavior in the two orientations of compression:

Orientation	O^+	$\perp m$
Major slip system	$\{2\bar{1}\bar{1}0\}$ $\langle c \rangle$	Duplex $\{10\bar{1}0\}$ $\langle a \rangle$
Creep curve shape	Incubation stage	
	Accelerating creep rates	No significant incubation stage
	Hardening stage	
	Decelerating creep rates	Hardening stage only
	Tertiary stage	
	Failure only for $T = 600$-$620°C$	No tertiary stage
Relative creep rates	High	Low
Activation energies		
For creep, kcal mol^{-1}		
α-quartz field	22 ± 1	51 ± 3
β-quartz field	22 ± 2	28 ± 2
Stress exponent, n		
α-quartz field	3.0 ± 0.2	5.3 ± 0.5
β-quartz field		3.4 ± 0.3

Differences in the development of the incubation stage are likely to be caused by differences in dislocation multiplication mechanisms. We believe that the strong contrasts in the absolute creep rates and in the activation energies for creep in the two suites of experiments are caused by an anisotropy in diffusion of impurities in quartz, most likely in the diffusion of water or some species related to water.

Introduction

Remarkably small concentrations of structurally bound water (0.001 to 0.100 wt %) greatly reduce the plastic yield strengths of quartz crystals [*Griggs and Blacic*, 1964, 1965; *Heard and Carter*, 1968; *Blacic*, 1971; *Baëta and Ashbee*, 1970a, b; *Hobbs* et al., 1972]. This hydrolytic weakening has considerable potential for application to the strength of rocks in equilibrium with water in the earth's crust. Much work on the phenomenon has focused on the effects of the direction of compression and of the temperature on the yield strengths and slip mechanisms in synthetic quartz crystals deformed in short-term constant strain rate experiments [*Griggs*, 1967; *Blacic*, 1971, 1975; *Baëta and Ashbee*, 1969, 1970a; *Hobbs* et al., 1972; *Morrison-Smith* et al., 1976; *Kekulawala* et al., 1978]. Creep and constant strain rate tests performed in the past 7 years have also explored the effects of the time duration of loading on synthetic crystals [*Balderman*, 1974; *Kirby*, 1975, 1977; *Kirby and McCormick*, 1979]. All of these long-term tests were performed on crystals compressed at $45°$ to $[2\bar{1}\bar{1}0]$ and to $[0001]$, and slip on $\{2\bar{1}\bar{1}0\}$ $\langle 0001 \rangle$ was the primary plastic deformation mechanism. Slip parallel to $\langle a \rangle$ on various slip systems also is known to be an important aspect of the plasticity of synthetic quartz [*Christie and Green*, 1964; *Baëta and Ashbee*, 1969, 1970a, 1973;

Fig. 1. (a) Dead-load creep tester used in the deformation experiments. The specimen is heated by the miniature cylindrical split-tube furnace. The load, applied by steel weights, is transmitted through a 10:1 lever arm. Vertical displacement of loading column is measured independently by a dial gage and a displacement transducer. The moving element of the transducer rests on a micrometer head used in routine calibrations. Temperature is regulated by an on-off-type controller that senses the output of a thermocouple whose hot junction lies within the furnace windings. The temperature of the specimen is monitored with an additional thermocouple whose output is corrected to 0°C with an electronic reference junction. The hot junction of this thermocouple is positioned within the ceramic base anvil at the bottom of the specimen. (b) View of specimen and loading column with the front half of furnace removed. The hemispherical dome is just visible above the specimen. The white ceramic centering rings aid in initial alignment of the specimen.

Hobbs et al., 1972; *Twiss,* 1974, 1976; *Blacic,* 1975; *Morrison-Smith et al.,* 1976]. We have compressed samples taken from one synthetic quartz crystal cut in two orientations: one to activate ⟨c⟩ slip, the other to activate ⟨a⟩ slip. Our objective was to compare the rheologies of synthetic crystals that deform by slip in these two directions.

Experimental Details

Apparatus

We performed all experiments in a 10:1 lever-actuated dead-load creep apparatus, an improved version of a prototype used in earlier studies [*Kirby,* 1975, 1977; *Kirby and*

McCormick, 1979]. This apparatus (Figure 1a) provided a constant compressive force on specimens held at nominally constant temperature at atmospheric pressure. Noteworthy features of this tester include a miniature split-tube furnace that uses three independently powered heating elements and a self-aligning Al_2O_3 loading column. A ground and polished hemispherical dome was mated to a hemispherical seat in a two-part piston (Figure 1b). This design prevented failure by fracture upon loading caused by misalignment of the column. The load-bearing faces of the samples are still parallel after testing. We maintained relatively constant specimen temperature by using

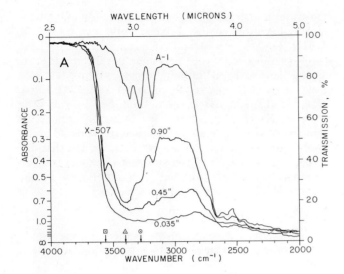

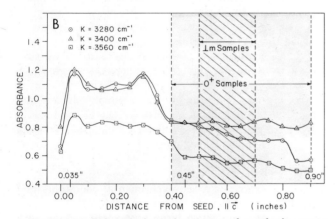

Fig. 2. (a) Infrared absorption curves of synthetic quartz crystal X-507, with aperture positioned at three different distances from the seed plate (0.035", 0.45", and 0.90") and the dry natural crystal A-1 used as a reference standard. Unpolarized radiation, 13.9-mm-thick polished plates oriented parallel to $(0\bar{1}\bar{1}0)$, spectra taken at room temperature. Greater absorbance near the seed (0.035") than near the crystal's edge (0.90") indicates that the center of the crystal is 'wetter' than the outer region. The three symbols located on the abscissa near 3500 cm^{-1} identify the wave numbers used in the fixed wave number profiles of Figure 2b. (b) Profiles of IR absorbance of X-507 at wave numbers 3280, 3400, and 3560 cm^{-1}, showing the region of nearly uniform absorbance from which the prisms for deformation experiments were taken. As indicated in (a), this quartz crystal is 'wetter' near the seed.

an electronic controller connected to a chromel-alumel thermocouple located in the upper furnace windings. A second chromel-alumel thermocouple, whose hot junction touched the bottom of the specimen and whose wires passed through an axial hole in the Al_2O_3 base anvil, was used to monitor the specimen temperature. The recorded temperatures were corrected for temperature gradients within the specimens, based on calibrations with hollow specimens. Specimen shortening was determined by using

an LVDT (linear voltage differential transformer) that monitored the motion of the yoke assembly (see Figure 1a) with respect to the mounting plate.

The following experimental uncertainties apply to our tests: (1) Temperature variation within specimens: $\pm\frac{1}{2}$% to 1% of the nominal temperature in °K; (2) Temperature drift: typically less than 2°C; (3) Stress (taken as the axial force divided by the room temperature cross sectional area, uncorrected for lateral expansion of the specimens): ±0.4% of the nominal stress; (4) Strain (expressed as the ratio of the shortening Δl to the original room temperature length l_0): ±0.01% ± 0.2% of the total nominal strain; corrections were applied for thermal expansion effects caused by temperature drift; direct measurement of specimen strain after testing agrees with the instrumental measurements to within ±0.2% total strain; (5) Time under load: ±0.5 min ±0.05% of nominal time.

Samples

We prepared oriented-right prisms of nearly square section from the Z-growth zone of a very well characterized synthetic quartz crystal provided by Bell Telephone Laboratories (see table below).

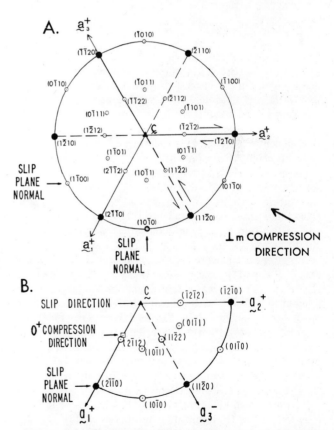

Fig. 3. Stereographic pole figure of alpha quartz showing orientations of $\perp$m (a) and O^+ (b) compression axes and the slip systems activated.

Growth History, Composition, and Physical Properties of Crystal X-507

Seed plate orientation*	5° to (0001)
	('Z' growth)
Hydrothermal solution*	1.0 m NaOH
Growth chamber temperature*	338°C
Hydrothermal pressure*	138 MPa
Growth rate*	1.6 mm/day
Infrared analysis†	240–520 $OH^-/10^6$ Si atoms
Absorption integration limits: 2800–3700/cm; see text and Figure 4	
Attenuation parameter, Q*	3.9×10^5

Trace element analysis, ppm*‡			
Na	50	Ag	<5
Li	<40	Ti	38
K	20	Co	ND
Fe	63	Al	67

Alpha-beta inversion temperature¶	574.8°C	
Unit cell dimensions¶	a	4.9131 Å
	c	5.4046 Å

Structurally right-handed (space group $P3_12$), based on sense of optical activity in basal orientation

Data sources:
* Bell Telephone Laboratories.
† This work.
‡ Analytical Laboratories, U.S. Geological Survey, Menlo Park, CA.
¶ *Ghirso et al.* [1979].

The hydroxyl concentration, as determined by infrared absorption in the 3-μm region [*Blacic*, 1975], is inhomogeneous in this crystal (Figure 2a), varying from 240 to 520 $OH^-/10^6$ Si atoms. The growth zone from which the specimens were taken is fairly uniform in infrared absorbance at the wavelengths of the principal hydroxyl absorption modes (Figure 2a, b), indicating a fairly uniform hydroxyl concentration of about 370 $OH^-/10^6$ Si atoms.

Prisms of approximate dimensions 11.2 mm × 4.4 mm × 4.4 mm were cut in two orientations (Figure 3): the O^+ orientation with long dimensions at 45° to [$2\bar{1}\bar{1}0$] and [0001] and prism faces parallel to (01$\bar{1}$0) and ~($\bar{2}$112); the $\perp m$ orientation with long dimensions perpendicular to (01$\bar{1}$0) and prism faces parallel to (0001) and (2$\bar{1}\bar{1}$0). The prism orientations were referenced to well-developed $m = \{10\bar{1}0\}$, $r = \{10\bar{1}1\}$, and $z = \{01\bar{1}1\}$ crystal faces. The polarity of the $\langle a \rangle$ axes was established from the larger development of r-rhombohedral faces relative to the less developed z-rhombohedra (see *Coe and Paterson* [1969] for orientation convention). The orientations of the prism compression faces are likely to be accurate to within 1°. Most of the prisms were ground with 600 grit SiC. Subsequent polishing had no significant effect on creep or on the deformation features (Figure 7b).

Previous work on other crystals has shown that compression in the O^+ orientation promotes $\{2\bar{1}\bar{1}0\}$ $\langle c \rangle$ slip [*Baëta and Ashbee*, 1969, 1970a; *Kirby*, 1975, 1977; *Kirby and McCormick*, 1979]. We selected the $\perp m$ orientation in order to have zero-resolved shear stress coefficients S_0 on

the systems with slip directions parallel to $\langle c \rangle$ and to have high values of S_0 on the systems $\{10\bar{1}0\}$ $\langle a \rangle$, $S_0 = 0.43$; $\{10\bar{1}1\}$ $\langle a \rangle$, $S_0 = 0.34$; $\{01\bar{1}1\}$ $\langle a \rangle$, $S_0 = 0.34$. These three systems with slip directions parallel to $\langle a \rangle$ are among those activated in short-term constant strain rate tests on synthetic quartz [*Baëta and Ashbee*, 1969, 1970a; *Hobbs et al.*, 1972; *Blacic*, 1971, 1975; *Twiss*, 1974, 1976; *Morrison-Smith et al.*, 1976]. We therefore anticipated that the O^+ and $\perp m$ compression orientations would exclusively promote $\langle c \rangle$ and $\langle a \rangle$ slip, respectively, and that differences in rheologies of test specimens in these two orientations could be associated with differences in the kinetics of slip in the $\langle c \rangle$ and $\langle a \rangle$ directions.

Experimental Procedure

We performed the experiments in the following way:
1. Apparatus assembled with sample aligned in position.
2. Sample slowly heated to desired test temperature. Heating time: 60–90 min.
3. Displacement transducer calibrated after apparatus reached thermal equilibrium.
4. Load applied over a period of less than 10 s by carefully lowering the scissors jack that bore the weights (see Figure 1a).
5. Continuous recording of thermocouple emf and displacement transducer output during creep of specimen.
6. Load removed and specimen air quenched to room temperature in less than 2 min.

Experimental Results

Table 1 summarizes the experimental conditions and creep histories of the 44 tests on crystal X-507. The specimens generally sustained large permanent strains with only minor microcracking, most of which occurred during unloading and cooling. In the O^+ specimens taken to large strain (>3%), some axial cracking occurred in association with stress concentrations at the boundaries of broad deformation bands (Figure 4e). Both O^+ and $\perp m$ specimens were cloudy after deformation. This condition, observed in almost all tests on synthetic quartz [*Blacic*, 1971, 1975; *Hobbs*, 1968; *Baëta and Ashbee*, 1969, 1970a, 1973; *Morrison-Smith et al.*, 1976; *Kirby*, 1975, 1977; *McCormick*, 1977; *Kirby and McCormick*, 1979], is caused by precipitation of water vapor into submicroscopic bubbles [*Dodd and Fraser*, 1965, 1967]. The influence of precipitation on the creep behavior of synthetic quartz is examined in Appendix 2.

Deformation Mechanisms

In Table 2 we summarize the observations that pertain to the mechanisms of deformation in the two suites of samples. We emphasize below several contrasting features in the plasticity of crystal X-507, which is compressed in the two orientations:
1. All of the observations on the O^+ specimens are consistent with more extensive evidence from a 'wetter' crys-

TABLE 1. Tabulated Creep Data for Crystal X-507

Experiment	Orientation	Stress, bar	Temperature, °C*	Strain, %†	Time Under Load, min	Strain Rates§, s⁻¹ Maximum	2% Strain	4% Strain	Incubation Time, min	Comment
LC-18	O⁺	1,400	543	4.29	377	4.11 − 6	2.92 − 6	1.18 − 6	77	
LC-19	O⁺	1,222	508	4.52	1,445	1.34 − 6	9.67 − 7	4.19 − 7	401	
LC-20	O⁺	1,400	512	4.60	931	2.03 − 6	1.58 − 6	7.22 − 7	291	
LC-21	O⁺	1,400	450	0.46	834	Failed before reaching maximum strain rate				?
LC-22	O⁺	1,400	452	4.46	3,144	5.24 − 7	5.24 − 7	2.10 − 7	1,150	
LC-24	O⁺	1,397	397	4.67	9,238	1.89 − 7	1.89 − 7	9.88 − 8	3,860	
LC-25	O⁺	1,400	550	4.20	370	3.49 − 6	2.78 − 6	1.19 − 6	62	
LC-26	O⁺	1,400	703	4.16	58	2.78 − 5	1.79 − 5	8.29 − 6	5	
LC-27	O⁺	1,400	650	4.25	76	1.66 − 5	1.34 − 5	6.67 − 6	9	
LC-28	O⁺	1,400	600	3.49	60	1.64 − 5	1.43 − 5			X
LC-29	O⁺	1,400	600	4.51	100	1.06 − 5	1.01 − 5	6.33 − 6	18	X
LC-30	O⁺	1,000	511	4.86	2,577	8.81 − 7	5.56 − 7	2.39 − 7	464	
LC-31	O⁺	1,600	512	4.70	553	2.94 − 6	2.60 − 6	1.59 − 6	175	
LC-32	O⁺	800	512	4.49	6,066	3.54 − 7	2.01 − 7	8.23 − 8	1,030	
LC-33	O⁺	1,000	448	4.77	7,110	2.55 − 7	2.27 − 7	1.34 − 7	2,590	
LC-35	O⁺	1,000	549	4.48	1,459	1.27 − 6	7.92 − 7	3.13 − 7	198	
LC-37	⊥m	1,000	509	1.00	17,327	2.23 − 8	4.36 − 9‡		2,810	
LC-40	⊥m	1,600	509	3.78	20,105	2.32 − 7	2.86 − 8		271	
LC-41	⊥m	1,400	702	4.15	221	8.54 − 6	3.07 − 6	2.00 − 6	3	
LC-43	⊥m	1,200	509	2.53	30,290	3.15 − 8	9.66 − 9		59	
LC-44	⊥m	1,400	651	4.03	589	2.84 − 6	1.52 − 6	6.60 − 7	5	P
LC-45	⊥m	1,400	509	2.20	15,936	6.81 − 8	1.84 − 8		521	P
LC-46	⊥m	1,400	550	3.98	11,578	5.15 − 7	5.82 − 8		42	
LC-47	⊥m	1,600	650	3.25	258	5.68 − 6	1.77 − 6	1.20 − 6¶	5	
LC-48	⊥m	1,000	649	4.58	2,886	1.01 − 6	2.98 − 7	1.57 − 7	18	
LC-49	⊥m	1,400	600	4.25	2,759	1.22 − 6	2.68 − 7	1.51 − 7	2	
LC-50	⊥m	1,200	650	4.88	1,339	2.88 − 6	6.51 − 7	4.30 − 7	14	
LC-51	⊥m	1,400	650	4.53	671	4.10 − 6	1.19 − 6	7.43 − 7	8	
LC-52	⊥m	1,400	454	2.49	87,962	9.78 − 9	3.82 − 9		None	
LC-53	⊥m	1,400	748	10.07	477	1.51 − 5	5.26 − 6	3.83 − 6	5	
LC-54	⊥ m	800	653	4.38	6,999	5.53 − 7	1.24 − 7	6.23 − 8	43	
LC-55	O⁺	1,400	751	4.14	40	3.90 − 5	2.00 − 5	9.30 − 6	2	
LC-56	O⁺	1,400	602	3.66	62	1.20 − 5	1.20 − 5		12	X
LC-58	O⁺	1,400	623	4.06	53	1.40 − 5	1.40 − 5	1.40 − 5	11	X
LC-59	O⁺	1,400	810	6.04	33	9.28 − 5	4.27 − 5	2.51 − 5	1	
LC-60	O⁺	1,000	397	4.44	19,850	7.98 − 8	7.71 − 8	5.33 − 8	9,080	
LC-61	⊥m	1,400	552	6.20	17,111	6.39 − 7	7.63 − 8	5.46 − 8	43	
LC-62	⊥m	1,400	807	9.75	306	2.46 − 5	9.21 − 6	6.00 − 6	1	
LC-63	⊥m	1,400	508	0.44	1,028	7.58 − 8			None	P
LC-64	⊥m	2,000	512	2	420	1.61 − 6				X
LC-65	⊥m	1,800	510	5.07	15,508	4.26 − 7	5.57 − 8	3.51 − 8	165	
LC-67	⊥m	1,400	512	0.05	205	Unloaded before reaching maximum strain rate				P
LC-68	⊥m	1,400	511	4.44	33,068	1.08 − 7	2.31 − 8	1.65 − 8	123	
LC-77	O⁺	1,400	804	4.62	27	9.09 − 5	3.53 − 5	1.72 − 5	1	

*Temperature at maximum strain rate, ±2°C drift.
†Instrumental strain at end of test.
‡At 1.00% strain.
¶Extrapolated from 3.25% strain.
§Multiplier and exponent of 10, i.e., 4.11 − 6 = 4.11 × 10⁻⁶.
?—Failure not reproducible.
X—Severe cracking leading to tertiary creep and failure.
P—Specimen surfaces polished prior to deformation test.

tal (X-0) reported earlier [Kirby, 1975, 1977; McCormick, 1977; Kirby and McCormick, 1979] and with observations on O⁺ specimens of intermediate OH⁻ concentration (S. H. Kirby and J. W. McCormick, unpublished data, 1976–1979). The identification of the major slip system, {2̄110} ⟨0001⟩, is unequivocal. The style of the plasticity of O⁺ specimens is lamellar, i.e., the {2̄110} ⟨0001⟩ dislocations tend to concentrate in localized slip bands approximately parallel to (2̄110), causing the stress-optical effects characteristic of deformation lamellae. The fact that the slip bands on O⁺ specimens of X-507, revealed as surface markings on (2̄112) surfaces, are not as straight as in O⁺ tests on wetter crystals [Kirby and McCormick, 1979], suggests a greater tendency for cross slip to (101̄0) and

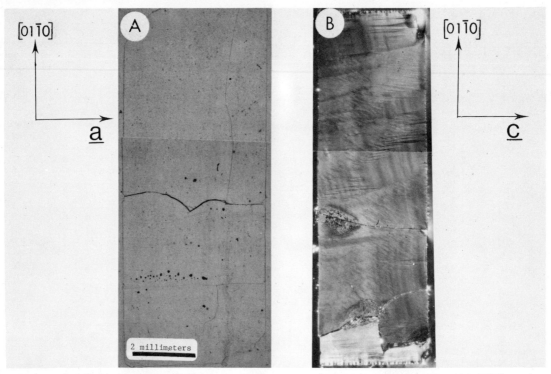

Fig. 4(a,b). Photomicrographs of LC-40: ⊥m orientation, 510°C, 1600 bar, 3.8% shortening: (a) (0001) section, plane-polarized light; (b) (2̄1̄10) section, crossed polars. Faint lamellar features parallel to specimen axis are caused by inhomogeneities in hydroxyl concentration parallel to the growth interface. Faint lamellar features parallel to [0001] are not identified.

(1̄100). Lastly, the operation of a single slip system in prisms constrained to maintain coaxial ends leads to the formation of a broad deformation band approximately parallel to (0001). In this region, {2̄1̄10} ⟨0001⟩ slip is concentrated, and external rotations of the slip plane and direction occur [*Turner*, 1962; *Kirby and McCormick*, 1979]. The marked barreling of the O⁺ prisms is caused by this inhomogeneity of slip and by additional frictional constraints that the ceramic pistons place on the distribution of plastic strain.

2. The ⊥m specimens remained right prism shaped to axial strains of 10% (Figure 4). The anisotropy in radial strain is extreme, with nearly zero strain in the [0001] direction (Figure 6). Thus the complete finite strain tensor, in terms of the axial strain ϵ_{11} and referred to the coordinate system in Figure 6, is

$$[\epsilon_{ij}] = \begin{bmatrix} \epsilon_{11} & 0 & 0 \\ 0 & \left(\dfrac{1}{1+\epsilon_{11}} - 1\right) & 0 \\ 0 & 0 & \sim 0 \end{bmatrix} \quad (1)$$

where compressive strain is positive. Consistent with the form of strain tensor given, the plastic deformation does not change the orientation of the specimen optic axis (Figure 4c, d). Unlike the O⁺ specimens, the ⊥m suite does

not contain strongly developed deformation bands, and the deformation is generally much more homogeneous. Preliminary transmission electron microscopy (by J. McCormick) on sample LC-40 has not revealed any prominent slip band arrays as were present in O⁺ specimens of wetter crystals [*McCormick*, 1977; *Kirby and McCormick*, 1979]. The form of the strain tensor (1) and the lack of rotation of the optic axis require that the primary slip systems have slip plane normals and slip directions perpendicular to [0001]. The most prominent slip bands, as revealed by surface markings (Figure 5) and by selective chemical etching [*Christie and Ord*, 1979], are approximately parallel to (101̄0) and (1̄100). The primary slip systems must therefore be (101̄0) [1̄2̄10] and (1̄100) [1̄1̄20]. *Christie and Ord* [1979] note some secondary slip on several other slip systems; the plastic strain tensor (Figure 6 and equation (1)) indicates that these systems (Table 2) could not have contributed significantly to the total plastic strain.

In summary, the observations collected in Table 2 show that the two suites of specimens deformed by different slip systems, namely, for O⁺, {2̄1̄10} ⟨c⟩, and for ⊥m, duplex {101̄0} ⟨a⟩ slip. Differences in the creep properties of samples in these two orientations of compression, described below, are to be linked to these differences in deformation mechanisms.

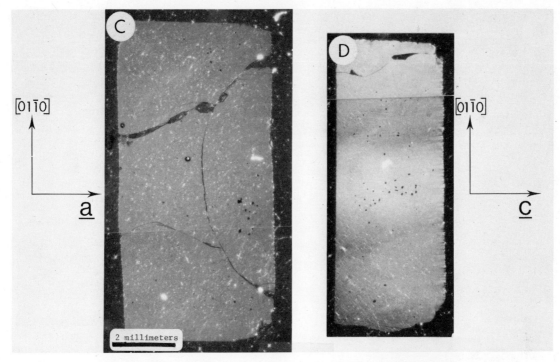

Fig. 4(c,d). Photomicrographs of LC-53: ⊥m orientation, 750°C, 1400 bar, 10.1% shortening; crossed polars. (c) (0001) section, polars slightly off normal. (d) (2$\overline{1}$10) section.

The Creep Curves

Duplicate tests generally showed excellent reproducibility (Figure 7a). Creep experiments on ground and on carefully polished specimens indicate that surface condition has little effect on the creep behavior (Figure 7b).

Creep curves for all tests listed in Table 1, plotted on appropriate scales, are available on microfiche.* We plot representative creep curves in Figure 8 for the O^+ and ⊥m orientations in both the α- and β-quartz stability fields.

The two-stage form of the O^+ creep curves (Figure 8a, c) is identical to that developed in earlier tests on crystals with higher hydroxyl concentration [*Kirby*, 1975, 1977; *Kirby and McCormick*, 1979]:

1. An incubation stage in which creep rates increase with time.

2. A hardening stage in which creep rates decrease continuously with time.

The duration of the incubation stage in the O^+ experiments systematically increases with decreasing temperature and stress (see Table 1). A subsequent hardening stage developed in all but four experiments. In the four experiments in the temperature interval 600°C–620°C, the hardening stage is absent or poorly developed, and a tertiary stage of accelerating creep rates leads to failure by fracture at strains of less than 5% (Figure 10c). Above

---

* Creep curves are available with entire article on microfiche. Order from AGU; document GM 80-001, price $1.00. Payment must accompany order.

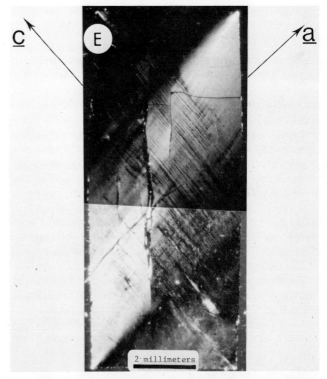

Figure 4e. Photomicrograph of deformed specimen LC-25, viewed between crossed polars: O^+ orientation, 550°C, 1400 bar, 4.2% shortening. Note (2$\overline{1}$10) deformation lamellae, broad basal deformation band, and axial cracks.

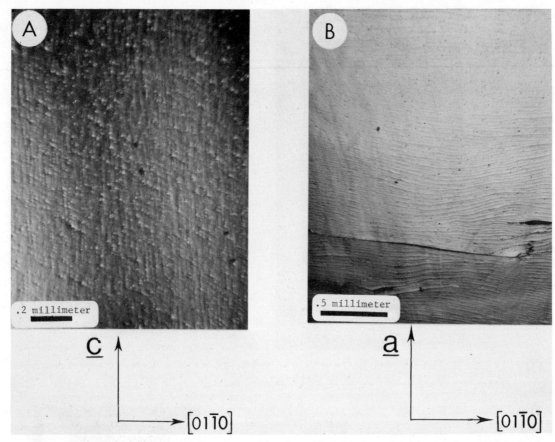

Fig. 5. Photomicrographs of polished surfaces with reflected light. Deformed specimen LC-44: 650°C, 1400 bar, 4.3% shortening. Compression direction was parallel to [01$\bar{1}$0]. (*a*) (2$\bar{1}$10) surface. Note traces of slip bands approximately parallel to [0001]. (*b*) (0001) surface. Note lack of slip bands and the presence of shallow axial cracks.

and below this temperature interval, normal ductility and hardening occur in the O$^+$ experiments.

Creep experiments on ⊥m specimens do not exhibit prominent incubation stages as those on the O$^+$ specimens. The incubation times t_i (defined in Figure 8*a*), where detectable, do not vary systematically with experimental conditions (Table 1). Hardening occurs in all ⊥m tests. The hardening rates ($d\dot{\epsilon}/d\epsilon$) are not strongly dependent on experimental conditions, except for an unusually high hardening rate at $T = 550°$C and $\sigma = 1400$ bar (note the differences in the strain rates at maximum value and at 2%, indicated in Figure 13*b*).

Another important difference between the two suites of experiments is that the ⊥m specimens crept at much lower rates than O$^+$ specimens deformed under the same conditions (compare the time scales of the creep curves in Figures 8, 9, and 10, and see the tabulated creep rates in Table 1).

Creep Rate Variations With Experimental Conditions

All of the creep curves exhibited transient behavior, i.e., the creep rates varied with strain and time. Previous work on another crystal cut in the O$^+$ orientation showed that

dislocations rapidly multiply during the incubation stage, reaching maximum values at about the maximum strain rate [*Kirby and McCormick*, 1979]. In the hardening stage the dislocation density remained constant, while progressively more structurally bound water precipitated into spherical bubbles. This led to dehydration hardening [*Kirby and McCormick*, 1979]. Since the densities and configurations of dislocations were steady state once the maximum strain rate was reached, the maximum strain rate best represents the steady state strain rate caused by dislocation flow. We present evidence in Appendix 2 that this same conclusion applies to the present suite of experiments. However, the contrasts in the mechanical behavior of these two suites of experiments are so great that comparisons are independent of where on the creep curves the strain rates are picked. Accordingly, we have determined the standard empirical rheological constants that describe the effects of stress and temperature on maximum strain rates.

Stress Effects. Creep rates increase systematically with increasing applied stress for both orientations (Figures 9 and 12). The sensitivity of strain rate to changes in applied stress is not greatly influenced by the α- and β-

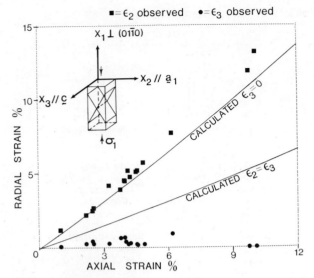

■ = ϵ_2 observed   ● = ϵ_3 observed

Fig. 6. Shape change of the deformed $\perp m$ prisms. Filled squares—radial strain parallel to [2$\bar{1}$10]. Filled circles—radial strain parallel to [0001]. Both are plotted against the axial shortening of the prisms in percent strain.

quartz transformation (Figure 12b) or by crystal orientation (Figure 12a, b). Fits of the creep rate and stress data to laws of the form

$$\dot{\epsilon} = B\sigma^n$$

give $n = 4.2 \pm 1.2$ for these tests (Table 3). This is consistent with all previous tests on 'wetter' crystals in the O^+ orientation [Kirby and McCormick, 1979].

Temperature Effects. Interpretation of the effects of temperature on the creep curves (Figures 10 and 11) is more complex than for the stress effects for three reasons:

1. The smooth and systematic effect of temperature on the creep rates of the O^+ specimens is interrupted in the interval 600°C–620°C by a tertiary creep stage which leads to failure by fracture. Above and below this temperature interval, the normal incubation and hardening stages develop. The creep curves for three of the tests (LC-29, -56, and -58) that exhibited a tertiary stage are shown in Figure 10c. The samples from two tests that were stopped before failure were severely cracked. We argue in Appendix 1 that an interaction between plastic deformation and microfracturing which involves the alpha-beta transition was responsible for the creep instability. The creep rates in tests over the interval $T = 600°C–620°C$ are therefore not included in our analysis of the temperature effects on creep rates.

2. O^+ specimens tested at lower temperatures and higher stresses were found to develop more cracks than those tested at the higher temperatures and lower stresses. Although the contribution to the total permanent strain was small compared to the slip strain, we were concerned that microfracturing was indirectly affecting strain rates by modifying internal stresses, thereby bias-

ing the comparison between the O^+ and $\perp m$ suites of runs. A series of tests at $\sigma = 1000$ bar, in which microfractures were less numerous than in the $\sigma = 1400$-bar runs, indicates that neither the minor microcracking nor applied stress have large effects on the temperature sensitivity of creep rate.

3. Previous work on O^+ crystals with higher hydroxyl concentrations has shown that the sensitivities of creep rates to changes in temperature are systematically lower in the β-quartz stability field than in the α-quartz field [Kirby, 1975, 1977; Kirby et al., 1977]. In contrast, our O^+ tests on crystal X-507 indicate that the temperature sensitivity of creep rate is not greatly different in the α- and β-quartz fields (Figure 13a). However, in the $\perp m$ data, rather large changes in temperature sensitivity are apparent, especially in the maximum strain rate data (Figure 13a).

Table 3 presents the results of least squares fits of the maximum strain rates ($\dot{\epsilon}_{max}$) to a flow law of the form:

$$\dot{\epsilon} = A\sigma^n \exp(-E_c^*/RT) \qquad (2)$$

where A, n, and E_c^* are material constants. In view of the uncertain effect of the α-β transition in the O^+ data, we also provide fits to the data over the entire temperature range for these experiments.

Anisotropy in the Steady State Rheology of Synthetic Quartz

Specimens in the O^+ and $\perp m$ orientations differ in rheological behavior in two significant ways:

1. Over our range of experimental conditions ($400 \leq T \leq 800°C; 800 \leq \sigma \leq 1800$ bar), the $\perp m$ crystals creep at considerably lower rates than the O^+ crystals. The contrasts in axial creep rates illustrated in Figures 12 and 13a are actually less than the contrasts in the glide shear strain rate, since duplex $\{10\bar{1}0\}$ $\langle a \rangle$ slip occurs in the $\perp m$ specimens. Neglecting both a small difference in resolved shear stress coefficient and any slip interference effects, the creep rates of the $\perp m$ specimens should be halved in any comparison with O^+ specimens. Thus the shear creep rates for the two orientations of compression at $T = 450°C$ and $\sigma = 1400$ bar actually differ by more than a factor of 100.

2. The creep rates of $\perp m$ specimens are much more sensitive to variations in temperature than those of O^+ specimens (Figure 13a). This conclusion is independent of interpretations of the creep curves and of what effects the α-β transition has on the rheology. The activation energies for creep are consistently higher in the $\perp m$ specimens (Table 3).

Why do the $\perp m$ and O^+ crystals behave so differently? It is generally agreed that diffusion plays an important if not controlling role in the hydrolytic weakening process. When the preliminary results of this work showed that the activation energies for creep varied by a factor of 2 with orientation [Linker and Kirby, 1978], we initially questioned the importance of diffusion in water-weakened

TABLE 2. Deformation Features in Crept Specimens of Synthetic Quartz Crystal X-507

Compression Orientation	O^+ (Figure 3b)		$\perp m$ (Figure 3a)	
Prism face orientation	$\sim(\bar{2}112)$	$(01\bar{1}0)$	$(2\bar{1}10)$	(0001)
Radial strain normal to prism faces	$\epsilon_r \simeq \left(\dfrac{1}{\epsilon_a+1}\right)-1$	~ 0	$\epsilon_r \simeq \left(\dfrac{1}{\epsilon_a+1}\right)-1$ (Fig. 6)	~ 0 (Fig. 6)
Barreling normal to prism faces?	Pronounced (Figure 4e)	None	None (Figure 4a, c)	None (Figure 4b, d)
Deformation features on polished faces	Prominent wavy slip bands $\sim \parallel [01\bar{1}0]$. Evidence for cross slip.	No slip bands. Short, shallow axial cracks at strain inhomogeneities.	Faint short slip bands $\sim \parallel [0001]$ (Figure 5a).	No slip bands. Shallow, continuous axial cracks (Figure 5b).
Features revealed by NH_4HF_2 etching*	ND†	ND†	Prominent wavy bands of etch pits at $\sim 90°$ to $[01\bar{1}0]$. Fainter bands at 55 to 61° to $[10\bar{1}0]$.	Prominent continuous bands of etch pits at 55° to 67° to $[01\bar{1}0]$. Fainter discontinuous bands of etch pits at 42° to 51° to $[01\bar{1}0]$.
Observations in thin sections cut parallel to prism faces				
Plane polarized light	Inclined deformation lamellae.	Deformation lamellae $\sim \parallel (2\bar{1}10)$.	No features observed	No features observed (Figure 4a, c)
Between crossed polars	Relatively uniform extinction position.	Deformation lamellae. Broad deformation band $\sim \parallel (0001)$ in which c-axis is rotated (Figure 4e)	Subtle variation of extinction position in bands $\sim \parallel [0001]$ in α-quartz samples only (Figure 4b, d).	Sections at extinction, no change in optic axis orientation.
Major slip system	$\{2\bar{1}10\}\langle 0001\rangle$ (Fig. 3B)		Duplex slip on $\{10\bar{1}0\}\langle a\rangle$: $(10\bar{1}0)[1\bar{2}10]$ and $(\bar{1}100)[\bar{1}\bar{1}20]$ (Fig. 3A)	
Secondary slip systems	Cross slip to $\{hki0\}\langle 0001\rangle$		Some secondary slip on systems of the type: $\{10\bar{1}1\}\langle c\pm a_1\rangle$, $\{\bar{1}011\}\langle c\pm a_3\rangle$, $\{11\bar{2}2\}\langle c\pm a_3\rangle$, and $\{\bar{1}2\bar{1}2\}\langle c-a_2\rangle^*$.	

*Unpublished report, *Christie and Ord* [1979].
†ND: Not determined

quartz. However, a subsequent review of the impurity diffusion data for quartz, summarized below, indicates that the diffusion of impurities varies with direction in ways similar to the creep anisotropy. This suggested to us that an anisotropy in the diffusion of water (or some related species) is responsible for the observed creep anisotropy. *Blacic* [1971, 1975] demonstrated that slip parallel to [0001] is strongly preferred over (0001) $\langle a\rangle$ slip in synthetic quartz crystals compressed in the O^+ and $(10\bar{1}1)$ directions and proposed that this preference stems from a higher water diffusivity in the [0001] direction. No details were given linking the two phenomena. Following our review of impurity diffusion in quartz, we review and extend the key concepts of hydrolytic weakening and show that for simple dislocation geometries, diffusion anisotropy leads to creep anisotropy.

The diffusivities of most impurities in quartz are strikingly higher in the [0001] direction than in directions perpendicular to [0001]. Electrical conduction in quartz occurs by ionic migration. The electrical conductivity of quartz is typically many orders of magnitude higher parallel to [0001] than normal to [0001] [*Sosman*, 1927; *Mortley*, 1969]. The diffusion rates of impurities are so low in directions perpendicular to [0001] that measurements in that direction are often not even attempted [*Vogel and Gibson*, 1950; *Verhoogen*, 1952]. The generally accepted explanation of this diffusion anisotropy is that the open channels parallel to [0001] in the α- and β-quartz structures allow easy migration of impurity ions.

Direct experimental data on diffusion in the two principal directions are available for three impurities: $^{22}Na^+$, $^3H_2O-^1H_2O$, and $H_2^{18}O$. *Frischat* [1970] reports for $^{22}Na^+$

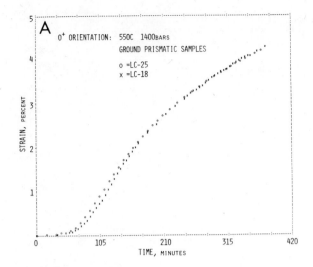

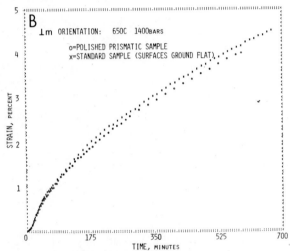

Fig. 7. The reproducibility of the creep curves. (a) Two O⁺ tests run at the same conditions. (b) Two ⊥m tests with polished and unpolished prism specimens.

at room pressure, $D_{\parallel c}/D_{\perp c} \simeq 10^4$ at 440°C and $D_{\parallel c}/D_{\perp c} \simeq 200$ at 790°C (see Figure 13b). The activation energy for ^{22}Na⁺ diffusion in the $\perp[0001]$ direction is double that for diffusion in the [0001] directions. The activation energies are also systematically lower in the α-quartz stability field. *Sang-Hwang* [1975] did not observe a significant anisotropy of H⁺ diffusion in ^{3}H$_2$O–^{1}H$_2$O interdiffusion experiments on quartz. *Choudhury et al.* [1965] measured the uptake of H$_2$^{18}O water into quartz crystals subject to hydrothermal treatment at $T = 667$°C and a pressure of 820 bar. The diffusivity in the [0001] direction was a factor of 40 higher than in the $\perp[0001]$ direction.

Specifically, how can a diffusion anisotropy cause a creep anisotropy? We do not have the insight and data to put forward a quantitative explanation of the phenomenon. We can, however, apply existing ideas on hydrolytic weakening to simple dislocation loop geometries and predict a creep anisotropy that is caused by a diffusion anisotropy.

To do so, we need to briefly review current understanding of the hydrolytic weakening process.

The quartz structure is made up of [SiO$_4$]$^{4-}$ tetrahedra that share corners in a three-dimensional polymeric structure. Plastic deformation of high-purity quartz thus requires the breaking of strong Si-O bonds to disrupt the Si-O-Si bridges between silicon atoms. *Griggs and Blacic* [1964, 1965] proposed that hydrolysis chemically degrades the strength of the Si-O-Si bridges.

$$\text{Si-O-Si} + \text{H}_2\text{O} \rightarrow \text{Si-O}\begin{array}{c}\text{H}\cdots\\\cdots\\\text{H}\cdots\end{array}\text{O-Si} \qquad (3)$$

Only the weak secondary (hydrogen) bonds between hydrogens and oxygens (dotted) need be broken to disrupt the Si to Si bridges in subsequent deformation. Since even in the 'wettest' quartz crystal less than 1% of all Si-O-Si bridges are hydrolyzed at any given time, this chemical degradation of strength must be localized at the moving dislocations that produce the plastic strain. The rate of diffusion of H-O-H (the structurally bound counterpart of the H$_2$O molecule) controls rates of hydrolysis of the Si-O-Si bridges just ahead of a moving dislocation. After hydrolysis, the dislocation can advance by hydrogen bond exchange [*Griggs*, 1967; Figure 14]. Therefore, the dislocation velocity is controlled by how fast H-O-H can diffuse to Si-O-Si sites ahead of a dislocation. Even though the H-O-H concentration necessary to saturate the 'dangling' bonds at dislocations is small (~1 ppm at a dislocation density of 10^9 cm/cm^3), the strength of synthetic quartz crystals decreases smoothly with increasing OH⁻ concentration [*Griggs*, 1967, 1974; *Kirby et al.*, 1977; *Kirby and Linker*, 1979]. This shows that the water between dislocations plays a key role in the hydrolytic weakening process. It is well known that impurities tend to segregate to the vicinity of dislocations in metallic alloys and that the segregation is driven by the elastic interaction between the stress field of a dislocation and the structural distortion created by an impurity defect. Free energy potential wells for impurities thus exist around dislocations, and the nature of these wells depends on the geometries of the elastic stress fields and of the impurity distortion. *Blacic* [1971] and *Griggs* [1974] proposed that such potential wells exist for H-O-H in quartz and that consequently clouds of H-O-H accompany dislocations in water-weakened quartz. These clouds are local reservoirs of H-O-H that increase the probability of hydrolysis of Si-O-Si bridges adjacent to the dislocations. As a dislocation moves, the cloud of H-O-H's moves with it.

Plastic strain occurs by the expansion of dislocation loops, and this reduces the concentration per dislocation line length of H-O-H in the cloud and in the dislocation core region. *Griggs* [1974] considered two processes for H-O-H replenishment: radial diffusion of H-O-H from the interdislocation region and pipe diffusion parallel to the dislocation lines in the core and cloud regions. Enhanced diffusion along dislocations, well documented in many

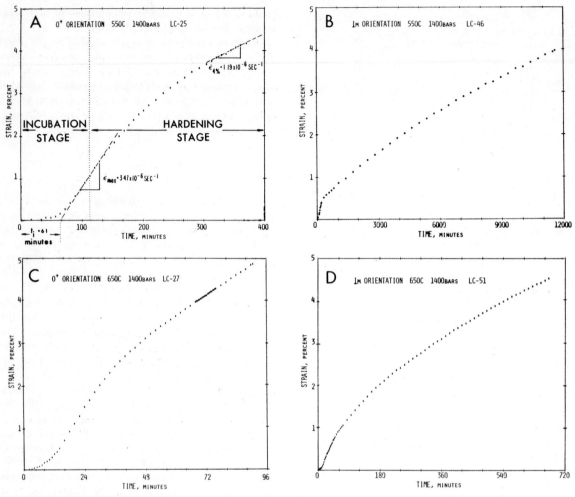

Fig. 8.   Four typical creep curves for the O$^+$ (a and c) and $\perp m$ (b and d) compression of alpha quartz (top row) and beta quartz (bottom row). Definitions of parameters used in the determination of flow law constants are shown in (a).

materials, would tend here to equilibrate H-O-H concentration along a dislocation loop. Thus the rate of loop expansion, proportional to the dislocation velocity and directly related to creep rate, is controlled by the rates that the H-O-H clouds move with the dislocations and by how fast the clouds and cores are replenished after loop expansion. Both of these processes are in turn controlled by the rates of H-O-H diffusion.

Before considering the effects of diffusion anisotropy on the rates of expansion of dislocation loops with simple geometries, we need to know how H-O-H interacts with dislocations in quartz. The equilibrium concentration c of a solute species is given by

$$c = c_0 \exp \; (-E/kT) \qquad (4)$$

where c_0 is the solute concentration in stress-free crystal, E the reversible work done in inserting the solute species into a given position in the field of the dislocation k the Boltzmann's constant, and T the temperature in °K

[*Hirth and Lothe*, 1968]. The interaction energy E is generally considered to be dominated by the elastic work done in introducing the distortion of the solute species $\Delta\epsilon_{ij}$ into the stress field of the dislocation σ_{ij}:

$$E \sim \sigma_{ij}\Delta\epsilon_{ij} \qquad (5)$$

Although the elastic stress fields for straight dislocations in anisotropic quartz have been calculated [*Heinisch et al.*, 1975], we do not have a specific model for the distortions associated with the H-O-H defect. We do know that hydrostatic pressure increases the solubility of water in quartz at low pressure [*Kats et al.*, 1962; *Shaffer et al.*, 1974; *Choudhury et al.*, 1965; *Sang-Hwang*, 1975; *Paterson and Kekulawala*, 1979]. A negative pressure-volume work term, then, is apparently important in E. Lacking detailed data on the structure of the H-O-H defect, we assume that this term dominates. The segregation of H-O-H to dislocations is thus dependent on the hydrostatic stress component of the dislocation stress field. In quartz, screw

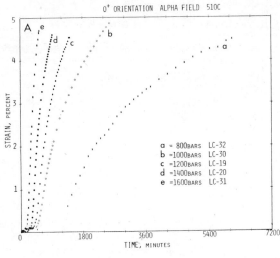

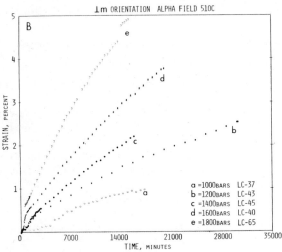

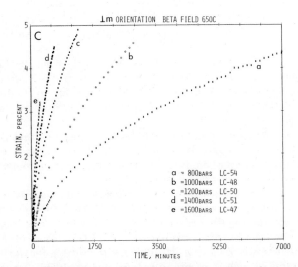

Fig. 9. Collected creep curves, showing the effects of stress: (a) O+ orientation; (b,c) ⊥m orientation. Alpha-quartz stability field (top row, a and b). Beta-quartz stability field (bottom row, c).

dislocations have very small hydrostatic components, as compared to edge dislocations [*Heinisch et al.*, 1975], thus indicating that radial gradients in solubility and radial diffusion are probably not important for screw dislocations. In contrast, edge dislocations in quartz have strong gradients in the hydrostatic components of their stress fields, and therefore radial diffusion is likely to be important.

In Figure 15 we explore the consequences of a strong diffusion anisotropy on the expansion rates of dislocations with simple sqaure-loop geometries. Consider an edge dislocation with an equilibrium distribution of H-O-H. Glide movement of this edge dislocation causes the solubility distribution to change. The chemical potential gradients caused by this movement are primarily parallel to the slip direction (the Burgers vector direction). A high diffusion rate parallel to the Burgers vector should facilitate the movement of the cloud, thereby increasing the glide velocity under stress. Screw dislocations, lacking large potential wells produced by hydrostatic pressure, are not expected to exhibit such cloud drag anisotropy effects. Replenishment of H-O-H, following depletion caused by dislocation loop expansion, occurs by radial diffusion toward edge dislocations. A high diffusion rate radial to the edge segments increases the rate of replenishment. Lastly, screw dislocations, lacking this hydrostatic pressure driving potential for diffusion, are probably replenished by pipe diffusion. A high bulk crystal diffusion rate parallel to the screw dislocation should favor loop expansion facilitated by pipe diffusion.

The effects of diffusion anisotropy predicted here are in harmony with our observed creep anisotropy. The high creep rates associated with $\{2\bar{1}\bar{1}0\}$ $\langle 0001 \rangle$ slip are likely to result from the high diffusion rate for H-O-H parallel to the slip direction (Figure 15a):

1. Edge segments move in the high-diffusivity direction, facilitating the movement of the H-O-H clouds.

2. The high diffusivity direction is radial to the edge segments, facilitating H-O-H replenishment by radial diffusion after loop expansion.

3. Replenishment of H-O-H for screws is facilitated by pipe diffusion parallel to the high-diffusivity direction.

Creep by $\{10\bar{1}0\}$ $\langle a \rangle$ slip should be difficult by the same arguments (Figure 15b):

1. Edges move in the low-diffusivity direction, inhibiting cloud movement.

2. Radial diffusion to edges is in low-diffusivity directions.

3. Pipe diffusion along screws is in the low-diffusivity direction. We are uncertain how more complex loop geometries than considered in Figure 15 affect the above predictions.

If the Blacic-Griggs hypothesis of H-O-H segregation to dislocations is correct, then what happens when cloud-bearing dislocations reach the surface of a single crystal? Presumably, the water in this H-O-H cloud is lost to the atmosphere. This hardening process, caused by slip-in-

TABLE 3. Experimental Flow Law Parameters, Synthetic Quartz Crystal X-507

Compression Orientation	$\log_{10}A$, bar^{-n}s^{-1}	Activation Energy for Creep, E_c^*, kcal/mol			Stress Exponent, n	
		α-quartz field	β-quartz field	α & β-quartz field	α-quartz field	β-quartz field
O$^+$	−8.9	E_c^* determined at σ = 1400 bar			3.0 ± 0.2	
		22.3 ± 1.0	21.7 ± 1.8	21.8 ± 0.4		
O$^+$		E_c^* determined at σ = 1000 bar				
		20.4 ± 0.8				
⊥m	−9.5	E_c^* determined at σ = 1400 bar			5.3 ± 0.5	3.4 ± 0.3
	−9.7	50.9 ± 3.0	27.5 ± 1.8	34.6 ± 1.8		

Strain rates picked at the maximum strain rate.

duced water loss, may be important in synthetic quartz deformation tests at low temperature, where recovery does not reduce the number of dislocations reaching the surface.

Recent data on the high-temperature rheology of quartz-bearing rocks suggest that the rate-controlling process for creep is also H-O-H diffusion. The activation energy E_c^* for steady state creep for nominally 'dry' quartzite is 36.0 ± 4.0 kcal/mol [*Christie et al.*, 1979] and 38.2 ± 5.1 kcal/mol in the presence of water [*Koch et al.*, 1980]. The steady state flow of Westerly granite (approximately 30% quartz) indicates that E_c^* = 30.5 kcal/mol [*Carter et al.*, 1981, this volume]. Our range of E_c^* values for alpha quartz (22 ± 1 kcal/mol for O$^+$ specimens and 51 ± 3 kcal/mol for ⊥m specimens) bracket the above values. The activation energies for creep of anhydrous refractory silicates and oxides typically exceed 80 kcal/mol [*Tullis*, 1979; *Kirby and McCormick*, 1981; *Bretheau et al.*, 1979]. The low values of E_c^* for quartz-bearing rocks, even nominally under anhydrous conditions, and the comparison with our results on synthetic crystals are highly suggestive that a single process rate controls, namely the diffusion of H-O-H.

Summary and Conclusions

We have compressed synthetic quartz crystals at constant load in two crystallographic directions: One orientation of compression activated single slip parallel to ⟨c⟩ and the other activated duplex slip parallel to ⟨a⟩. Specimens that deformed by ⟨c⟩ slip showed strong initial transient creep behavior and crept at much higher rates than the specimens that flowed by ⟨a⟩ slip, which exhibited little initial transient creep behavior. Creep rates were more sensitive to temperature in ⟨a⟩ slip samples than in ⟨c⟩ slip samples. The initial transient stage is a stage of dislocation multiplication, and the differences in the initial transients in the two suites of samples are caused by expected differences in the dislocation multiplication mechanisms. Both suites showed a hardening stage which resulted from the loss of structural water caused by nonequilibrium processes during the tests. Differences in absolute creep rates and in temperature sensitivity of

creep rates in the two suites of tests are mimicked by similar anisotropy effects for impurity diffusion in quartz. We speculate that diffusion anisotropy of H-O-H (or some related hydroxyl defect) is responsible for the anisotropy in rheology.

Appendices: Interpretation of the Creep Curves

Appendix 1: The Creep Instability

We review the observations pertinent to the interpretation of the creep instability.

1. The instability, which leads to failure by fracture, occurs only in the relatively dry O$^+$ crystals and not in the ⊥m tests on X-507. Previous experiments on crystals with higher hydroxyl concentration have not developed the instability [*Kirby*, 1977; *Kirby et al.*, 1977; *Kirby and McCormick*, 1979].

2. Failure by fracture is heralded by a tertiary creep stage in which creep rates increase with time and strain (Figure 10c). In all but one test, the hardening stage did not develop.

3. Unstable creep is only observed in the 600°C to 620°C tests. Experiments at 550°C and below and at 650°C and above show the normal incubation and hardening stages.

4. Although creep curve reproducibility is not good in the temperature interval of unstable creep, the average creep rates at low strain (<1%) are in keeping with interpolation of the creep data from temperatures above and below (Figure 10c).

5. Specimens unloaded before termination through failure by fracture exhibit prominent optical evidence for plastic deformation (slip bands and deformation bands) and also are severely microfractured.

6. Permanent plastic strain must precede the instability. Natural quartz crystals with very low hydroxyl concentrations neither creep nor fail by fracture under these same experimental conditions [*Kirby*, 1975]. These observations rule out a strictly brittle fracture origin for the instability, such as occurs in natural quartz crystals loaded to higher stresses [*Scholz*, 1972].

It is obvious that the enhanced creep rates in the tertiary stage are due, directly or indirectly, to micro-

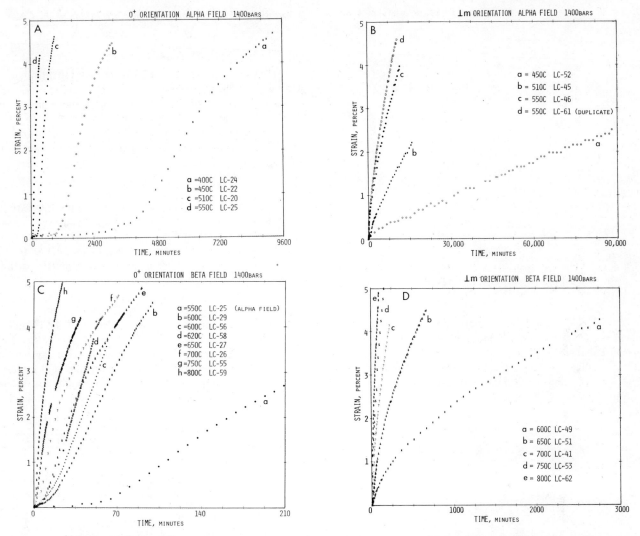

Fig. 10. Collected creep curves, showing the effects of temperature at 1400-bar compressive stress: (*a*, *c*) O⁺ orientation; (*b*, *d*) ⊥*m* orientation. Alpha-quartz stability field (top row, *a* and *b*). Beta-quartz stability field (bottom row, *c* and *d*).

fracturing. A permanent plastic strain is required for the instability to develop. The lack of instability in the ⊥*m* experiments (which deform much more homogeneously than the O⁺ specimens) suggests that stress concentrations at plastic strain inhomogeneities in the O⁺ specimens are important in the instability. Plastic strain inhomogeneities develop in all of the O⁺ tests, most of which creep stably. Thus it appears that some additional phenomenon that occurs over a narrow temperature range is also an important factor. We suggest that the phenomenon is the α-β phase change. *Coe and Paterson* [1969] reviewed the large changes in the physical properties that accompany the α-β quartz transition. In particular, the heat capacity, elastic constants, and thermal expansion coefficients undergo large changes within ±20°C of the room pressure hydrostatic transition temperature of 573 ± 4°C. Also, profuse small-scale Dauphiné twins occur at, or somewhat below,

the nominal transition temperature. *Coe and Paterson* [1969] proved that differential compressive stress raises the transition temperature by 5°C to 11°C per kilobar, depending on the orientation of the compression axis. If the α-β transition is somehow involved in the instability at temperatures as high as 620°C, then compressional stress concentrations of 3 to 7 times the applied stress (σ = 1400 bar) are required.

Thermal shocking near the α-β transition is a commonly reported phenomenon in quartz crystals heated or cooled rapidly [*Sosman*, 1927]. We believe that the creep instability is related to this thermal shock phenomenon. Given the relatively large effects of nonhydrostatic stress on the transition temperature, and the large changes in heat capacity, thermal expansion coefficients, and elastic constants near the transition, small changes in the stress distribution caused by plastic strain inhomogeneities can

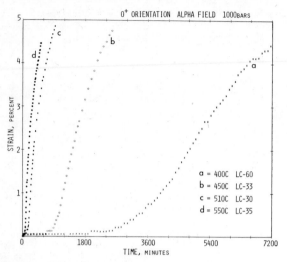

Fig. 11. Collected creep curves, showing the effects of temperature at 1000-bar compressive stress; O⁺ orientation, alpha-quartz stability field.

lead to large changes in temperature and thermoelastic strains. This interesting phenomenon merits further work. The instability may have importance in the mechanics of emplacement of crustal plutons into quartz-bearing rocks since microfracturing can degrade the fracture strength of the wall rock.

Appendix 2: Micromechanical Interpretation of the Creep Curves

The Incubation Stage

O^+ *Orientation.* Kirby and McCormick [1979] demonstrated that the incubation stage of creep in crystal X-0 (O^+ orientation, ~4000 ppm OH^-) is a consequence of rapid multiplication of $\{2\bar{1}\bar{1}0\}$ $\langle 0001 \rangle$ dislocations. This increase in dislocation density produces an increase in the glide shear strain rate $\dot{\gamma}$ through the Taylor-Orowan relation

$$\dot{\gamma} = Nb\bar{v} \qquad (A1)$$

where N is the mobile dislocation density, b the magnitude of the Burgers vector, and v the average dislocation velocity. Since b is not dependent on strain or time, an increase in N not compensated for by decreases in $\bar{v}$ leads to increasing creep rates with time and strain. A multiplication law of the form

$$dN/dt = \delta(\delta N) \qquad (A2)$$

where δ is a breeding coefficient, satisfied all the observational constraints from O^+ experiments on crystal X-0 [Kirby and McCormick, 1979]. Multiplication laws with the form of (A2) generally apply to materials that are oriented for single slip but that tend to cross slip out of the active slip plane [Gilman, 1969]. Multiple cross glide is a

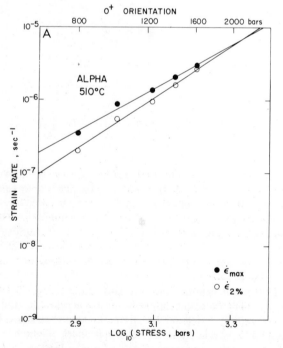

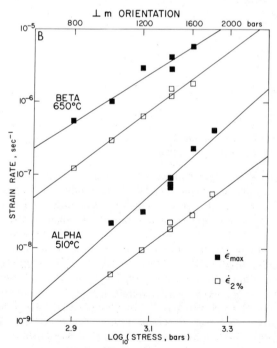

Fig. 12. Stresses and strain rates plotted on logarithmic scales. The slopes of the fitted least squares lines are equal to the n values in equation (3). Solid symbols indicate maximum strain rates, open symbols indicate strain rates at 2% strain. The uncertainties in the values of the stresses and strain rates are smaller than the dimensions of the symbols. (a) O^+ orientation, alpha-quartz stability field, 510°C. (b) $\perp m$ orientation, alpha-quartz stability field (510°C, lower pair of lines), beta-quartz stability field (650°C, upper pair of lines).

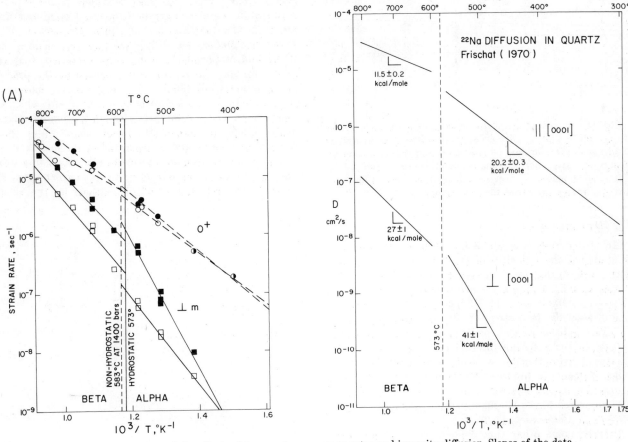

Fig. 13. Arrhenius plots of the effects of temperature on creep rates and impurity diffusion. Slopes of the data are proportional to the activation energies of the processes. (*a*) Creep rate variation with temperature. (*b*) Sodium diffusion variation with temperature. Note the remarkable way that this diffusion data mimics the creep data of (*a*).

mechanism whereby double cross slip from the primary slip plane produces paired superjogs that can serve as Frank-Read dislocation multiplication sources [*Gilman*, 1969, p. 114]. The form of (A2) follows directly from this geometry. There is clear evidence for cross slip in synthetic crystals compressed in the O⁺ orientation [*Baëta and Ashbee*, 1969, 1970*a*; *Kirby and McCormick*, 1979].

Many constant strain rate tests on synthetic quartz crystals compressed in the O⁺ orientation exhibit a two-stage plastic yield behavior (yield deflection from an elastic curve followed by a stress drop) [*Baëta and Ashbee*, 1970*a*; *Hobbs et al.*, 1972; *Balderman*, 1974; *Morrison-Smith et al.*, 1976; *McCormick*, 1977]. This yield behavior is the constant strain rate counterpart of the incubation stage of our O⁺ creep tests and has been successfully modeled by *Hobbs et al.* [1972] and *Griggs* [1974], assuming the multiplication law of (A2). Crystals of many materials that have low initial dislocation density and that are oriented for single slip show yield drop and incubation stage behavior in constant strain rate and creep tests. Apparently this same dislocation multiplication mechanism occurs in a wide range of crystalline materials.

⊥*m Orientation.* We stated earlier that the incubation stage is absent or poorly developed in our ⊥*m* tests. Most ⊥*m* runs reached maximum strain rate very early in the tests. The measured incubation time t_i (defined in Figure 8*a*) does not vary systematically with experimental conditions as it does in the O⁺ tests (Table 1). The lack of significant yield drop behavior in constant strain rate ⊥*m* tests [*Baëta and Ashbee*, 1970*a*; *Hobbs et al.*, 1972] is consistent with the poorly developed incubation stage. We believe that the incubation stages in ⊥*m* and O⁺ tests are different because of differences in the dislocation multiplication mechanisms. Simultaneous (10$\overline{1}$0) [1$\overline{2}$10] and ($\overline{1}$100) [$\overline{1}$120] slip inevitably produces dislocation jogs by intersections of dislocation loops on the two primary glide planes. Jogs of intersection origin can then serve as new Frank-Read dislocation sources. The combination of multiplication by multiple cross glide and multiplication at jogs of intersection origin constitutes a more explosive breeding of dislocations than solely the simple multiple cross glide mechanism that leads to (A2). Electron microscopy now in progress will test this explanation of high initial creep rates in ⊥*m* specimens.

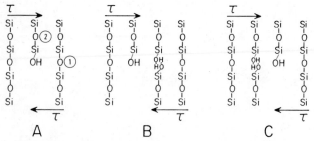

Fig. 14. Sketch of hydrolytic weakening mechanism [after *Griggs*, 1967]. Glide of edge dislocation facilitated by hydrolysis of adjacent Si-O-Si bridges (at position 1) and hydrogen bond exchange with saturated 'dangling bonds.' This process can also facilitate climb motion by hydrolysis at position 2 and diffusion of $Si(OH)_4$ groups away from the dislocation core.

The Hardening Stage

An earlier study of crystal X-0 (~4000 ppm OH) demonstrated that the reduction of strain rate with time and with strain in that particular crystal cannot be caused by the classical hardening mechanisms associated with strain [*Kirby and McCormick*, 1979]. Specifically, neither reduction of resolved shear stress on the operating slip system by strain-induced rotation or by lateral expansion nor increases in the resisting internal stresses caused by increases in dislocation density could account for the magnitudes of the hardening rates. Expanding on this last point, dislocation density was essentially constant in the O^+ samples deformed to the hardening stage [*McCormick*, 1977; *Kirby and McCormick*, 1979].

Clearly, some other time-dependent physicochemical process provides most of the hardening in creep tests at atmospheric pressure. *Kirby and McCormick* [1979] concluded that the precipitation of molecular water into submicroscopic bubbles was the principal hardening process for crystal X-0. In that crystal, TEM observations indicated that the amount of water contained by bubbles increased progressively in the hardening stage. Infared absorption spectra of the deformed samples confirmed that absorption caused by 'structurally bound' water decreased and that caused by molecular water increased. Heat treatment alone can cause precipitation of water [*Dodd and Fraser*, 1965, 1967; *Bambauer et al.*, 1969], and heat treatment at atmospheric pressure prior to deformation testing renders synthetic quartz stronger [*Kirby*, 1975; *Kirby and McCormick*, 1979; *Kekulawala et al.*, 1978]. Simultaneous heating and deformation at low confining pressure promote precipitation [*Baëta and Ashbee*, 1970a, 1973; *Balderman*, 1974; *Morrison-Smith et al.*, 1976; *Kirby*, 1975, 1977; *Kekulawala et al.*, 1978; *McCormick*, 1977; *Kirby and McCormick*, 1979]. The process is time dependent and thermally activated. Apparently, synthetic crystals that precipitate water and become cloudy are supersaturated with respect to water at the test conditions.

Our specimens became cloudy during creep; this phenomenon is a macroscopic manifestation of light scattering by water bubbles. Preliminary TEM on a few of our

deformed specimens confirmed the existence of submicron-sized bubbles like those so widely reported in experimentally deformed synthetic quartz [*Baëta and Ashbee*, 1973; *Morrison-Smith et al.*, 1976; *McCormick*, 1977; *Kirby and McCormick*, 1979]. We therefore believe that Kirby and McCormick's earlier interpretation, that the hardening stage in crystal X-0 is caused by the precipitation of water into bubbles, also applies to the hardening stage in both O^+ and $\perp m$ tests on crystal X-507, which we report here. The specific precipitation hardening mechanism is simply the loss of structurally bound water. Since this water is no longer incorporated in the quartz structure and cannot promote the hydrolytic weakening process, creep rates decrease. A less effective hardening mechanism is the retardation of dislocations by the elastic

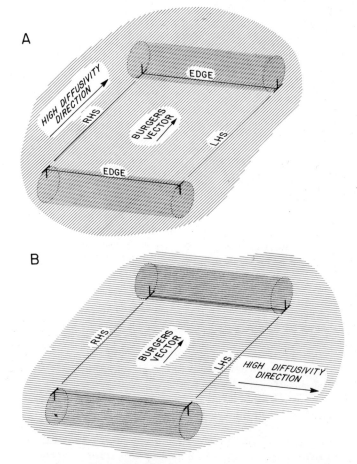

Fig. 15. Schematic diagram of the effects of anisotropy of diffusion of H-O-H on the rates of expansion of rectangular dislocation loops in quartz. The H-O-H clouds on the edge segments (shown as grey cylinders) are in the regions of compression, since pressure increases the solubility of H-O-H in quartz (see text). LHS—left-handed-screw dislocation segment. RHS—right-handed-screw dislocation segment. See text for additional explanation. (*a*) Slip direction parallel to high-diffusivity direction. (*b*) Slip direction and slip plane normal perpendicular to the high-diffusivity direction.

interaction with the bubbles, which can move only by means of diffusion [*Kirby and McCormick*, 1979].

We have not directly measured the volume fraction occupied by bubbles in our X-507 specimens, so we are unable to estimate the amount of precipitation and to rigorously test our interpretation of the hardening stage. However, the degree of cloudiness and the magnitude of the hardening with strain (as indicated by the ratio ($\dot{\epsilon}_{2\%}/\dot{\epsilon}_{max}$)) are consistent with the precipitation hardening mechanism. First, the degree of cloudiness increases with increasing strain, particularly at low strain. Secondly, the specimens that show the most hardening (the 550°C, 1400 bar, $\perp m$ specimens, Figures 8*b* and 13*a*) are also the cloudiest. Lastly, the $\perp m$ runs tend to show greater hardening to strains of 2% than experiments on O$^+$ specimens tested at the same conditions (compare the separations of the $\dot{\epsilon}_{max}$ and $\dot{\epsilon}_{2\%}$ data in Figure 12 (*a*, *b*) and those in Figure 13*a*). We think this difference stems from the much longer time that the $\perp m$ specimens are at temperature and under load. The creep rates in the two suites at $\sigma = 1400$ bar nearly converge at $T = 800$°C; supporting this explanation, the initial hardening rates at $T = 800$°C are approximately the same.

To summarize our interpretation of the creep curves: By analogy with TEM and theoretical work on wetter crystals, the incubation stage represents a time of rapid dislocation multiplication. Dislocation density maximizes at the maximum strain rate and remains stabilized to steady state values throughout the hardening stage. Hardening results from the precipitation of molecular water into bubbles. All of our observational data support this analogy.

Since our major interest is in the steady state rheology of these crystals, we should focus our attention on the effects of temperature, stress, and orientation on the maximum strain rates, because they represent creep rates where steady state dislocation density is achieved and where the precipitation hardening effects are smallest. This is to say, the $\dot{\epsilon}_{max}$ data most closely approach the steady state creep rates in the absence of precipitation.

Acknowledgments. We thank John Christie for loan of equipment and John Christie and Alison Ord for permission to use unpublished optical data on the $\perp m$ suite of specimens. John McCormick performed some initial electron microscopy and built the data processing and display systems used in this work. R. A. Laudise and E. D. Kolb of Bell Telephone Laboratories of Murray Hill, New Jersey generously donated crystal X-507 to us and provided data on its growth history and some of its properties. Charles E. Hildebrand's assistance in design and his craftsmanship in making the components made possible the successful construction of the apparatus. Robert Coe provided helpful comments on the B.A. (University of California, Santa Cruz) thesis of M.F.L. on which this paper is based. Jim Blacic, Barry Raleigh, and John Christie improved the manuscript by furnishing helpful reviews.

References

Baëta, R. D., and K. H. G. Ashbee, Slip systems in quartz, I, Experiments, *Am. Mineral., 54,* 1551–1573, 1969.

Baëta, R. D., and K. H. G. Ashbee, Mechanical deformation of quartz, I, Constant strain-rate compression experiments, *Philos. Mag., 22,* 601–623, 1970*a*.

Baëta, R. D., and K. H. G. Ashbee, Mechanical deformation of quartz, II, Stress relaxation and thermal activation parameters, *Philos. Mag., 22,* 625–635, 1970*b*.

Baëta, R. D., and K. H. G. Ashbee, Transmission electron microscopy studies of plastically deformed quartz, *Phys. Status Solidi, 18,* 155–170, 1973.

Balderman, M. A., Relationship of yield stress and strain-rate in hydrolytically weakened synthetic quartz, M.Sc. thesis, 119 pp., Univ. of Calif., Los Angeles, 1972.

Balderman, M. A., The effect of strain rate and temperature on the yield point of hydrolytically weakened synthetic quartz, *J. Geophys. Res., 79,* 1647–1652, 1974.

Bambauer, H. U., G. O. Brunner, and F. Laves, Light scattering of heat-treated quartz in relation to hydrogen containing defects, *Am. Mineral., 54,* 718–724, 1969.

Blacic, J. D., Hydrolytic weakening of quartz and olivine, Ph.D. thesis, 205 pp., Univ. of Calif., Los Angeles, 1971.

Blacic, J. D., Plastic deformation mechanisms in quartz: The effect of water, *Tectonophysics, 27,* 271–294, 1975.

Bretheau, P. T., J. Castaing, J. Rabier, and P. Veyssiere, Mouvment des dislocations et plasticité à haute temperature des oxydes binaires et ternaire, Adv. Phys., *28,* 835–1014, 1979.

Carter, N. L., D. A. Anderson, F. D. Hansen, and R. L. Kranz, Creep and creep-rupture of granitic rocks, this volume, 1981.

Choudhury, A., D. Palmer, G. Ansel, H. Curier, and P. Baruch, Study of oxygen diffusion in quartz by using the nuclear reaction O^{18} (p,α) N^{15}, Solid State Commun., *3,* 119–122, 1965.

Christie, J. M., and H. W. Green, Several new slip systems in quartz, *Eos Trans. AGU, 45,* 102, 1964.

Christie, J. M., and A. Ord, Microstructural study of deformed synthetic quartz crystals, A preliminary report, *Contract Rep. P.O. 60202,* 11 pp., U.S. Geol. Surv., Reston, Va., 1979.

Christie, J. M., P. S. Koch, and R. P. George, Flow law of quartzite in the alpha field, *Eos Trans. AGU, 60,* 948, 1979.

Coe, R. S., and M. S. Paterson, The alpha-beta inversion in quartz: A coherent phase transition under nonhydrostatic stress, *J. Geophys. Res., 74,* 4921–4948, 1969.

Dodd, D. M., and D. B. Fraser, The 3000–3900 cm^{-1} absorption bands and anelasticity in crystalline quartz, *J. Phys. Chem. Solids, 26,* 673–686, 1965.

Dodd, D. M., and D. B. Fraser, Infrared studies of the variation of H-bonded OH$^-$ in synthetic α-quartz, *Am. Mineral., 52,* 149–160, 1967.

Frischat, G. H., Sodium diffusion in natural quartz crystals, *J. Am. Ceram. Soc., 53,* 357, 1970.

Ghirso, M. S., I. S. E. Charmichael, and L. K. Moret, Inverted high temperature quartz: Unit cell parameters and properties of the α-β inversion, *Contrib. Mineral. Petrol.*, *68*, 307-323, 1979.

Gilman, J. J., *Micromechanics of Flow in Solids*, 294 pp., McGraw-Hill, New York, 1969.

Griggs, D. T., Hydrolytic weakening of quartz and other silicates, *Geophys. J. R. Astron. Soc.*, *14*, 19-32, 1967.

Griggs, D. T., A model of hydrolytic weakening in quartz, *J. Geophys. Res.*, *79*, 1655-1661, 1974.

Griggs, D. T., and J. D. Blacic, The strength of quartz in the ductile regime (abstract), *Eos Trans.* AGU, *45*, 102-103, 1964.

Griggs, D. T., and J. D. Blacic, Quartz: Anomalous weakness of synthetic crystals, *Science*, *147*, 292-295, 1965.

Heard, H. C., and N. L. Carter, Experimentally induced 'natural' intragranular flow in quartz and quartzite, *Am. J. Sci.*, *266*, 1-42, 1968.

Heinisch, H. L., G. Sines, J. W. Goodman, and S. H. Kirby, Elastic stresses and self-energies of dislocations of arbitrary orientation in anisotropic media: Olivine, orthopyroxene, calcite, and quartz, *J. Geophys. Res.*, *80*, 1885-1896, 1975.

Hirth, J. P., and J. Lothe, *Theory of Dislocations*, 780 pp., McGraw-Hill, New York, 1968.

Hobbs, B. E., Recrystallization of single crystals of quartz, *Tectonophysics*, *6*, 353-401, 1968.

Hobbs, B. E., A. C. McLaren, and M. S. Paterson, Plasticity of single crystals of synthetic quartz, in *Flow and Fracture of Rocks*, Geophys. Monogr. Ser., vol. 16, edited by H. C. Heard et al., AGU, Washington, D. C., 1972.

Kats, A., Hydrogen in alpha quartz, *Philips Res. Rep.*, *17*, 133-195, 1962.

Kekulawala, K. R. S. S., M. S. Paterson, and J. N. Boland, Hydrolytic weakening in quartz, *Tectonophysics*, *46*, T1-T6, 1978.

Kirby, S. H., Creep of synthetic alpha quartz, Ph.D. thesis, Univ. of Calif., Los Angeles, 1975.

Kirby, S. H., The effects of the alpha-beta phase transformation on the creep properties of hydrolytically-weakened synthetic quartz, *Geophys. Res. Lett.*, *4*, 97-100, 1977.

Kirby, S. H., and M. F. Linker, Creep of hydrolytically weakened synthetic quartz crystals at atmospheric pressure: Effects of hydroxyl concentration, *Eos Trans. AGU*, *60*, 949, 1979.

Kirby, S. H., and J. W. McCormick, Creep of hydrolytically weakened synthetic quartz crystals oriented to promote $(2\bar{1}\bar{1}0)$ [0001] slip: A brief summary of work to date, *Bull. Minéral.*, *102*, 124-137, 1979.

Kirby, S. H., and J. W. McCormick, The inelastic properties of rocks and minerals: Strength and rheology, in *Physical Properties of Rocks and Minerals*, edited by R. S. Carmichael, CRC Press, West Palm Beach, Fla., in press, 1981.

Kirby, S. H., J. W. McCormick, and M. Linker, The effect of water concentration on creep rates of hydrolytically weakened synthetic quartz single crystals, *Eos Trans. AGU*, *58*, 1239, 1977.

Koch, P. S., J. M. Christie, and R. P. George, Flow law of 'wet' quartzite in the alpha-quartz field, *Eos Trans. AGU*, *61*, 376, 1980.

Linker, M. F., and S. H. Kirby, Creep of hydrolytically weakened synthetic quartz: Tests with crystals oriented to promote duplex $\{10\bar{1}0\}$ ⟨a⟩ slip, *Eos Trans. AGU*, *59*, 1185, 1978.

McCormick, J. W., Transmission electron microscopy of experimentally deformed synthetic quartz, Ph.D. thesis, 171 pp., Univ. of Calif., Los Angeles, 1977.

McLaren, A. C., and J. A. Retchford, Transmission electron microscope study of the dislocations in plastically deformed synthetic quartz, *Phys. Status Solidi*, *33*, 657-668, 1969.

Morrison-Smith, D. J., M. S. Paterson, and B. E. Hobbs, An electron microscope study of plastic deformation in single crystals of synthetic quartz, *Tectonophysics*, *33*, 43-79, 1976.

Mortley, W. S., Electrical conductivity of quartz, *Nature*, *221*, 359, 1969.

Paterson, M. S., and K. R. S. S. Kekulawala, The role of water in quartz deformation, *Bull. Minéral.*, *102*, 92-98, 1979.

Sang-Hwang, J. S., Diffusion and solubility of water in quartz, Ph.D. thesis, 174 pp., Case West. Reserve Univ., Cleveland, Ohio, 1975.

Scholz, C. H., Static fatigue in quartz, *J. Geophys. Res.*, *77*, 2104-2114, 1972.

Shaffer, E. W., J. S.-L. Sang, A. R. Cooper, and A. H. Heuer, Diffusion of tritiated water in α-quartz, in Geochemical Transport and Kinetics, edited by A. W. Hoffman, et al., *Publ. 634*, pp. 121-138, Carnegie Inst. of Wash., Washington, D. C., 1974.

Sosman, R. B., *The Properties of Silica*, Sci. Monogr. 37, Am. Chem. Soc., Washington, D. C., 1927.

Tullis, J. A., High-temperature deformation rocks and minerals, *Rev. Geophys. Space Phys.*, *17*, 1137-1154, 1979.

Turner, F. J., Rotation of the crystal lattice in kink bands, deformation bands, and twin lamellae of strained crystals, *Proc. Nat. Acad. Sci. USA*, *48*, 955-963, 1962.

Twiss, R. J., Structure and significance of planar deformation features in synthetic quartz, *Geology*, *2*, 329-332, 1974.

Twiss, R. J., Some planar deformation features, slip systems, and submicroscopic structures in synthetic quartz, *J. Geol.*, *84*, 701-704, 1976.

Verhoogen, J., Ionic diffusion and electrical conductivity in quartz, *Am. Mineral.*, *37*, 637-655, 1952.

Vogel, R. C., and G. Gibson, Migration of sodium ions through quartz plates in an electric field, *J. Chem. Phys.*, *18*, 490-494, 1950.

An Experimental Study of the Role of Water in Quartz Deformation

K. R. S. S. KEKULAWALA,[1] M. S. PATERSON, AND J. N. BOLAND[2]

Research School of Earth Sciences, Australian National University, Canberra, A.C.T., 2600 Australia

Basic equilibrium and kinetic properties of water in quartz are elucidated in a combined mechanical, infrared absorption, and electron microscope study on a hydroxyl-bearing synthetic quartz crystal. Changes in yield strength and nature of infrared absorption observed after heating specimens at 900°C and various pressures are correlated with the precipitation or redissolution of water in bubbles visible in the electron microscope. From these observations it is concluded that the solubility of water in quartz at 900°C and 300 MPa pressure is between 200 and 400 H per 10^6Si and that the diffusivity of the dissolved hydroxyl-bearing species is very low, probably less than 10^{-19} $m^2 s^{-1}$ at 900°C in initially dry quartz but increasing significantly with hydroxyl concentration. These results explain the sluggishness and associated experimental difficulties in studies on hydrolytic weakening in this pressure range. They indicate that equilibrium may not be attained in many such experiments and pose pertinent questions concerning extrapolation to geological conditions.

Introduction

The phenomenon of 'hydrolytic weakening' in quartz, discovered by *Griggs and Blacic* [1964], was first observed in natural single crystals deformed in an environment containing water from dehydrating talc. Soon afterward, a similar effect was demonstrated in synthetic crystals that initially show a broad hydroxyl band in the infrared absorption spectrum [*Griggs and Blacic*, 1965; *Griggs*, 1967]. Evidently a water-related hydroxyl species within the quartz structure is involved. However, it has remained to characterize this species more fully and to identify the factors influencing its incorporation in the quartz in a form that is mechanically effective.

The mere presence of water is in itself insufficient to ensure easy deformability of quartz in laboratory experiments. This insufficiency has been shown up by experiments in which strong natural quartz was heated in the presence of water at confining pressures of 300–500 MPa; no obvious hydrolytic weakening was found after long periods of heating [*Kekulawala et al.*, 1978]. Also, quartzite containing abundant water in fluid inclusions is found to be relatively strong compared to OH-containing synthetic crystals when tested at similar confining pressures. Thus it would appear that in the Griggs and Blacic experiments the high level of confining pressure used (around 1500

MPa) is an important factor. A similar conclusion follows from observations of lower strength in quartzite at higher pressures by *Tullis et al.* [1979]. The reasons for the pressure effect have not yet been elucidated, but pressure is clearly an important variable.

The present study, combining mechanical, infrared absorption, and electron microscope approaches, attempts to separate and to identify some basic equilibrium and kinetic aspects of the role of OH in the properties of quartz. Preliminary accounts of the findings have been given in *Kekulawala et al.* [1978] and in *Paterson and Kekulawala* [1979]. The latter paper also gives a review setting out the background in some detail, which therefore need not be repeated here. We shall now describe observations on the mechanical properties of synthetic quartz crystals subjected to various histories of heat treatment and relate these observations to infrared absorption spectra measured at low temperature and to electron microscope observations on bubbles. The observations will then be discussed in terms of the solubility and diffusivity of the kinetically effective hydroxyl species responsible for hydrolytic weakening. Finally, some implications for the mechanical properties of quartz under other conditions are considered.

Apparatus and Procedures

Deformation tests were carried out in an argon gas-medium high pressure apparatus, previously described [*Paterson*, 1970]. The apparatus has an internal furnace and an internal load cell. Displacements are determined with an externally mounted LVDT, correcting for apparatus

---

[1]Now at the Department of Physics, University of Sri Lanka, Colombo 3, Sri Lanka.

[2]Now at Instituut Voor Aardwetenschappen, Rijksuniversiteit, Utrecht, The Netherlands.

distortion. Temperatures are measured with a Pt/Pt-Rh thermocouple within the hollow loading piston, 2–3 mm from the end of the specimen. The temperature gradient is determined in calibration runs with a hollow dummy specimen and is normally less than 5°–10°C over the length of the specimen. Uncertainty in stress determination in these tests is thought to be about 1–2 MPa and that in strains to be less than about 0.2% strain. In most cases the specimens were prisms of 5 mm square cross sections, 13 mm long. They were mechanically sealed within 7 mm diameter copper jackets of 0.25 mm wall thickness which collapse around the prisms when pressure is applied. The loading direction is $\perp m$ prism.

Infrared absorption spectra were measured at 4°K with a double-beam Perkin-Elmer 180 spectrophotometer with LiF optics fitted with a continuous-flow liquid helium cryostat. An unpolarized beam was transmitted parallel to the c axis of the quartz either using the prismatic specimens described above, in which case the thickness traversed was 5 mm, or using 0.5 to 2 mm thick slices cut normal to the c axis. In all cases the faces through which the beam passed were polished. By using the relationship $i_t = i_0 10^{-\epsilon c t}$ where i_0, i_t are the incident and transmitted intensities, ϵ the molar absorption coefficient, c the molar concentration of the absorbing species, and t the thickness traversed, it follows that

$$c = \Delta/I \qquad (1)$$

where Δ is the area under the curve of absorption coefficient $(\log_{10}(i_0/i_t)/t)$ plotted against wave number (n), and I is the integral molar absorption coefficient $(\int \epsilon dn)$ for the given interval in wave number over which absorption by the species is being considered. By using an effective value of $I \approx 24000 \, l(\text{mol OH})^{-1} \text{cm}^{-2}$ suggested by a recent review of the calibration of OH determination (M. S. Paterson, manuscript in preparation, 1980), the relation (1) becomes

$$c = 0.94\Delta \qquad (2)$$

when c is expressed in H per 10^6Si and Δ is in cm^{-2}.

Electron microscope observations were made with a 200 KV JEOL microscope on ion-thinned specimens. Bubble sizes and densities were determined by taking a series of 10–20 20,000X to 30,000X photographs at random for each specimen and measuring or counting on the photographs. For the density determinations the thicknesses of the sections were variously obtained by using the intersection of dislocations or twins of known crystallographic orientation or the extinction contours in 2-beam situations; the thicknesses were about 0.5 μm.

Heating experiments at atmospheric pressure were done in argon in a muffle furnace. The quartz specimens were placed in a hollow block of copper, and the temperature was raised and lowered very slowly, especially in the vicinity of the alpha-beta transition, to avoid cracking. Heating experiments under elevated pressure were done in the gas-medium deformation apparatus mentioned

above, wrapping the polished specimens in platinum foil and then sealing them in copper jackets.

Specimen Material and Preparation

Most experiments were done on specimens cut from the Z-growth zone of a synthetic quartz crystal designated W1, supplied by D. W. Rudd of Western Electric (cf., crystals used by *Hobbs et al.* [1972]; *Morrison-Smith et al.* [1976]; *Twiss* [1974, 1976]). The hydroxyl content of crystal W1 decreases monotonically in the c axis direction from the seed to the outside. Therefore, the crystal was cut into 5 mm thick slabs normal to the c axis and $5 \times 5 \times 13$ mm prismatic specimens were cut from the slab with their long axes lying in the slab perpendicular to the $m\{10\bar{1}0\}$ crystallographic prism face and with the lateral faces normal to the a and c axes (Figure 1). In this way, specimens from a given slab or layer have approximately uniform OH contents that are reproducible from specimen to specimen. Infrared absorption spectra for three successive layers outward from the seed are shown in Figure 1. By using relation (2), these spectra correspond to average total hydroxyl contents of about 800, 400, and 200 H per 10^6Si for layers 1, 2, and 3, respectively. As far as possible, infrared measurements were made on each specimen tested mechanically and studied in the electron microscope.

Also shown in Figure 1 is the absorption spectrum for a high quality synthetic crystal, A6-13, supplied by B. Sawyer. This crystal appears to be essentially devoid of all forms of hydroxyl and has been used as a standard to deduct the absorption effects intrinsic to quartz itself when determining total absorption owing to hydroxyl in other specimens. Other crystals used in work mentioned in this paper are a synthetic quartz, E-1-1, also supplied by Sawyer which has a lower hydroxyl content than W1, ranging from 50 to 200 H per 10^6Si, and a clear natural rock crystal designated N2, of negligible hydroxyl content except that giving rise to a few sharp absorption bands.

Mechanical Properties: Influence of Heat Treatment

Mechanical tests were carried out under standardized conditions of 800°C 300 MPa confining pressure and 10^{-5} s^{-1} strain rate in order to be able to compare the mechanical properties after various heat treatments at 900°C. Lowering the temperature to 800° avoids significantly prolonging the heat treatment period. Attention was focused on the initial yield stress as a measure of resistance to plastic deformation, and in general any deformation involved did not exceed a few percent.

An initial survey of a wide variety of quartz crystals showed that many types were very strong and exhibited no plastic deformation under the above conditions; that is, their yield stresses were in excess of about 1000–1400 MPa, the limit set by fracture of the crystals or excessive flow in the loading pistons. Except for the very dry synthetic crystal A6-13, the very strong crystals were natural crystals and included optically clear, colorless crystals

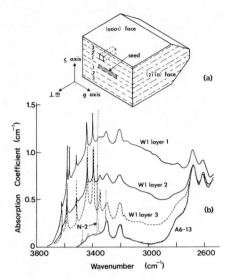

Fig. 1.   (a) Sketch of synthetic quartz crystal W1 showing the mode of slicing into layers and the orientation of specimens cut from the layers. (b) The infrared absorption spectra for the three layers of crystal W1. Also shown below are the spectra for synthetic crystal A6-13 (solid line) and natural crystal N2 (dotted).

(rock crystal), cloudy crystals with high fluid inclusion content, cloudy amethysts, a smoky quartz, and a clear citrine.

All optically clear crystals of this group had very low hydroxyl contents (less than a few tens H per 10^6Si), and the cloudy crystals gave low temperature infrared spectra dominated by ice bands. In contrast, other crystals showed initial yield stresses of the order of 100 MPa at 800°C. The latter group includes, as is now well known, the synthetic crystals with grown-in hydroxyl contents in the range of roughly a hundred to several thousand H per 10^6Si and, in addition, as reported by *Kekulawala et al.* [1978], natural amethyst crystals of optically clear quality with similar infrared absorption spectra. Some milky vein quartzes were found to have intermediate properties.

Synthetic crystal W1 was selected for more detailed study, particularly of the effects of various heat treatments (Table 1). The stress-strain behavior at small strains before heat treatment is shown in Figure 2a for the three layers of different initial OH contents (Figure 1). The strength clearly decreases as the OH content increases. When such crystals are heated at 900°C at atmospheric pressure for durations of the order of 10 hours before testing, they become turbid and the initial yield stress is raised to around 300-400 MPa. The influence of the duration of heating was reported earlier [*Kekulawala et al.*, 1978] for the crystal E-1-1, showing that the degree of strengthening continued to increase, but with diminishing rate, with increasing heating time up to 100 hours, when a yield strength of 1200 MPa was reached.

In contrast to the marked effect at atmospheric pressure, heating under elevated hydrostatic pressure pro-

duces less change in yield strength. Figure 2b shows that for a set of specimens from layer 1 (Figure 1) heated at 900°C for the same length of time, increase in yield strength is smaller the higher the pressure applied during heating. Similarly, Figure 2c, for a specimen from layer 2 heated 25 hours at 300 MPa pressure and 900°C, shows that there has been very little change in yield strength although there appears to be a slight change in the subsequent course of the stress-strain curve. Specimens heated at 300 MPa pressure did not develop appreciable turbidity.

Finally, it has been found that the strengthening induced by heating at atmospheric pressure can be reversed by subsequent heating at 300 MPa pressure (Figure 3a). Increasing the duration of subsequent heating under pressure increases the degree of this reweakening, but complete restoration of the original level of yield strength was not achieved within the same period as was used in the initial atmospheric-pressure heat treatment. However, curiously, if after a certain interval of heating under pressure the specimen is lightly deformed (1-2%) and then the heating under pressure continued, a moderate degree of secondary or rehardening was observed to occur (Figure 3b); this effect was repeatable, but it has not been explored in detail.

Reweakening of synthetic quartz previously partially strengthened by heat treatment, as just described, is the only case of actually inducing hydrolytic weakening that we have been able to demonstrate. All attempts to induce hydrolytic weakening by heating initially strong quartz in water at pressures up to 500 MPa and temperatures up to 900°C for periods up to 10 hours or so have produced no detectable weakening or change in the infrared absorption spectrum.

Some indications of the mechanisms of deformation have been obtained from observations on slip traces on the prepolished faces of the W1 synthetic crystal specimens. Thus, at small strains, slip lines are always visible parallel to the trace of m {$10\overline{1}0$} planes on the (0001) faces but are absent on the {$2\overline{1}\,\overline{1}0$} faces, indicating that the main mechanism of deformation is slip on the $\{m\}\langle a\rangle$ system with Schmid factor 0.43. This conclusion is supported by the absence of any change in the cross-sectional dimension normal to the (0001) faces, up to about 2% strain. However, the slip traces are somewhat wavy, suggesting that there is also some cross slip on other planes containing the same $\langle a\rangle$ slip direction. In a few cases (3005, 3015/16, 3021/25, 3030/31, 3032/33, 3035/36) where the straining was continued to about 4%, a small change in the cross-sectional dimension normal to the (0001) faces was detected, and additional slip traces appeared on the lateral faces of both orientations, corresponding to traces of the s {$2\overline{1}\,\overline{1}1$} planes. It is therefore concluded that in these cases there is a small activity of the $\{s\}\langle c + a\rangle$ slip systems with Schmid factor 0.46. Finally, in view of this indication of slip with a $\langle c + a\rangle$ Burgers vector, it is interesting to record that in two experiments (3012, 3055) on clear ame-

TABLE 1. Heat-Treatment Experiments on Synthetic Crystal W1

		Heat Treatment (900°C)			
Number	Crystal Layer	Hours	Pressure (MPa)	Yield Stress, MPa	Remarks
3067	3	—	—	100	virgin condition; lowest OH
2980	2	—	—	75	virgin condition; intermediate OH
3005	2	—	—	80	virgin condition; intermediate OH
3084	1	—	—	60	virgin condition; highest OH
3088	1	—	—	60	virgin condition; highest OH
3124	1	—	—	50	virgin condition; highest OH
3044	1	—	—	50 }	virgin condition; highest OH
		+7	atmos.	>500 }	
2982	2	9	atmos.	>240	fractured without yielding
2983	2	9	atmos.	>300	fractured without yielding
2985	2	10	atmos.	>270	fractured without yielding
3079	1	10	atmos.	>350	fractured
3080	1	10	7.5	290	
3076	1	10	60	185	
3075	1	10	114	160	
3074	1	10	200	115	
3081	1	10	290	70	
2981	2	8	300	90	
2987	2	11½	300	—	infrared only
2988	2	25	300	100	
2989	2	6½	300	—	E. M. study only
3062	2	25	300	70	
3007/8	2	21	atmos.	390 }	
		+10	300	320 }	
3015/16	2	21	atmos.	390 }	
		+18	300	200 }	
3027/28	2	5	atmos.	400 }	fractured
		+2	300	— }	
3035/36	1	15	atmos.	550 }	
		+24	300	300 }	
3041	2	7	atmos.	— }	infrared only
		+20	300	— }	
3042	2	16	atmos.	— }	infrared only
		+24	300	— }	
3091	2	24	atmos.	— }	infrared only
		+9	300	— }	
3092	2	15	atmos.	— }	infrared only
		+16	300	— }	
3006	2	6½	atmos.	— }	heated in argon
		6½	300	205 }	
3030/31	2	6	atmos.	250 }	heated in air
		+8 → 15	300	200 → 300 }	
3032/33	1	15	atmos.	>540 }	
		+14 → 24	300	200 → 280 }	
3021/25	2	60	atmos.	>740 }	fractured at end of run at
		+10 → 19	300	>740 → >740 }	860 MPa (upper yield point
		→33 → 43		→680 → 800 }	at 700 MPa)

The plus sign signifies further heat treatment following that given in the previous line. The arrow signifies a continuation of heating under the same conditions up to the total duration next listed. The greater than sign specifies that the specimen was loaded to this stress without yielding.

thyst faint wavy slip lines were observed on the (0001) faces parallel to the traces of $r\{10\bar{1}1\}$ planes; these lines possibly represent $\{r\}\langle c + a \rangle$ slip (Schmid factor 0.46) since the resolved shear stress on the $\{r\}\langle a \rangle$ systems was close to zero.

Infrared Absorption Observations

Information about the nature of the hydroxyl- or water-related impurity content of quartz crystals can be obtained from infrared absorption spectra in the 3-μm region, especially when measured at low temperatures (78°K is suitable but 4°K has been used here for reasons of equipment set-up). Three principal categories of absorption at low temperature can be distinguished [Kekulawala et al., 1978]:

1. Sharp, dichroic absorption bands, varying considerably from one type of quartz to another, none of which can be uniquely correlated with hydrolytic weakening.

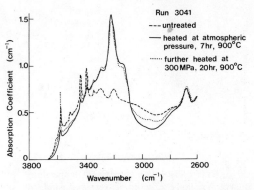

Fig. 4. Infrared absorption spectra for a specimen from synthetic quartz W1, layer 2, in the as-received condition, then after heating at 900°C, atmospheric pressure for 7 hours, and finally after further heating at 900°C, 300 MPa pressure for 20 hours.

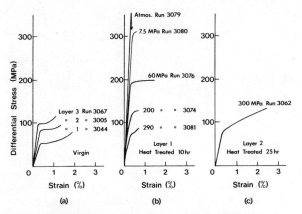

Fig. 2. Stress-strain curves measured at 800°C, 300 MPa confining pressure for specimens from synthetic crystal W1 for various conditions: (a) Specimens from the three layers (see Figure 1a) in initial condition. (b) Specimens from layer 1 after heating for 10 hours at 900°C at pressures shown. (c) Specimen from layer 2 after heating for 25 hours at 900°C, 300 MPa pressure.

2. A very broad band of isotropic absorption extending from about 3600 to 2600 cm^{-1} or even beyond, underlying such sharp bands as may occur; the hydroxyl species responsible for hydrolytic weakening appears to absorb in this band.

3. A complex absorption band of intermediate width, peaking at 3200 cm^{-1} and corresponding to absorption by ice presumably formed by the freezing of water in fluid inclusions; the presence of this absorption is not correlated with deformability of quartz at high temperatures, at least in laboratory experiments at confining pressures up to a few hundred megapascals.

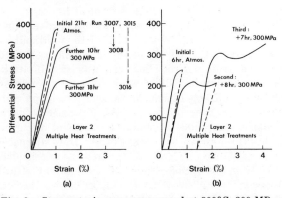

Fig. 3. Stress-strain curves measured at 800°C, 300 MPa confining pressure for synthetic quartz W1, layer 2, with multiple heat treatment: (a) Two specimens, each first heated at 900°C, atmospheric pressure for 21 hours (upper curve, runs 3007 and 3015), then heated at 900°C, 300 MPa pressure for 10 and 18 hours (runs 3008 and 3016), respectively. (b) A single specimen, heated first at 900°C, atmospheric pressure for 6 hours (run 3030), then at 900°C, 300 MPa for 8 hours (run 3031) and finally, after about 2% strain, at 900°C 300 MPa for a further 7 hours (run 3031 continued).

When specimens of synthetic quartz W1 are heated at 900°C at atmospheric pressure, raising the yield stress as just described, there are distinctive changes in the infrared spectrum involving all three categories of absorption (Figure 4). The sharp bands are markedly reduced in strength, the underlying very broad band is reduced to a lesser but still substantial extent (as is evident especially in the otherwise clear region around 2900 cm^{-1}), and a prominent ice band appears. Except for some continuing gradual diminution of the sharp bands, these changes appear to approach completion fairly early in the course of heating, perhaps by about 5 hours. Thus the height of the ice peak and the lowering of absorption around 2900 cm^{-1} are almost as well developed after 7 hours as after 60 hours of heating (cf., the observation of *Jones* [1975] that 'equilibrium' is established in about 12 hours at a given temperature). However, in spite of the relative changes in the spectrum, the total integral absorption over the whole 3600-2600 cm^{-1} region does not change very much with heating; there is possibly a slight increase but not by more than the order of 10%. The changes in the infrared spectrum with heating are accompanied by development of an optical turbidity or milkiness (also a slight brownish tinge if the heating is done in air), but this does not appear to affect significantly the background level of infrared absorption in the 3-μm region. Prior deformation of the specimen at 300 MPa confining pressure does not modify the effect of heating at atmospheric pressure.

In contrast to the effects at atmospheric pressure, when specimens of W1 are similarly heated at 300 MPa hydrostatic pressure, relatively little change is observed in the infrared spectrum (Figure 5); the sharp bands remain largely unchanged, while the very broad band is only slightly reduced and a corresponding slight rise appears at the position of the ice band. There is similarly very little change in the spectrum after a deformation experiment at this pressure. At intermediate pressures the amount of change with heating is intermediate, the effect increasing with decreasing pressure as shown in Figure 6.

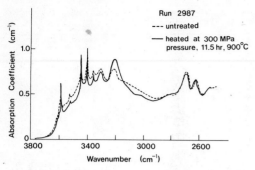

Fig. 5.   Infrared absorption spectra for a specimen from synthetic quartz W1, layer 2, in the as-received condition and then after heating at 900°C, 300 MPa pressure for 11.5 hours.

When specimens of W1 are heated first at atmospheric pressure and subsequently at 300 MPa, the initial changes in the spectrum are partially reversed (Figure 4). Both the sharp bands and the very broad band (as indicated by the trough around 2900 cm^{-1} and the region above 3400 cm^{-1}) are partially restored at the expense of some reduction in the ice band. However, it is evident that full recovery of the original spectrum, if achievable, would take a much longer time at 300 MPa pressure than the initial changes take at atmospheric pressure. Reversal of the initial changes is also found to be slower when the initial duration of heating at atmospheric pressure is longer.

We also record here (Figure 7) the infrared absorption spectrum of a specimen of clear natural quartz (N2) that was heated in the presence of water from dehydrating talc for 12 hours at 900°C under 1500 MPa confining pressure in a solid medium apparatus (by courtesy of R. Coe).

Electron Microscope Observations

Synthetic quartz crystal W1 is initially free of any features visible in the optical microscope. In transmission

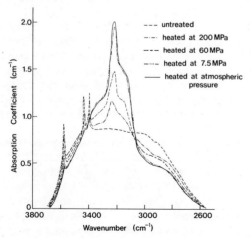

Fig. 6.   Infrared absorption spectra for hydroxyl only (the spectrum for crystal A6-13, Figure 1, has been subtracted from the measured spectra) for a series of specimens from crystal W1, layer 1, heated for 10 hours at 900°C at the pressures shown.

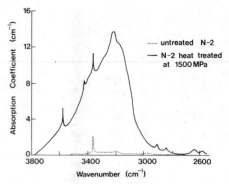

Fig. 7.   Infrared absorption spectrum for a specimen of natural quartz after heating at 900°C, 1500 MPa pressure for 12 hours in the presence of water from dehydrating talc; the spectrum in the original condition is shown dotted.

electron microscopy, no dislocations are seen (initial dislocation density is around 10^7 m^{-2} [Hobbs et al., 1972]). However, there is a high density (up to $10^{18} - 10^{19}$ m^{-3}) of small contrast features (Figure 8a) similar to those noted by *Morrison-Smith et al.* [1976], *Jones* [1978], and *Kirby and McCormick* [1979] in other synthetic quartzes (it is notable, however, that such features are completely absent in the amethysts studied by us). These features consist of a region of diffraction contrast of 30–80 nm across, with a diametral zone of no contrast generally normal to the diffraction vector **g**. In the latter respect, they may differ somewhat from the features in crystal W2 described by *Morrison-Smith et al.* [1976] and correspond to those observed in the synthetic quartz X-0 and interpreted by McCormick as arising from spherical inclusions of 8.5 ± 0.5 nm diameter acting as sources of hydrostatic pressure [*McCormick and Kirby*, 1979].

After deformation of several percent at 10^{-5} strain rate at 800°C and 300 MPa confining pressure or after simple heating for 5 hours or more at 900°C with or without confining pressure, the density of the above-mentioned features is markedly reduced. However, other precipitate-like features now appear. Under weak diffracting conditions, in which all diffracted beams have large extinction errors s, polyhedral or spherical precipitates are observed having less-than-background intensity in the print (brighter-than-background as seen on the electron-microscope screen), as shown in Figure 8b. The contrast is considered to arise from absorption effects and is consistent with the features being cavities [*Amelinckx*, 1970]; we therefore shall refer to them as 'bubbles.' Under diffraction conditions in which one or more reflections is more strongly excited, the bubbles are again of similar appearance except that they sometimes show greater-than-background contrast if suitably situated within the thickness of the foil, but now additional small dark features are revealed as shown in Figure 8c. The latter features resemble defect clusters, and the image width is strongly dependent on diffraction conditions [cf., *Wilkens*, 1970]; they have not been included in the category of bubbles in the following text and in Table 2. The density of the bubbles themselves is not directly correlated with the density of the initial contrast features; in particular, it is much higher in W1 than in E-1-1 specimens in spite of similar densities of initial contrast features. We now describe in more detail the bubble development under the various kinds of heat treatments.

Sizes and densities of bubbles formed during heating at 900°C at atmospheric pressure for various times are given in Table 2. These bubbles have polyhedral or 'negative crystal' form (Figure 8b), with approximately constant aspect ratio when viewed in profile, suggesting that the form is an equilibrium one [cf., *McLaren and Phakey*, 1966]. The 'sizes' quoted are the diameters of spheres of the same volume as the bubbles (in practice, approximately 1.75 times the minimum dimensions of the bubbles). In most cases, the bubble population has been divided subjectively into two groups, 'large' and 'small,' during the measuring, although the dividing line between the groups becomes rather arbitrary for the shorter heating durations. In spite of wide scatter, the data in Table 2 appear to define a definite trend for the density of bubbles to increase with annealing time, this increase arising mainly from the appearance of bubbles of smaller size, so that the mean size decreases as the population grows. However, the total volume of bubbles does not increase noticeably, within uncertainty of measurement, since the volume contributed by later-appearing smaller bubbles is relatively small (after about 100 hours, the majority of new bubbles appear to be only of the order of 10 nm in size); this observation is consistent with relative constancy of the infrared absorption spectrum after the first few hours of annealing (see earlier). The volume fraction of bubbles in specimen 3079 from layer 1 is higher than that in specimens from layer 2, corresponding to its higher OH content (Figure 1; the apparent difference between ratio of volume fractions of bubbles and ratio of OH contents between the two layers is probably not significant in view of the uncertainties in the determinations of the volume fractions). In addition to observations on bubbles, it has also been noted that after annealing the dislocation density rises appreciably, roughly to the order of $10^{12}m^{-2}$ from 10^7 m^{-2}, and that the larger bubbles, but not generally the smaller ones, are often linked by dislocations [cf. *McLaren and Phakey*, 1966].

When annealing is done under pressure, there are several changes in the characteristics of the bubble population (Table 2):

1. Although up to 60 MPa the negative crystal form persists, at 300 MPa the bubbles formed are in general spherical (Figure 8c), similar to those previously observed in synthetic quartz deformed at 300 MPa [*Morrison-Smith et al.*, 1976].

2. With increasing annealing time at 300 MPa, the bubbles increase in size and decrease in density, indicating that some bubbles grow at the expense of others.

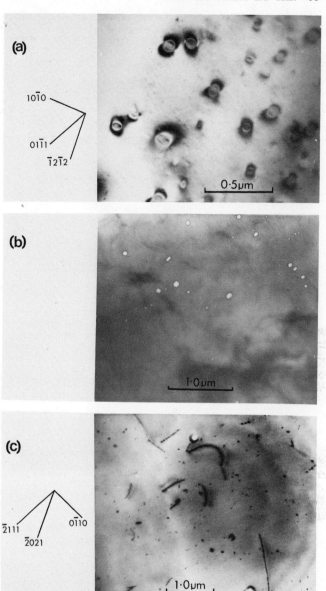

Fig. 8. Electron micrographs (bright field) of specimens of synthetic quartz crystal W1. (a) Initial condition; $g = \bar{1}2\bar{1}2$. (b) After heating for 96 hours at 900°C, atmospheric pressure (sample Y). (c) After heating for 25 hours at 900°C, 300 MPa pressure (run 3062; this specimen has also been strained about 2% at 800°C after the heat treatment).

3. The volume fraction of bubbles formed decreases as the pressure increases.

4. In contrast to the situation at atmospheric pressure, the dislocations appearing in specimens annealed under pressure (whether subsequently deformed or not) do not in general link the bubbles to a great extent.

When specimens previously annealed at atmospheric pressure are further heated under 300 MPa hydrostatic pressure (Table 2), the characteristics of the bubble popu-

TABLE 2. Bubble Sizes and Densities after Heat Treatment at 900°C

Speci-mens	Crystal Layer	Pressure, MPa	Heating Time, hour	Large Bubbles			Small Bubbles			All Bubbles		
				Mean Size, nm	Density, 10^{18} m^{-3}	Volume Fraction, ppm	Mean Size, nm	Density, 10^{18} m^{-3}	Volume Fraction, ppm	Mean Size, nm	Density, 10^{18} m^{-3}	Volume Fraction, ppm
S	2	atmos.	5	64 ± 6	2.6 ± 1.5	360	55 ± 8	2.4 ± 1.4	210	60	5	570
W	2	atmos.	23	62 ± 8	3.8 ± 3.0	480	41 ± 7	4.0 ± 2.8	140	51	8	620
X	2	atmos.	48	59 ± 9	5.2 ± 2.7	560	31 ± 5	6.8 ± 4.0	110	43	12	570
Y	2	atmos.	96	57 ± 7	6.0 ± 3.1	570	19 ± 6	10.1 ± 7.2	30	33	16	600
3079	1	atmos.	10							60 ± 12	18 ± 12	2040
3080	1	7.5	10							63 ± 9	14 ± 10	1790
3076	1	60	10							73 ± 16	5.2 ± 3.0	1040
2989	2	300	6.5							35 ± 6	5.8 ± 4.0	130
3062	2	300	25							105 ± 30	0.3 ± 0.25	180
3027/28	2	atmos/300	5/2							64 ± 10	2.6 ± 1.8	360
3030/31	2	atmos/300	6½/15							83 ± 13	0.7 ± 0.4	210
3021/25	2	atmos/300	60/43							74 ± 15	2.5 ± 1.5	530
3032/33	1	atmos/300	15/24							123 ± 30	0.3 ± 0.2	290

The specimens S, W, X, Y were not deformed after the heating, whereas the other specimens were slightly deformed after heating in order to determine the yield stresses; it is thought that this deformation does not significantly change the bubble size and density. The plus or minus figures given are for two standard deviations.

lation fall between those for the individual treatments alone and reflect various degrees of sluggishness in reversing the initial changes. When the initial annealing has been of short duration, there is a clear trend for the bubble density and volume fraction to decrease and the size of the remaining bubbles to increase as the subsequent annealing time at 300 MPa is increased (3027/28 and 3030/31; cf., S), but reversal of the precipitation after 60 hours at atmospheric pressure is evidently more sluggish (3021/25). Changes in the specimen from the OH-richer layer appear to be more easily reversible, as indicated by the larger bubble size and smaller volume fraction (3032/33; cf., 3079). Dislocation configurations in the latter specimen are typical of a 'high temperature' regime [cf., Morrison-Smith et al., 1976] and the bubbles are largely spherical, whereas in the others the dislocations are crystallographically aligned and more numerous, as for a 'low-temperature' regime, and the bubbles, at least the larger ones, retain their negative crystal form. In light of the discussion later, this suggests that the initially OH-richer specimen still has more OH in a kinetically effective form after the double heat-treatment.

Some observations were also made on the specimen of clear natural quartz that was heated in the presence of water from dehydrating talc at 1500 MPa confining pressure (Figure 7). This specimen now contains spherical bubbles ranging in size from about 70 to 120 nm with a density of around $0.5 \cdot 10^{18}$ m^{-3}. There is also a dislocation density of up to $5 \cdot 10^{12}$ m^{-2} and some cracking. The bubbles tend to be linked by dislocations, but the majority of the dislocations are not associated wth bubbles, and it is not clear whether the dislocations have been generated owing

to nonhydrostatic stresses in the solid pressure medium or simply in the heating cycle as noted for the other specimens. Generally, the appearance is similar to that of a milky vein quartz but with lower densities of both bubbles and dislocations. A specimen of the same quartz heated at 900°C in the presence of water at 300 MPa shows no bubbles or dislocations within it, and no change in infrared spectrum is observed.

Discussion

General

We have previously pointed to the correlation between hydrolytic weakening and the presence of a broad absorption band in the 3600–2600 cm^{-1} region of the low-temperature infrared spectrum, thus associating this absorption 'signature' with the particular hydroxyl species responsible for the weakening [Kekulawala et al., 1978]. The present observations are consistent with this correlation although some qualification is needed. In addition, they permit some conclusions to be drawn about the kinetic and equilibrium properties of the hydroxyl in the quartz. We now discuss these aspects in turn.

At first sight the mechanical, infrared, and electron microscope observations can each be explained by the precipitation, as water, of the 'broadband' or 'gel-type' hydroxyl responsible for the hydrolytic weakening. Thus the strengthening of the synthetic quartz on heating at atmospheric pressure is accompanied by the formation of bubbles which presumably contain water, the freezing of which gives rise to the ice band that appears in the low temperature infrared spectrum at the same time as the very broad band is diminished in intensity. The strengthening is therefore correlated with removal of gel-type hydroxyl from the quartz structure. The same conclusion has been reached from creep tests by *Kirby and McCormick* [1979], who also set out arguments against the hardening being mainly due to the precipitates themselves. It is consistent with this picture that heat treatment at 300 MPa pressure which does not affect the yield strength also does not appreciably change the broad infrared absorption; evidently, the relatively small number of bubbles that now appear do not represent the major part of the hydroxyl initially in the quartz, which is still not precipitated. In all cases, regardless of the pressure, the approximate constancy of the total infrared absorption in the 3-μm region shows that the water is not lost from the crystal as a whole to any appreciable extent during the heat treatments.

On closer examination, some qualifications to the above general conclusions are needed:

1. Even after prolonged heating at atmospheric pressure, when it might be expected from the observed strength increase that almost all the hydroxyl would have been precipitated as water, the ice band in the infrared spectrum only makes up about one half of the total absorption, indicating that as much as one half of the precipitated hydroxyl is in a form that is not freezable as ice but which again gives a very broad absorption, including in the trough region around 2900 cm^{-1} (cf., the lack of correlation between development of milkiness and change in infrared absorption noted by *Chakraborty and Lehmann* [1977]). It is likely that this nonfreezable component occurs in films of monolayer or similar thickness on the surfaces of the bubbles, the films consisting mainly of water absorbed on or reacting with the quartz but possibly including some of the silica (amounting to up to a few percent) which would have been in solution in the bubbles at 900°C; in fact, the films may be similar to those discussed by *Langer and Flörke* [1974] in relation to hydroxyl absorption in opal. Calculation shows that for a water-filled bubble of 60 nm diameter with a bubble pressure of 50 MPa at 900°C (see below), a monolayer of water on the surface of the bubble could accommodate roughly one third of the water in the bubble; at 180 MPa bubble pressure, this fraction would be about one tenth. Thus, correlation between the very broad infrared absorption and hydrolytic weakening may not always be as strong as was initially thought, and in particular, significant absorption around 2900 cm^{-1} may not necessarily indicate low strength at high temperatures, especially when small bubbles are present.

2. The observations at atmospheric pressure show that there are quite different rates of change in mechanical properties and infrared spectrum during heat treatment. However, in spite of the lack of change in the infrared spectrum after a few hours, precipitation or segregation of water continues to occur, as indicated by the electron microscopy observations on bubbles; but as time progresses the new bubbles that become visible are smaller and smaller if the present observations are significant.

The change in strength thus evidently reflects the continuing precipitation or segregation, while the apparent lack of change in the infrared absorption may arise from two factors: first, the total amount of water appearing as later-formed bubbles is relatively small because of their small sizes and, second, a larger proportion of it may be in nonfreezable form in surface films, because of the larger surface to volume ratio in smaller bubbles, and therefore not be distinguished from the original unprecipitated hydroxyl.

3. The pressures that can be deduced for the bubbles when formed may need further study and rationalization. If a mole fraction X of hydroxyl is precipitated into a volume fraction x of bubbles at 900°C, the pressure p (in MPa) in the bubbles will be given approximately by $p = 220X/x$ if X/x is less than about 1.5 (it is assumed that the specific volume of water at 900°C is given by $0.55/p$ m^3 kg^{-1}, an expression that fits the actual p-v-T data to within better than 10% from atmospheric pressure to almost 300 MPa; see table in *Clark* [1966, p. 379]). Taking $X = 400$ ppm and $x = 600$ ppm as representative of the precipitation at atmospheric pressure (Table 2), the calculated pressure in the bubbles when formed at atmospheric pressure is about 150 MPa (or about 10% less if there is a monolayer of water on the surface of the bubbles). This figure is somewhat higher than corresponds to a purely mechanical equilibrium between surface energy and internal pressure as expressed in the formula $p = 4\gamma/d$, where γ is the surface energy and d the bubble diameter; with $d = 60$ nm (Table 2) and the theoretical estimate of $\gamma = 0.7$ J m^{-2} at 900°C [*Davidge*, 1979, p. 77; see also *Lawn and Wilshaw*, 1975, p. 77] this formula gives $p \approx 50$ MPa. However, it is possible that the higher pressure can be rationalized through a thermodynamical treatment, as set out by *Russell* [1972], in which the chemical potentials of the OH species in the bubble and in the quartz are equilibrated at a transient stage when the crystal is still supersaturated in hydroxyl. However, the case of precipitation at 300 MPa presents a more immediate problem. A volume fraction of up to 180 ppm of bubbles has been observed (Table 2). Since it could be expected that pressure in the bubbles must at least slightly exceed the external pressure of 300 MPa for the bubbles to form, the amount of hydroxyl required to fill this volume fraction of bubbles would have to be at least 250 ppm. On the other hand, the infrared observations suggest that not more than a small fraction of the initial 400 ppm hydroxyl could have been precipitated. Possibly, the discrepancy arises from experimental inaccuracies, especially in the volume fraction determinations owing to considerable uncertainty in thickness determination and inadequate characterization of the bubble size distribution. Analogous measurements reported by *Jones* [1978] do not present the same problem. Thus, Jones observes an average volume fraction of 2500 ppm of bubbles for the presumably nearly complete precipitation of 1000° ppm hydroxyl at atmospheric pressure, giving a calculated bubble pressure at 850°C of about 90 MPa,

while for precipitation in 10 min at 1000 MPa pressure and 850°C he observes an average volume fraction of bubbles of 230 ppm which would require about half of the initial 1000 ppm hydroxyl to give a bubble pressure equal to the internal pressure (using the actual p-v-T data for 1000 MPa).

Kinetic Aspects—Diffusivity

The observed coarsening of the bubble pattern during heating at 300 MPa indicates that a transfer or interchange of material is occurring on the scale of the bubble spacing. In an earlier, brief discussion of these observations [*Paterson and Kekulawala*, 1979], it was implicitly assumed that this transfer occurred by direct bulk diffusion and that the rate-controlling process was the diffusion of water from bubble to bubble, from which it was concluded that the diffusivity of the water-related species in quartz containing about 400 H/10^6Si at 900°C was of the order of 10^{-18} m^2 s^{-1}. It is doubtful whether such a conclusion is valid, for two reasons. First, the coarsening may not occur by direct bulk diffusion but rather by the migration and mutual coalescence of bubbles, as suggested for radiation-damaged materials where various factors have been proposed as rate-controlling, including surface diffusion, nucleation of new surface layers, and bulk diffusion [e.g., *Gruber*, 1967; *Gulden*, 1967; *Beeré and Reynolds*, 1972]. Second, even if bulk diffusion is the rate-controlling process, either in case of direct transfer of material or in bubble migration, it is possible and even likely that it is the diffusivity of the Si or O of the quartz that is concerned and to which the above value of 10^{-18} m^2 s^{-1} would apply. This value would then have to be taken as a lower limit for the diffusivity of the water-related species in this quartz. On the other hand, the failure to weaken strong milky quartz through re-solution of water from bubbles with spacings of the order of 1 μm during many hours of heating at 300 MPa suggests that the corresponding OH diffusivity in a relatively dry quartz matrix at 900°C may be substantially less than 10^{-18} m^2 s^{-1}. In the case of the heat-treated synthetic quartz, in which the bubble spacing is more on the scale of 0.1 μm, the ability to reweaken these specimens would be consistent with an OH diffusivity in them of the order of 10^{-19} m^2 s^{-1}.

During heating at atmospheric pressure, in contrast to the situation at 300 MPa, the early-formed coarser bubbles do not appear to grow much, but rather the population of visible bubbles appear to be augmented by successive addition of smaller and smaller bubbles of closer and closer spacing. Thus, if this is the case, the range over which material is effectively mobile evidently becomes less as the major part of the water is precipitated. By applying the same argument as before, it can be concluded that the controlling step, probably self-diffusion of Si or O in the quartz, is dependent on the concentration of OH in the quartz. It may not follow directly that diffusivity of OH is itself concentration-dependent, as previously asserted; nevertheless, it is likely that diffusivity of OH is in fact concentration-dependent in view of the diffusivity limits discussed in the previous paragraph and the change in scale of bubble spacing with time in the atmospheric pressure heat treatment.

Equilibrium Aspects—Solubility

The considerations in the previous section suggest that the specimens from layer 2 heated at 900°C and 300 MPa pressure for the longer periods would have been brought to a state of equilibrium between the hydroxyl in solid solution and that in the bubbles. Thus the hydroxyl remaining in solid solution will represent the solubility of OH in quartz at this temperature and pressure. The absence of significant change in initial yield strength and in infrared absorption during heating point to the solubility being not much less than the 400 H/10^6 Si initially present, although the bubble pressure considerations point to a substantial fraction of this hydroxyl having possibly been precipitated. Thus the best estimate we can give at present for hydroxyl solubility is that it is between 200 and 400 H/10^6Si at 300 MPa. The specification of 300 MPa calls for some further comment. This pressure is the total external pressure applied to the specimen. The internal pressure of water in the bubbles may be slightly higher, at least by about 30 MPa because of the $4\gamma/d$ term and perhaps somewhat more if there is any residual influence of a higher driving force for precipitation from initial supersaturation. However, the latter effects probably can be neglected to a first approximation, allowing the above solubility figure to be taken as applying to 300 MPa water pressure as well as total pressure.

It is difficult to make further deductions on solubility. Clearly, the solubility at 900°C at atmospheric pressure is very low, but the reduction in diffusivities during precipitation and the consequent greater sluggishness in approaching equilibrium may make it very difficult to determine the actual solubilities at low pressures. On the other hand, *Jones'* [1978] observations quoted in the previous section would indicate roughly 500 H/10^6Si as an upper limit for the solubility at 850°C, 1000 MPa (subject to uncertainties about how closely equilibrium is approached in 10 min, whether the assumption of 1000 MPa in the bubbles is valid and whether the pressure in the quartz itself is maintained in the experimental assembly during quenching). Such a value, taken to be approximately applicable also to 900°C, would be more nearly consistent with the solubility being proportional to the square root of the pressure rather than to the pressure itself, but such a conclusion is clearly very tentative. The result shown in Figure 7 suggests at first sight a much higher solubility at 1500 MPa, but the presence of the high bubble density in this specimen makes this interpretation questionable; the extraordinary difference in behavior at 1500 MPa is at present highly enigmatic.

No conclusions about the temperature dependence of the solubility can be drawn from the present work.

Practical Implications

It is widely accepted that water plays an important role in determining the rheological properties of quartz. However, it is now clear that it is only effective if it is incorporated in the quartz in a particular solid solution form (so far characterized, probably inadequately, as gel-type on the basis of its infrared absorption) or at least if it is dispersed throughout the quartz on a very fine scale, a scale certainly finer than is resolvable in the optical microscope if it is to be effective in experiments at geologically realistic pressures on a laboratory time scale.

An understanding of both the kinetic and equilibrium aspects of water in quartz is therefore essential in assessing whether laboratory experiments refer to specimens in equilibrium with the experimental environment and so whether it is valid to extrapolate such results to other conditions of temperature, pressure, water fugacity, and strain rate. The conclusions drawn in the two previous sections represent some limits that can be present be deduced concerning the solubility and diffusivity of water or hydroxyl in quartz. Clearly, further studies are needed. However, the indications are that sluggishness of approach to equilibrium is a general and serious problem in experimentation at pressures below 1000 MPa, especially at lower temperatures; only in experiments above 1000 MPa, such as those by *Griggs and Blacic* [1964], has the movement of water through quartz on the scale of the whole specimen been demonstrated.

Many of the experiments carried out so far on synthetic quartz, especially those at atmospheric pressure, have obviously been done on specimens that are not in equilibrium in respect to their hydroxyl content, as shown by the formation of bubbles during deformation [*Morrison-Smith et al.*, 1976; *Kirby and McCormick*, 1979]. To do experiments that can be reliably extrapolated to geological conditions, it is evidently necessary to use a confining pressure and also to avoid supersaturation of OH at the chosen confining pressure. The amount of hydroxyl in solid solution then should be chosen to be that corresponding to the geological conditions contemplated or, if higher OH contents are chosen to accelerate strain rates in the laboratory, a knowledge of the dependence of the flow law on hydroxyl content will be needed in extrapolating to geological conditions.

Acknowledgments. The authors have benefited greatly from discussions with many colleagues, especially J.-C. Doukhan, A. C. McLaren, D. H. Mainprice, G. Smith, and R. W. T. Wilkinson, and they are grateful to G. R. Horword, P. J. Percival, P. E. Willis, and T. W. White for technical assistance. They are also grateful for access to the infrared spectrometer in the Department of Solid State Physics, Australian National University through the kindness of W. A. Runciman, N. B. Manson, and E. R. Vance.

References

Amelinckx, S., The study of planar interfaces by means of electron microscopy, in *Modern Diffraction and Imaging Techniques in Material Science*, edited by S. Amelinckx, R. Gevers, G. Remaut and J. Van Landuyt, pp. 257-294, North-Holland, Amsterdam, 1970.

Beeré, W., and G. L. Reynolds, Rate controlling nucleation and diffusion processes in faceted inert gas bubbles and voids, *Acta Metall., 20*, 845-848, 1972.

Brunner, G. O., H. Wondratschek, and F. Laves, Ultrarot untersuchungen über den Einbau von H in naturlichen Quarz, *Z. Elektrochem., 65*, 735-750, 1961.

Chakraborty, D., and G. Lehmann, Infrared studies of X-ray irradiated and heat treated synthetic quartz single crystals, *N. Jahrbuch Mineral., Monatshefte*, 289-298, 1977.

Clark, S. P., Jr. (Ed.), *Handbook of Physical Constants*, Mem. Geol. Soc. of Am., *97*, 1966.

Davidge, R. W., *Mechanical Behaviour of Ceramics*, Cambridge University Press, Cambridge, 1979.

Griggs, D. T., Hydrolytic weakening of quartz and other silicates, *Geophys. J. R. Astron. Soc., 14*, 19-31, 1967.

Griggs, D. T., and J. D. Blacic, The strength of quartz in the ductile regime (abstract), *Eos Trans. AGU, 45*, 102, 1964.

Griggs, D. T., and J. D. Blacic, Quartz: Anomalous weakness of synthetic crystals, *Science, 147*, 292-295, 1965.

Gruber, E. E., Calculated size distributions for gas bubble migration and coalescence in solids, *J. Appl. Phys., 38*, 243-250, 1967.

Gulden, M. E., Migration of gas bubbles in irradiated uranium dioxide, *J. Nucl. Mater., 23*, 30-36, 1967.

Hobbs, B. E., A. C. McLaren, and M. S. Paterson, Plasticity of single crystals of synthetic quartz, in *Flow and Fracture of Rocks, Geophys. Monogr. Ser.*, vol. 16, edited by H. C. Heard, I. Y. Borg, N. L. Carter, and C. B. Raleigh, pp. 29-53, AGU, Washington, D. C., 1972.

Jones, M. E., Water weakening of quartz, and its application to natural rock deformation, *J. Geol. Soc. London, 131*, 429-432, 1975.

Jones, M. E., The influence of hydrostatic pressure on the precipitation of structure-bound water in micro-inclusions in quartz, *Philos. Mag. Part A, 37*, 703-706, 1978.

Kekulawala, K. R. S. S., M. S. Paterson, and J. N. Boland, Hydrolytic weakening in quartz, *Tectonophysics, 46*, T1-T6, 1978.

Kirby, S. H., and J. W. McCormick, Creep of hydrolytically weakened synthetic quartz crystals oriented to promote $\{2\bar{1}\bar{1}0\}\langle0001\rangle$ slip: A brief summary of work to date, *Bull. Minéral., 102*, 124-137, 1979.

Langer, K., and O. W. Flörke, Near infrared absorption spectra (4000-9000 cm^{-1}) of opals and the role of 'water' in these $SiO_2 \cdot nH_2O$ minerals, *Fortschr. Mineral., 52,* 17-51, 1974.

Lawn, B. R., and T. R. Wilshaw, *Fracture of Brittle Solids,* Cambridge University Press, Cambridge, 1975.

McLaren, A. C., and P. P. Phakey, Transmission electron microscope study of bubbles and dislocations in amethyst and citrine quartz, *Aust. J. Phys., 19,* 19-24, 1966.

Morrison-Smith, D. J., M. S. Paterson, and B. E. Hobbs, An electron microscope study of plastic deformation in single crystals of synthetic quartz, *Tectonophysics, 33,* 43-79, 1976.

Paterson, M. S., A high-pressure, high-temperature apparatus for rock deformation, *Int. J. Rock Mech. Mining Sci., 7,* 517-526, 1970.

Paterson, M. S., and K. R. S. S. Kekulawala, The role of water in quartz deformation, *Bull. Minéral., 102,* 92-98, 1979.

Russell, K. C., Thermodynamics of gas-containing voids in metals, *Acta Metall., 20,* 899-907, 1972.

Tullis, J., G. L. Shelton, and R. A. Yund, Pressure dependence of rock strength: Implications for hydrolytic weakening, *Bull. Minéral., 102,* 110-114, 1979.

Twiss, R. J., Structure and significance of planar deformation features in synthetic quartz, *Geology, 2,* 329-332, 1974.

Twiss, R. J., Some planar deformation features, slip systems, and submicroscopic structures in synthetic quartz, *J. Geol., 84,* 701-724, 1976.

Wilkens, M., Identification of small defect clusters in particle-irradiated crystals by means of transmission electron microscopy, in *Modern Diffraction and Imaging Techniques in Material Science,* edited by S. Amelinckx, pp. 233-256, North-Holland, Amsterdam, 1970.

Creep and Creep Rupture of Granitic Rocks

Neville L. Carter, Douglas A. Anderson, Francis D. Hansen, and Robert L. Kranz

Center for Tectonophysics, Texas A&M University, College Station, Texas 77843

A review is given of recent experimental studies of flow properties and processes of granitic rocks deformed over a wide range of physical conditions. Preliminary new creep data for Westerly granite, deformed to low strains, were obtained in dry compression tests in a Griggs solid pressure medium apparatus at 1.0 GPa confining pressure, temperatures from 470° to 765°C, constant stress differences of from 0.6 to 1.2 GPa, all in the α quartz stability field. High-temperature transient creep data fit an exponential decay flow law very well and were also fit to a power law, for comparison with previous work, with the result $\epsilon_t = 7 \times 10^{-5} \sigma^{2.2} t^{0.5} \exp(-30.5/RT \cdot 10^{-3})$ for stress in MPa, where time is in seconds and E is in kcal/mole. Steady state creep results fit a power law $\dot{\epsilon}_s = 1.4 \times 10^{-9} \exp(-25.3/RT \cdot 10^{-3}) \sigma^{2.9}$. Because of experimental uncertainties, differences in the activation energies and stress exponents for ϵ_t and $\dot{\epsilon}_s$ are regarded as insignificant. The experiments and analyses indicate that high-temperature transient creep gives way to steady state creep at strains less than 1% and in short times. Steady state flow should thus dominate natural creep of granitic rocks at moderate to high temperatures. Preliminary optical and TEM analyses of the specimens indicate that these low creep strains are accommodated primarily by quartz, secondarily by micas, and little, if at all, by feldspars, as seems to be true also of several naturally deformed granitic rocks examined petrographically. The close accord of the activation energies for creep of granitic rocks observed here and in previous studies with those found for steady state creep of quartzite [Koch et al., 1980] also suggests that deformation of quartz controls the creep rate of granitic rocks. Activation energies for creep of feldspars, under most favorable conditions for low energies [Tullis and Yund, 1979b], appear to be too high to account for the results.

Introduction

The creep behavior of granitic rocks at shallow to moderate crustal depths is important to a wide variety of national concerns: geothermal energy extraction, radioactive waste isolation, and earthquake hazards reduction. Near the earth's surface, the behavior is elastic brittle and has been studied extensively experimentally. However, at moderate depths, at elevated temperature and pressure, the behavior is most probably semibrittle in which thermally activated processes contribute to the creep strain and fracture state. At such depths, at shallow depth at elevated temperature, and, perhaps, under most crustal physical conditions, granitic and similar crystalline rocks are expected to flow both in transient and steady state creep regimes. Despite their apparent importance, transient creep and semibrittle behavior of crystalline rocks at elevated temperature and pressure have received very little attention. Thus we have launched an intensive investigation of this research topic, initially on high-temperature transient and steady state creep of granitic rocks, and the preliminary results for Westerly granite are presented here. For background, and continuity, it will be necessary first to review briefly results of studies of low-temperature creep of granite. Because of space limitations, this review cannot be comprehensive and we must focus only on the most salient features of recent experimental work.

Creep of Granitic Rocks
General Statement

Extensive creep (constant stress) and constant strain rate tests have shown that the creep strain of any material is given by

$$\epsilon = \epsilon_e + \epsilon_p + \epsilon_t + \epsilon_s + \epsilon_a \qquad (1)$$

where ϵ_e is the instantaneous elastic strain ($\Delta\sigma/E$) upon loading, ϵ_p is the instantaneous plastic strain produced

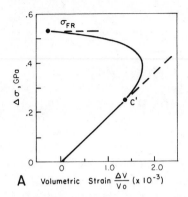
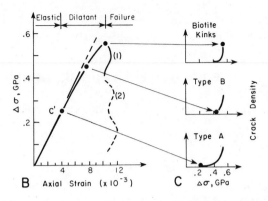

Fig. 1. (a) Volumetric strain as a function of differential stress of Westerly granite loaded to failure. Stress-strain curve of (b) Westerly granite loaded to failure and beyond with (c) initiation of microcracks of various types at various stresses and their associated density as a function of stress (after *Tapponnier and Brace* [1976]).

during loading, ϵ_t is the transient or primary creep strain, ϵ_s is the secondary or steady state creep strain, and ϵ_a is the accelerating or tertiary creep strain. Inasmuch as we are interested primarily in appreciable creep strains, ϵ_e and ϵ_p (generally $<10^{-2}$) may be neglected although substructural changes may be associated with ϵ_p. In rocks the accelerating creep strain ϵ_a is generally observed at stresses above one half the short-term breaking strength in unconfined creep tests and in low-temperature, low-pressure triaxial creep tests. Under these conditions the accelerating creep stage is generally attributed to microfracturing leading to macroscopic failure by faulting. As applied to granite, this type of behavior may be important in natural deformations and in experimentally deformed natural aggregates and will be discussed in context subsequently. For general deformations of crustal rocks, the contributions of transient and steady state creep are expected to constitute the bulk of the creep strain. Steady state creep of rocks has been reviewed extensively recently by *Carter* [1976], *Heard* [1976], *Tullis* [1979], and *Handin and Carter* [1980] and warrants no further treatment here, although the steady state results for Westerly granite [*Tullis and Yund*, 1977] will be summarized below. Rather, we shall concentrate on transient creep of granite over a large range in temperature as this type of creep is expected to be important over most crustal conditions.

Creep at Low Temperatures

Low-temperature creep of rocks has been researched vigorously in recent years because of its possible importance in association with earthquake precursors [*Martin*, 1979; *Byerlee*, 1978; *Carter and Kirby*, 1978; *Kranz*, 1979a, c]. Nonrecoverable creep in metals at temperatures below $0.3\ T_m$ and at stresses high enough to promote significant dislocation mobility generally follows a transient creep law of the form

$$\epsilon_t = \alpha \log (1 + \nu t) \qquad (2)$$

where α and ν are constants that depend on test conditions and material properties [*Garofalo*, 1965; *Weertman and*

Weertman, 1970]. There is some evidence for analogous low-temperature logarithmic creep of rocks and minerals that show significant plasticity at low temperature. Minerals behaving in this way (e.g., halite, calcite, gypsum, layered silicates, pyroxenes) are few, however, and high-confining pressures and stresses are required to produce very limited creep strain [*Carter and Kirby*, 1978]. Most rock-forming silicates are brittle at low temperature and deform permanently by microfracturing at stresses greater than about half their short-term breaking strength. This permanent strain is accompanied by a volume increase (dilatancy) with acoustical emissions associated with microfracturing [e.g., *Handin et al.*, 1963; *Brace et al.*, 1966; *Scholz*, 1968a, b, c; *Hardy et al.*, 1969; *Lockner and Byerlee*, 1977]. Other changes associated with dilatancy, reviewed recently by *Byerlee* [1978], and currently under investigation as possible phenomena useful for earthquake prediction, include seismic velocities, electrical resistivity, magnetic moment, and radon emission. Again, because of space limitations, we shall henceforth confine our discussion largely to direct observations of the deformational processes.

Tapponnier and Brace [1976] have recently summarized the mechanical behavior of Westerly granite compressed at 23°C, a constant strain rate of $10^{-5}/s^{-1}$, and a confining pressure of 50 MPa. The chief emphasis of the study was to observe the nature of crack development and interaction with preexisting natural cavities, and they chose to study, using scanning electron microscopy (SEM), ion-polished sections of the deformed material according to the method described by *Brace et al.* [1972] and *Sprunt and Brace* [1974a]. The pressure of 50 MPa was chosen as a compromise to avoid undue damage of natural cavities in the starting material [*Sprunt and Brace*, 1974b] and of end effects in unconfined tests. Volumetric strains were determined as a function of stress for the samples compressed to varying fractions of the fracture strength, and the results are reproduced in Figure 1a. It is seen that the stress at onset of dilatancy, C', occurs at about 250 MPa, or about one half of the compressive strength of Westerly granite under these conditions.

MICROFRACTURES

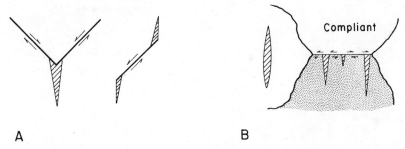

A   B

Fig. 2.   Crack models for brittle rock. Grain boundary shearing models for (*a*) initiating extension cracks and (*b*) isolated crack and elastic mismatch models (after *Brace et al.* [1966] and *Tapponnier and Brace* [1976]).

Tapponnier and Brace [1976] also examined some of the earlier samples of Westerly granite deformed by *Wawersik and Brace* [1971] beyond the peak stress into the postfailure regime. These tests were done in a specially designed stiff testing machine at the same temperature and strain rate but at different pressures (3.5 to 150 MPa) from those of Tapponnier and Brace. The characteristic type of stress-strain curve is as shown in Figure 1*b*. Optical examination of specimens compressed at pressures near 50 MPa [*Wawersik and Brace*, 1971] indicated the presence of extensive transgranular cracking followed by faulting in region (1), at and beyond the peak stress, and complete faulting associated with unloading cracks in region (2). A summary of the more recent observations of *Tapponnier and Brace* [1976], for the prefailure regime, is depicted schematically in Figures 1*b* and 1*c*. Here, the onset of dilatancy is associated with type A cracks. At about 450 MPa, type B microcrack density accelerates sharply, and near the peak stress, about 550 MPa, exten-

EN ECHELON   EN PASSANT   CRACK-PORE

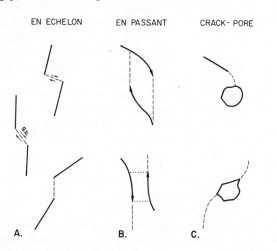

A.   B.   C.

Fig. 3.   Various types of crack and cavity interactions observed by *Kranz* [1979*b*, Figure 1]. En echelon arrays are most common, and of these, that shown on far left, not illustrated by *Kranz* [1979*b*] is most frequent [e.g., *Dunn et al.*, 1973].

sive plastic strain of biotite is noted, as indicated by the kinking on the {001} slip planes of suitably oriented grains. Type A cracks are associated with arrays of natural grain boundary, low aspect ratio cavities, with previously induced, partially healed transgranular cracks and with cleavage cracks. Type B cracks are transgranular cracks emanating from interior pores in grains, preexisting grain boundary cavities and from high angle grain boundaries of different minerals, especially when biotite and magnetite are involved. Most of these cracks are oriented subparallel to σ_1 and are termed precursive microfractures by *Friedman* [1975] because they coalesce to produce macroscopic shear failures. At or near the peak stress, most grain boundaries are continuously cracked, regardless of their orientation with respect to the compression axis. The crack density in the radial direction nearly doubled at peak stress, roughly in accord with lateral strain determinations and with the results of *Peng and Johnson* [1972] on Chelmsford granite and of *Kranz* [1979*a*] on Barre granite, these granites being appreciably coarser grained than the Westerly.

Tapponnier and Brace [1976] observed that nearly all stress-induced cracks were extensile in nature (a few exceptions were noted as also were observed by *Friedman et al.* [1970]. This results simply because these brittle materials are very strong in shear and comparatively weak in tension or extension, as observed previously by many others. They discussed their results in terms of crack models proposed by *Brace et al.* [1966], as shown in Figure 2*a* and of the elastic mismatch model shown in Figure 2*b*. The highly favored (because they readily explain hysteresis effects) grain boundary shear models depicted in Figure 2*a* were rejected on the basis of the observations noted above. Axial cracks of the type shown in Figure 2*b* (left) could contribute to the hysteresis provided that shear stress is transmitted across them that could arise if the cracks are inclined somewhat to σ_1, if crack edges interpenetrate, or if the cracks are closed and lie at the interface of two different minerals. *Tapponnier and Brace* [1976] also suggested that the hysteresis could arise from

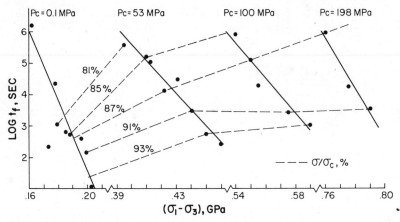

Fig. 4.   Time to failure under constant stress difference at various confining pressures for dry Barre granite. Solid lines are linear least squares regression of data for each confining pressure and dashed lines connect points at constant σ/σ_c [*Kranz*, 1979*c*].

sliding along boundaries between minerals of different elastic properties, such as are illustrated in Figure 2*b* (right) and especially if preexisting cavities occurred along those boundaries, as is commonly the case. They suggested further that minor constituents, such as biotite, which deforms plastically at or near the peak stress, causing cracking in adjacent grains, and magnetite, which is unusually stiff, might control fracture stability and hence brittle strength of Westerly granite and possibly other rocks.

Kranz [1979*a*, *c*], in a comprehensive study of creep of Barre granite compressed normal to the rift plane, confirmed the tensile nature of stress-induced cracks as observed by *Tapponnier and Brace* [1976]. In Kranz's experiments, the granite specimens were loaded to a stress equal to 87% of their fracture strength, and the development and growth of cracks, as a function of time, was investigated by SEM. New cracks were continuously generated with time and the average crack length increased although the width remained unchanged. Crack interaction with other cavities increased with time until accelerating creep became important, whereupon coalescence of precursive microfractures was suggested to be more important than individual crack growth. *Kranz and Scholz* [1977] have obtained evidence that a critical amount of inelastic volumetric strain must be achieved during crack coalescence in tertiary creep before instability leading to macroscopic failure ensues and that the time to failure is logarithmically dependent on stress although the critical inelastic strain is stress independent. This is equivalent to stating that a critical crack density [*Cruden*, 1974] must be achieved for the onset of tertiary creep and, ultimately, for fracture instability. They showed further that the magnitude of the inelastic volumetric strain at the onset of tertiary creep seemed to depend on the brittle fracture strength of the quartzite and granite specimens studied.

Kranz [1979*b*] reviewed crack growth, interaction, and coalescense in general and gave examples of each type of

interaction from his SEM work. He distinguished between three types of interaction, en echelon, en passant, and crack-pore, as illustrated in Figure 3, where it is assumed that a uniaxial compressive stress is applied vertically at a remote boundary. The most commonly observed en echelon arrays (Figure 3*a*, left) are extensile fractures parallel to σ_1 and are generally linked by shear along grain boundaries [e.g., *Dunn et al.*, 1973]; these were not illustrated by Kranz. Other en echelon interactions, observed by Kranz (Figure 3*a*, right) involve linking of extension fractures slightly inclined to σ_1 (Figure 3*a*, top) and of highly inclined fractures, by extension cracks (Figure 3*a*, bottom). En passant interactions (Figure 3*b*) involve cracks on parallel planes that, because of changes in the stress fields, first deviate from and then bend sharply toward each other (Figure 3*b*, top) or linkage is achieved by strain relief (dotted lines, Figure 3*b*, bottom). Figure 3*c* shows deflection of the path of a crack into a void (top) and cracks originating at sharp void boundaries (bottom).

Kranz [1979*a*] has listed the sequence of events that he believes occurs in low temperature constant strain rate and creep tests. For constant strain rate tests, the sequence envisioned is (1) closure of cracks at large angles to the maximum compressive stress, (2) rupturing of material bridging cavities at grain boundaries and partially healed cracks, (3) production of new cracks associated with old cracks or grain boundaries and opening of cavities oriented at a low angle to σ_1, (4) appearance of transgranular cracks connecting pores, grain boundaries, and other cracks, (5) acceleration of crack production accompanied by kinking in suitably oriented micas, (6) formation of a fault by rapid coalescence of cracks through and along grain boundaries.

For creep tests the specimens are loaded very rapidly to about stage (4) in the sequence outlined above. From that point there appears to be a slow production of new cracks, growth of existing cracks, and crack coalescence. The fundamental difference, then, is that in constant strain rate tests, crack growth and development are limited by the rate at which stress is applied to the rock, whereas in creep tests, crack growth and development are limited by the rate at which corrosive agents can decrease the crack tip strength. It should be noted here that many of the observations of *Tapponnier and Brace* [1976] and of *Kranz* [1979a, b] have been made previously by many others in experimental and optical studies of deformed crystalline rocks [e.g., *Borg and Handin*, 1966; *Friedman*, 1975].

Anisotropy of granite also has to be taken into account. Thus, for example, *Brace* [1965] measured the linear compressibility of Westerly and Stone Mountain granites and found them to be markedly anisotropic with the most compliant direction normal to the direction of longest grain dimension, as is to be expected for grain boundary cracks. *Douglas and Voight* [1969] showed, by fabric analysis, that a similar anisotropy in the Barre, Stanstead, and Laurentian granites was related to microfractures in the quartz grains. *Friedman and Bur* [1974] studied the effects of residual stress and of fabric on the orientations of fractures induced by point loading Charcoal granodiorite using optical and sonic techniques. They found that pre-existing microcracks (especially at long grain boundaries), exsolution lamellae and prestress all affected the preferred orientation of newly induced fractures thus emphasizing the importance of anisotropy on fracture orientations under conditions in which $\sigma_2 = \sigma_3$ across the boundaries of the specimens. More recently, *Siegfried and Simmons* [1978] have used differential strain analysis, at confining pressures to 200 MPa, to determine the zero pressure strain tensor of Westerly granite. They, too, found the crack distribution to be anisotropic (most cracks parallel to the rift plane) with the principal strain axes reflecting principal axes of crack distribution, at pressures less than 50 MPa. At pressures above 50 MPa, they suggested that the strain was due mainly to deformation of minerals and of pores that had not yet closed.

It has long been recognized that pressure has profound effects on the creep deformation of granite at low temperature, and several of these effects have been articulated recently by *Kranz* [1980]. Increases in pressure at constant stress difference increase the time required for creep rupture (Figure 4); conversely, for a given time to rupture, higher stress differences are required at higher pressures. In addition, pressure increases the creep rupture resistance above the increase in fracture strength, as indicated by the general trend of the dashed lines in Figure 4. Those lines connect points at different pressures at constant σ/σ_c (applied stress/room temperature fracture strength at the corresponding pressure); they should be horizontal if the increase in the stress required for time dependent rupture was due entirely to an increase of fracture strength with pressure.

The creep-inhibiting effect of pressure is also evident in Figure 5a, which shows the volumetric strain rates for different samples deformed in creep at $\sigma/\sigma_c = 87\%$ and at different pressures. The higher the pressure, the lower the

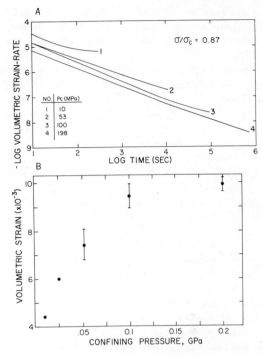

Fig. 5. (a) Volumetric strain rate as a function of time for creep tests on Barre granite. Stress applied was 87% of the short term fracture strength at the corresponding pressure. Curves end just prior to the onset of tertiary creep, culminating in creep rupture (data from *Kranz* [1979c]). (b) Dilatant volumetric strain at the onset of tertiary creep as a function of pressure. Individual dots indicate a single test and dots with error bars represent mean values of several constant stress tests whose differences in $(\sigma_1 - \sigma_3)$ ranged from 30 to 70 MPa. These variations in stress difference at constant confining pressure did not affect the volumetric strain in any systematic way. Error bars are the standard deviation about the mean [*Kranz*, 1979c].

strain rate at any given time after loading. However, lower strain rates at higher pressures do not mean lower creep strains as indicated in Figure 5b, which shows the inelastic volumetric strain prior to the onset of accelerating creep as a function of confining pressure. Higher pressures require more volumetric strain to accumulate prior to the onset of the instability and this correlates with increases in number and average lengths of cracks.

Kranz [1980] has suggested several possible sources for this fatigue-inhibiting effect of pressure, involving chemical and mechanical effects of pressure on processes at crack tips and on crack arrays, hypotheses that require further careful testing. These include (1) an increase in the energy barrier for continued propagation, (2) closing of extensile cracks (the vast majority) inhibiting migration rates of corrosive fluids and gases to crack tips, and (3) reduction of the rate of crack linking, perhaps by diminishing tensile and shearing stress concentrations around individual cracks by the superposition of confining pressure. Finally, the most important mechanical and chemical effects of pore fluids on such brittle rock properties as fracture and frictional strength, creep rate, and static fatigue, among many others, can not be over-emphasized; this topic has been reviewed thoroughly by *Martin* [1979].

The foregoing has constituted a brief summary of static fatigue; it is clear that cyclic fatigue, at least above some threshold stress level, must also be taken into account. *Attewell and Farmer* [1973] showed for dolomite, as had previously been observed for other materials, that cyclic fatigue is a function of peak stress level and of the amplitude and frequency of stress cycles. Specifically, the number of cycles that a rock can withstand increases exponentially as the peak stress is lowered and the higher the frequency at the same peak and mean stress, the higher the number of cycles the rock can sustain prior to failure, although the strain is about the same. *Haimson et al.* [1973] and *Haimson* [1974] recorded acoustic emissions which indicated that the first stress cycle (at frequencies of 1–5 cycles/s) damaged the material most, followed by a stage of decelerating crack development with cycling and ending with a stage at which an increasing number of cracks initiate and/or propagate. An investigation of the processes indicated grain boundary separation during the first cycle, followed by intragranular cracking in succeeding cycles, ending in crack extension, coalescence, widening, and fracture. *Scholz and Kranz* [1974], during slow uniaxial cycling of Westerly granite, showed that each cycle produced more dilatancy than the previous one, that the onset of dilatancy began at a lower stress with each cycle, and that most of the dilatant strain is recoverable with a time dependent character.

The above mentioned studies were carried out at atmospheric or very low confining pressures. *Zoback and Byerlee* [1975] and *Hadley* [1976] repeated the work of *Scholz and Kranz* [1974], also carrying out tests at confining pressures to 200 MPa. Both studies concluded that at the higher pressures, dilatancy cycling is stable, increasing very little from cycle to cycle and without lowering the stress of the onset of dilatancy with each cycle. *Scholz and Koczynski* [1979] reinvestigated the problem, taking into account the effects of dilatancy anisotropy, peak stress per cycle, cycle amplitude, and rate of loading. They found, as in previous studies, that those specimens loaded at higher rates undergo more cycles prior to failure and that the lower the rate of loading, the greater the dilatant strain at the same stress level. Contrary to the results of *Zoback and Byerlee* [1975] and *Hadley* [1976], *Scholz and Koczynski* [1979] found no stabilization of fatigue for confining pressures up to 300 MPa. They suggested that an insufficient number of cycles had been carried out in the former two studies to observe the fatigue effect. Scholz and Koczynski propose that three types of damage occur in these cyclic fatigue tests: (1) stress-induced cracking, accomplished on initial loading to peak stress, (2) time dependent, stress-corrosion cracking upon further cycling, and (3) fatigue cracking as the result of continuous opening and closing of existing cracks.

Creep at Elevated Temperature

The most obvious effect of changing the temperature of polycrystalline aggregates, beyond a threshold temperature that varies with material, is introduction of extension fractures resulting from stresses arising from differential thermal expansion or contraction of adjacent grains. Thermal cracking of crystalline rock has recently been studied theoretically [e.g., *Johnson et al.*, 1978; *Bruner*, 1979] and experimentally [e.g., *Richter and Simmons*, 1974; *Simmons and Cooper*, 1978; *Kern*, 1978; *Johnson et al.*, 1978; *Friedman and Johnson*, 1978; *Bauer and Johnson*, 1979 and *Friedman et al.*, 1979]. The thermal cracks observed by *Freidman et al.* [1979] occurred along, and at high angles to, crystal boundaries and they are invariably shorter and less regular than stress-induced cracks. The studies cited above have shown that the degree of thermal cracking in crystalline rocks depends on mineralogy, fabric, grain size, effective confining pressure, residual strain, and prior thermal and stress history. It seems evident that the type of bonding between grains, including second phases or impurities at grain boundaries, would also play an important role. Mineralogy is obviously important as the coefficients of thermal expansion differ for different minerals. Because these coefficients vary as a function of direction within the crystals, it might be expected, and has been shown experimentally [*Kern*, 1978], that little thermal cracking would occur in monomineralic rocks with strong preferred orientations. Apparently, rocks with finer grain size show less thermal cracking than do those with coarser grains [*Richter and Simmons*, 1974]. Increasing the effective confining pressure inhibits thermal cracking, as expected [*Kern*, 1978; *Friedman et al.*, 1979]. In their work on Sioux quartzite, *Friedman and Johnson* [1978] showed that the residual strain state in the rock influenced the orientation of thermally induced intragranular cracks.

Simmons and Cooper [1978] found, during thermal cycling experiments on Westerly granite, that the volume of new cracks depends exponentially on the maximum temperatures; the porosity doubled for each 140°C increment. For repeated cycling to the same temperature, they found a steady 10% increase in crack porosity for at least the first few cycles. These results differ somewhat from those of *Friedman et al.* [1979] on Charcoal granodiorite (discussed below) who observed a linear increase in thermal cracking with temperature and from other results (B. Johnson, personal communication, 1980) that indicate that new thermal cracking does not occur until the previous maximum temperature is exceeded. *Simmons and Cooper* [1978] showed that cracks produced by thermal cycling increase the thermal expansion over the value predicted for the constituent minerals but that the presence of cracks decreases thermal expansion of the rock by allowing some minerals to expand into the cracks. The tensor thermal expansion for Westerly and Chelmsford granites showed an inverse correlation between linear crack porosity and lin-

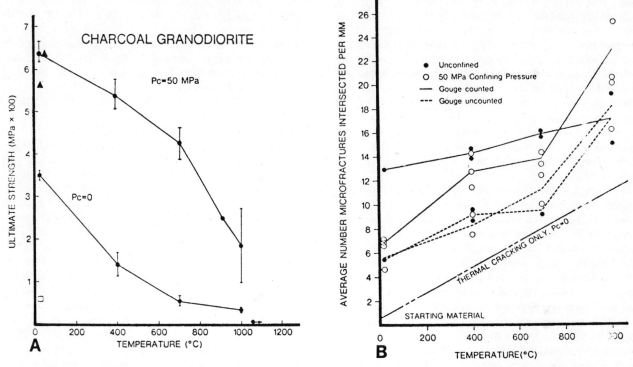

Fig. 6.   (a) Strength versus temperature of Charcoal granodiorite at 0 and 50 MPa and (b) microcrack density as a function of temperature for thermal and stress induced cracks [*Friedman et al.*, 1979].

ear thermal expansion at temperatures between 25° and 400°C.

Another effect of temperature was discovered by *Wu and Thomsen* [1975], who deformed Westerly granite at atmospheric pressure in creep to 150°C at stresses from 190 to 225 MPa. They monitored microfracturing by signals from a piezoelectric transducer and in some experiments measured radial and axial strain. At 150°C there was an initial increase in microfracturing but then this activity decayed sharply, effectively strengthening the granite. They attributed this strengthening effect to driving off water so that it is no longer available for stress-aided corrosion or water weakening [*Griggs and Blacic*, 1965; *Griggs*, 1967; *Blacic*, 1975]. The importance of water as a corrosive agent was also demonstrated in several other types of experiments [*Wu and Thomsen*, 1975].

A most significant result was obtained by *Summers et al.* [1978] in their study of flow of water through Westerly granite in the temperature range 100° to 400°C. In their creep experiments up to 17 days at 50 MPa confining pressure, stress differences of 0 to 350 MPa, inlet pressures were 27.5 MPa and outlet pressures were 0.1 MPa. The initial permeability was observed to be higher at elevated temperature by one to two orders of magnitude over 25°C permeability and this was attributed mainly to thermal cracking. However, the permeability decreased significantly during the first half day of H_2O flow. At elevated temperature, quartz and plagioclase dissolved near the

pore pressure inlet and precipitated down the pore pressure gradient causing a reduction in H_2O flow rate. The final flow rate after 10 days was less at 300°C than at lower temperature and measurable flow stopped in most 400°C experiments after 1.5 days. Thus the permeability was shown to be both temperature and time dependent, but there was no reproducible dependence on stress.

Friedman et al. [1979], in their experiments on charcoal granodiorite at a strain rate of 10^{-4}/s, temperatures to about 1000°C and confining pressures of 0.1 and 50 MPa, distinguished carefully between thermal cracks and stress-induced cracks. The marked decrease in ultimate strength with increasing temperature is as shown in Figure 6a; as was expected, the strengths are much lower in unconfined tests. In unconfined tests, there appears to be a linear increase in thermal cracking with temperature (Figure 6b) as opposed to the exponential increase observed for Westerly granite by *Simmons and Cooper* [1978]. Abundance of stress-induced cracks is reasonably constant with increasing temperature (Figure 6b). An important observation by *Friedman et al.* [1979] is that, at 50 MPa confining pressure, raising the temperature to 1000°C does not affect the room temperature strength of the rock, whereas this strength is lowered appreciably by similar heat treatment in unconfined tests (Figure 6a). This indicates that the weakening as a function of temperature at pressure is not due to thermal cracking but must result from the temperature sensitivity of stress-induced fracturing process. Although failure occurred by macroscopic faulting at less than 3% strain and brittle failure was observed throughout, the temperature effect on fracturing at pressure suggests that thermally activated semibrittle processes may have contributed to the failure. Precursive stress-induced extension fractures are observed in every specimen and commonly are concentrated in an en echelon manner along the path of the throughgoing fault (Figure 3a, left), as was observed indirectly by *Soga et al.* [1978]. Some of these cracks were produced prior to faulting, but others are feather fractures induced during displacement along the fault [*Friedman*, 1975].

In contrast to the *Friedman et al.* [1979] results at low pressure and at a high strain rate of 10^{-4}/s, *Tullis and Yund* [1977] observed a change from brittle to ductile deformation in their dry compression tests on Westerly granite at 150 to 1,500 MPa confining pressure and at the relatively low strain rate of 10^{-6}/s. At low pressures and temperatures, irregular faults are also formed in the same manner, but rather than one major fault, several parallel faults with small offset are formed. Under these conditions at high temperature, sharp faults occur by the linking up, with biotite grains and grain boundaries, of cracks induced by dislocation tangles along straight deformation bands in the quartz and feldspar. Quartz and feldspar grains adjacent to those high temperature faults show high dislocation densities and cell structures.

At the highest pressures (1.0 to 1.5 GPa) and low temperatures, throughgoing faults resulted from microcrack linkage but- only at high differential stresses. At intermediate temperatures, grain scale faults occurred, as a result of cracks nucleated from dislocation tangles, but no throughgoing faults developed. At the highest temperatures employed (to 1000°C), dislocations in both quartz and feldspar were sufficiently mobile to promote steady state flow with no associated fracturing. *Tullis and Yund* [1977] found that the transition from dominantly microfracturing to dominantly dislocation motion occurred over the temperature ranges 300° to 400°C for quartz and 550° to 650°C for feldspar at pressures above 500 MPa. Thus, the behavior of granite under most conditions of this study was transitional between brittle and ductile; that is, semibrittle processes dominated.

Tullis and Yund [1978] added 0.1 to 0.2 wt. % H_2O, by vacuum impregnation, to Westerly granite and, after sealing the specimens mechanically and preheating at pressure at 700°C for 24 hours, deformed them at 300° to 600°C, pressures of 1.0 to 1.5 GPa at a strain rate of 10^{-6}/s. All of the specimens were weaker than their dry counterparts and at temperatures of 400°C and above, the quartz showed much more evidence for deformation and recovery than in the dry specimens, whereas 500° to 600°C was required for feldspar to show similar behavior. *Tullis et al* [1979] have also adduced evidence that pressure has an important effect on hydrolytic weakening of silicates. In their nominally dry experiments on Westerly granite, En-

field aplite, Hale albite, and Heavitree quartzite, they found that at 900°C, at a strain rate of 3×10^{-6}/s, the higher pressures in the range 500 to 1500 MPa produced lower yield strengths than did the lower pressure. They attribute this effect to increased solubility of H_2O in the crystal structure at high pressure that, in turn, promotes plastic flow and recovery.

Tullis and Yund [1979a, b] and *Yund and Tullis* [1980] have also investigated the effects of H_2O, pressure and strain on the Al/Si order-disorder kinetics of feldspars. In general, they have shown that the presence of H_2O, high pressure, and mobile dislocations at high temperature all enhance the disordering rate and lower the activation energy for disordering. Again, increased equilibrium concentrations of H_2O at the higher pressures are called upon to account for this hydrolytic weakening effect and the presence of mobile dislocations may aid by permitting pipe diffusion. *Yund and Tullis* [1980] showed that the activation energy for disordering in albite decreased from 87 kcal/mole for samples disordered in air to about 67 kcal/mole for those disordered at 1.0 GPa pressure. *Tullis and Yund* [1979b] determined activation energies for creep of fine-grained monomineralic aggregates of albite and anorthosite at 1.5 GPa confining pressure, 300 MPa differential stress, and temperatures between 850° and 1100°C. The samples contained about 0.1 wt. % H_2O and creep activation energies obtained were 55 kcal/mole for the albite rock and 60 kcal/mole for anorthosite. These important studies have shed considerable light on the physics of deformation of feldspars, and the authors have suggested that diffusion of tetrahedral Al and Si ions, rather than oxygen, may be the rate-controlling process for deformation of feldspar-rich rocks.

The studies discussed above were carried out primarily with a view toward determining the effects of temperature and/or pressure on the flow processes and no attempts were made to determine creep flow laws. As was pointed out by *Carter and Kirby* [1978], two principal types of creep equations have been applied to high temperature transient creep:

$$\epsilon_t = \epsilon_T \left[1 - \exp\left(-t/\tau\right)\right] \tag{3}$$

where ϵ_T is the total transient creep strain and τ is a relaxation time for transient creep and

$$\epsilon_t = \beta t^m = \beta_0 \sigma^n t^m \exp\left(-E/RT\right) \tag{4}$$

where β_0 is a constant, σ is the stress, and n is the stress exponent; t is time (seconds) and m is the time exponent; E is the activation energy for the rate-controlling process and R and T (°K) have their usual significance. Equation (3) is to be preferred because [*Carter and Kirby* [1978] (1) it has theoretical support from the observation that the exponential decay law follows a first-order rate equation, (2) equation (4) breaks down at high and low stresses [*Garofalo*, 1965], (3) experimental results from a large number of metals and alloys best fit this equation,

TABLE 1. Oxygen Diffusion in Feldspars [*Giletti et al.*, 1978]
Diffusion Distance (Å) for 5.7 s ($x = 2(Dt)^{1/2}$

	500°C	600°C	700°C	800°C
Adularia	24	63	135	251
Albite	22	50	93	155

and (4) experimental results for rocks (dunite, clinopyroxenite, and halite) deformed to sufficiently high transient creep strains (>0.01) to distinguish between (3) and (4), fit (3) very satisfactorily.

The power law transient creep equation (4) has, however, been applied universally to the relatively abundant low strain, high-temperature transient creep data available for rocks. *Carter and Kirby* [1978] have challenged the utility of this relation for the reasons given above and because (1) the creep strains achieved experimentally are extremely small (10^{-3} to 10^{-5}) where anelastic creep should also operate; (2) steady state creep is not achieved, and extrapolations according to (4) underestimate actual creep rates when the transient creep rates predicted by (4) fall below the steady state strain rates; and (3) void formation and boundary sliding are likely to take place in the uniaxial compression experiments commonly employed and these processes may have little relevance to rheology of the earth's crust [*Kirby and Raleigh*, 1973]. Nonetheless, most of the published transient creep data for granite have been given and interpreted in tems of (4) and, following *Carter and Kirby* [1978] are summarized below.

Misra and Murrell [1965] conducted creep tests in uniaxial compression on a microgranodiorite from Penmaenmaur, Caernarvonshire, at temperatures to 750°C. More recently, *Murrell and Chakravarty* [1973] have extended those tests to 1045°C at stress levels ranging from 70 MPa at 830°C to 6.5 MPa at 1045°C. Creep strains at the lower temperature after 5 hours deformation were in the range 2.5×10^{-3} to 4×10^{-4} and at the higher temperature, from 8.5×10^{-3} to 3×10^{-1}; steady state flow was not achieved and the deformation mechanisms were not determined. *Murrell and Chakravarty* [1973] found the mechanical behavior at the higher temperatures to fit well equation (4) with $n = 1$, $m = 0.37$ and $E = 176$ kJ/mol for creep stresses near 100 MPa [*Murrell*, 1976]; β_0 is calculated from their data to be 3×10^2 for σ, the longitudinal differential stress, expressed in MPa and t in seconds as will be the practice hereafter.

A low activation energy of 6 kJ/mol is calculated from *Rümmel*'s [1969; Figure 4] data on granite deformed under uniaxial compression at a stress of 87 MPa, $t = 4000$ s in the temperature range of 100 to 400°C. After 3600 s duration at 400°C, $n = 1.35$ for stress levels between 40 and 87 MPa [*Rümmel*, 1969; Figure 4] and $m = 0.25$ at 400°C, 87 MPa in the time interval 10^2 to 10^4 seconds [*Rümmel*, 1969; Figure 8]. The material constant β_0 is calculated to be 2×10^{-6} at a stress of 87 MPa, $\epsilon = 0.023$, $T = 400$°C and $t = 4000$ s. It should be noted that these constants may not be especially significant because most of Rümmel's Flossenberg granite deformed in anelastic logarithmic creep.

The studies discussed above were all conducted at atmospheric pressure and the only careful study of high temperature transient creep of granitic rocks under confining pressure is that of *Goetze* [1971] on Westerly granite. His experiments were carried out at temperatures to melting at 500 MPa confining pressure and 100 MPa internal H_2O pressure. Constant differential stresses applied were between 12.8 and 64.3 MPa and creep strains ranged from 10^{-3} to 10^{-5}; the creep deformation was clearly transient in nature. The parameters n, m, and E were determined to be 1.7, 0.49 and 159 kJ/mol, respectively [*Goetze and Brace*, 1972], the latter value having been obtained at 64 MPa stress, corrected to a temperature of 711°C. Using (4) and the units employed in this paper, β_0 is calculated to be 3.9.

It is seen from the calculations given above that the values of the stress and time exponents, n and m, vary within narrow limits but the material constant β_0 diverges widely in value. However, the most significant feature to observe is the large increase in activation energy in the higher temperature tests, as is observed for metals [*Weertman and Weertman*, 1970]. The low activation energy, 6 kJ/mol, obtained in the temperature range 100° to 400°C, is very likely to be that for thermally activated microfracturing, especially in view of somewhat similar results from experiments on basalt [*Lindholm et al.*, 1974]. The central difficulty in assessing the physical meaning of these creep data is that the deformation processes responsible for the transient creep were not determined.

Current Results

Westerly granite cylinders have been deformed dry in compression to low strains in a Griggs-type solid confining pressure medium apparatus [*Green et al.*, 1970]. The cylindrical cores, 6.35 mm in diameter by 17.8 mm long, were backed by Al_2O_3-TiO endpieces and surrounded by 0.076-mm thick platinum jackets, both to reduce the thermal gradient and to isolate the specimen chemically from the talc confining medium. All samples were dried overnight at 300°C in air and were then placed in a dessicator after cooling. Experiments were run at constant differential stresses of 0.6 to 1.2 GPa, temperatures in the range 470° to 765°C, and longitudinal strains from 0.002 to 0.02, all at a constant confining pressure of 1.0 GPa. Thus all experiments were carried out in the α quartz stability field [*Koch et al.*, 1980].

The experiments were terminated at low creep strains because we are concerned here primarily with the high-temperature transient creep behavior of granite, although the steady state was achieved in many of the runs. In conventional creep experiments, steady state flow properties are generally the object and are assumed to be path independent, but in attempting to determine transient creep properties, it is the path toward a steady state substructure that we wish to evaluate. This has required two modifications of the conventional loading procedure, in addition to more sensitive displacement measurements (a displacement of $5 \times 10^{-4''}$ is recorded as 2 cm on the recorder, with a sensitivity of $5 \times 10^{-5''}$). The first modification was to raise the stress quickly, after equilibrating at the desired temperature and pressure, to minimize the time available for plastic strain and recovery to take place during loading. The specimens were loaded at about 140 MPa/s manually, typically reaching the desired stress difference in about 6 s. *Giletti et al.* [1978] have determined oxygen diffusion parameters for adularia and albite in a hydrothermal environment; this should place approximate upper bounds on the diffusion-controlled recovery in two dominant minerals in Westerly granite. The relation $x = 2(Dt)^{1/2}$ [*Shewmon*, 1963] gives x, the distance that oxygen will diffuse in time t for a given diffusion coefficient D. Combined with the results of *Giletti et al.* [1978], diffusion distances at various temperatures are given in Table 1; for example, diffusion distances at typical loading times for adularia and albite at 700°C are 135 and 93 Å, respectively, and hence should not allow significant recovery. The second modification is to control the drive motor speed manually at the beginning of the experiment to avoid oscillations due to insufficiently rapid damping of the creep servo. By using this procedure, differential stresses are maintained constant to about 10 MPa (ca. 1%); runs in which the stress level varied by more than 25 MPa were rejected.

It must be emphasized that the results presented here are of a very preliminary nature; we are just beginning to learn how to do transient creep experiments and how to interpret them. Finally, early exploratory tests were run at a constant strain rate of 10^{-6}/s to strains near 10% in order to estimate the constant stress levels that would be required for the creep tests.

Analyses of the strain time data obtained from these high-temperature creep tests are comprised of three parts: fitting the transient portion to an appropriate functional form, determining variations of functional coefficients with variations in stress and temperature, and comparing transient creep parameters with those determined from steady state creep. Fundamental to this procedure is the judicious choice of a functional form which, among several possibilities, has both a sound physical basis and is amenable to empirical verification. *Carter and Kirby* [1978] have discussed the various functional forms and so only a summary of the topic need be given here.

Andrade [1910] originated the concept that the strain time relationship at high temperature is the sum of the instantaneous elastic and plastic strain during loading ($\epsilon_e + \epsilon_p$), the transient creep strain (ϵ_t), and the steady state creep strain ($\dot{\epsilon}_s t$):

$$\epsilon(t) = (\epsilon_e + \epsilon_p) + \epsilon_t(t) + \dot{\epsilon}_s t \qquad (5)$$

On the basis of empirical results, Andrade found that, for high-temperature transient creep, the expression given in (4) fits the data satisfactorily over a wide range of stresses and temperatures, with $0 < m < 1$; generally m is 1/3 to 1/2. Equation (5) can thus be written as

$$\epsilon = \epsilon_0 + \beta\, t^m + \dot{\epsilon}_s t \qquad (6)$$

and has been employed by many workers since.

More recently, *Webster et al.* [1969] have presented an analysis based on first order kinetics that resulted in an equation of the form

$$\epsilon = \epsilon_0 + \epsilon_T[1 - \exp(-t/\tau)] + \dot{\epsilon}_s t \qquad (7)$$

where, again, ϵ_T is the total transient creep strain and τ is a relaxation time for the rate-controlling transient creep process. As was noted above, (7) has a much firmer empirical and theoretical basis than has (6) and is to be preferred. Furthermore, the fundamental observation by *Dorn* [1954] of the equalities of activation energies for high-temperature transient creep, steady state creep, and self-diffusion is generally interpreted to mean that the same process controls creep rates in both regimes, namely, diffusion through recovery processes. Above a minimum stress level, for which grain boundary sliding processes are dominant in uniaxial tests, τ and $\dot{\epsilon}_i$ are found to be proportional to $\dot{\epsilon}_s$, where $\dot{\epsilon}_i$ is the initial strain rate upon loading. *Amin et al.* [1970] show that

$$\epsilon = \epsilon_0 + [(\dot{\epsilon}_i - \dot{\epsilon}_s)/(K\dot{\epsilon}_s)] \cdot [1 - \exp(-K\dot{\epsilon}_s t)] + \dot{\epsilon}_s t \qquad (8)$$

where $K\dot{\epsilon}_s = 1/\tau$ and $(\dot{\epsilon}_i - \dot{\epsilon}_s)/(K\dot{\epsilon}_s) = \epsilon_T$. K, ϵ_T, and $\dot{\epsilon}_i/\dot{\epsilon}_s$ should be independent of stress and temperature. $\dot{\epsilon}_i$ may be calculated by differentiating the basic equation with respect to time, and evaluating at $t = 0$

$$\dot{\epsilon}_i = \dot{\epsilon}_s + \epsilon_T/\tau = \dot{\epsilon}_s + 4\epsilon_T/t_s \qquad (9)$$

where t_s is the time to steady state creep and $\dot{\epsilon}_i/\dot{\epsilon}_s$ is a constant [*Carter and Kirby*, 1978]. *Amin et al.* [1970] showed that if the total time dependent strain ($\epsilon - \epsilon_0$) is plotted against the steady state contribution to the creep strain, $\dot{\epsilon}_s t$, creep data taken at different temperatures and stresses plot on the same curve in the transient and steady state regimes. This indicates that temperature and stress effects on transient creep rates are accounted for by the effects of these parameters on steady state creep. Finally, the starting dislocation substructure has a marked influence on such transient creep parameters as τ, t_s, $\dot{\epsilon}_i$, and ϵ_T. It follows then that the above relationships between transient and steady state creep parameters are not fundamental material properties but must be referred to specific initial defect states of the material [*Carter and Kirby*, 1978].

Determinations of the various parameters, $\dot{\epsilon}_i$, $\dot{\epsilon}_s$, ϵ_T, and τ, were made to test the applicability of the analysis to Westerly granite. In addition, a least squares fit to the general equation (7) has been carried out by fitting the data to four parameters, ϵ_0, $\dot{\epsilon}_s$, ϵ_T, and $r = 1/\tau$, by using a nonlinear technique developed by A. F. Gangi for analysis of stress variation in constant strain rate tests. We minimize the mean square fractional error of the actual data

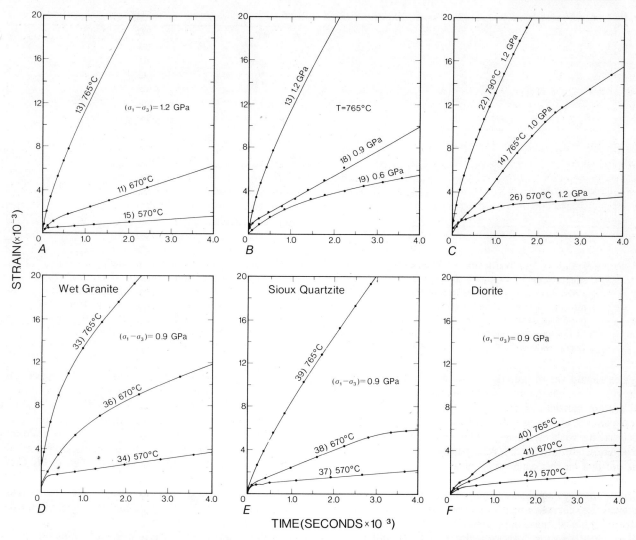

Fig. 7. (*a–c*) Strain time curves for representative creep experiments on dry Westerly granite; dots are reduced data points. Curves in (*a*) and (*b*) are fits to equation (7). The curves in (*c*) are an eyeball fit to the data points and emphasize the peculiar behavior discussed in the text. (*d–f*) 'Wet' granite (*d*), dry Sioux quartzite (*e*) and dry diorite (*f*). Curves are all hand fit to the data points and are not fit to any equation. Temperatures stated are based on calibrations at low pressure using a hollowed core of Westerly granite in place of the sample; corrections have not been made for the effect of pressure on the emf of the Pt-Pt/Rh thermocouple.

set with respect to a data set generated by initial estimates of the parameters. This then linearizes a nonlinear procedure (by approximating the desired parameters) and then permits iteration to find the best values for the parameters. Sample data sets (dots) and the fits generated (curves) are presented in Figures 7*a* and *b*, which show the expected increase of strain with time at constant stress and increasing temperature (Figure 7*a*) and a similar increase with increasing stress at constant temperature (Figure 7*b*). It is apparent that the fits to (7) are quite satisfactory, though the data are few. It should be noted that because of the very rapid loading rate and inability to scale the recorders properly in this short interval, ϵ_0 determinations were erratic and were eliminated from both data and fits in Figures 7*a–f*.

As for independent determination of the constancy of the various parameters listed above, the results are not entirely satisfactory. ϵ_T appears to be independent of stress and temperature, varying within a factor of about two of 0.003, a value lower than was anticipated. $\dot{\epsilon}_i$ tends to increase with increasing stress and temperature (with a few irregularities) and shows generally consistent results with values calculated from (9). The ratio $\dot{\epsilon}_i/\dot{\epsilon}_s$ shows some variation also, though the significance of this variation (up to a factor of 4) is inconclusive. However, the major inconsistency with the theory appears in the relaxation time τ,

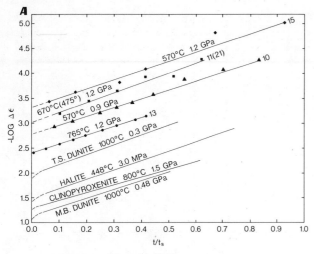

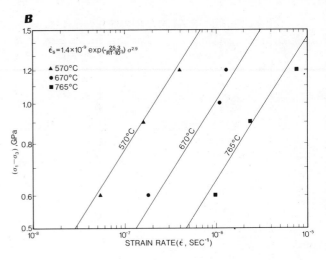

Fig. 8. (a) Test of fit of high-temperature transient creep data to equation (3). Slopes on $-\log\Delta\epsilon$ versus t/t_s should equal -1.74, where $\Delta\epsilon = \epsilon_T - \epsilon_t$ and t_s is the time required to achieve steady state creep [Carter and Kirby, 1978]. Curves in the lower part of the diagram are from Carter and Kirby [1978, Figure 2] with Twin Sisters dunite (T.S.) from data of Goetze and Brace [1972], halite polycrystals from Burke [1968], clinopyroxenite, from unpublished data by S. H. Kirby (1980) and Mt. Burnet dunite (M.B.) data from Post [1973]. Data points and eyeball fit lines in upper part of diagram are for five representative specimens of Westerly granite from this study. (b) Least squares fit to steady state creep data for dry Westerly granite using (11) in text.

which should vary regularly with stress and temperature as should $\dot{\epsilon}_s$, but in an inverse manner (equation (8)). When normalized to $\dot{\epsilon}_s$ for each test, K should be constant, independent of temperature and stress; it is not, generally decreasing with increasing stress and temperature. Furthermore, our data for different stresses and temperatures do not plot on the same curve in $(\epsilon - \epsilon_0)$ versus $\dot{\epsilon}_s t$ space.

Attempts to improve the fit to (7) as our technique improved merely complicated the matter further. Figure 7c shows creep curves in which two transient portions, with a steady state region sandwiched between, occur in three representative runs. This behavior is not yet understood and may indicate that better data and/or more refined data analysis are needed, or that two or more deformation mechanisms, with different rate constants, are operative [Gangi, this volume; Parrish and Gangi, this volume]. However, attempts to fit the data by using the multiple mechanism fits also proved to be unsuccessful. The anomalous behavior of granite shown in Figure 7c is reminiscent of synthetic quartz crystals deformed at a constant strain rate [Hobbs et al., 1972; Balderman, 1974; Morrison-Smith et al., 1976] and has been analyzed in terms of the Haasen microdynamical theory of yielding [Hobbs et al., 1972; Griggs, 1974]. To our knowledge, however, this behavior has not been observed in natural quartz or feldspar crystals; evidently the matter requires further careful investigation.

Figures 7d and 7e show the results of some preliminary experiments on 'wet' Westerly granite, Sioux quartzite and diorite (locality unknown) carried out merely for a comparison with the behavior of the dry granite. Curves for these figures, as well as Figure 7c, are hand fits to the data points (dots) and do not represent fits to any equa-

tion. Following the Tullis and Yund [1978] procedure, wet granite specimens were prepared by vacuum impregnation of distilled H_2O (0.1 wt. % added), were dried at room temperature and then encapsulated in platinum jackets. Wet granite samples were then soaked for 30 min at the ambient conditions of the experiment prior to deformation at the conditions shown in Figure 7d. Comparisons with curves of Figures 7a–c for the nominally dry granite indicate that the wet granite deforms much more readily, under comparable conditions, as expected [Goetze, 1971; Tullis and Yund, 1978].

Both the quartzite and diorite were deformed under dry conditions. The mechanical behavior of the quartzite (Figure 7e) appears to lie between that of wet and dry granite and some of the peculiar behavior observed in Figure 7c also seems to be present in the quartzite. This appears to be true also of the diorite (Figure 7f) but only at very low strains and times and thus may not be significant. The diorite, evidently, is much more creep resistant, under comparable conditions, than the other materials and deformed only in relatively low temperature transient creep throughout the strain time (to 8×10^3 s) range employed.

Returning now to an analysis of the dry granite data, the only functional difference between (6) and (7) is the expression for the transient portion of the creep curve, as given in (3) and (4). As pointed out by Carter and Kirby [1978], most high temperature transient creep data for rocks, primarily under uniaxial conditions, have been analyzed in terms of (4). The few data for rocks deformed to sufficiently high transient creep strains to distinguish between the two equations fit the physically more reasonable equation (3) very satisfactorily. This is shown in Figure 8a, which is a plot of log $\Delta\epsilon$ versus t/t_s where $\Delta\epsilon$ is $\epsilon_T - \epsilon_t$

and t_s is the time required to reach the steady state. These transient data should be linear on such a plot, with a slope equal to -1.74 and an intercept at log $\Delta\epsilon_T$ [*Carter and Kirby*, 1978, p. 821]. The data for other rocks are shown in the lower portion of the figure and were taken from Carter and Kirby's Figure 2. Data for five of our experiments on high-temperature creep of Westerly granite are shown by the four curves with data points in the upper part of the diagram. Two other experiments (16 at 670°C, 1.0 GPa and 23 at 570°C, 1.0 GPa) could have been plotted on the figure, but are so close to experiments 10 and 11 as to have cluttered the diagram. The slopes of the seven lines range from -1.2 to -2.0, with an average value of -1.72. Thus the high temperature transient creep data for Westerly granite, as well as other rock types for which comparable data are available, fit (3) very well.

Our transient creep data have also been hand fit to the power law equation (4) for comparison with similar data for granitic rocks, all of which have been cast in these terms. The value of the stress exponent n (= d log ϵ/d log σ) is 2.2, as determined at 765°C, 500 s, at stresses between 0.6 and 1.2 GPa. The time exponent, m (= d log ϵ/d log t) varies from 0.22 at 570°C, 0.9 GPa, through 0.51 at 670°C, 1.2 GPa to an average of 0.69 at 765°C, 0.9 and 1.2 GPa. Thus m increases with increasing temperature, as observed for halite [*Hansen and Carter*, 1980] and other materials and so we have arbitrarily adopted 0.5 as representative of the average behavior. The activation energy, E [$= -Rd \ln \epsilon/d$ $(1/T)$] is 30.5 kcal/mole (128 kJ/mole) at

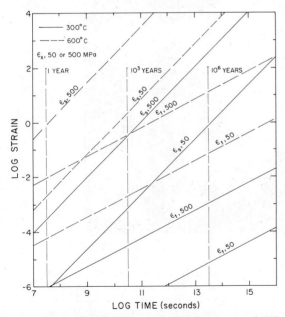

Fig. 9.   Extrapolation in log strain-log time space of (10) and (12) for high-temperature transient and steady state creep, respectively, of dry Westerly granite. Extrapolations are for 300° and 600°C and stress differences of 50 and 500 MPa. Steady state creep gives the higher creep rate, at comparable stress temperature conditions, swamping transient creep in the strain time interval.

1.2 GPa, and 500 s between 670° and 765°C; this is our best estimate. Using these values for n, m, and E, β_0 is estimated to be 7×10^{-5} (ranges between 5×10^{-5} and 9×10^{-5} for the various tests). Therefore our best estimate of the power law high temperature transient creep equation is:

$$\epsilon_t = 7 \times 10^{-5}\sigma^{2.2}t^{0.5} \exp(-30.5/RT \cdot 10^{-3}) \qquad (10)$$

The steady state creep data were fit to the well-substantiated power law that adequately represents the high-temperature creep behavior of most materials over a wide range of physical conditions:

$$\dot{\epsilon}_s = A \exp(-Q_c/RT)\sigma^n \qquad (11)$$

where A is a nearly temperature insensitive material constant, Q_c is the activation energy for creep, and the remaining parameters have been specified previously. The least squares fit of the data to this flow law is shown in Figure 8b and, although the data are few at present, they appear to fit (11) quite well. The resulting steady state creep equation is

$$\dot{\epsilon}_s = 1.4 \times 10^{-9}(\pm 3.5 \times 10^{-9})$$
$$\cdot \exp(-25.3 \pm 2.2/RT \cdot 10^{-3})\sigma^{2.9\pm0.3} \qquad (12)$$

The activation energy, given here in kcal/mole, is slightly lower than that estimated above for power law transient creep, but the difference is well within estimated errors and is regarded as insignificant. The exponent on stress, 2.9, is also close to the 2.2 exponent found for the high-temperature transient estimate. Where they are comparable, these steady state creep data are compatible with the constant strain rate data of *Tullis and Yund* [1977]. At a strain rate of 10^{-6}, they found that at 600° and 700°C, the steady state flow stresses were about 1.2 and 0.8 GPa, respectively. Under comparable creep conditions, we find a steady state strain rate of 1.1×10^{-6}/s at 570°C, 1.2 GPa and 1.3×10^{-6}/s at 670°C, 0.8 GPa.

In Figure 9, the power law transient and steady state creep behavior of Westerly granite deformed in our experiments are compared, arbitrarily, at 300° and 600°C and differential stresses of 50 and 500 MPa. The comparison is made in log strain versus log time space by extrapolating (10) and (12) over the time range 10^7 s (0.32 year) to 10^{16} s (3.17×10^8 years) and a strain interval 10^{-6} to 10^4. At any given set of conditions, steady state creep gives the higher creep rate over the entire strain time range shown and hence completely dominates the creep strain. The slopes of the lines representing the two creep regimes are such that they converge and intersect to the left and, in some instances, below the diagram. For example, at 600°C and 500 MPa, transient and steady state curves intersect at log $t =$ 3.5 and log $\epsilon = -4.0$; other intersections are at lower strains. These intersections (not shown) mark the time at which transient and steady state creep contribute equally to the total creep strain and at which the creep strain rates are identical. To the left of the intersections, transient creep dominates and to the right, steady state creep

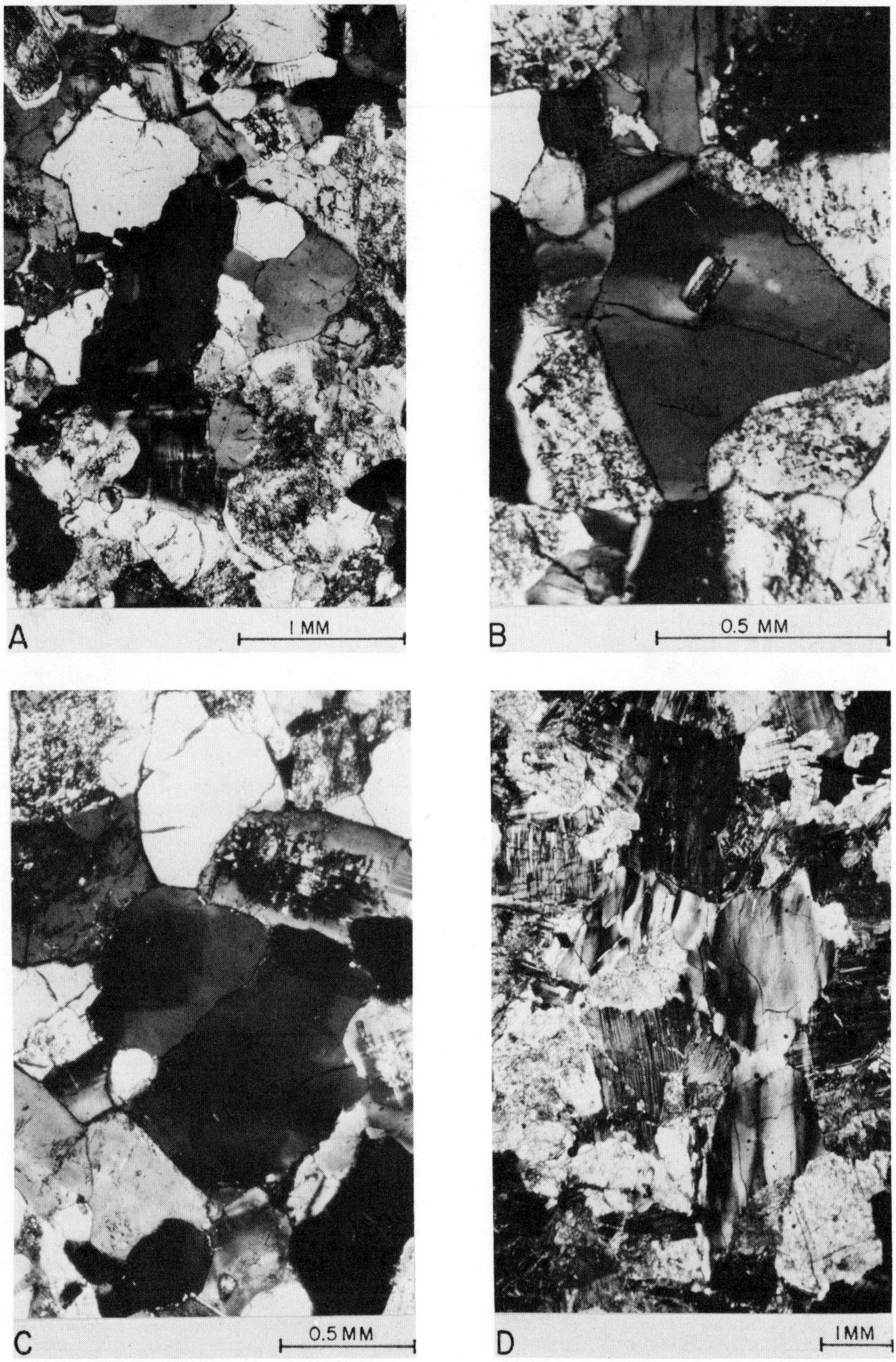

Fig. 10. (*a–c*) Optical photomicrographs of 'undeformed' Westerly granite and (*d*) of Barre granite. (*a*) Typical texture at low magnification. (*b*) Strain around a zircon (?) inclusion in quartz. (*c*) A slightly deformed, well-recovered quartz grain. (*d*) This section shows that quartz in the coarser-grained Barre granite is much more highly deformed than in the Westerly. A few micas are kinked in both specimens but feldspars appear largely undeformed although some bending and possible mechanical twinning is observed sporadically in the Barre.

dominates; over the strain time range shown in Figure 9, the contribution due to transient creep is negligible.

We must now discuss the deformational processes observed in the experiments to the extent to which they have

been investigated; analyses again are preliminary. The texture, mineralogical constitution, and microstructures for the Westerly granite block used are as shown in Figures 10*a–c*. The average grain size is about 0.5 mm, and

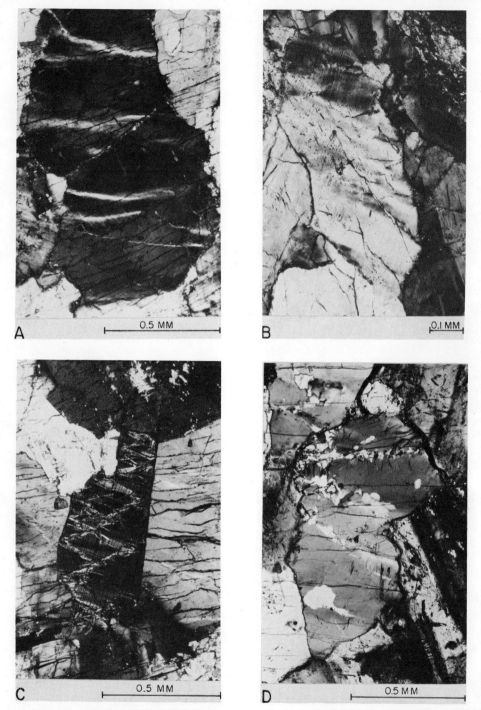

Fig. 11.   Microstructures induced in initial constant strain-rate tests (10^{-6}/sec.) at 1.0 GPa confining pressure, various temperatures, deformed to strains near 20%. Pervasive cracks evident in (a), (c), and (d) are unloading fractures. (a) Irregular kink bands (white horizontal bands) in quartz (spec. 2, 635°C). (b) Undulatory extinction and subbasal deformation lamellae (north-south trending linear features) in quartz (spec. 1, 740°C). (c) Kinks in biotite grain (spec. 2, 635°C). (d) Recrystallized zones in quartz grain (spec. 6, temperature unknown, but high).

modal analyses [*Brace*, 1965] give quartz 27.5%, k feldspar 42%, plagioclase (An_{17}) 31.4%, and micas and accessories 4.9%. Figure 10a shows, at low magnification, the interlocking texture of quartz and feldspar, the degree of alter-

ation of the feldspars, and a few microcracks. Strain in a quartz grain produced by the presence of an inclusion, apparently zircon, is evident in Figure 10b and Figure 10c shows the occasional well-recovered (polygonized) quartz

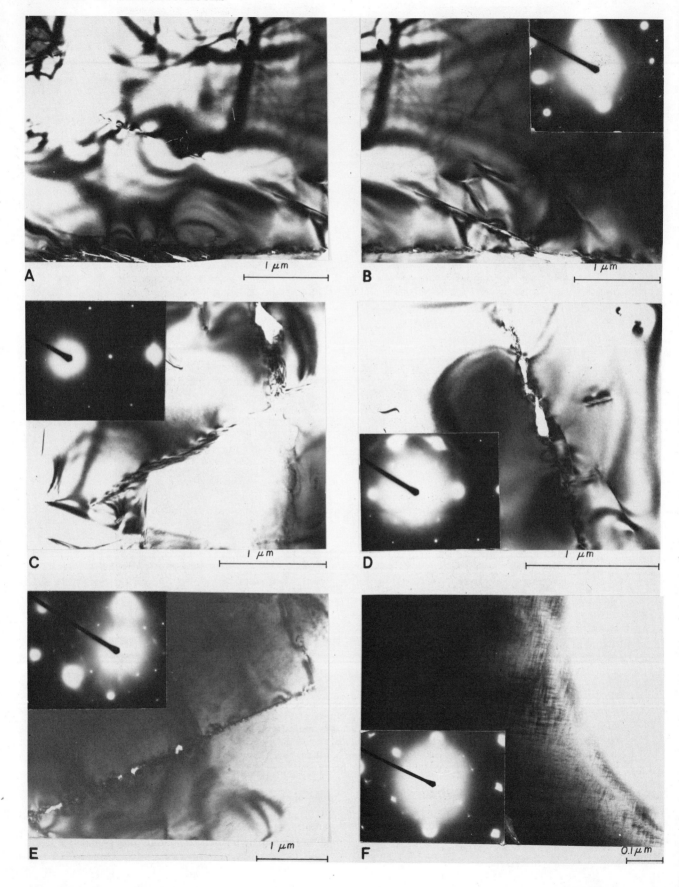

TABLE 2. Transient Creep Parameters for Granitic Rocks

Material	Temperature Range, °C	Effective Pressure, MPa	Stress, MPa	$\varepsilon_t = \beta_0 \sigma^n t^m \exp(-E/RT)$				Source
				β_0	n	m	E, kJ/mole	
Granite	20–400	0	87	2×10^{-6}	1.35*	0.25*	6	*Rümmel* [1969]
Granite	711	400	64	3.9	1.7	0.49	159	*Goetze* [1971]
Granite	470–765	1000	600–1200	7×10^{-5}	2.2	0.51	128	this study
Granodiorite	830–1045	0	100	3×10^{2}	1	0.37	176	*Murrell and Chakravarty* [1973]; *Murrell* [1976]

β_0 is calculated for longitudinal differential stress expressed in MPa and time in seconds.
*These values are of doubtful significance as the material deformed dominantly by logarithmic creep.

grain with very small disorientations across polygon boundaries. This latter feature may be compared with deformed quartz in the much coarser-grained Barre granite (Figure 10d) which, although fairly well-recovered, typically shows large strains and occasional subbasal deformation lamellae. Some of the favorably oriented micas (those with {001} slip planes at low to moderate angles to σ_1) in both granites are kinked, but the feldspars show little optical evidence of internal deformation. One gets the impression from studying these and other granitic rocks petrographically that the plastic creep strain is taken up dominantly by the quartz and secondarily by the low volume percentage of micas; the feldspars appear to act merely as pistons transmitting boundary stresses.

This same observation seems to hold for the experimentally deformed Westerly granite specimens, as shown in Figure 11a-d, at least under the conditions employed in our experiments. Ignoring the horizontal unloading cracks in Figure 11a (as well as in Figures 11c and 11d), the main features are gentle bending of the crystal structure and thin, irregular deformation bands showing intense rotations. These features have been observed previously in quartz [e.g., *Heard and Carter*, 1968; *Ave'Lallemant and Carter*, 1971; *Tullis et al.*, 1973; *Parrish et al.*, 1976; *Tullis and Yund*, 1977]. Fracturing is commonly associated with these features, especially in single crystals [*Heard and Carter*, 1968] and it appears in such instances as if localized plastic flow precedes the faulting, indicating semibrittle behavior. While these thin, irregular deformation bands are the most common microstructures observed in quartz in our specimens, gentle, regular undulatory extinction subparallel to the c axis, associated with subbasal deformation lamellae (Figure 11b), are also produced in

Fig. 12. (opposite) TEM micrographs of (a-e) specimens 8 (700°C, 1.0 GPa confining pressure, 0.8 GPa constant stress, 1% strain) and (f) 10 (600°C, 1.0 GPa confining pressure, 0.9 GPa constant stress, 0.3% strain). Diffraction patterns are shown as insets and phase determinations were by EDAX. (a,b) Overlapping micrographs showing fine en echelon arrays of cracks at bottom, a straight sharp crack near the center and isolated dislocations above this crack. (c) Northeast-trending array of en echelon cracks or of wall of dislocation or both. (d) North-northwest-trending en echelon crack array intimately associated with dislocations with free dislocations to the right and dislocation dipoles in upper right corner. (e) Two sets of cleavage cracks in microcline, no dislocations. (f) Very fine grid-twinning in microcline, no dislocations or fractures.

some grains with {0001} planes in orientations of high shearing stress. Highly kinked micas (Figure 11c) are also very common in the experimentally deformed specimens, much more so than in the starting material. Finally, recrystallized zones (Figure 11d), presumably of highly strained regions such as those shown in Figure 11a, were observed in one specimen whose temperature soared accidentally beyond the dehydration temperature of the talc confining medium. While the examples of the microstructures shown in Figure 11 are from relatively highly strained specimens, and have been described previously for the highly strained Westerly granite samples of *Tullis and Yund* [1977], this same type of deformation, in its incipient (and less photogenic) development is also observed optically in samples having undergone low creep strain.

Transmission electron micrographs (Figure 12) were obtained from ion-milled foils of specimens 8 and 10 deformed, respectively, to 1% strain at 670°C, 0.8 GPa and to 0.3% strain at 570°C, 0.9 GPa. Diffraction patterns are shown as insets in the micrographs and phase identification was accomplished by energy dispersive x ray analysis (EDAX). Figures 12a-d show TEM micrographs of quartz (spec. 8) and Figures 12e and f are micrographs of alkali feldspars (specs. 8 and 10, respectively). Figures 12a and b are overlapping micrographs of the same general area as can be seen by matching bend contours in the upper parts of the micrographs, or the microcracks in the lower parts. The most significant feature in these micrographs is the en echelon array of small cracks giving rise to the long straight fracture oriented nearly horizontally at the bottom of the figures. By contrast, the straight, sharp WNW-trending fracture shows no evidence of en echelon arrays, yet this long fracture is subparallel to the smaller array below. Isolated dislocations are observed above this crack (Figure 12b) and fringes at its tip (Figure 12a, left center) and adjacent to the horizontal fracture are associated with rotations due to strain along the cracks.

An apparent en echelon array is also observed trending northeast in Figure 12c, but it could not be ascertained whether this array is composed of cracks or a wall of dislocations or both. The en echelon crack array shown in Figure 12d is certainly intimately associated with dislocations. Free dislocations are also observed to the right of the array and dislocation dipoles are seen in the upper right corner. Contrary to the quartz, dislocations were not observed in alkali feldspars in the two specimens examined to date. The two sets of fractures in Figure 12e are

TRANSIENT CREEP OF GRANITIC ROCKS

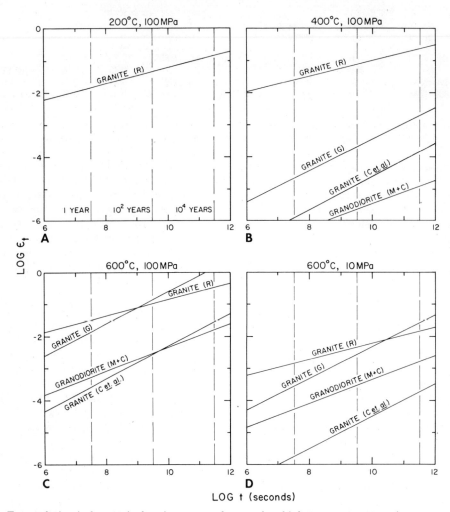

Fig. 13. Extrapolation in log strain–log time space of power law high-temperature transient creep equation (4), using parameters given in Table 2, and various temperatures and stresses: (R) *Rümmel* [1969], (G) *Goetze* [1971], (M&C) *Murrell and Chakravarty* [1973], and (C. et al.) this study.

cleavage cracks and Figure 12*f* shows only fine grid-twinning in microcline.

These few TEM observations to date are consistent with the more extensive petrographic analyses in that they suggest, in specimens deformed to low strains, that quartz deforms by a combination of plastic flow and fracturing whereas, at least, the alkali feldspars are only slightly fractured or are totally undeformed. The intimate association of dislocations with some of the en echelon crack arrays in quartz is regarded as an indication of semibrittle behavior, also in accord with conclusions based on optical studies.

Summary, Discussion, and Conclusions

The form of the inelastic flow law for low-temperature brittle creep of crystalline rocks is not yet firmly established [*Kranz and Scholz*, 1977]. The flow process has, however, been well-documented as extension fracturing with macroscopic shear fractures most commonly pro-

duced by linking of en echelon arrays of precursive extension fractures, generally by shearing along grain boundaries. The rate-controlling process for microfracturing is thought to be chemical corrosion of stressed crack tips by fluids and gases, leading to time dependent cracking. Increases in pressure may inhibit fluid migration rate, increase the energy for crack propagation and/or inhibit crack linkage, thus increasing fracture resistance. Cycling the stress at low pressures and frequencies, above a threshold stress level, produces more dilatancy with each cycle and the onset of dilatancy occurs at a lower stress level with each cycle. At moderate confining pressure, the results are equivocal, but it appears that pressure inhibits the effects observed during uniaxial cycling, at least for a large number of cycles. Finally, anisotropy plays an important role in the creep properties of the several granites. Preexisting cracks strongly affect the creep strain and the orientation of new cracks, dominantly extensile in nature, are affected by preexisting cracks and flaws and prestress as well as by the applied stress.

Increasing the temperature, above a threshold value, introduces thermal cracks that arise from differential thermal expansion of adjacent crystals in most instances. Application of pressure inhibits thermal cracking, but stress-induced cracks are temperature sensitive, possibly indicating the presence of thermally activated or semibrittle flow processes, although the material may appear to be brittle from a macroscopic viewpoint. Microcracks formed under such conditions have the same en echelon form as those at low temperature, but slip dislocations are commonly involved in the fracturing process. At sufficiently high pressures and temperatures and low strain rates, brittle and semibrittle processes are suppressed entirely and only ductile flow ensues. Both high effective pressure and the presence of H_2O enhance ductility by increasing the equilibrium concentration of H_2O in the crystal structure, thereby effecting hydrolytic weakening. These same parameters, in addition to creep strain, apparently also enhance Al-Si disordering in feldspars and lower the activation energy for the process.

Forms of the flow laws for high-temperature transient and steady state creep of granitic rocks have a somewhat firmer basis than for low-temperature creep, but they are not yet entirely satisfactory. Creep parameters for the power law transient equation (4) are presented in Table 2. In Figure 13 these various equations have been extrapolated, for comparative purposes, in log ϵ_t − log t space for the temperature interval 200° to 600°C at a differential stress of 100 MPa (Figure 13a–c) and at a stress level of 10 MPa at 600°C (Figure 13d). At 200°C, 100 MPa, only Rümmel's [1969] data plot in the strain time interval chosen, yielding strains in the range 7×10^{-3} to 0.85. Because of the low activation energy, creep strains estimated from Rümmel's data do not change appreciably with increasing temperature. At 400°C, all transient creep equations are represented, but the highest creep strain (apart from Rümmel's curve) is 3×10^{-3}, obtained from Goetze's [1977] data, and this requires 3×10^4 years. At 600°C, 100 MPa, all equations predict appreciable transient creep strains in the time interval of 1–10^4 years, with Rümmel's and Goetze's data yielding unreasonably high strains at the longer times. Of course, lowering the differential stress by one order of magnitude (Figure 13d) reduces the creep strain at any given time by one to two orders of magnitude, depending on the stress exponent. We emphasize that these plots are for comparative purposes only, showing the effects of stress and temperature on high-temperature transient creep behavior, and are suspect in view of the questionable validity of (4). In any instance, extrapolations to the higher strains are useful only in that they generally place lower bounds on the total creep strain [Figure 9; Carter and Kirby, 1978].

Our experimental results, if they are representative, indicate that high-temperature transient creep of granite gives way to steady state creep at strains less than 1% and in fairly short times. Referring to Figure 9, and the geologically more reasonable stress level of 50 MPa, it is evident that appreciable steady state strains can take place at times between 10^3 and 10^6 years at 300°C and in shorter times at higher temperatures. Thus it would appear that if significant solid state flow occurs in granitic rocks, as is certainly the case in orogenic regions, then this flow takes place predominantly by steady state creep. This tentative conclusion is contrary to the view expressed earlier by Carter and Kirby [1978], a view based on Carter's prejudice that in turn was based on lack of evidence for appreciable flow of feldspars in naturally deformed rocks.

We now inquire as to the flow processes that give rise to steady state creep of granitic rocks. Information on this topic comes from studies of naturally induced microstructures, those produced experimentally, and the empirically derived flow parameters. Unfortunately, microstructures in naturally deformed granitic rocks have received only cursory attention to date, except perhaps in relatively atypical environments, such as major fault and mylonite zones. Our own petrographic observations on several different granitic rocks indicate that the creep strains are not large and that they are accommodated primarily by flow of quartz, secondarily by kinking of micas, and little, if at all, by deformation of the feldspars. We are aware of no evidence to contradict this generalization but we emphasize that a systematic detailed study of microstructures of selected granitic rocks from different structural environments should be carried out to confirm or repudiate this assertion and to ascertain the validity of all experimental work on granitic rocks.

Our preliminary experiments at low to moderate strains appear to be consistent with the observations above although, whereas most of the deformed quartz grains in the naturally deformed granitic rocks that we have examined are well recovered, those in the experiments show significant recovery only at the highest temperatures. Our experiments were nominally dry and Tullis and Yund [1978] have adduced evidence that the presence of H_2O enhances recovery rates in quartz in granite, as had been observed earlier for quartzites [e.g., Parrish et al., 1976]. The quartz grains in most of our experiments show the thin, irregular deformation bands described above that we take as indicative of semibrittle behavior. These features, both on optical and electron microscope levels, have not been reported for naturally deformed granitic rocks, but then they have not been searched for. Finally, although Tullis and Yund [1977, 1978] found evidence for appreciable plasticity and recovery of feldspar (enhanced also by the addition of H_2O) in their experiments at high confining pressure, high temperature, large strains, and high differential stresses, we are aware of no counterparts in naturally deformed feldspars in granitic rocks. Plastic flow, polygonization, and recrystallization features very similar to those in quartz have been observed by us in feldspars in rare anorthosites, and so this type of deformation should be easily recognized optically.

As for the empirically determined creep flow parameters, the activation energy for creep is most significant. The value that we obtain, 25 kcal/mole is the only one available for steady state creep of granite and is, within experimental error, the same as our estimated value for high-temperature transient creep (30.5 kcal/mole). *Goetze*'s [1971] value for high-temperature transient creep, 38 kcal/mole, is somewhat higher but his experiments were carried out at lower effective pressure (400 MPa) and in the presence of 100 MPa H_2O pressure. His value, 38 kcal/mole, is identical to that found for steady state flow of wet quartzite by *Koch et al.* [1980] in the temperature interval 750 to 900°C. Their value for steady state flow of dry quartzite was somewhat lower, about 36 kcal/mole, and values of the stress exponent n are 2.4 (wet) and 2.9 (dry). By contrast, activation energies for creep of albite rock and anorthosite, under the most favorable conditions for low energies, are 55 and 60 kcal/mole, respectively [*Tullis and Yund*, 1979*b*]. These observations, in conjunction with preliminary analyses of naturally and experimentally induced microstructures, suggest that deformation of quartz may control the creep rate of granitic rocks.

We conclude, tentatively, that creep of granitic rocks at moderate to high temperatures takes place predominantly in the steady state regime and is rate limited by flow processes in quartz.

Acknowledgments. We are especially pleased to dedicate this article in honor of our good friend John Handin whose brilliance and leadership in the field of geological rock mechanics has illuminated a pathway of excellence, from which we continue to stray. The experimental part of this research was supported by DOE contract DE-ASOS-79ER10361. The remainder was supported in part by the DOE contract, by U.S.G.S. contract 14-001-17716, and by RE/SPEC, Inc. We are grateful for this support and also wish to acknowledge constructive critical comments on the manuscript by M. Friedman and P. Gnirk.

References

Amin, K. E., A. K. Mukherjee, and J. E. Dorn, A universal law for high-temperature diffusion-controlled transient creep, *J. Mech. Phys. Solids*, *18*, 413-426, 1970.

Andrade, C. N., CaC., Viscous flow in metals, *Proc. R. Soc. London Ser. A*, *84*, 1-12, 1910.

Attewell, P. B., and I. W. Farmer, Fatigue behavior of Rock, *Int. J. Rock Mech. Mining Sci.*, *10*, 1-9, 1973.

Ave'Lallemant, H. G., and N. L. Carter, Pressure dependence of quartzite plastic flow mechanisms, *Am. J. Sci.*, *270*, 203-218, 1971.

Balderman, M. A., The effect of strain rate and temperature on the yield point of hydrolytically weakened synthetic quartz, *J. Geophys. Res.*, *79*, 1647-1652, 1974.

Bauer, S., and B. Johnson, Effects of slow uniform heating on the physical properties of igneous rocks, paper presented at 20th Symp. on Rock Mechanics, University of Texas, Austin, Tex., June 1979.

Blacic, J. D., Plastic deformation mechanisms in quartz: The effect of water, *Tectonophysics*, *21*, 271-294, 1975.

Borg, I. Y., and J. W. Handin, Experimental deformation of crystalline rocks, *Tectonophysics*, *3*, 249-348, 1966.

Brace, W. F., Some new measurements of linear compressibility of rocks, *J. Geophys. Res.*, *70*, 391-398, 1965.

Brace, W. F., B. W. Paulding, and C. Scholz, Dilatancy in the fracture of crystalline rocks, *J. Geophys. Res.*, *71*, 3939-3953, 1966.

Brace, W. F., E. Silver, K. Hadley, and C. Goetze, Cracks and pores: A closer look, *Science*, *178*, 162-164, 1972.

Bruner, W. M., Crack growth and thermoelastic behavior of rocks, *J. Geophys. Res.*, *84*, 5578-5590, 1979.

Burke, P. M., High temperature creep of polycrystalline sodium chloride, Ph.D. thesis, Stanford Univ., Stanford, Calif., 1968.

Byerlee, J., A review of rock mechanics studies in the United States pertinent to earthquake prediction, *Pure Appl. Geophys.*, *116*, 586-602, 1978.

Carter, N. L., Steady state flow of rocks, *Rev. Geophys. Space Phys.*, *14*, 301-360, 1976.

Carter, N. L., and S. H. Kirby, Transient creep and semibrittle behavior of crystalline rocks, *Pure Appl. Geophys.*, *116*, 807-839, 1978.

Cruden, D. M., The static fatigue of brittle rock under uniaxial compression, *Int. J. Rock Mech. Mining Sci.*, *11*, 67-73, 1974.

Dorn, J. E., Some fundamental experiments on high-temperature creep, *J. Mech. Phys. Solids*, *19*, 77-83, 1954.

Douglas, P. M., and B. Voight, Anisotropy of granites: A reflection of microscopic fabric, *Geotechnique*, *19*, 376-398, 1969.

Dunn, D. E., L. J. LaFountain, and R. E. Jackson, Porosity dependence and mechanisms of brittle fracture in sandstones, *J. Geophys. Res.*, *78*, 2403-2417, 1973.

Friedman, M., Fracture in rocks, *Rev. Geophys. Space Phys.*, *13*, 352-357, 1975.

Friedman, M., and T. R. Bur, Investigations of the relations among residual strain, fabric, fracture and ultrasonic attenuation and velocity in rocks, *Int. J. Rock Mech. Mining Sci.*, *11*, 221-234, 1974.

Friedman, M., and B. Johnson, Thermal cracks in unconfined Sioux quartzite, paper presented at 19th Symposium on Rock Mechanics, University of Nevada, Reno, Nevada, June 1978.

Friedman, M., R. D. Perkins, and S. Green, Observation of brittle deformation features at the maximum stress of Westerly granite and Solenhofen limestone, *Int. J. Rock Mech. Mining Sci.*, *7*, 297-306, 1970.

Friedman, M., J. Handin, N. G. Higgs, and J. R. Lantz, Strength and ductility of four dry igneous rocks at low pressures and temperatures to partial melting, paper presented at 20th Symposium on Rock Mechanics, University of Texas, Austin, Tex., June 1979.

Gangi, A. F., A constitutive equation for one-dimensional transient and steady state flow of solids, this volume.

Garofalo, F., *Fundamentals of Creep and Creep-Rupture in Metals*, 258 pp., MacMillan, New York, 1965.

Giletti, B. J., M. P. Semet, and R. Yund, Studies in diffusion, 3, Oxygen in feldspars: An ion microprobe determination, *Geochim. Cosmochim. Acta*, *42*, 45-51, 1978.

Goetze, C., High temperature rheology of westerly granite, *J. Geophys. Res.*, *76*, 1223-1230, 1971.

Goetze, C., and W. F. Brace, Laboratory observations of high-temperature rheology of rocks, *Tectonophysics, 13,* 583-600, 1972.

Green, H. W., D. T. Griggs, and J. M. Christie, Syntectonic and annealing re-crystallization of fine-grained quartz aggregates, in *Experimental and Natural Rock Deformation*, edited by P. Paulitsch, pp. 272-335, Springer, New York, 1970.

Griggs, D. T., Hydrolytic weakening of quartz and other silicates, *Geophys. J. R. Astron. Soc., 14,* 19-31, 1967.

Griggs, D. T., A model of hydrolytic weakening in quartz, *J. Geophys. Res., 79,* 1655-1661, 1974.

Griggs, D. T., and J. D. Blacic, Quartz: Anomalous weakness of synthetic crystals, *Science, 147,* 292-295, 1965.

Hadley, K., The effect of cyclic stress on dilatancy: Another look, *J. Geophys. Res., 81,* 2471-2474, 1976.

Haimson, B. C., Mechanical behavior of rock under cyclic loading, in *Advances in Rock Mechanics*, vol. 2, pt. 1, National Academy of Science, Washington, D. C., 1974.

Haimson, B. C., C. M. Kim, and T. M. Tharp, Tensile and compressive cyclic stresses in rock, paper presented at 14th Symposium on Rock Mechanics, Pennsylvania State Univ., University Park, Pa., June 1973.

Handin, J. W., and N. L. Carter, Rheology of rocks at high temperature, paper presented at 4th International Congress on Rock Mechanics, Montreux, Switzerland, 1980.

Handin, J. W., R. V. Hager, Jr., M. Friedman, and J. N. Feather, Experimental deformation of sedimentary rocks under confining pressure: Pre-pressure tests, *Am. Assoc. Pet. Geol. Bull., 47,* 717-755, 1963.

Hansen, F. D., and N. L. Carter, Creep of rocksalt at elevated temperature, paper presented at 21st Symposium on Rock Mechanics, Univ. of Missouri, Rolla, Mo., May 1980.

Hardy, H. R., Emergence of acoustic emission, microseismic activity as a tool in geomechanics, in *Proceedings of the First Conference on Acoustic Emission/Microseismic Activity in Geologic Structures and Materials*, Trans. Tech. Publ., Clausthal, West Germany, 1977.

Hardy, H. R., R. Y. Kim, R. Stetanko, and Y.-T. Wang, Creep and microseismic activity in geological materials, paper presented at 11th Symposium on Rock Mechanics, University of California, Berkeley, 1969.

Heard, H. C., Comparison of the flow properties of rocks at crustal conditions, *Phil. Trans. R. Soc. London Ser. A, 283,* 173-186, 1976.

Heard, H. C., and N. L. Carter, Experimentally induced 'natural' intragranular flow in quartz and quartzite, *Am. J. Sci., 266,* 1-42, 1968.

Hobbs, B. E., A. C. McLaren, and M. S. Paterson, Plasticity of single crystals of synthetic quartz, *Flow and Fracture of Rocks, Geophys. Monogr. Ser.*, vol. 16, edited by H. C. Heard, I. Y. Borg, N. L. Carter, and C. B. Raleigh, pp. 29-54, AGU, Washington, D. C.,1972.

Johnson, B., A. F. Gangi, and J. Handin, Thermal cracking of rock subjected to slow, uniform temperature changes, paper presented at 19th Symposium on Rock Mechanics, University of Nevada, Reno, June 1978.

Kern, H., The effect of high temperature and high confining pressure on compressional wave velocities in quartz-bearing and quartz-free igneous and metamorphic rocks, *Tectonophysics, 44,* 185-204, 1978.

Kirby, S. H., and C. B. Raleigh, Mechanisms of high-temperature, solid-state flow in minerals and ceramics and their bearing on creep behavior of the mantle, *Tectonophysics, 19,* 165-197, 1973.

Koch, P. S., J. M. Christie, and R. P. George, Jr., Flow law of 'wet' quartzite in the α-quartz field (abstract), *Eos Trans. AGU, 61,* 346, 1980.

Kranz, R. L., Crack growth and development during creep of Barre granite, *Int. J. Rock Mech. Mining Sci., 16,* 23-35, 1979a.

Kranz, R. L., Crack-crack and crack-pore interactions in stressed granite, *Int. J. Rock Mech. Mining Sci., 16,* 37-47, 1979b.

Kranz, R. L., The static fatigue and hydraulic properties of Barre granite, Ph.D. thesis, Columbia Univ., New York, N.Y., 1979c.

Kranz, R. L., The effects of confining pressure and stress difference on static fatigue of granite, *J. Geophys. Res., 85,* 1854-1866, 1980.

Kranz, R. L., and C. Scholz, Critical dilatant volume of rocks at the onset of tertiary creep, *J. Geophys. Res., 82,* 4893-4898, 1977.

Lindholm, U. S., L. M. Yeakley, and A. Nagy, The dynamic strength and fracture properties of Dresser basalt, *Int. J. Rock Mech. Mining Sci., 11,* 181-192, 1974.

Lockner, D., and J. Byerlee, Acoustical emission and creep in rocks at high confining pressure and differential stress, *Seis. Soc. Am. Bull., 67,* 243-258, 1977.

Martin, R. J., III, Pore pressure effects in crustal processes, *Rev. Geophys. Space Phys., 17,* 1132-1137, 1979.

Misra, A. K., and S. A. F. Murrell, An experimental study of the effect of temperature and stress on the creep of rocks, *Geophys. J. R. Astron. Soc., 9,* 509-535, 1965.

Morrison-Smith, D. T., M. S. Paterson, and B. E. Hobbs, An electron microscope study of plastic deformation in single crystals of synthetic quartz, *Tectonophysics, 33,* 43-79, 1976.

Murrell, S. A. F., Rheology of the lithosphere—Experimental indications, *Tectonophysics, 36,* 5-24, 1976.

Murrell, S. A. F., and S. Chakravarty, Some new rheological experiments on igneous rocks at temperatures up to 1120°C, *Geophys. J. R. Astron. Soc., 34,* 211-250, 1973.

Parrish, D. K., and A. F. Gangi, A nonlinear, least-squares technique for deforming multi-mechanism, high-temperature creep flow laws, this volume.

Parrish, D. K., A. L. Krivz, and N. L. Carter, Finite element folds of similar geometry, *Tectonophysics, 32,* 183-207, 1976.

Peng, S., and A. M. Johnson, Crack growth and faulting in cylindrical specimens of Chelmsford granite, *Int. J. Rock Mech. Mining Sci.*, *9*, 37-86, 1972.

Post, R. L., Jr., The flow laws of Mt. Burnett dunite, Ph.D. thesis, Univ. of Calif., Los Angeles, 1973.

Richter, D, and G. Simmons, Thermal expansion behavior of igneous rocks, *Int. J. Rock Mech. Mining Sci.*, *11*, 403-411, 1974.

Rümmel, F., Studies of time dependent deformation of some granite and eclogite rock samples under uniaxial constant compressive stress and temperatures up to 400°C, *Z. Geofiz.*, *35*, 17-42, 1969.

Scholz, C. H., Microfracturing and inelastic deformation of rock in compression, *J. Geophys. Res.*, *73*, 1417-1432, 1968*a*.

Scholz, C. H., Experimental study of the fracture process in brittle rock, *J. Geophys. Res.*, *73*, 1447-1454, 1968*b*.

Scholz, C. H., Mechanism of creep in brittle rock, *J. Geophys. Res.*, *73*, 3295-3302, 1968*c*.

Scholz, C. H., and R. L. Kranz, Notes on dilatancy recovery, *J. Geophys. Res.*, *79*, 2132-2135, 1974.

Scholz, C. H., and T. A. Koczynski, Dilatancy anisotropy and the response of rock to large cyclic loads, *J. Geophys. Res.*, *84*, 5525-5534, 1979.

Shewman, P. G., *Diffusion in Solids*, 272 pp. McGraw-Hill, New York, 1963.

Siegfried, R., and G. Simmons, Characterization of oriented cracks with differential strain analysis, *J. Geophys. Res.*, *83*, 1269-1278, 1978.

Simmons, G., and H. W. Cooper, Thermal cycling in three igneous rocks, *Int. J. Rock Mech. Mining Sci.*, *15*, 145-148, 1978.

Soga, N., H. Mizutani, H. Spetzler, and R. J. Martin, The effect of dilatancy on velocity anisotropy in Westerly granite, *J. Geophys. Res.*, *83*, 4451-4458, 1978.

Sprunt, E., and W. F. Brace, Direct observation of micro-cavities in crystalline rocks, *Int. J. Rock Mech. Mining Sci.*, *11*, 139-150, 1974*a*.

Sprunt, E., and W. F., Brace, Some permanent changes in rocks due to pressure and temperature, Paper presented at International Congress on Rock Mechanics, sponsor, Denver, Month, 1974*b*.

Summers, R., K. Winkler, and J. Byerlee, Permeability changes during flow of water through Westerly granite at temperatures of 100° to 400°C., *J. Geophys. Res.*, *83*, 339-344, 1978.

Tapponnier, P., and W. F. Brace, Development of stress-induced microcracks in Westerly granite, *Int. J. Rock Mining Sci.*, *13*, 103-112, 1976.

Tullis, J., High temperature deformation of rocks, *Rev. Geophys. Space Phys.*, *17*, 1137-1154, 1979.

Tullis, J., and Y. A. Yund, Experimental deformation of Westerly granite, *J. Geophys. Res.*, *82*, 5705-5718, 1977.

Tullis, J., and Y. A. Yund, An experimental study of the rheology of crustal rocks, *Prog. Summ. Tech. Rep. 6*, U.S. Geol. Survey Nat. Earthquake Hazards Red. Program, Menlo Park, Calif., 379-381, 1978.

Tullis, J., and Y. A. Yund, An experimental study of the rheology of crustal rocks, *Prog. Summ. Tech. Rep. 8*, U.S. Geol. Survey Nat. Earthquake Hazards Red. Program, Menlo Park, Calif., 517-520, 1979*a*.

Tullis, J., and Y. A. Yund, An experimental study of the rheology of crustal rocks, *Prog. Summ. Tech. Rep. 9*, U.S. Geol. Survey Nat. Earthquake Hazards Red. Program, Menlo Park, Calif., 550-554, 1979*b*.

Tullis, J., J. M. Christie, and D. T. Griggs, Microstructures and preferred orientations in experimentally deformed quartzites, *Geol. Soc. Am. Bull.*, *84*, 297-314, 1973.

Tullis, J., G. L. Shelton, and Y. A. Yund, Pressure dependence of rock strength: Implications for hydrolytic weakening, *Bull. Mineral.*, *102*, 110-114, 1979.

Wawersik, W. R., and W. F. Brace, Post-failure behavior of granite and diabase, *Rock Mech.*, *3*, 61-85, 1971.

Webster, G. A., A. P. D. Cox, and J. E. Dorn, A relationship between transient and steady state creep at elevated temperatures, *Met. Sci. J.*, *3*, 221-225, 1969.

Weertman, J., and J. R. Weertman, Mechanical properties, strongly temperature-dependent, in *Physical Metallurgy*, edited by R. W. Cahn, 983-1010, 1970.

Wu, F. T., and L. Thomsen, Microfracturing and deformation of Westerly granite under creep conditions, *Int. J. Rock Mech. Mining Sci.*, *12*, 167-173, 1975.

Yund, Y. A., and J. Tullis, The effect of water, pressure, and strain on the Al/Si order-disorder kinetics in feldspar, *Contrib. Mineral Petrol.*, *72*, 297-302, 1980.

Zoback, M. D., and J. D. Byerlee, The effect of cyclic differental stress on dilatancy in Westerly granite under uniaxial and triaxial conditions, *J. Geophys. Res.*, *80*, 1526-1530, 1975.

Activation Volume for Steady State Creep in Polycrystalline CsCl: Cesium Chloride Structure

H. C. HEARD

Lawrence Livermore National Laboratory, Livermore, California 94550

S. H. KIRBY

U.S. Geological Survey, Menlo Park, California 94025

Annealed polycrystalline samples of CsCl have been deformed in triaxial compression at 150°–400°C (T), at 0.1–400 MPa confining pressure (P), and at constant strain rates ($\dot{\epsilon}$) ranging from 10^{-2} to 10^{-6} s^{-1}. Stress-strain data from these tests show mainly an elastic-plastic response, but with minor work hardening occurring at the lower temperatures and at the higher rates. At the values of T, P, and $\dot{\epsilon}$ investigated, the steady state flow stress σ ranged from 1.3 to 65 MPa. At similar T and $\dot{\epsilon}$, σ at 400 MPa was 2–4 times that at low P. This σ increase with P is related to intracrystalline plasticity, not to dilatancy. Measured values of $\dot{\epsilon}$, T, P, and σ from 39 tests were fitted to a flow law of the form $\dot{\epsilon} = B\sigma^N \exp[-(Q^* + PV^*)/RT]$, where Q^* is the activation energy and V^* the activation volume for steady state flow. A least squares fitting routine was used to calculate B, N, Q^*, and V^*. At one standard deviation and in SI units, log B is -19.87 ± 0.71, N is 4.42 ± 0.29, Q^* is 150 ± 9 kJ/mol, and V^* is $53 \pm 5.5 \cdot 10^{-6}$ m^3/mol. Calculation of V_c^* from an empirical relation based on T_m/T and measured dT_m/dP on melting of the CsCl phase yields a much larger activation volume: $83 \cdot 10^{-6}$ m^3/mol. Previously, the V_c^* has been calculated by Keyes' method based on the shear and bulk moduli in isotropic materials or on this method as corrected for cubic solids. The range of values for V_c^* (17–$74 \cdot 10^{-6}$ m^3/mol) is so broad that they are of little use in estimating V^*. Values for Q^* and V^* determined here are shown to be consistent with both the measured activation energy and a model activation volume for self-diffusion in CsCl based on a vacancy diffusion mechanism.

Introduction

It is widely held that convection associated with solid-state flow in the earth's mantle is responsible for motions of the lithospheric plates at the surface. Although transient creep in the upper mantle can be important (as in the surface rebound associated with crustal unloading) and tertiary creep can be dominant locally (as in active seismic zones near these plate margins), secondary creep or steady state flow is thought to be the most widespread phenomenon within the mantle. Secondary creep in metals, ceramics, and rock-forming minerals at atmospheric pressure is affected by such parameters as temperature, deviatoric stress, elastic moduli, grain size, and other physicochemical factors such as defect concentration and mobility and trace concentrations of other chemical species. In nearly all naturally occurring earth materials where secondary creep is important, the high temperature and low deviatoric stress levels necessary for these proc-

esses occur only at depth. The high superposed pressure associated with that depth and temperature adds further complications to the understanding of secondary creep. Although it has often been neglected in the past, pressure affects the creep behavior by increasing the elastic moduli as well as the activation enthalpy for flow. For deformation at relatively shallow depths, the magnitude of these pressure effects is small in comparison to the other terms in the flow equation that best fits most secondary creep data at $T/T_m > 0.4$ and $\sigma/\mu < \sim 10^{-3}$:

$$\dot{\epsilon} = A\,(\mu/T)\,(\sigma/\mu)^N \exp[-(Q^* + PV^*)/RT] \tag{1}$$

where T is the absolute temperature; T_m is the melting temperature; σ is the principal stress difference ($\sigma_1 - \sigma_3$); μ is the isothermal shear modulus; $\dot{\epsilon}$ is the strain rate; Q^* is an activation energy; P is pressure; V^* is an activation volume; and A, N, and R are material constants. But for flow at depth, as at the extreme pressures existing in the mantle, the associated changes in the moduli and activation enthalpy can exert significant effects. Many data are available on the elastic behavior of appropriate mineral phases at pressure and temperature [Bonner and Schock, 1980]. However, only recently has the pressure effect on the activation enthalpy ($Q^* + PV^*$) been assessed and V^* derived for a single common silicate mineral, olivine [Ross et al., 1979; Kohlstedt et al., 1980].

Direct measurements of V^* have been reported for two additional nonmetallic compounds: ice and AgBr [Christy, 1954; Weertman, 1970]. We report here new data determining V^* for a fourth nonmetal, CsCl, to an accuracy heretofore not attained. Such determinations of V^* are not only useful for direct application to a specific flow problem but also critical in constraining theories that seek to predict V^* from other thermophysical measurements. Because of the lack of data concerning V^* for secondary creep for virtually all common minerals, coupled with the difficulty of making accurate experimental determinations, most researchers desiring such information have resorted to assessing V^* by calculation, using either of two approximate methods. The first is related to the pressure effect on T_m:

$$V_c^* = Q^* \left(\frac{dT_m}{dP}\right)\left(T_m - P\frac{dT_m}{dP}\right)^{-1} \quad (2)$$

where V_c^* is the calculated activation volume [Weertman, 1970; Sherby et al., 1970; Sammis et al., 1977]. The second method is based on an elastic point defect model for migration:

$$V_c^* = \left[\left(\frac{\partial \ln \mu}{\partial P}\right)_T - \beta\right]\left[1 - \left(\frac{\partial \ln \mu}{\partial \ln T}\right)_P - \alpha T\right]^{-1} Q^* \quad (3)$$

where β is the isothermal compressibility and α is the volume coefficient of thermal expansion [Keyes, 1963; Sammis et al., 1980]. Both methods have limitations when applied to CsCl.

The purpose of the present contribution is to determine experimentally the activation volume for secondary creep in a material where V^* could reasonably be expected to be large and to compare these results with V_c^* based on (2) and (3). The V^* should be large in CsCl (bcc structure) because it has an exceptionally large dT_m/dP [Clark, 1959; Vaidya and Kennedy, 1971] and a very large dC_{44}/dP [Barsch and Chang, 1967]. We also chose CsCl for this study instead of an appropriate oxide or silicate mantle mineral because CsCl exhibits secondary creep and possesses a large σ dependence upon P within the P, T measurement and control capability of existing constant strain rate apparatus. A total of 39 compression experiments were performed on annealed polycrystalline CsCl. Pressures for these tests ranged from 0.1 to 400 MPa, temperatures from 150° to 400°C, strain rates from 10^{-2} to 10^{-6} s^{-1}, and resultant stresses from 1.3 to 65 MPa. If one could demonstrate from these results that $V^* = V_c^*$ for

this model material and if the silicate and oxide minerals existing at depth in the mantle behave similarly, then the pressure effect on secondary creep in the mantle can be evaluated by these methods with more confidence. We show, however, that neither (2) nor (3) gives an acceptable estimate of the true value of V^* for steady state flow in CsCl. Therefore, by analogy, use of these methods may yield misleading results when applied to other materials.

Starting Material and Experimental Apparatus

Two polymorphs of CsCl have been reported [Clark, 1959]. The stability fields of these are shown in Figure 1. For the purpose of the present study we wished to cover as wide a P, T range as possible without crossing phase boundaries, as from the bcc to the fcc structure, with the accompanying complications associated with the large volume change (~18%). We therefore limited our tests to those conditions in Figure 1 identified by circles. The extremes of T/T_m for these tests range from 0.45 to 0.73. For reference, σ/μ varied from $6 \cdot 10^{-3}$ to $2 \cdot 10^{-4}$, based on μ for the aggregate as calculated by the Voigt-Reuss average method.

Right-circular cylindrical samples were machined in two sizes, 25.4-mm diameter by 50.8-mm length and 19-mm diameter by 38-mm length, from billets hydrostatically pressed at 250 MPa and 20°C. The larger specimen size was used in the higher-temperature, lower-strain-rate experiments to improve experimental accuracy at low stress. The starting material was CsCl of ≥99.9% purity furnished in <40-mesh size. The density of the machined test cylinders was randomly checked by weighing and by micrometer measurement before annealing under pressure; porosity averaged about 2%. At each stage of the sample preparation process, all materials were stored in desiccators over anhydrous magnesium perchlorate to minimize water adsorption by the test material.

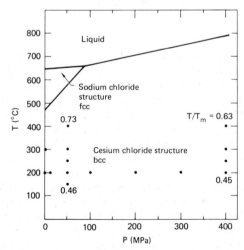

Fig. 1. Phase diagram for CsCl showing pressure and temperature conditions used in this work [from Clark, 1959]. Also indicated is the range of T/T_m values for the tests.

Before beginning compressive loading at the P, T, and $\dot{\epsilon}$ reported in Table 1, each sample was jacketed in 0.25-mm-thick 1100F aluminum and annealed at 400°C and 50 MPa for 1 hour. During the anneal as well as deformation, pore pressure in the sample was maintained at 0.1 MPa by a hole through the loading piston. Measurements of density after this anneal (but before straining) yielded porosities of <0.5%. Thus with this preparation and annealing procedure uniformly applied to all samples, we could expect that the resulting rheologic behavior of each sample resulted from the P, T, ϵ, $\dot{\epsilon}$, and σ unique to each test and was not due to the different pretest histories of the sample and jacket.

Reported values for σ (= $\sigma_1 - \sigma_3$) in Table 1 are corrected for porosity changes in the starting samples, for changes in cross-sectional area of the samples with strain, and for loads borne by the aluminum jackets.

The apparatus used in these tests has been described previously [Carter and Heard, 1970]. In the constant strain rate tests reported here, accuracies in temperature are 1°-2°C, and accuracies in pressures are 0.1-1.0 MPa, depending somewhat on the experimental conditions. Strain rates are calculated near the maximum strains measured (10%) and are constant to 1%. Accuracies in stress range from 0.1 to 0.2 MPa, depending on sample diameter.

Results

After correction for apparatus and jacket effects and the change in sample dimensions as a result of strain and porosity changes, values for σ were plotted versus axial strain ϵ. Resulting σ versus ϵ curves for the annealed CsCl were then compared as a function of P, T, and $\dot{\epsilon}$. Values for σ at 4, 6, 8, and 10% strain are summarized in Table 1

TABLE 1. CsCl Data Summary

Test	Pressure, MPa	Temperature, °C	Strain Rate, s⁻¹	Stress (σ) at Selected Strains (ϵ), MPa 4%	6%	8%	10%
996	0.1	200	$1.52\cdot10^{-4}$	25.7	26.2	26.4	26.6
1005	0.1	300	$1.55\cdot10^{-4}$	4.9	4.9	4.9	4.9
995	10	200	$1.52\cdot10^{-4}$	23.1	25.0	26.4	27.3
999	50	150	$1.51\cdot10^{-4}$	27.3	42.0	44.8	46.5
958	50	200	$1.52\cdot10^{-4}$	30.0	32.2	33.8	34.4
967	50	200	$1.52\cdot10^{-4}$	29.4	31.7	32.9	33.6
962	50	200	$1.54\cdot10^{-5}$	16.0	16.3	16.5	16.7
965	50	200	$1.54\cdot10^{-6}$	13.4	13.5	13.5	(13.5)
1000	50	250	$1.54\cdot10^{-4}$	17.7	18.4	18.4	18.4
946	50	300	$1.83\cdot10^{-2}$	12.0	13.2	13.9	14.3
926	50	300	$1.55\cdot10^{-3}$	7.9	8.4	8.6	8.7
924	50	300	$1.55\cdot10^{-4}$	6.4	6.5	6.5	6.5
957	50	300	$1.15\cdot10^{-4}$	5.5	(5.7)	5.8	(5.9)
957	50	300	$1.15\cdot10^{-5}$	(4.0)	4.1	(4.2)	4.3
966	50	300	$1.16\cdot10^{-6}$	2.2	2.3	2.3	2.3
944	50	400	$1.85\cdot10^{-2}$	5.6	5.9	6.1	6.2
929	50	400	$1.56\cdot10^{-3}$	2.6	2.7	2.7	2.8
953	50	400	$1.16\cdot10^{-4}$	1.0	1.1	1.2	1.3
956	50	400	$1.16\cdot10^{-4}$	2.0	(2.1)	2.1	2.1
956	50	400	$1.16\cdot10^{-5}$	(1.3)	1.3	(1.3)	(1.3)
994	100	200	$1.52\cdot10^{-4}$	29.1	30.6	31.5	32.2
993	200	200	$1.52\cdot10^{-4}$	36.4	38.8	40.2	40.7
992	300	200	$1.52\cdot10^{-4}$	43.6	47.3	48.6	49.0
968	400	200	$1.49\cdot10^{-4}$	56.0	60.1	62.6	64.5
970	400	200	$1.52\cdot10^{-5}$	(47.4)	(48.4)	(48.9)	(49.5)
970	400	200	$1.52\cdot10^{-6}$	38.5	(39.5)	(40.1)	(40.6)
998	400	250	$1.52\cdot10^{-4}$	29.1	31.3	32.8	33.2
948	400	300	$1.89\cdot10^{-2}$	35.2	39.3	41.6	43.0
969	400	300	$1.82\cdot10^{-2}$	25.7	27.4	28.3	28.7
927	400	300	$1.55\cdot10^{-3}$	25.4	26.0	26.3	26.4
925	400	300	$1.56\cdot10^{-4}$	12.3	12.4	12.5	12.6
971	400	300	$1.56\cdot10^{-4}$	15.2	15.8	16.3	16.8
972	400	300	$1.15\cdot10^{-5}$	9.4	(9.8)	9.9	(10.1)
972	400	300	$1.15\cdot10^{-6}$	(5.6)	(5.9)	(6.1)	6.2
949	400	400	$1.38\cdot10^{-2}$	16.1	16.5	16.7	16.9
973	400	400	$1.85\cdot10^{-2}$	13.8	14.1	14.3	14.4
928	400	400	$1.56\cdot10^{-3}$	7.0	7.3	7.5	7.6
931	400	400	$1.17\cdot10^{-4}$	3.0	3.3	3.4	3.6
932	400	400	$1.17\cdot10^{-5}$	1.7	1.9	2.1	2.2

Parentheses indicate interpolated or extrapolated value.

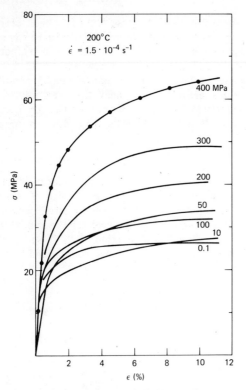

Fig. 2. Stress-strain curves for CsCl tested at 0.1–400 MPa, $\dot{\epsilon} = 1.5 \cdot 10^{-4}$ s^{-1}, and 200°C. The calculated values of stress and strain plotted on the 400-MPa curve show typical data scatter and are representative of all σ–ϵ curves.

along with the P, T, and $\dot{\epsilon}$ conditions for each test. In Figure 2 we compare this σ-ϵ behavior as P is increased from 0.1 to 400 MPa, but with $\dot{\epsilon}$ and T held constant at $1.5 \cdot 10^{-4}$ s^{-1} and 200°C, respectively. Although there are minor inconsistencies at the lower pressures, in general, σ increases with P.

When testing brittle materials at $T/T_m < {\sim}0.3$ and at pressure, it is usual to observe such an increase in σ with P. At sufficiently high pressure, depending on the material, the σ dependence on P becomes small. This increase of σ with P in brittle media is caused by the internal friction characteristics of the material and is associated with tensile and shear fractures, zones of intense local deformation (Lüder's bands), and dilatancy [*Griggs and Handin*, 1960; *Heard*, 1967]. When the flow stress becomes only weakly influenced by pressure, the material is said to be perfectly plastic and approximates a Von Mises solid. In materials undergoing such a brittle-ductile transition with pressure, a plot of fracture or flow stress versus P usually shows a parabolic envelope (concave downward) at low P, followed by a linear region at higher P. At a still higher P the slope progressively decreases until the stress is nearly independent of P. However, the point to note is that over the brittle-ductile transition range there is a conspicuous decrease in the derivative of stress with pressure. The trend

of the CsCl results shown in Figure 2 suggests that fracture and dilatancy are not important over our range of test conditions.

In Figure 3 we plot σ at $\epsilon = 10\%$ from Figure 2 plus one additional test from Table 1, all at 200°C and $\dot{\epsilon} = 1.5 \cdot 10^{-4}$ s^{-1}. For the conditions investigated here, this T and $\dot{\epsilon}$ are among the most favorable for brittle behavior. The absence of any obvious decrease in slope in Figure 3, together with the complete absence of any serrate shape or negative slope of the σ-ϵ curves (Figure 2), as is usual with brittle or transitional behavior in nonmetallic materials, argues against a frictional mechanism. No fractures, faults, or Lüder's bands were observed in any sample from these tests or any other test summarized in Table 1. The concave upward curve through the experimental data is consistent with an increased strength with P for high-temperature plastic flow (equation (1)). In addition, Figure 3 also shows the porosity ϕ of each sample after straining, for comparison with the porosity after annealing at 50 MPa and 400°C but before any strain (<0.5%). After strain, these test data indicate that porosity remains constant at about 0.5% except at the lowest test pressures, 0.1 and 10 MPa. Initially, this suggested that some dilatancy may have occurred at low P. However, the test procedure that was uniformly followed was to cool after deformation at the indicated test pressure. Therefore any increase in porosity could result from either dilatancy from strain or from differential thermal contraction in the sample-jacket-piston assembly during cooling. In Figure 3 the single point, depicted by the cross, at $\phi = 1.9\%$ and 0.1 MPa is from an undeformed sample annealed as usual but cooled at 0.1 MPa. These results suggest that most, if not all, of the indicated porosity increase at low P is not due to deformation-related dilatancy but to the cooling history.

In Figure 4 we show typical σ-ϵ data for the CsCl compressed at several temperatures from 150° to 400°C and at a constant P of 50 MPa and a constant $\dot{\epsilon}$ of 1.2–1.5 $\cdot 10^{-4}$ s^{-1}. The curves at 400 MPa appear similar in shape but are elevated in σ. Strain hardening, a characteristic of transient creep, is apparent at the lower temperatures to strains of about 8%. As T is increased, transient behavior diminishes until at $T > 250$°C, CsCl approximates an elas-

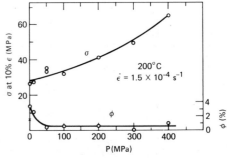

Fig. 3. Stress at 10% ϵ and porosity ϕ versus P for CsCl tested at $\dot{\epsilon} = 1.5 \cdot 10^{-4}$ s^{-1} and 200°C.

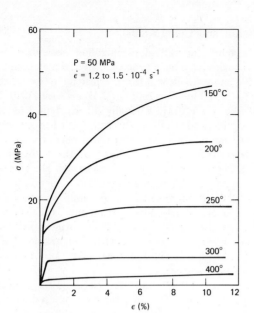

Fig. 4. Stress-strain curves for CsCl tested at 150°–400°C, $\dot{\epsilon}$ = 1.2–1.5 · 10⁻⁴ s⁻¹, and 50 MPa.

tic-plastic solid. It is obvious here and from Table 1 that moderate increments in T strongly lower the flow stress at any given strain and pressure. At other values of $\dot{\epsilon}$, curves are again much like those in Figure 4; σ and work hardening increase with $\dot{\epsilon}$.

In Figure 5, more σ-ϵ results are compared, but in this case over a broad range of strain rate but at constant T of 300°C and constant P of 50 MPa. The $\dot{\epsilon}$ was maintained constant during the test, as indicated near each curve, except in the case of the irregular curve centered at 5 MPa. In this instance the CsCl sample was loaded at a $\dot{\epsilon}$ of 1.2 · 10⁻⁵ s⁻¹ to a strain of 1.8%; then the $\dot{\epsilon}$ was increased to 1.2 · 10⁻⁴ s⁻¹. The higher rate was continued to a strain of 4.8%, lowered by a factor of 10 to 6% strain, raised to 10⁻⁴ s⁻¹ to a strain of 8.8%, and then again lowered by a factor of 10 until the test was terminated at 10.6% total strain. This procedure allowed several strain rates to be investigated for the same sample. Connection between the σ values at each $\dot{\epsilon}$ (dashed curves in Figure 5) then allowed inference of the complete σ-ϵ curve for both. The behavior in the few tests accomplished in this manner is consistent with the main body of data taken at a single constant strain rate for each sample. In Figure 5, transient creep is apparent to an ϵ of about 8% at the highest strain rate and becomes negligible at ϵ > ~1% at $\dot{\epsilon}$ < 10⁻⁴ s⁻¹. All stress-strain curves indicate that σ decreases with $\dot{\epsilon}$. Curves for σ versus ϵ at other pressures or temperatures are quite similar except for the σ increase with P and σ decrease with T.

Comparison of all results summarized in Table 1 together with those data selected for presentation in Figures 2, 4, and 5 indicates that steady state creep predominates at 10% strain at all P, T, and $\dot{\epsilon}$ investigated. At lower strains the importance of transient flow depends

mainly on T and $\dot{\epsilon}$ and less on P. Only secondary creep occurs at a strain of >1% at intermediate and higher T and at intermediate and lower $\dot{\epsilon}$. The equivalence of T and $\dot{\epsilon}$ in affecting the steady state flow stress suggests that the process responsible for creep is thermally activated. Furthermore, for the range of T/T_m and σ/μ investigated here this process is likely to be diffusion controlled. All evidence presented so far on the rheological behavior of CsCl deformed over our range of conditions is consistent with a steady state flow equation of the type of (1).

Data from the 39 tests summarized in Table 1 were fitted to (1) by a least squares routine, which minimizes the deviation in log $\dot{\epsilon}$. Values for $\dot{\epsilon}$, T, P, σ (at 10% ϵ), and μ were the input, and best values were calculated for A, Q^*, V^*, and N. Because we did not determine μ for our polycrystalline aggregates at P, T, we calculated the Voigt-Reuss average values using Barsch and Chang's [1967] CsCl data for C_{11}, C_{12}, and C_{44} together with their respective pressure and temperature derivatives. In addition, we also fitted our results to a similar steady state flow equation, which is simpler and involves fewer parameters and assumptions, namely,

$$\dot{\epsilon} = B\sigma^N \exp\left[-(Q^* + PV^*)/RT\right] \qquad (4)$$

In those instances where (1) could be compared with (4) for other materials, available secondary creep data indicate that (4) is marginally better for marble and halite but (1) seems favored for several metals and ceramics [Bird et al., 1969; Heard, 1972; Heard and Raleigh, 1972].

Table 2 lists the values for B, N, Q^*, and V^* calculated from (4). These values are in SI units, and uncertainties have been determined at one standard deviation. Also

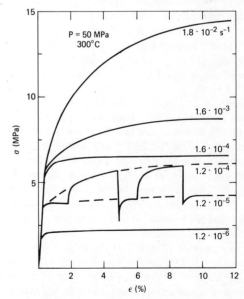

Fig. 5. Stress-strain curves for CsCl tested at $\dot{\epsilon}$ = 1.8 · 10⁻² to 1.2 · 10⁻⁶ s⁻¹, 300°C, and 50 MPa. (See text for a discussion of multiple $\dot{\epsilon}$ test.)

TABLE 2. Calculated Values (in SI Units) for B, N, Q^*, and V^* from Equation (4)

Parameter	Value
$\log B$	-19.87 ± 0.71
N	4.42 ± 0.29
Q^*	150 ± 9 kJ/mol
V^*	$53.1 \pm 5.5 \cdot 10^{-6}$ m^3/mol

Equation (4) is $\dot{\epsilon} = B\,(\sigma)^N \exp\left[-(Q^* + PV^*)/RT\right]$. The standard deviation between $\log \dot{\epsilon}$ and $\dot{\epsilon}_c$ is 0.42.

shown is the standard deviation between the measured and the calculated strain rate $\dot{\epsilon}_c$ based on these values. We also calculated A, N, Q^*, and V^* as above but from (1), using both the Voigt and the Reuss average values for μ. Resulting values using the Voigt average μ are $\log A = 16.78 \pm 1.67$, $N = 4.45 \pm 0.29$, $Q^* = 149 \pm 9$ kJ/mol, and $V^* = 48.9 \pm 5.3 \cdot 10^{-6}$ m^3/mol. Comparison of these values with those in Table 2 shows N, Q^*, and V^* to be virtually identical. Standard deviations in each quantity were equal or only slightly higher using (4) as compared to (1). Little or no difference could be seen when the Voigt-calculated μ was used as compared to the Reuss-derived μ. Because of the closeness of fit between (1) and (4) and because (4) is simpler than (1), not requiring terms in T and μ, we prefer the analysis using (4).

In fitting the CsCl data from Table 1 to (4) there is the inherent assumption that B, Q^*, V^*, and N are constant over the range of the $\dot{\epsilon}$, T, P, and σ parameters investigated. For the stress exponent N we tested linearity by plotting $\log \sigma$ versus $\log \dot{\epsilon} - \log B + (Q^* + PV^*)/2.3RT$ for each test, using values listed in Table 2. In Figure 6 we compare the results of these calculations with N equal to 4.42. Inspection of the results indicates that N is constant over the range of σ investigated and that deviations from N^{-1} are about constant at any position. Comparison of both the high-pressure and the low-pressure data sets with the line of slope equal to 4.42^{-1} demonstrates that N is unaffected by P.

Similarly, in Figure 7 we illustrate the temperature dependence by plotting T^{-1} versus $\log \dot{\epsilon} - \log B - N \log \sigma$. As noted, the slopes of the two lines plotted at 50 and 400 MPa are $[(Q^* + PV^*)/2.3R]^{-1}$ and are derived from best fit parameters for (4) listed in Table 2. Calculated values for each test shown at both pressures indicate a linear trend in $1/T$ but with more dispersion than is shown in Figure 6. The slight divergence of the two isobars over this temperature interval is a result of the pressure effect on the activation enthalpy.

The constancy of V^* with P is shown in Figure 8 by plotting the calculated values for $\log \dot{\epsilon} - \log B + N \log \sigma + Q^*/2.3RT$ for all tests at 200°C. Comparison of these points with the line based on (4) from Table 2 indicates that V^* is constant with P within the stated deviation. Similar results can be demonstrated by comparing test data at other temperatures from Table 1.

Discussion and Conclusions

To account for diffusion-controlled secondary creep, many theories have been proposed, all of which take essentially the same functional form. These have been most recently reviewed by *Weertman and Weertman* [1970, 1975] and *Weertman* [1974]. Virtually all are based on a dislocation glide and climb mechanism with the assumption that diffusion of single or groups of atoms or vacancies is the rate-controlling process. None of these theories are widely accepted for all types of materials. Most lead to a stress exponent (N) in the range 3–6, depending on the specific dislocation mechanism assumed to prevail. The stress exponent of 4.4 deduced from our CsCl tests is consistent with these models. In addition, other alkali halides also show power law behavior with comparable N values, 5.5 for NaCl [*Burke*, 1968; *Heard*, 1972] and 4.5 for KBr [*Yavari and Langdon*, 1981].

Sherby et al. [1970] and *Weertman* [1970] calculated values for Q_c^* and V_c^* mainly for metals based on T_m and dT_m/dP and compared the results with diffusion measurements. *Sammis et al.* [1977, 1980] extended these comparisons further to include ice, olivine, and several alkali halides. Predicted values for Q_c^* seem to be in good agreement with those determined from both diffusion and creep data. On the basis of (2), calculated values for V_c^* also seem to be reasonably consistent with measurement for the metals, ice, and olivine but not for the alkali halides. The cause for this discrepancy may be that the pressure effect on the creep rate for alkali halides is not adequately described by (2). However, recognizing these problems, especially as applied to CsCl, we calculate V_c^* using (2) for comparison to V^*, $53 \cdot 10^{-6}$ m^3/mol. Taking a value for Q^* of 150 kJ/mol (Table 2, equation (4)), dT_m/dP of 0.483°C/MPa (from *Clark* [1959]: bcc phase, 95–800 MPa), and the average P, T_m from 95 to 400 MPa (Figure 1) of

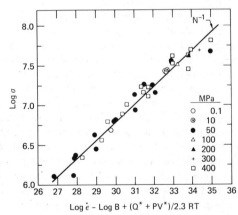

Fig. 6. Plot of $\log \sigma$ as a function of $\log \dot{\epsilon} - \log B + [(Q^* + PV^*)/2.3RT]$ calculated for each test summarized in Table 1 and based on best fit values for B, N, Q^*, and V^* from Table 2, equation (4). The line indicated has a slope of $(4.42)^{-1}$.

248 MPa and 724°C, respectively, we calculate $V_c{}^*$ to be $83 \cdot 10^{-6}$ m³/mol.

Sammis et al. [1980] extensively reviewed and discussed the various models proposed to calculate the $V_c{}^*$ associated with defect migration in elastic solids. Equation (3) estimates $V_c{}^*$ in an isotropic solid if the strain energy is entirely shear [*Keyes*, 1963]. However, if the strain energy is assumed to be purely dilational, then μ should be replaced by β^{-1}. When cubic solids are considered, then μ in (3) becomes a function of C_{11}, C_{12}, and C_{44}. *Sammis et al.* [1980] thus calculate $V_c{}^*$ to range between 17 and $74 \cdot 10^{-6}$ m³/mol.

In alkali halides with the CsCl *bcc* structure, cations diffuse more slowly than anions [*Harvey and Hoodless*, 1967; *Hoodless and Turner*, 1972a]. The slowest diffusing species generally controls the rate of bulk diffusion because of the electroneutrality constraint. In the case of CsCl we would expect that the activation energy for self-diffusion of Cs⁺ ($E_{Cs}{}^*$) best correlates with the activation energy for creep Q^*. Inspection of Table 3 shows that $E_{Cs}{}^* \sim Q^*$. However, the absolute diffusion rates and activation energies for Cs⁺ and Cl⁻ are so closely similar that we cannot rule out creep controlled by Cl⁻ self-diffusion.

Although there are no high-pressure data on self-diffusion or ionic conductivity for CsCl, room pressure data are useful to compare with the creep data. Studies of electrical conductivity of pure and doped alkali halides, including CsCl, indicate that Schottky defects (paired anion and cation vacancies) predominate. The equilibrium concentrations of these vacancies are given by

$$\eta_c \eta_a = \exp\left(-E_s{}^*/RT\right) \qquad (5)$$

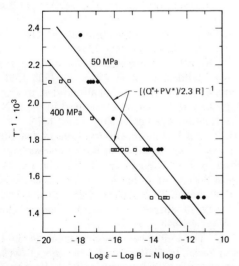

Fig. 7. Plots of T^{-1} as a function of $\log \dot{\epsilon} - \log B - N \log \sigma$ calculated for tests at 50 and 400 MPa from Table 1 and based on best fit values for B, N, Q^*, and V^* from Table 2, equation (4). The indicated lines at both pressures are proportional to $Q^* + PV^*$ (see text).

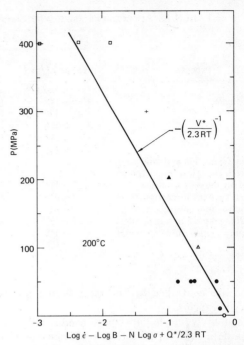

Fig. 8. Plot of P as a function of $\log \dot{\epsilon} - \log B - N \log \sigma + Q^*/2.3RT$ for all tests at 200°C from Table 1, based on best fit values for B, N, Q^*, and V^* from Table 2, equation (4). Comparison of V^* from individual tests with the line indicates that V^* is constant with P.

where η_c and η_a are the molar concentrations of cation and anion vacancies (which are equal in pure crystals) and $E_s{}^*$ is the formation energy for Schottky defects. The rates of transport of Cs⁺ and Cl⁻ are controlled by the concentrations of the corresponding vacancies and by the mobilities of the ions jumping into these vacancies. Measurement of the electrical conductivity of pure and doped CsCl crystals permits the separation of the formation and migration enthalpies for conduction by ionic transport:

$$H^* = H_m{}^* + \tfrac{1}{2}H_s{}^* = E_m{}^* + \tfrac{1}{2}E_s{}^* + PV_m{}^* + P\left(\tfrac{1}{2}V_s{}^*\right) \qquad (6)$$

where $E_m{}^*$ and $V_m{}^*$ are the migration activation energy and activation volume, respectively, and $E_s{}^*$ and $V_s{}^*$ are the formation energy and volume for Schottky defects, respectively. At low temperatures or in very impure crystals the defect concentration is fixed by the concentrations of impurities with valences different from those of the matrix species, and thus the ionic conductivity and self-diffusion rates are controlled by the mobility terms $D \propto \exp\left[-(E_m{}^* + PV_m{}^*)/RT\right]$. This is the extrinsic (impurity controlled) regime for ionic conductivity and self-diffusion. The low values of $E_m{}^*$ for both Cs⁺ and Cl⁻ ion vacancies (Table 3) indicate that if self-diffusion controls creep rates in our experiments, it cannot be impurity assisted. The good agreement between the activation energies for self-diffusion and those predicted from the conductivity data supports the belief that the rates of self-

TABLE 3. Activation Parameters for Creep, Self-Diffusion, and Vacancy Formation and Migration in CsCl

Process	Activation Energy		Activation Volume	
	Value, kJ/mol	Source	Value, × 10⁻⁶ m³/mol	Source
Creep	150 ± 9	this work	53 ± 5.5	this work
Self-diffusion	$Cs^+ = 148$	*Harvey and Hoodless* [1967]		
	$Cs^+ = 130 \pm 5$	*Hoodless and Turner* [1972a]		
	$Cl^- = 122$	*Harvey and Hoodless* [1967]		
	$Cl^- = 124 \pm 3$	*Hoodless and Turner* [1972a]		
	$Cl^- = 130$	*Fredericks* [1975]		
Vacancy formation and migration (from electrical conductivity)				
Schottky defect formation	$E_s{}^* = 171$	*Hoodless and Turner* [1972b]	$V_s{}^* = 48\text{--}84$	*Yoon and Lazarus* [1972]
Cation vacancy migration	$E_m{}^* = 61$	*Hoodless and Turner* [1972b]	$V_m{}^* = 11\text{--}14$	*Keyes* [1963]
Anion vacancy migration	$E_m{}^* = 34$	*Hoodless and Turner* [1972b]	$V_m{}^* = 6\text{--}8$	*Keyes* [1963]
Self-diffusion parameters predicted from vacancy diffusion model	$E_m{}^* + \frac{1}{2}E_s{}^*$ $Cs^+ = 147$ $Cl^- = 120$	*Henderson* [1972]	$V_m{}^* + \frac{1}{2}V_s{}^*$ 36–56 30–50	

diffusion are controlled by the rates of ion jumps into Schottky vacancies.

Unfortunately, no high-pressure conductivity or self-diffusion data exist for CsCl that would allow us to determine $V_m{}^*$ and $V_s{}^*$ directly. The Keyes relation (equation (3)) has been very successful in predicting the activation volume for ion migration in *fcc* alkali halides [*Yoon and Lazarus*, 1972]. Depending on whether shear or dilatational strains are contemplated, $V_m{}^*$ ranges from 11 to $14 \cdot 10^{-6}$ m³/mol for Cs^+ and from 6 to $8 \cdot 10^{-6}$ m³/mol for Cl^- (Table 3). These values are far lower than the activation volume determined for creep, again indicating that extrinsic diffusion does not control the creep process. There is no generally accepted theory for predicting the Schottky defect volume $V_s{}^*$. In the alkali halides the formation volumes for Schottky defects are 1.1–1.9 times the molar volume. For CsCl under our test conditions the molar volume is $44 \cdot 10^{-6}$ m³/mol, and $V_s{}^*$ is $48\text{--}84 \cdot 10^{-6}$ m³/mol if it follows the pattern for other alkali halides. Combining $V_s{}^*$ with our estimates of $V_m{}^*$, the predicted activation volume for either Cs^+ or Cl^- diffusion ranges from 30 to $56 \cdot 10^{-6}$ m³/mol (Table 3), consistent with our experimentally determined activation volume for creep. We believe, however, that only electrical conductivity or diffusion measurements at pressure will provide a definitive test of the relationship between creep and self-diffusion and also will determine the likeliest diffusing species governing creep.

Acknowledgments. We thank R. N. Nimbletoes for a penetrating review. This work was performed under the auspices of the U.S. Department of Energy by Lawrence Livermore National Laboratory under contract W-7405-Eng-48.

References

Barsch, G. R., and Z. P. Chang, Pressure derivatives of the elastic constants, II, Adiabatic, isothermal and inter-mediate pressure derivatives of the elastic constants for cubic symmetry, *Phys. Status Solidi, 19,* 139–151, 1967.

Bird, J. E., A. K. Mukherjee, and J. F. Dorn, Correlations between high temperature creep behavior and structure, in *Quantitative Relation Between Properties and Microstructure,* edited by D. Brandon and A. Rosen, pp. 255–342, Israel University Press, Jerusalem, 1969.

Bonner, B. P., and R. N. Schock, Physical properties of rocks and minerals, in *McGraw-Hill/CINDAS Data Series on Material Properties,* edited by Y. S. Touloukian and C. Y. Ho, McGraw-Hill, New York, in press, 1980.

Burke, P. M., High temperature creep of polycrystalline sodium chloride, Ph.D. thesis, Stanford Univ., Stanford, Calif., 1968.

Carter, N. L., and H. C. Heard, Temperature and rate dependent deformation of halite, *Am. J. Sci., 269,* 193–249, 1970.

Christy, R. W., Creep of silver bromide at high temperature, *Acta Metall., 2,* 284–295, 1954.

Clark, S. P., Jr., Effect of pressure on the melting points of eight alkali halides, *J. Chem. Phys., 31,* 1526–1531, 1959.

Fredericks, W. J., Diffusion in alkali halides, in *Diffusion in Solids, Recent Developments,* edited by A. S. Nowick and J. J. Burton, Academic, New York, 1975.

Griggs, D. T., and J. Handin, Observations on fracture and a hypothesis of earthquakes, in *Rock Deformation, Mem. 79,* edited by D. T. Griggs and J. Handin, pp. 347–364, Geological Society of America, Boulder, Colo., 1960.

Harvey, P. J., and I. M. Hoodless, Self-diffusion and ionic conductivity in single crystals of caesium chloride, *Philos. Mag., 16,* 543–551, 1967.

Heard, H. C., The influence of environment on the brittle failure of rocks, in *Failure and Breakage of Rock, Proceedings of the 8th Symposium of Rock Mechanics,* pp. 82–93, American Society of Mechanical Engineers, New York, 1967.

Heard, H. C., Steady state flow in polycrystalline halite at pressure of 2 kilobars, in *Flow and Fracture of Rocks,*

Geophys. Mono. Ser., vol. 16, edited by H. C. Heard, I. Y. Borg, N. L. Carter, and C. B. Raleigh, pp. 191-209, AGU, Washington, D. C., 1972.

Heard, H. C., and C. B. Raleigh, Steady state flow in marble at 500° to 800°C, *Geol. Soc. Am. Bull., 83*, 935-956, 1972.

Henderson, B., *Defects in Crystalline Solids*, Crane, Russak, New York, 1972.

Hoodless, I. M., and R. G. Turner, Self diffusion in single crystal CsCl, *J. Phys. Chem. Solids, 33*, 1915-1919, 1972a.

Hoodless, I. M., and R. G. Turner, Ionic conductivity of single crystals of CsCl, *Phys Status Solidi A, 11*, 689-694, 1972b.

Keyes, R. W., Continuum models of the effect of pressure on activated processes, in *Solids Under Pressure*, edited by W. Paul and D. M. Warschauer, pp. 71-99, McGraw-Hill, New York, 1963.

Kohlstedt, D. L., H. P. K. Nichols, and P. Hornack, The effect of pressure on the rate of dislocation recovery in olivine, *J. Geophys. Res., 85*, 3122-3130, 1980.

Ross, J. V., H. G. Ave'Lallemant, and N. L. Carter, Activation volume for creep in the upper mantle, *Science, 203*, 261-263, 1979.

Sammis, C. G., J. G. Smith, G. Schubert, and D. A. Yuen, Viscosity-depth profile of the earth's mantle: Effects of polymorphic phase transitions, *J. Geophys. Res., 82*, 3747-3761, 1977.

Sammis, C. S., J. G. Smith, and G. Schubert, A critical assessment of estimation methods for activation volume, submitted to *J. Geophys. Res.*, 1980.

Sherby, O. D., J. L. Robbins, and A. Goldberg, Calculation of activation volumes for self-diffusion and creep at high temperature, *J. Appl. Phys., 41*, 3961-3968, 1970.

Vaidya, S. N., and G. C. Kennedy, Effect of pressure on the melting of cesium chloride, *J. Phys. Chem. Solids, 32*, 2301-2304, 1971.

Weertman, J., The creep strength of the earth's mantle, *Rev. Geophys. Space Phys., 8*, 145-168, 1970.

Weertman, J., High-temperature creep produced by dislocation motion, in *Dorn Memorial Symposium*, edited by J. C. M. Li and A. K. Mukherjee, Plenum, New York, 1974.

Weertman, J., and J. R. Weertman, Mechanical properties, strongly temperature dependent, in *Physical Metallurgy*, 2nd ed., edited by R. W. Cahn, pp. 983-1010, North-Holland, Amsterdam, 1970.

Weertman, J., and J. R. Weertman, High temperature creep of rock and mantle viscosity, *Annu. Rev. Earth Planet. Sci., 3*, 293-314, 1975.

Yavari, P., and T. G. Langdon, The influence of grain boundaries on creep of KBr, in *Proceedings of International Conference on Surfaces and Interfaces in Ceramics and Ceramic-Metal Systems, Berkeley, California, August 1980*, edited by A. G. Evans and J. A. Pask, Plenum, New York, in press, 1981.

Yoon, D. N., and D. Lazarus, Pressure dependence of ionic conductivity in KCl, NaCl, KBr, and NaBr, *Phys. Rev. B, 5*, 4935-4945, 1972.

Stiff Testing Machines, Stick Slip Sliding, and the Stability of Rock Deformation

NEVILLE G. W. COOK

Department of Materials Science and Mineral Engineering, University of California, Berkeley, California 94720

For more than a century, geologists and geophysicists have been interested in the transition from brittle to ductile deformation of rock. In most laboratory experiments on brittle rock the system comprising the testing machine and rock specimen becomes unstable at or near the peak of the stress-strain curve, resulting in violent failure, because of the rapid release of the relatively large amount of strain energy stored in the machine. This has obscured the study of the 'work-softening,' or brittle deformation of rocks, in which increasing deformation leads to decreasing resistance to deformation, so that any local concentration of deformation tends to become the focus of further deformation leading to the formation of a fracture. Only recently have stiff, or servo-controlled, testing machines, which obviate this instability, become available to enable the progressive loss of strength of brittle materials with increasing deformation to be studied. Brittle materials exhibit also dilatancy and a tendency to form extension fractures in triaxial compression which have a direction parallel to that of the maximum compressive stress. The processes of brittle deformation, dilatation, and fracture are slow in relation to theoretical velocities of dynamic crack propagation. Work-softening deformation, in which resistance is a decreasing function of displacement, appears to be related closely to the phenomenon of 'stick slip' observed in experiments on frictional sliding. The conditions of confining stresses and temperatures at which the transition from brittle to ductile deformation occurs in rock appear to be related to those separating stable sliding from stick slip sliding. Ductile deformation, which takes place without loss of resistance, or 'work-hardening' deformation, is inherently stable and aseismic, whereas brittle deformation may be conditionally stable or intrinsically unstable and seismic.

The hardest stones alone give way to crushing at once, without warning. All others begin to crack or split under a load less than that which finally crushes them....

Rankine [1894]

Introduction

Laboratory studies of the mechanical behavior of rock have been concerned to a great extent with elucidating observations of deformations in crustal rocks. Most such studies have involved the axial compression of right circular cylinders of rock under various conditions of radial confining stresses, temperatures, and pore fluid pressures.

It has been known qualitatively for over a century that if the radial expansion in a compression test is resisted, the rock can sustain a greater axial stress than it can in the absence of such resistance, and that the deformation of many rocks becomes ductile at sufficiently high radial confining stresses and at elevated temperatures [*Becker,* 1893]. Early experiments [*Adams,* 1912] made use of closely fitted metal cylinders to obtain the radial resistance or confining stress. The classic triaxial compression tests, in which radial confining stresses are applied to the cylindrical surfaces of the test specimen through an impermeable jacket, were done first by *von Karman* [1911] and *Boker* [1915]. The compressive strength of rocks, that is, the maximum ordinate of the stress axis as a function of compressive strain parallel to this stress, increases with increasing confining stresses, as is illustrated in Figure 1. When the ability of rock to resist load decreases with increasing deformation, as in the first three curves of this

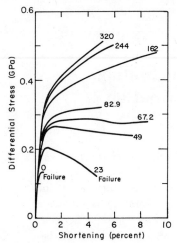

Fig. 1. Axial stress-strain curves for Carrera marble in triaxial compression tests at room temperature [after *von Karman*, 1911] showing the effect of confining stress on the transition from brittle to ductile deformation. The numbers adjacent to each curve indicate the confining stress in megapascals.

figure for zero and low confining stresses, it is considered to be brittle [*Jaeger and Cook*, 1979]. Curves for greater confining stresses in this figure exhibit ductile deformation, that is, the rock undergoes permanent deformation without loss of ability to resist load [*Jaeger and Cook*, 1979]. Ductile is defined in a phenomenological sense as deformation which results in total strains greater than a few percent but may include elastic and rotational deformations as well as dilatancy within the rock. The conditions of confining stresses, temperature, and pore fluid pressures at which rocks undergo a transition from brittle to ductile deformation have been a topic of continued research [*Handin and Hager*, 1957, 1958; *Robinson*, 1959; *Griggs et al.*, 1960; *Heard*, 1960]. In general, both the strength and ductility of rocks increase with increasing confining stress, whereas the strength decreases and the ductility increases with increasing temperature, as is illustrated in Figure 2 [*Heard*, 1976].

Ductile deformation is essentially stable, whereas brittle deformation is potentially unstable [*Whitney*, 1943]. Recently, with the advent of stiff, or servo-controlled, testing machines [*Barnard*, 1964; *Cook and Hojem*, 1966; *Wawersick and Fairhurst*, 1970; *Hudson et al.*, 1971; *Hojem et al.*, 1975], brittle deformation has been studied.

Stick slip is a frictional phenomenon resulting from a decrease in the coefficient of sliding friction with displacement [*Bowden and Leben*, 1939; *Jaeger and Cook*, 1979]. In this sense it is related to 'work-softening,' or brittle deformation. The stability or instability of brittle deformation and frictional sliding is of great interest in connection with aseismic and seismic deformation of the crust [*Reid*, 1910; *Griggs and Handin*, 1960; *Orowan*, 1960; *Brace and Byerlee*, 1966; *Dieterich*, 1974].

The simple theory of stick slip based on the concept of constant coefficients of static and dynamic friction [*Blok*, 1940; *Morgan et al.*, 1941; *Jaeger and Cook*, 1971, 1979] accords with many of the observations of this phenomenon made in laboratory experiments of sliding friction between rock surfaces. However, sliding does not always occur by stick slip; under some conditions it is stable.

The difference between brittle and ductile behavior of rock is of fundamental importance. Ductile behavior both in laboratory tests and in geological deformation is intrinsically stable. Brittle behavior, both in laboratory tests and in geological deformation, is conditionally stable. Brittle behavior can give rise to violent fracture and stick slip sliding. The theory of brittle fracture [*Griffith*, 1921; *Berry*, 1960a, b; *Jaeger and Cook*, 1979] can be related to violent fracture and stick slip sliding.

Stiff and Servo-Controlled Testing Machines

The first definitive explanation of the effect of the stiffness of a testing machine on the failure test specimens appears to have been given by *Whitney* [1943]. In discussing some stress-strain curves for concrete in a paper by *Jensen* [1943], he stated that

the vertical descent of the stress-strain curves . . . is without basis . . . High strength cylinders fail suddenly but as a matter of fact during the failure the elastic movement of the testing machine head imposes a large additional strain on the concrete. This strain may be considerably greater than the total strain up to the time that failure starts. At that time a large amount of elastic energy stored in the cylinder and in the machine (about three times as much in the latter as in the former) and the release of this energy causes (sic) the breakdown of the cylinder.

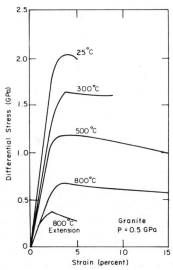

Fig. 2. Axial stress-strain curves for granite in triaxial compression tests at 0.5-GPa confining stress [after *Griggs et al.*, 1960] showing the effect of temperature on the transition from brittle to ductile deformation and on the strength.

Today, there are essentially two approaches to testing the strengths of materials, including rock and concrete. The first, traditional approach is to apply a progressively increasing load, or stress, to a specimen and to observe the resulting deformation, or strain. This approach is epitomized by the 'deadweight' testing machine (Figure 3a). The second approach is to apply a progressively increasing deformation, or strain, to a specimen and to observe the load, or stress, generated by the reaction of the specimen to this deformation. This approach involves the use of what may be called a 'deadbeat' testing machine, that is, a testing machine in which ideally, the displacement is independent of the load (Figure 3b).

Deadweight testing machines can be used to measure the properties only of elastic and ductile 'work-hardening' materials, that is, those materials in which the slopes of the load deformation of stress-strain curves are always positive. This is illustrated in Figure 3a, which shows how every increase in load produces an increase in deformation

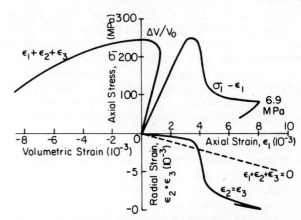

Fig. 4. Axial stress-strain, axial stress-volumetric strain, and axial strain-radial strain curves obtained in a laboratory triaxial test, on a specimen of argillaceous quartzite [after *Cook*, 1979].

until the increased resistance of the specimen to the applied load balances this load. Machines such as this cannot be used to measure the work-softening properties of brittle materials, that is, those in which the slope of the load deformation or stress-strain curves becomes negative. Test systems involving specimens of such materials in a deadweight machine become unstable beyond the peak of the curve, where the characteristic of the material changes from work-hardening to work-softening.

All testing machines have a finite stiffness, so that their characteristics fall between those of ideal deadweight and deadbeat testing machines. The ratio between displacement and load in a testing machine defines its stiffness k. The elastic energy U stored in a testing machine of stiffness k and recoverable from it as a load S decreases to zero is

$$U = S^2/2k \qquad (1)$$

Although the nature of work-softening deformation of brittle materials had been recognized [*Griggs and Handin*, 1960], it could not be studied until sufficiently stiff, or servo-controlled, testing machines became available. In general, any system is stable if additional work is required to effect a virtual displacement of that system, whereas if any virtual displacement of the system results in the release of energy, that system is potentially unstable. A machine-specimen system becomes unstable beyond the point on the load deformation, or stress-strain, curve where the line representing the stiffness of the machine becomes tangential to the slope of this curve (Figure 3c). However, with a machine so stiff that the slope of the load deformation, or stress-strain, curve is always flatter than that of the line representing the stiffness of the machine (Figure 3d), the work-softening deformation of brittle materials is stable and can be studied.

Most published results suggest that brittle rocks in a triaxial compression test behave in a manner described by the stress-strain curves illustrated in Figure 4 [*Brace*,

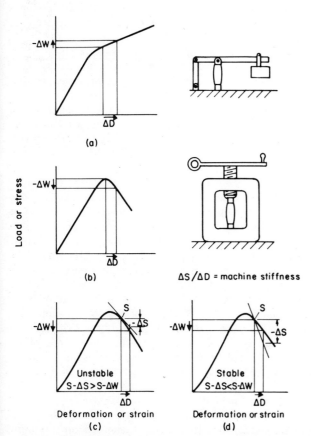

Fig. 3. Load displacement or stress-strain curves for (a) an elastic, work-hardening material in a deadweight testing machine, (b) an elastic, brittle work-softening material in a deadbeat testing machine, (c) unstable deformation of a work-softening material in a 'soft' testing machine, and (d) stable deformation of a work-softening material in a 'stiff' testing machine.

1971; *Bieniawski*, 1971; *Jaeger and Cook*, 1979]. At least three features of importance to the stability of rock deformation have emerged from studies of brittle behavior in stiff or servo-controlled testing machines.

First, at a value of the maximum compressive stress of between a half and two thirds of their strength, brittle rocks become dilatant, that is, their volume increases with increasing compressive deformation in relation to that of an elastic material [*Brace et al.*, 1966; *Schock et al.*, 1973]. At and beyond the peak of the maximum compressive stress-strain curve, brittle rocks become so dilatant that their volume actually increases in comparison with their original volume prior to being subjected to stress. Such dilatancy must be associated with the opening of cracks and voids in brittle rocks subjected to deviatorial compressive stresses. These cracks and voids will have an important effect on transient pore fluid pressures and hence the state of 'effective triaxial stresses' [*Duba et al.*, 1974]. Such effects have been observed in laboratory tests by

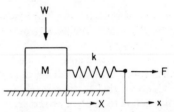

Fig. 6. A simple lumped parameter mechanical model for stick slip oscillations, comprising a mass M, pressed against a surface by a normal load W, and driven parallel to this surface by a force F applied through a spring of stiffness k.

Robinson [1959] and *Heard* [1960] but were not explained at that time. Transient reductions in pore pressure resulting from dilatation increase the effective confining stress, making the rock less brittle, as is illustrated in Figure 5 [*Frank*, 1965].

Second, for rock in a brittle, or work-softening, condition, any increase in deformation results in a decrease in the resistance of the rock to further deformation. Accordingly, any local concentration of deformation becomes work-softened, and further deformation is focused into this weakened region, which thereby extends to form a fracture. Conversely, for rock in a work-hardening condition, any increase in deformation results in an increase in the resistance of the rock to further deformation, so that deformation tends to spread throughout the rock.

Third, according to most theories the propagation of brittle fractures is a rapid phenomenon with terminal fracture velocities approaching the speed of propagation of Rayleigh waves [*Berry*, 1960a, b]. Although it is true that cracks can be propagated in rocks at these speeds, *Wawersick and Fairhurst* [1970] were able to control the brittle deformation of rock specimens in an inadequately stiff testing machine by manually decreasing the load as brittle deformation leading to fracture progressed. This shows that the processes of brittle deformation, dilatation, and fracture occur over periods of time of the order of a second (as evidenced by manual reaction times) in typical laboratory specimens, with dimensions of the order of 100 mm, rather than in periods of tens of microseconds, as suggested by theory. Were it not for this phenomenon, which is far from understood, the study of brittle deformation using servo-controlled testing machines of inadequate fundamental stiffness could not be done, because the hydraulic and mechanical reaction time of such machines is significantly greater than the times predicted theoretically for brittle fracture propagation.

Finally, the primary mode of brittle fracture in triaxial compression appears to be by extension fractures tending parallel to the direction of the maximum (compressive) principal stress. The mechanism of fracture is not clear but probably is related to the phenomenon of dilatation [*Fairhurst and Cook*, 1966; *Cook*, 1979]. However, such fractures do not allow for kinematic failure of the machine-specimen system, and the study of them has been ob-

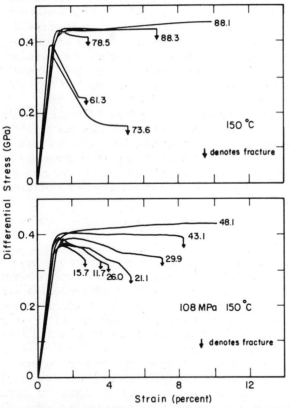

Fig. 5. Stress-strain curves for specimens of Solenhofen limestone tested in triaxial compression without (above) and with (below) interstitial pore water pressure [after *Heard*, 1960], showing increases in ductility in the latter resulting from transient changes in the 'effective confining stress.' The nominal, static effective confining stresses, that is, the differences between the confining stresses applied externally and the interstitial pore water pressures (if any), in megapascals, are indicated by the numbers adjacent to each curve.

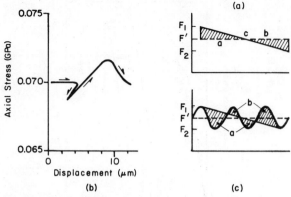

Fig. 7. (a) A graphical representation of one stick slip oscillation for the model illustrated in Figure 6, assuming that sliding is controlled by constant coefficients of static and dynamic friction. (b) An example of 'stable stick slip,' where the resistance to sliding is a monotonic function of displacement [after *Jaeger and Cook*, 1971]. The 'stick' was initiated by a small reversal of loading. (c) An illustration of stick slip where the resistance to sliding is an oscillatory function of displacement, showing the equivalence between this case and that in Figure 7a, where the coefficients of friction are constants.

scured by subsequent fractures which do allow for such failure occurring during the violent termination of tests of brittle rock in conventional testing machines, except where shear fractures have been observed in triaxial extension tests [*Brace*, 1964].

Stick Slip

The term stick slip has been applied to a wide variety of relaxation oscillations associated with sliding friction using materials as different as metals [*Bowden and Leben*, 1939] and rocks [*Brace and Byerlee*, 1966] and in many different experimental configurations from simple riders [*Bowden and Tabor*, 1950], through triaxial tests [*Jaeger*, 1959; *Handin and Stearns*, 1964] to *Bridgman*'s [1936, 1937] high-pressure anvils rotating relative to one another.

The simplest model of stick slip oscillations is a lumped parameter model comprising a mass M, pressed against a surface by a normal load W, and driven parallel to this surface by a force F, applied through a spring of stiffness k (Figure 6) [*Jaeger and Cook*, 1971]. Initially, suppose that the resistance to sliding movement between the mass and the surface can be characterized by a static coefficient of friction μ and a dynamic coefficient of friction μ' such that $\mu' < \mu$ and both have constant values independent of displacement. Let the force F be applied at a constant velocity V.

At the point of slipping, the spring must have extended a distance x_1 such that

$$kx_1 = F_1 = \mu W \qquad (2)$$

Slipping ceases at some extension of the spring, x_2, such

that

$$kx_2 = F_2 \qquad (3)$$

The reduction in elastic energy stored in the spring as a result of this movement is $k(x_1^2 - x_2^2)/2$. If V is very small in relation to the velocity during slipping, this reduction in the stored energy is virtually equal to the work done against friction during slipping, that is,

$$k(x_1^2 - x_2^2)/2 = \mu' W(x_1 - x_2) \qquad (4)$$

so that

$$k(x_1 + x_2) = 2\mu' W \qquad (5)$$

Therefore

$$F_2 = kx_2 = (2\mu' - \mu) W \qquad (6)$$

$$F_1 - F_2 = K(x_1 - x_2) = 2(\mu - \mu') W \qquad (7)$$

Equation (7) suggests that the difference between the maximum and minimum shear forces in a stick slip oscillation is twice as great as the difference between the force required to start slipping, $F_1 = \mu W$, and that during slipping, $F' = \mu' W$; this difference is independent of the stiffness k of the system [*Byerlee and Brace*, 1968]. Stick slip oscillations repeat themselves at intervals of displacement, $X = x_1 - x_2$ with periods of $T = X/V$.

A graphical representation of this solution is given in Figure 7a. Note that the area a, representing the kinetic energy acquired by the mass, must equal the area b, representing the loss of this kinetic energy against friction as the mass overshoots the force equilibrium position given by the point c.

A more refined analysis of stick slip oscillations, taking into account the equations of motion of the mass, has been given by several authors [*Blok*, 1940; *Jaeger and Cook*, 1979] and has been applied to seismic faulting by *Nur* [1974]. It is interesting to note that to damp such oscillations, a viscous dashpot in parallel with the spring with a damping coefficient $k_d = 2(Mk)^{1/2}$ would be needed.

A model as simple as that described above is not likely to be a good representation of frictional sliding between

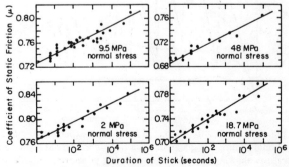

Fig. 8. The coefficient of static friction as a function of time between sandstone surfaces separated by gouge [after *Dieterich*, 1972].

rock surfaces. The weakest assumption in this respect is that the resistance of these surfaces to sliding can be represented by two constants, μ and μ'. In reality, the resistance to sliding is likely to be a function of both displacement and time.

Considering first the dependence on displacement, two conditions may exist as represented in Figures 7b and 7c. In Figure 7b the transition from static to sliding resistance is illustrated as a monotonic function of displacement. Stick slip oscillations will occur in this case if the stiffness of the spring, k, has a slope flatter than the curve representing the change in sliding resistance as a function of displacement. If the slope of k is steeper than this curve, then stable sliding can be made to occur as is illustrated in Figure 7b. In the case where stick slip occurs, an analysis similar to that described above can be used to determine the period and displacement of the stick slip oscillations. Another possibility suggested by *Brace and Byerlee* [1968] is that sliding resistance is an oscillatory function of displacement, as is illustrated in Figure 7c, and that the result of this is essentially the same as that given by the simple theory based on displacement invariant values for the coefficients of static and dynamic friction. This can be seen by comparing Figures 7a and 7c and noting that the mass comes to rest in both cases when all the kinetic energy of the mass, represented by area a, has been absorbed by frictional resistance, represented by the area b, as the mass overshoots its force equilibrium position. In both the simple and oscillatory theories the change in force between the initiation and cessation of sliding is independent of the stiffness of the applied force, provided that the slope representing this stiffness is much less than that of the steep portions of the oscillatory variations in sliding resistance.

Dieterich [1972] has shown that for surfaces separated

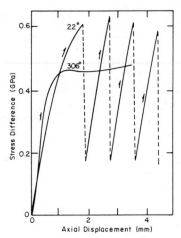

Fig. 10. The results of triaxial laboratory experiments on frictional sliding across a clean, dry saw cut in granite, showing stick slip deformation at room temperature and stable deformation at a temperature of 306°C [after *Brace*, 1972].

by gouge, the coefficient of static friction is an increasing function of the period of time for which the surfaces are in stationary contact, as illustrated in Figure 8. The period for which surfaces are in stationary contact during stick slip oscillations is given by $T = X/V$, or

$$T = 2(\mu - \mu')\,W/kV \qquad (8)$$

so that this phenomenon may be expected to exacerbate stick slip oscillations once they start.

Typical important results concerning the frictional behavior of rocks, obtained from laboratory triaxial compression tests, are illustrated in Figures 9 and 10 [*Summers and Byerlee*, 1977; *Brace*, 1972]. At low values of the confining stress, stable sliding occurs. Above some critical value of the confining stress, a short period of stable sliding is followed by violent stick slip sliding. As is illustrated in Figure 10, an increase in temperature promotes stable sliding. A particularly striking feature of the stick slip sliding is the very sudden drop in stress during the slip, indicating that little energy is dissipated in this process, apparently less than that stored during the preceding increase in load while the surfaces are stuck. This phenomenon is reminiscent of brittle fracture, and Griffith's theory of brittle fracture may be extended to a situation similar to this. Briefly, closed Griffith cracks, subjected to normal stress and a shear stress sufficient to cause sliding, affect both the modulus of deformation and the strength of a material [*Cook*, 1965; *Jaeger and Cook*, 1979], which can be represented by a Griffith locus [*Berry*, 1960a, b] in the stress-strain plane (Figure 11). This has a vertical tangent corresponding to a critical crack length per unit volume, less than which the elastic strain energy in the volume influenced by the crack is sufficient to cause intrinsic instability by crack extension once the stress equals the strength. The expression for the critical length

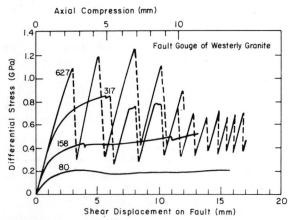

Fig. 9. The results of triaxial laboratory experiments on dry frictional sliding of a gouge-filled saw cut in granite at room temperature, showing the effects of confining pressure on stable and stick slip deformation [after *Summers and Byerlee*, 1977]. The numbers adjacent to each curve indicate the confining stress in megapascals.

of a closed crack is

$$C_c = \left\{ \frac{4\,(\mu^2 + 1)^{1/2}}{6\pi L\,[\,(\mu^2 + 1)^{1/2} - \mu\,]} \right\}^{1/2} \qquad (9)$$

where C_c is the critical length of crack, L is the crack length per unit volume, and μ is the coefficient of Coulomb friction.

Adhesive contact between adjacent rock surfaces is only partial, and it seems reasonable to assume that the degree of contact would increase with increased confining pressure. If the portions of nonadhesive contact are thought of as closed Griffith cracks, L will be large at low confining stresses. As the confining stress increases, so will L decrease, causing also an increase in the critical crack length. It can be argued that at low confining stresses the actual cracks are large and critical crack lengths needed for intrinsic instability are small, whereas at high confining stresses, actual crack lengths are reduced and critical crack lengths increase in comparison with those at low confining stresses. From this it follows that stable sliding commences from large, weak cracks at low confining pressures and that intrinsically unstable, stick slip sliding commences from strong cracks of subcritical length at high confining pressures, as illustrated by paths I and II in Figure 11, respectively. It is interesting to note how closely the Griffith locus at high confining stress approximates the concept of a static coefficient of friction followed by a drop, virtually without displacement, to a near-constant dynamic coefficient of friction.

The use of brittle fracture theory to explain the effects of temperature, which has been observed to reduce strength so markedly, is less obvious. Perhaps the effects of temperature on sliding are to preclude high stress concentrations giving rise to instability by promoting plastic deformation in regions of stress concentration.

In order to apply (9) to a single, intrinsically unstable earthquake fault of length C_c in a volume of crustal rock with a representative side length l, $C_c \leq L/l^3$, so that

$$\left(\frac{C_c^2 L}{l^3} \right)^{1/3} = C_c' \leq \left\{ \frac{4\,(\mu^2 + 1)^{1/2}}{6\pi\,[\,(\mu^2 + 1)^{1/2} - \mu\,]} \right\}^{1/3} \qquad (10)$$

where C_c' is a dimensionless fault length and the other symbols are as defined above. The values of C_c' range from 0.6 to 0.9 as μ ranges from zero to unity. If the volume of crustal rock influenced by a crack in the form of a fault is of the order of a right square prism extending downward the same depth as the fault, any fault with a length in the range given above becomes intrinsically unstable at a sufficiently great value of the difference between the shear stress tending to cause sliding and the friction tending to resist it.

Jaeger and Cook [1979] also have derived an expression relating the difference between the shear and normal stresses on such a fault to the Griffith strength

$$(|\tau| - \mu\sigma_n) = (2\alpha E/\pi C)^{1/2} = T_g \qquad (11)$$

where

$|\tau|$	modulus of the shear stress acting across the fault;
μ	Coulomb friction on the fault;
σ_n	normal stress across the fault;
α	apparent surface energy of fracture;
E	Young's modulus of the rock;
C	length of the fault;
T_g	Griffith tensile strength.

In laboratory experiments on sliding friction the value of C is of the order of millimeters and that of $(|\tau| - \mu\sigma_n)$ is of the order of 10 MPa. In an earthquake fault, the value of C is of the order of kilometers, so that the value of $(|\tau| - \mu\sigma_n)$ should, according to (11), be of the order of 10 kPa. Similarly, in laboratory experiments the stress drop during stick slip resulting from intrinsic instability is of the order of 500 MPa, and the stress drop should scale as $C^{1/2}$ according to (11). Therefore the stress drop on an earthquake fault with a length of the order of kilometers should be of the order of 0.5 MPa [*Jamison and Cook*, 1980].

Finally, the requirements for conditionally stable defor-

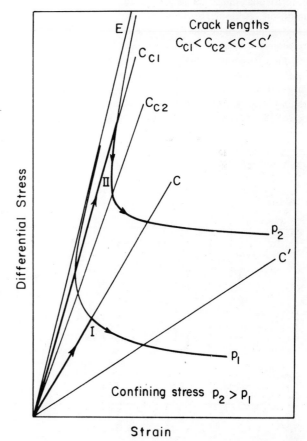

Fig. 11.   A sketch showing Griffith loci for low (p_1) and high (p_2) confining stresses, critical crack lengths C_{c1} and C_{c2} for these conditions, and elastic loading followed by a conditionally stable sliding path I and an intrinsically unstable sliding path II.

mation of an earthquake fault should be examined. It can be shown [*Aki*, 1978; *Jaeger and Cook*, 1979] that the relationship between the change in shear stress on a fault and the average shear displacement across the fault is given by

$$\tau/G = u/C \qquad (12)$$

where

τ change in shear stress;
G modulus of rigidity of the crust;
u average shear displacement across the fault;
C minimum dimension of the fault in its plane.

The average stiffness with which the crust loads the fault, that is, the ratio of the change in shear stress to the displacement across the fault accompanying it, is

$$\tau/u = G/C \qquad (13)$$

where the symbols are as defined above. From (13) it is clear that the length of faults which could slide in a conditionally stable manner is of the order of a meter if this stiffness is to be steeper than the slope of the shear stress-shear displacement curve measured in laboratory experiments on friction. However, the ratio of shear stress to shear displacement has the units of specific force, which are of no obvious physical relevance to this problem. Accordingly, it seems more appropriate to convert the shear displacement to a shear strain across the thickness of the gouge by dividing the displacement by the thickness. This results in a ratio between shear stress and shear strain with units of stress or specific energy, which is used extensively in comminution [*von Rittinger*, 1867; *Rosin and*

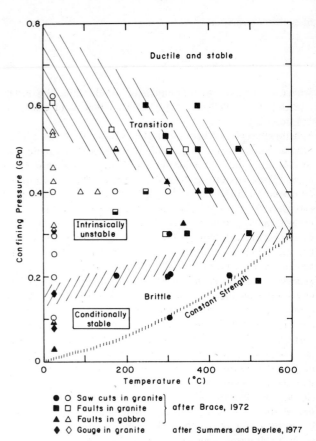

Legend:
● ○ Saw cuts in granite
■ □ Faults in granite   } after Brace, 1972
▲ △ Faults in gabbro
◆ ◇ Gouge in granite   after Summers and Byerlee, 1977

Fig. 13.   A sketch delineating the transition from brittle to ductile deformation as functions of confining stresses and temperatures (mainly after *Griggs et al.* [1960]), together with lines indicating the transition from conditionally unstable to intrinsically unstable sliding and constant strength. Superimposed on this are the results of triaxial laboratory experiments on frictional sliding in granite from *Brace* [1972] and *Summers and Byerlee* [1977]. Open points indicate stick slip, solid points stable sliding, and half-solid points a period of stable sliding followed by stick slip.

Rammler, 1933], and increases the length of a fault which can slide in a conditionally stable manner in proportion to the thickness of the gouge.

Conclusion

A laboratory test apparatus or a geologic fault system is mechanically stable or unstable depending upon whether or not any virtual displacement within the system requires the addition of work from outside the system or results in the release of some of the energy stored in the system, respectively.

The concepts of brittle and ductile deformation of rock are linked closely to stable and unstable deformation as defined above. *Griggs and Handin* [1960] represented the spectrum from brittle to ductile deformation as in Figure 12. Their first and second cases correspond to conditions of intrinsic instability, as suggested by the Griffith locus, in

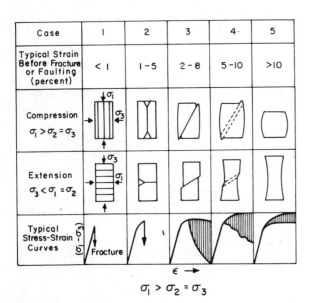

Fig. 12.   The spectrum of rock deformation and fracture from quite brittle and intrinsically unstable, to fully ductile and stable [after *Griggs and Handin*, 1960] showing how work-softening deformation results in shear fractures.

which the elastic strain energy stored in the rock specimen is sufficient to cause brittle fracture without the outside addition of work. Their third and fourth cases correspond to conditionally unstable work-softening, brittle deformation. Whether such deformation is unstable or stable depends upon the stiffness with which load or stress is applied. With a sufficiently stiff system, stable brittle deformation commences at some weakness, and continued deformation is focused into a 'work-softened' plane until a brittle fracture forms. Their fifth case corresponds to work-hardening, ductile deformation. This is intrinsically stable, and deformation spreads throughout the specimen.

Similar considerations appear to apply to frictional sliding. At high confining stresses and low to moderate temperatures, sliding is intrinsically unstable and occurs by violent stick slip, irrespective of the stiffness of the loading system. Such sliding can be made intrinsically stable by an increase in temperature, and possibly confining pressure, sufficient to make the rock ductile and work-hardening. At low confining stresses, sliding is conditionally stable depending upon the stiffness of the loading system.

In Figure 13 an attempt has been made to portray the properties of intact laboratory specimens of a typical hard rock, such as granite, in terms of temperature and confining stress and ductility as defined in the introduction. Using a more rigorous definition of ductility, *Schock et al.* [1973] and *Schock and Heard* [1974] have suggested that granite is brittle to stresses with values as high as 0.2 GPa. On this same figure the results of laboratory sliding experiments on similar rock [*Brace*, 1972; *Summers and Byerlee*, 1977] have been plotted. It appears that there may be a relationship between stick slip and intrinsically brittle deformation.

References

Adams, F. D., An experimental contribution to the question of the depth of the zone of flow in the Earth's crust, *J. Geol., 20,* 97-118, 1912.

Aki, K., A quantitative model of stress in a seismic region as a basis for earthquake prediction, in Proceedings of Conference III: Fault Mechanics and its Relation to Earthquake Prediction, *Geol. Surv. Open File Rep. U.S., 78-380,* pp. 7-13, 1978.

Barnard, P. R., Research into the complete stress-strain curve for concrete, *Magn. Concr. Res., 16,* 203-210, 1964.

Becker, G. F., Finite homogeneous strain, flow and rupture of rocks, *Geol. Soc. Am. Bull., 4,* 13-90, 1893.

Berry, J. P., Some kinetic considerations of the Griffith criterion for fracture, I, Equation of motion at constant force, *J. Mech. Phys. Solids, 8,* 194-206, 1960a.

Berry, J. P., Some kinetic considerations of the Griffith criterion for fracture, II, Equations of motion at constant deformation, *J. Mech. Phys. Solids, 8,* 207-216, 1960b.

Bieniawski, Z. T., Deformational behavior of fractured rock under multi-axial compression, in *Structure, Solid Mechanics and Engineering Design,* edited by M. Te'Eni, pp. 589-598, Interscience, New York, 1971.

Blok, H., Fundamental aspects of boundary lubrication, *J. Soc. Automot. Eng., 46,* 54-68, 1940.

Boker, R., Die Mechanik der bleibenden Formanderung in kristallinische aufgebauten Korpern, *Ver. Dtsch. Ing. Mitt. Forsch., 175,* 1-51, 1915.

Bowden, F. P., and L. Leben, The nature of sliding and the analysis of friction, *Proc. R. Soc., Ser. A, 169,* 371-391, 1939.

Bowden, F. P., and D. Tabor, *The Friction and Lubrication of Solids,* vol. 1, Oxford at the Clarendon Press, London, 1950.

Brace, W. F., Brittle fracture of rocks, in *State of Stress in the Earth's Crust,* edited by W. R. Judd, pp. 111-174, Elsevier, New York, 1964.

Brace, W. F., Micromechanics in rock systems, Proceedings of the Civil Engineering Materials Conference, Southampton, in *Structure, Solid Mechanics and Engineering Design,* part 1, edited by M. Te'Eni, pp. 187-204, Interscience, New York, 1971.

Brace, W. F., Laboratory studies of stick-slip and their application to earthquakes, *Tectonophysics, 14*(3/4), 189, 1972.

Brace, W. F., and J. D. Byerlee, Stick-slip as a mechanism for earthquakes, *Science, 153,* 990-992, 1966.

Brace, W. F., B. W. Paulding, Jr., and C. Scholz, Dilatancy in the fracture of crystalline rocks, *J. Geophys. Res., 71,* 3939, 1966.

Bridgman, P. W., Shearing phenomena at high pressure of possible importance to geology, *J. Geol., 44,* 653-669, 1936.

Bridgman, P. W., Shearing phenomena at high pressures, particularly in inorganic compounds, *Proc. Am. Acad. Arts Sci., 71,* 387-460, 1937.

Byerlee, J. D., and W. F. Brace, Stick-slip stable sliding, and earthquakes—Effect of rock type, pressure, strain rate, and stiffness, *J. Geophys. Res., 73*(18), 6031-6037, 1968.

Cook, N. G. W., The failure of rock, *Int. J. Rock Mech. Mining Sci., 2,* 389-403, 1965.

Cook, N. G. W., Rock fracture: Observations and interpretations, paper presented at the Fourth Tewksbury Symposium on Fracture at Work, Univ. of Melbourne, Melbourne, Australia, Feb. 1979.

Cook, N. G. W., and J. P. M. Hojem, A rigid 50-ton compression and tension testing machine, *S. A. Mech. Eng., 16,* 89-92, 1966.

Dieterich, J. H., Time-dependent friction in rocks, *J. Geophys. Res., 77,* 3690-3694, 1972.

Dieterich, J. H., Earthquake mechanisms and modeling, *Annu. Rev. Earth Planet. Sci., 2,* 275-301, 1974.

Duba, A. G., H. C. Heard, and M. L. Santor, Effect of fluid content on the mechanical properties of Westerley granite, *Rep. UCRL 51626,* Lawrence Livermore Lab., Livermore, Calif., 1974.

Fairhurst, C., and N. G. W. Cook, The phenomenon of rock splitting parallel to a free surface under compressive stress, paper presented at the First Congress of the International Society of Rock Mechanics, Lisbon, 1966.

Frank, F. C., On dilatancy in relation to seismic sources, *Rev. Geophys. Space Phys.*, *3*, 485, 1965.

Griffith, A. A., The phenomena of rupture and flow in solids, *Phil. Trans. R. Soc. London, Ser. A*, *221*, 163-198, 1921.

Griggs, D. T., and J. Handin, Observations on fracture and an hypothesis of earthquakes, in Rock Deformation, *Geol. Soc. Am. Mem.*, *79*, 347-364, 1960.

Griggs, D. T., F. J. Turner, and H. C. Heard, Deformation of rocks at 500° to 800°C, in Rock Deformation, *Geol. Soc. Am. Mem.*, *79*, 39-104, 1960.

Handin, J., and R. V. Hager, Jr., Experimental deformation of sedimentary rocks under confining pressure: Tests at room temperature on dry samples, *Am. Assoc. Pet. Geol. Bull.*, *41*, 1-50, 1957.

Handin, J., and R. V. Hager, Jr., Experimental deformation of sedimentary rocks under confining pressure: Tests at high temperature, *Am. Assoc. Pet. Geol. Bull.*, *42*, 2892-2934, 1958.

Handin, J., and D. W. Stearns, Sliding friction of rock (abstract), *Eos Trans. AGU*, *45*, 103, 1964.

Heard, H. C., Transition from brittle to ductile flow in Solenhofen limestone as a function of temperature, confining pressure and interstitial fluid pressure, in Rock Deformation, *Geol. Soc. Am. Mem.*, *79*, 193-226, 1960.

Heard, H. C., Comparison of the flow properties of rocks at crustal conditions, *Phil. Trans. R. Soc. London, Ser. A*, *283*, 173-186, 1976.

Hojem, J. P. M., N. G. W. Cook, and C. Heins, A stiff, two meganewton testing machine for measuring the 'work-softening' behaviour of brittle materials, *S. A. Mech. Eng.*, *25*, 250-270, 1975.

Hudson, J. A., E. T. Brown, and C. Fairhurst, Optimising the control of rock failure in controlled laboratory tests, *Rock Mech.*, *3*, 217-224, 1971.

Jaeger, J. C., The frictional properties of joints in rock, *Geofis. Pura Appl.*, *43*, 148-158, 1959.

Jaeger, J. C., and N. G. W. Cook, Friction in granular materials, Proceedings of the Civil Engineering Materials Conference, Southampton, in *Structure, Solid Mechanics and Engineering Design*, part 1, edited by M. Te'Eni, pp. 257-266, Interscience, New York, 1971.

Jaeger, J. C., and N. G. W. Cook, *Fundamentals of Rock Mechanics*, 3rd ed., Chapman and Hall, London, 1979.

Jamison, D. B., and N. G. W. Cook, Note on measured values for the state of stress in the earth's crust, *J. Geophys. Res.*, *85*, 1833-1838, 1980.

Jensen, V. P., The plasticity ratio of concrete and its effect on the ultimate strength of beams, *J. Am. Concr. Inst.*, *39*, 565-582, 1943.

Morgan, F., M. Muskat, and D. W. Reed, Friction phenomena and the stick-slip process, *J. Appl. Phys.*, *12*, 743-752, 1941.

Nur, A., Tectonophysics: The study of relations between deformation and forces in the earth, Proceedings of the Third Congress of the International Society for Rock Mechanics, Denver, in *Advances in Rock Mechanics*, vol. 1, part A, pp. 243-317, National Academy of Sciences, Washington, D. C., 1974.

Orowan, E., The mechanism of seismic faulting, *Geol. Soc. Am. Mem.*, *79*, 232, 1960.

Rankine, W. J. M., *A Manual of Civil Engineering*, 19th ed., Charles Griffin and Co., Ltd., London, 1894.

Reid, H. F., The mechanics of the earthquake, in *The California Earthquake of April 18, 1906, Report of the State Earthquake Investigation Committee*, Carnegie Institution, Washington, D. C., 1910.

Robinson, L. H., Jr., Effect of pore and confining pressure on the failure process in sedimentary rocks, *Colo. Sch. Mines Q.*, *54*, 177-199, 1959.

Rosin, P., and E. Rammler, The laws governing the fineness of powdered coal, *J. Inst. Fuel*, *7*, 1933.

Schock, R. N., and H. C. Heard, Static mechanical properties and shock loading response of granite, *J. Geophys. Res.*, *79*, 1662-1666, 1974.

Schock, R. N., H. C. Heard, and D. R. Stephens, Stress-strain behavior of a granodiorite and two graywackes on compression to 20 kilobars, *J. Geophys. Res.*, *78*, 5922-5941, 1973.

Summers, R., and J. Byerlee, A note on the effect of fault gouge composition on the stability of frictional sliding, *Int. J. Rock Mech. Mining Sci. Geomech. Abstr.*, *14*(3), 155-160, 1977.

von Karman, Th., Festigkeitsversuche unter allseitigem Druck, *Z. Ver. Dtsch. Ing.*, *55*, 1749-1757, 1911.

von Rittinger, P., *Lehrbuch der Aufbereitungskunde*, Ernst and Korn, Berlin, 1867.

Wawersick, W. R., and C. Fairhurst, A study of brittle rock fracture in laboratory compression experiments, *Int. J. Rock Mech. Mining Sci.*, *7*, 561-575, 1970.

Whitney, C. S., Discussion on paper by V. P. Jensen, *J. Am. Concr. Inst.*, *39*, 584-586, 1943.

Constitutive Properties of Faults With Simulated Gouge

U.S. Geological Survey, Menlo Park, California 94025

Direct shear experiments with a layer of simulated fault gouge consisting of crushed and sieved Westerly Granite sandwiched between intact blocks of Westerly Granite show a variety of competing time-, velocity-, and displacement-dependent effects similar to previous friction observations for slip on initially clean surfaces (Dieterich, 1979a). Gouge strength increases by approximately the logarithm of the time that the driving ram is held stationary. A change of sliding velocity results in an immediate but transient change in strength and a displacement-dependent residual change of strength. The transient and residual velocity-dependent effects result from the competition between a direct velocity-dependent process that operates continuously during slip and a process inversely dependent on velocity that becomes fully effective only following a finite displacement d_r at a specific velocity. The residual velocity-dependent effect results in strength changes that are of the same or opposite sign as the velocity change depending on the relative magnitude of the competing processes. The experiments with gouge show variations in the magnitude and sign of the residual velocity effect that appear to be correlated with the rate of comminution of the gouge. The displacement dependency measured by the parameter d_r scales by both gouge particle size and roughness of the surface in contact with the gouge. The parameter d_r is insensitive to thickness of the gouge layer, indicating that deformation in the gouge is highly localized, probably at the gouge-rock interface. A constitutive law of the general type proposed by Dieterich (1979a) adequately represents the range of behavior seen in these experiments. Because experimental results arise from interactions between the fault and the laboratory apparatus, a simple deterministic spring-slider numerical model that uses the constitutive law is employed to simulate the experiments. This single model reproduces in detail the full range of effects seen in these laboratory experiments. In addition, it produces a previously observed dependence of mode of slip (stick slip, stable slip, or oscillatory slip) on machine stiffness, normal stress, and the parameter d_r. Accelerating premonitory slip precedes all unstable slip events.

Introduction

Active faults exhibit a variety of deformation phenomena, the most obvious, of course, being earthquake slip. In addition, faults may undergo decelerating postseismic slip (afterslip), long-term stable slip (fault creep) in the absence of earthquake instability, and perhaps slow preseismic slip. Fault creep may be of the nearly constant velocity type, or it may be characterized by repeated alternating episodes of slow stable slip and rapid creep events. Parallel behavior has been demonstrated in laboratory experiments using simulated faults. Significantly, accentuation or inhibition of different slip characteristics can be brought about by changing the physical conditions of the experiment. An example that illustrates this point

is the demonstration that a change of the stiffness of the laboratory apparatus is sufficient to alter the mode of slip [*Dieterich*, 1979a]. Low stiffness favors unstable fault slip, and high stiffness results in stable slip. Other factors such as normal stress, strain rates, and surface characteristics of the fault also have been shown to significantly affect the mode of slip.

This range of slip behavior, in which the mode of slip can be changed simply by altering the characteristics of the loading apparatus, points to the inescapable conclusion that faults possess fairly complicated constitutive properties and that the observed mode of slip in both laboratory experiments and in nature can be regarded as the result of interactions between the constitutive properties of the fault and the total mechanical system that stresses the fault. In this context it seems reasonable to conclude that understanding of active faulting and the attainment of such practical goals as the prediction of the time and location of earthquake slip depend in large part on understanding both these factors.

Dieterich [1979a] discusses some of the general features of constitutive behavior implied by gross observations of modes of slip and within a general framework presents a specific formulation for a fault constitutive law. This is based on a limited series of experiments for slip on initially clean surfaces of granite. The general constitutive model at the minimum provides a qualitative explanation for the range of behavior observed, and the specific model accounts for the limited experimental data in detail. The principal purpose of the present study is to obtain experimental results for the more appropriate case of a fault with a layer of finely comminuted rock fragments, i.e., gouge, and to explore the extent to which the constitutive model can be applied to faults with simulated gouge.

A number of studies of faults with simulated gouge have been described in the literature [*Engelder et al.*, 1975; *Byerlee and Summers*, 1976; *Summers and Byerlee*, 1977; *Weeks and Byerlee*, 1978; *Byerlee et al.*, 1978; *Logan*, 1979]. In most of these studies it is difficult to extract specific quantitative constitutive parameters, because the measurements were designed to answer other questions. However, there are direct parallels between the modes of slip seen in the different gouge experiments and rock friction experiments without gouge. This leads one to expect analogous, if not quantitatively identical, constitutive properties for gouge layers and for rock friction.

Experimental Procedure

The experimental phase of this study examines the relationships between various physical parameters of simulated gouge layers and the response of the fault as a function of stress, displacement, and time when subjected to different types of load histories. The simulated gouge material used here consists of crushed and sieved Westerly Granite placed between blocks of intact Westerly Granite. Parameters that were varied are roughness of the rock surface in contact with the gouge, gouge thickness, and the size fractions of the gouge particles in the starting material. Surfaces with different roughnesses were prepared in a uniform manner by hand-lapping the surfaces using either 60-90, 240, or 600 mesh silicon carbide abrasive to produce surfaces designated throughout this paper as rough, medium, or fine, respectively. For identically prepared surfaces, *Dieterich* [1979a] gives surface roughnesses that were measured using a profilometer. Initial gouge layer thicknesses of 0.5 mm, 1.0 mm, and 2.0 mm were controlled using shims of appropriate thickness that were removed after the gouge and blocks were in place. Three particle size groups were prepared by sieving the crushed granite: (1) all size fractions of <250 μm (see Figure 1 for histogram), (2) all size fractions of <85 μm, and (3) the fraction consisting of 125-250 μm.

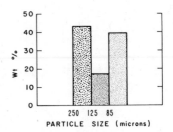

Fig. 1.   Histogram of the gouge particle sizes in the starting material <250 μm.

The sample assembly consists of a three-block, sandwich type direct shear configuration described previously [*Dieterich*, 1972] and shown schematically in Figure 2. Prior to loading, the loose, crushed granite is held in place using cellophane tape. The sample assembly includes a modification suggested by A. Ruina. Namely, the inner block has twice the thickness of the outer blocks instead of the previously used equal thicknesses for all blocks. The purpose of this geometry is to reduce differences in the vertical elastic displacements within the blocks induced by the vertical load. With these dimensions the vertical elastic displacements are approximately the same for the inner and outer blocks. This configuration reduces possible slip gradients that would arise from changes in the vertical load.

With this type of direct shear arrangement the vertical and horizontal hydraulic rams independently control the shear and normal stresses, respectively, on the sliding planes. Because of the large number of possible combina-

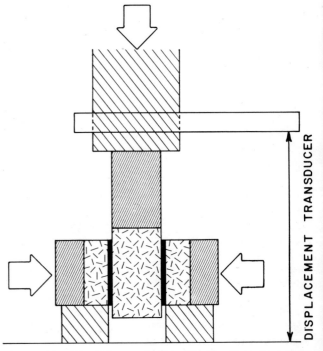

Fig. 2.   Diagram of the sample assembly. The dimensions of the sliding surface are 5.0 cm × 5.0 cm.

tions involving surface roughness, gouge thickness, and particle size to be tested, all experiments were conducted at a single normal stress: 10.0 MPa (100 bars). Normal stress was held constant during each test through the use of a hydraulic accumulator. Motion of the vertical ram was generally servo-controlled on displacement using a high-speed servo control valve. Reference signals for the servo controller were supplied from a computer, permitting pre-programing of the somewhat complicated velocity histories required for many of the experimental runs. Displacement feedback signals were obtained from a displacement transducer that spanned the sample. One end of the transducer was anchored on the loading column between the sample and the hydraulic cylinder, and the other end was mounted on the base that supports the sample assembly (see Figure 2). Resolution of displacements by the servo controller is variable, depending on amplifier gains and total displacement range. In all cases presented here it is at least as good as 0.2 μm and in most cases at least 0.1 μm. Although the servo valve has a response rate that can approach 1.0 KHz, mechanical resonance of the apparatus requires low-pass filtering of the feedback signal to remove frequencies above 30–40 Hz. Below those frequencies the effective stiffness of the vertical loading system expressed as shear stress at the slip surfaces divided by displacement was measured to be 650 MPa/cm.

Three types of tests were systematically conducted for the different combinations of gouge parameters: constant velocity, multiple velocity, and time dependence tests. The constant velocity tests were run at 2.5 μm/s and were done to examine the overall form of the stress/displacement curves and to permit reasonably direct comparisons to be made for the control of strength by the fault parameters. In the multiple-velocity tests the velocity was held constant for a predetermined displacement, then abruptly changed by a factor of 10, held constant for another displacement interval, then changed again, and so on. Slip velocities used for these tests were 0.25, 2.5, and 25.0 μm/s. The purpose of the multiple-velocity tests was to look for variations of strength as a function of velocity as seen in the earlier friction experiment of *Dieterich* [1979a]. Those earlier findings demonstrated two competing velocity effects. For time dependence tests, constant velocity slip is interrupted at specified displacements where the control displacement is held at zero for a specified time interval, followed by resumption of slip at constant velocity. Slip velocities of 2.5 μm/s and hold times of 1000, 100, 10, and 1 s were employed. The purpose of the time dependence test was to quantify the increase of strength with time of nominally stationary contact that was found to be closely related to the velocity dependence and is an essential part of the stick slip process [*Dieterich*, 1978a].

In addition, a few constant stress creep tests were tried. For these tests the shear stress was held constant at successively higher levels for an interval of 30 min, then increased to a higher level. The principal motivation for the creep experiments was to see if these measurements at 10.0-MPa normal stress could be related to the creep measurements reported by *Solberg et al.* [1978] at confining pressures to 450 MPa.

Experimental Results

In general, these experiments display time, velocity, and displacement effects that qualitatively agree with the effects reported by *Dieterich* [1979a] for friction of initially clean gouge-free surfaces. Quantitatively, these results show some differences from the previous experiments. Additionally, there are some features not specifically represented by the constitutive formulation of *Dieterich* [1979a]. The differences between the gouge-free and simulated gouge experiments may have significant implications for applications to natural faults and point to some difficulties that may be encountered in assessing the relevance to natural faulting of some types of experiments reported in the literature.

Condition of Gouge Following the Tests

During a test the gouge layer becomes compacted, and when examined macroscopically or using a hand lens, the gouge either displays a streaked and polished interface with the granite block or it contains well-defined oblique parting planes, or both. If present, the streaking and polishing at the interface indicate concentrated shear in a narrow zone adjacent to the rock. The streaking arises from mafic minerals that consisted originally of individual fragments that have been finely broken up and dragged out along the shear zone. Polishing of the rock surfaces also gives evidence for concentrated shear at the interface between the gouge and the intact rock. In some cases the gouge adheres very strongly to the rock with the result that the streaking cannot be observed, but when the adhering gouge is scraped away, the underlying rock exhibits a polishing down of the original roughness. The oblique parting planes apparently experienced shear also because mafic minerals could be seen to have been dragged out along these surfaces. These features correspond to the shear zones at the gouge-rock interface and the oblique shear planes previously documented in detail by the microscopic studies of deformed simulated gouge [*Tchalenko*, 1970; *Byerlee et al.*, 1978; *Logan*, 1979]. The oblique shears are consistently oriented at a low angle to both the gouge layer and the stress vector acting on the sliding surface as defined by the shear and normal stress. *Tchalenko* [1970] and *Logan* [1979] equate these features to Riedel shears. These features clearly provide some clues concerning the deformation process in the gouge layers. Although there must be some uncertainty in the present interpretations because a microscope was not used for the observations, it is very evident that the roughness of the interface and the particle dimensions of the gouge exert a strong control on the relative development of the shear structures. The most strongly polished and streaked surfaces along the gouge-rock interface were consistently seen with the coarsest gouge size fractions and the finely ground surfaces. Those combinations of parameters also produced the weakest development of the oblique shears. For the case of the coarsest gouge (125–250 μm) in contact with finely ground surfaces, no oblique shears could be seen, and the surfaces were polished, indicating that most deformation was concentrated slip at the gouge-rock interface. The oblique shears became more conspicuous as the roughness increased and the particle size was decreased. In addition, the oblique shears became more closely spaced as the gouge thickness was reduced.

The histograms of Figure 3 give the final size distributions for the gouge that initially consisted entirely of the 125- to 250-μm size fraction. Note that the greatest size reduction is found with the 0.5-mm layer in contact with the coarse surface and the least reduction in particle size is the 0.5-mm and 2.0-mm layers in contact with the finely ground surface. The former case had the strongest development of oblique shears, and the latter had no visible oblique shears and very prominent streaking and polish at the contact of the gouge with the rock.

In summary, the observations are consistent with the interpretation that fine interfaces enhance the development of concentrated shear zones at the contact of the gouge with the intact rock and reduce the development of the oblique shears. Rough surfaces enhance the development of oblique shears and result in greater deformation within the gouge layer as shown by greater grain size reduction in comparison to fine surfaces. Similarly, thin layers of gouge compared to thick layers result in more intense deformation within the gouge layer and show most highly developed oblique shears and greatest reduction in grain size.

Form of the Stress Versus Displacement Curves

When loaded at a constant velocity (2.5 μm/s), the stress plotted as a function of displacement usually shows an initial peak followed first by slow displacement weakening for 0.5 to 5.0 mm of slip and then stabilization at a residual strength, which does not change or increases very slowly with additional displacement (Figure 4). In ap-

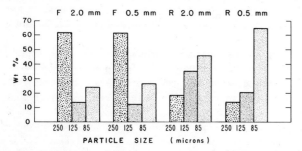

Fig. 3. Particle size histograms of the initial 125- to 250-μm gouge following deformation. The rough and fine interfaces are designated R and F, respectively. Layer thicknesses are 0.5 mm and 2.0 mm.

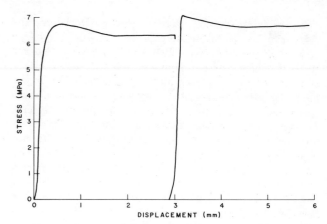

Fig. 4. Shear stress versus displacement for a 1-mm gouge layer consisting of the <250-μm size fraction and rough interface. Sliding velocity is 2.5 μm/s.

proximately 20% of the experiments the strength displays an asymptotic increase to the residual value without passing through a peak. Neither the occasional absence of the large-scale displacement weakening nor the displacement over which the weakening occurs appears to correlate with interface roughness, gouge thickness, or gouge particle size. The coarsest gouge (125–250 μm) consistently showed a less pronounced peak than the other size fractions.

Cycling of the shear stress to zero, even momentarily, after reaching the residual strength, acts to restore the peak in the stress-displacement curves (Figure 4). As seen in Figure 4, the cycling and/or the accompanying increase in total displacement results in an increase in the peak and residual strength. The slow increase in residual strength with displacement probably arises from increased comminution of the gouge fragments leading to better packing of the gouge particles. The relationship of the establishment of the transient peak strength to stress cycling probably involves gross rearrangement of the gouge fragments when the shear stress is dropped, possibly by the collapse of dilatant pore volume. The large-scale displacement weakening might then be related to the localization process whereby distinct slip planes become established in the gouge or at the gouge-rock contact. This interpretation is supported by the observations of *Tchalenko* [1970], *Logan* [1979], and *Dengo and Logan* [1979], who report similar stress-displacement curves. In those studies, oblique shears (Riedel shears) began to form at the peak strength, and the development of layer-parallel shears marked the beginning of slip at the residual strength.

The magnitudes of the residual strength plotted as a function of the gouge parameters are given in Figure 5. Throughout this paper, strength is expressed as the ratio of shear stress τ to normal stress σ (i.e., coefficient of friction μ). The letters R, M, and F designate rough, medium, and fine surface roughnesses, respectively, prepared by the lapping procedure noted above. As evident from Figure 5, gouge thickness, particle size, and perhaps to a lesser extent surface roughness significantly affected the final residual strength. Gouge with size fractions of <85 μm gives greater strength than gouge with size fractions of <250 μm, which has greater strength than the 125- to 250-μm gouge. Hence strength increases with the relative amount of the fine particle size fraction. These relationships of strength with gouge thickness and surface roughness correlate with the observations described above that relate the development of shear structures and grain size reduction. This suggests that strength is partially controlled by the amount or perhaps rate of comminution of the gouge. Hence experimental factors that are associated with enhancement of the development of oblique shears and increased comminution are also associated with increased shear strength. The slow increase of the residual strength with displacement noted above supports this interpretation, as do the observations of grain size reduction. In Figure 3, note that the greatest size reduction is found with the 0.5-mm gouge with the rough interface, which also has the highest strength of the 125- to 250-μm gouge experiments. Conversely, the 2.0-mm layer with a fine interface shows the least size reduction and the lowest strength. From these experimental data alone it cannot be determined unequivocally if the comminution process itself is directly linked to higher strength or if the shear strength is more directly linked to grain size which is changed by comminution. The large-scale displacement weakening which is apparently associated with the localization of slip and possibly reductions in rate of comminution suggests a possible direct dependence of strength on comminution.

Velocity Dependence

A step change of velocity results in transient and residual changes in strength that qualitatively resemble the features reported previously [*Dieterich*, 1979a] for the velocity-dependent slip on ground surfaces of granite. As shown by *Dieterich* [1979a], these two types of velocity dependence are distinct and apparently reflect competing underlying mechanical processes. Depending on the type of test, the different processes may appear relatively more or less important. The example given in Figure 6 is representative and illustrates the types of strength and displacement features observed for step changes in velocity. In general, if sliding at a constant velocity has proceeded for a sufficient distance for the strength to have stabilized at a residual value, an abrupt increase of velocity produces an immediate increase in strength followed by a decay in strength with displacement to a new residual value. The new residual strength is usually less than the residual strength for the previous slower sliding velocity. Conversely, an abrupt decrease of the slip velocity results in an immediate drop in strength followed by a displacement dependent rise in strength to a new residual strength that is greater than the previous strength. Hence there are two velocity-dependent effects. The first appears as a direct but transient velocity dependence; and the second, which is generally an inverse dependence with velocity, becomes fully evident only after a finite amount of slip has occurred. Throughout this paper the two types of velocity dependence will be referred to as the transient velocity effect and residual velocity effect. It must be emphasized here that this terminology is intended to describe the character of the observations and not the character of the

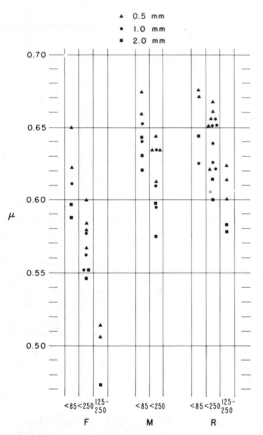

Fig. 5.   Residual strength for slip at 2.5 μm/s. The squares indicate 2-mm layers, circles indicate 1-mm layers, and the triangles indicate 0.5-mm layers. R, M, and F refer to rough, medium, and fine surfaces, respectively.

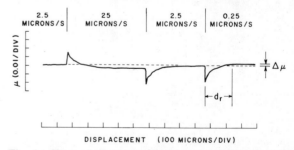

Fig. 6.   Shear stress versus displacement showing the typical effect of changes of slip velocity. The gouge is 1-mm thick, <250 μm in contact with rough surfaces.

underlying mechanisms. In fact, the jump in strength that marks the onset of the transient velocity effect is probably the result of a process related to comminution that operates continuously during slip. The reason for the decay of the strength jump with displacement giving the residual velocity effect is that other competing processes require a finite displacement to become fully evident.

Figure 6 defines parameters used to characterize the observations. The displacement required to reach fully the residual value following a factor of 10 change of velocity is designated as d_r. In the previous studies of *Dieterich* [1978a, 1979a, b] this experimentally observed parameter was designated d_c. However, d_c was also used directly as a constitutive parameter. Although principally controlled by constitutive properties, it is now clear that d_r also reflects some effects due to machine interactions. In this paper, d_r refers to the experimental observations, and d_c is retained to represent the related but not identical constitutive parameter. In Figure 7 the parameter B is the slope of the strength change $\Delta\mu$ with the logarithm of the change of velocity from V_1 to V_2:

$$B = \Delta\mu/\log_{10} (V_1/V_2) \qquad (1)$$

For the initial series of tests the velocity was changed only after the strength had stabilized at a residual value following the peak strength and gross displacement weakening described above. This preconditioning generally required 2.5–5.0 mm of slip. For those measurements the parameter B is negative, indicating that positive changes of velocity produced a decrease in residual strength. A summary of the residual velocity dependence B for the various combinations of parameters is given by Figure 7. The

most striking feature of Figure 7 is the relationship between B and surface roughness. Thickness and particle size show no systematic relationship to B within the scatter of the observations. The dependence of B on surface roughness indicates a possible connection with the observations noted above that relate surface roughness to the development of oblique and contact shears, the amount of grain size reduction in the gouge, and the magnitude μ. In comparison to fine surfaces the rough interfaces enhance the development of oblique shears, result in greater comminution of the gouge, have higher values of μ, and produce less negative values for the parameter B. This suggests that the process that gives less negative values of B is related to comminution processes in the gouge.

Overall, the result that B is negative agrees with the findings for residual velocity dependence for friction of initially clean granite surfaces [*Dieterich*, 1978a, 1979a] and for corundum on granite [*Scholz and Engelder*, 1976]. However, *Solberg and Byerlee* [1979] found a positive velocity dependence in triaxial experiments with a layer of crushed Westerly Granite along an inclined saw cut in a cylinder of Westerly Granite. There are two possible explanations for the different results seen in the latter experiments: First, fundamentally different mechanisms may operate at the higher confining pressures used for those experiments, or second, some aspect of the sample preparation or experimental procedure may result in the selective enhancement of one of the competing mechanisms that underly the rather complicated transient and residual velocity-dependent effects seen for the present experiment. The results obtained here favor the latter explanation. Figure 8 shows the results of a test that is similar to the usual velocity test except the velocity was cycled from the beginning of the run instead of first presliding the sample until the strength stabilized. This procedure more closely corresponds to the procedure used by *Solberg and Byerlee* [1979], who also cycled the velocity from the beginning of the run. Note that the velocity dependence as defined by parameter B is initially very strongly positive, as found by *Solberg and Byerlee* [1979] for the triaxial tests, but that B decreases with total displacement and stabilizes at a negative value. Hence the absence of preconditioning in the triaxial experiments appears to be an important causative factor for the positive velocity dependence. Because the purpose of these tests is to obtain better insight into the constitutive properties of natural faults that generally have very large displacements compared to laboratory tests, it was decided to conduct all subsequent velocity tests after first presliding the blocks to reach the residual strength. Under these conditions the velocity dependence appears to be relatively insensitive to additional displacement.

The agreement of the present results showing a direct residual velocity dependence at small displacements (Figure 8) with the triaxial results of *Solberg and Byerlee* [1979], the smooth change from positive to negative B with displacement, and the dependence of B on surface

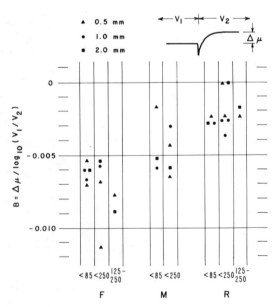

Fig. 7. Summary of the residual velocity dependence. All measurements followed at minimum 2.5 mm of slip to precondition the gouge.

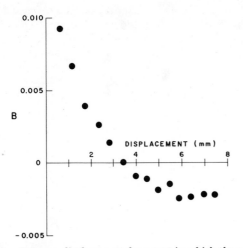

Fig. 8. *B* versus displacement for a test in which the velocity was cycled between 0.25, 2.5, and 25 μm/s from the beginning of the test. Note that the residual velocity dependence is initially positive and that it becomes negative with displacement. This experiment used 1 mm of gouge, <250 μm in contact with rough surfaces.

roughness (Figure 7) indicate that the different values of *B* seen in various experiments result from variations in the relative magnitudes of the two competing processes that alone give either positive or negative velocity dependence. Note, for example, that two tests in Figure 7 have *B* = 0. For those tests it is concluded that the process giving positive velocity dependence exactly offsets the process giving negative velocity dependence. The conclusion that competing velocity-dependent processes operate even when *B* = 0 is supported by the stress versus displacement records for those tests that show an initial but transient positive jump in μ when the velocity is increased followed by a decay to a zero net change of μ as the displacement increases. The point here is that the direct velocity effect is fully manifest immediately and causes the initial jump of μ. The competing velocity effect requires finite displacement to become fully evident and results in the progressive cancellation of the jump of μ as displacement increases. The final sign and magnitude of *B* reflect the magnitude of both velocity-dependent processes.

Recall from the earlier discussion of Figure 7 the conclusion that less negative values of *B* correlate with greater comminution of the gouge. This speculation is supported by the test illustrated by Figure 8, which displays positive values of *B* at the initial stages of slip where the rate of grain crushing is most likely a maximum. Later in the deformation, when *B* becomes negative and when shear zones are apparently well established, proportionately greater amounts of the deformation might take place by actual slip of particles past each other.

Figure 9 summarizes the relationship of the displacement required to reach the residual strength d_r to surface roughness, gouge thickness, and particle size. The variation of μ with displacement following a sudden veloc-

ity step was noted previously [*Dieterich*, 1978a, 1979a] and interpreted to be caused by a change in a velocity-dependent characteristic of the population of contacts between particles of gouge or across the slip surface. The two assumptions here are (1) there is some characteristic of the contacts that gives the inverse velocity dependence and (2) the displacement d_r required to reach a stable strength value following a velocity change is related to the displacement required to change completely the population of intergrain contacts established during slip at the previous velocity. A good correlation of d_r to surface roughness was found for the rock on rock tests supporting this conclusion [*Dieterich*, 1979a]. As shown by Figure 9, surface roughness and particle size both affect d_r in the present experiments. In addition, there is perhaps a weak indication that d_r may be larger for the thicker layers. The control of d_r by surface roughness and the weak or possibly nonexistent effect of thickness point again to the probable importance of concentrated shear zones that are strongly influenced by the roughness of the gouge-rock contact. The data in Figure 9 suggest that the coarser element of either the gouge or the surface roughness determines d_r. For example, with fine surfaces there are pronounced differences in d_r that appear to reflect the gouge particle size. For the rough surfaces the surface roughness is large in comparison to all but the coarsest particle sizes in the 125- to 250-μm gouge, and the d_r for the <85-μm gouge and the <250-μm gouge appear to be comparable.

Time Dependence

Time-dependent increase of the frictional strength of rocks has been observed in a variety of experiments and

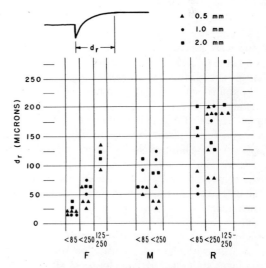

Fig. 9. Summary of the dispalcement d_r required to reach the residual strength following a factor of 10 change of velocity. Conditions for these measurements correspond to those of Figure 7. In addition, all measurements were made for decreases in velocity only. The upper curve shows an example of the stress versus displacement records from which the data were taken.

has been discussed at length in the literature [*Dieterich*, 1972; *Scholz et al.*, 1972; *Scholz and Engelder*, 1976; *Teufel and Logan*, 1978]. Time-dependent friction is related to observations of the residual velocity-dependent friction and directly underlies the causative mechanism of stick slip [*Dieterich*, 1978a]. The experimental curve of Figure 10 gives a stress versus displacement record that illustrates the characteristics of the time dependence tests. Note that the stress drops during the time the driving ram is held stationary (*a* to *b*). This is caused by slow continued slip of the surfaces at a rate that decays with time. Following the hold cycle the driving ram resumes motion at a constant velocity, and the gouge shows a transient increase in strength (*b* to *c*) that depends on the duration of the time that the ram was held stationary. Similar stress versus displacement records were found for the gouge-free tests reported by *Dieterich* [1979a]. The time-dependent increase in strength $\Delta\mu$ is taken to be defined by the stress maximum (level *c* minus level *a*). For all combinations of parameters with the present experiments, quantitatively similar increases in strength were observed when the driving ram was held stationary for a specific interval of time. The data of Figure 10 summarize the time dependence tests and give the peak increase of μ relative to the residual value of μ observed during slip at 2.5 μm/s that directly preceded and followed each hold cycle. Hold periods of 1, 10, 100, and 1000 s were used. The data appear to show a weak control of the time-dependent strength increase by the gouge particle size and the surface roughness. Within the scatter of the data, gouge layer thickness does not display a relationship to the magnitude of the time-dependent effect. The magnitude of the strength increases observed for the simulated gouge is within the range of values reported for nominally gouge-free surfaces [*Dieterich*, 1972].

A minimum displacement of 1.25 mm and a maximum of 6.0 mm were required to reach the large-scale residual strength preceding the hold cycles for the tests of Figure 10. A single exploratory experiment using 1 mm of gouge, <250 μm, and rough surfaces examined the effect of total displacement on the time dependence. This experiment (Figures 11a and 11b) employed 100-s hold cycles at successively greater displacements beginning at 0.25-mm total displacement. Sliding velocity between holds was 2.5 μm/s. Figure 11a gives the increase of strength, $\Delta\mu$, for the 100-s holds as a function of displacement. Figure 11b plots the displacement required to reach residual strength d_r following each hold cycle. The points P and R mark the approximate total displacements for the large-scale peak and residual strengths, respectively. There are two noteworthy features in Figure 11. First, the time-dependent effect becomes stable before completion of the large-scale displacement weakening and at relatively small total displacements (~1.75 mm) compared to the stabilization of the residual velocity effect in Figure 8 (5-6 mm). Second, d_r appears to decrease at the point where the large-scale displacement weakening effect is completed. The latter ob-

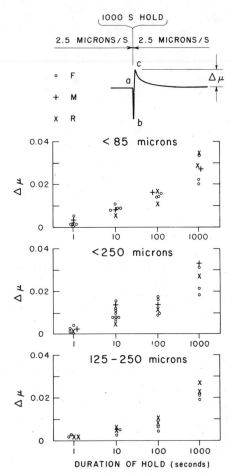

Fig. 10.   Summary of observations of time-dependent increase of the strength following holds of 1, 10, 100, and 1000 s. The circles, pluses, and crosses refer to fine, medium, and rough surfaces, respectively.

servations suggest that perhaps the deformation becomes fully localized at that displacement.

Creep Test

Solberg et al. [1978] and *Solberg and Byerlee* [1979] report on a series of creep tests on simulated gouge consisting of crushed Westerly Granite in contact with intact Westerly Granite. Those experiments employed a triaxial testing apparatus and were conducted at confining pressures to 450 MPa. In those tests the axial load was raised in small increments and held constant at each increment for 12 hours. They found a strong dependence for the rate of deformation on time and stress level. At the lowest stress levels at which permanent deformation was seen, the deformation rate decayed with time in a manner resembling transient creep. At higher stress levels the initial transient creep phase gave way to approximately steady state or secondary creep. The creep rates increased with the stress level, indicating a direct velocity dependence of the gouge strength. At still higher stresses the

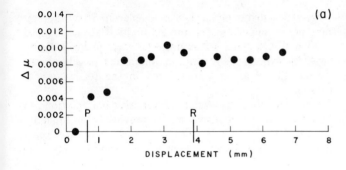

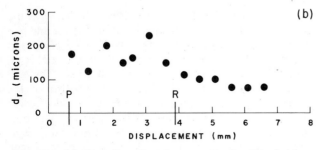

Fig. 11.   (a) Time-dependent increase of μ for 100-s cycles as a function of displacement. (b) Displacement d_r required to erase the increase in strength caused by the hold cycle.

creep rate accelerated and instability resulted. These results clearly suggest a relationship to the usual types of creep behavior seen in solids. The positive velocity dependence indicated by the steady state creep results may at first consideration appear to contradict the inverse velocity dependence of the residual strength seen in this and previous studies.

In an attempt to clarify the relationship of the present experiments and the previous constitutive models to the higher-pressure creep measurements a few creep tests were run at 10-MPa normal stress. The specific purpose was to see if creep results similar to those obtained by *Solberg et al.* [1978] could be obtained, and if so to establish the connection of the creep measurements to the constitutive model proposed by *Dieterich* [1979a]. The experiments employed the <250-μm gouge in 1-mm layers in contact with a rough ground surface. Figure 12 plots the displacement against time for a creep experiment within which the gouge layer has not been preconditioned with slip. Although the present experiment was not carried to such long time intervals as those of Solberg et al., the results of Figure 12 look very much like the effects reported by Solberg et al., and it is concluded that similar processes are active in each experiment. Because the displacements of the block are small throughout this experiment, the results from Figure 8 indicate that the residual velocity effect should show a positive dependence on velocity. Hence except for the difference in normal stress and duration of the test, the experimental conditions and results appear to be fully compatible with those of *Solberg et al.* [1978].

A second test was run using the same fault parameters,

but the gouge layer was first preconditioned by 5.5 mm of slip to determine the effect of negative residual velocity dependence on the creep results. From Figure 8, recall that the initial positive velocity dependence becomes negative following a few millimeters of displacement. For this creep test following the preconditioning the shear stress was not set to zero but was lowered to the first hold stress of 5.64 MPa. Figure 13 gives the results of the test. Although creep is observed over a small range of shear stresses and the total displacements are much less than those observed for the test without preconditioning, the general character of the results is similar to the results of Figure 12. It would appear therefore that the creep type of behavior observed in these experiments and the high-pressure triaxial results is enhanced by a positive residual velocity dependence but that it does not require it. Below it will be shown that the creep is made possible by the mechanism that causes the jump of strength seen for the transient velocity effect.

Constitutive Model

In a qualitative sense most of the experimental data described above agree with the results obtained previously

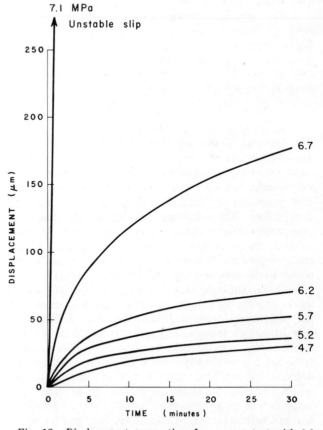

Fig. 12.   Displacement versus time for a creep test with 1.0 mm of <250-μm gouge in contact with a rough surface. This gouge layer was not preconditioned by prior slip. Labels on the curves give the value of the shear stress in megapascals.

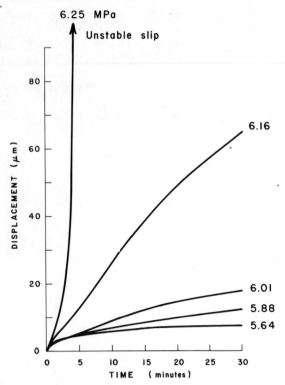

Fig. 13. Displacement versus time for a creep test with 1.0 mm of <250-µm gouge in contact with a rough surface. This gouge layer was preconditioned by 3.0 mm of slip. Note change of scale for displacements compared to Figure 11. Labels on the curves give shear stress in megapascals.

for nominally gouge-free surfaces of granite. In particular, following preconditioning, the overall form of the stress-displacement curves, the time dependence, the transient velocity effect, and the residual velocity effect are very similar to those observed previously. The magnitude of d_r is significantly greater for gouge than it is for gouge-free surfaces. The initial large-scale peak in the stress-displacement curves was not observed in the prior friction experiments and does not seem to be explainable by the fault friction constitutive model in its present state. The observations here and those of *Tchalenko* [1970] and *Logan* [1979] indicate that the large-scale displacement weakening from the peak strength at the beginning of a run is associated with the localization of the deformation into distinct shear zones. This probably results in a change in the dominant mode of deformation from breaking of interlocked grains throughout the gouge layer to localized deformation consisting more nearly of slip on surfaces within the gouge or at the gouge-rock contact. The large magnitude of the displacement required to reach stabilization at a residual strength and the sensitivity of the velocity-dependent parameter B to initial displacements may present problems for interpretation of experimental results in which the total displacements are limited. Because the peak appears to become reestablished only following a

large stress drop, the large-scale displacement weakening may not be important for natural faults if stress changes are small, i.e., small stress drops in earthquakes. At any rate, the large-scale stress maximum and possible localization effects are not considered in the following discussion.

The following discussion of constitutive relations is based on the constitutive model previously outlined by *Dieterich* [1979a]. The basis of that model is that the coefficient of friction μ can be expressed as the product

$$\mu = CF \qquad (2)$$

where C is proportional to the strength of the load-bearing asperities in response to the applied normal stress and F is inversely proportional to the strength of the asperities in response to the applied shear stress. Note that CF is dimensionless and independent of normal stress. The central notion of the model is that F, the strength of the asperities in shear, is a direct function of slip velocity and that C, which controls the size of the asperities, is a function of a history-dependent parameter θ. Several lines of evidence suggest that θ is the effective age of the asperities and that the observed time-dependent increase of fault strength occurs because the area of contact of the asperities increases with time [*Dieterich*, 1978a]. A specific empirical relationship for C that satisfactorily accounts for the increase in μ with time is

$$C = c_1 + c_2 \log_{10} (c_3\theta + 1) \qquad (3)$$

The parameters c_1, c_2, and c_3 are constants, and θ is proportional to the average age in seconds of the asperities. The dimensions of c_3 are s^{-1}. The mechanism for the increase in the size of the asperities is interpreted to be asperity creep in response to the normal stress [*Dieterich*, 1972, 1978a; *Scholz and Engelder*, 1976]. The observation of the transient velocity dependence (i.e., the jump in strength that occurs when velocity is suddenly changed) was interpreted to be evidence for a velocity dependence of the shear strength. The following empirical relationship was proposed by *Dieterich* [1979a, b]:

$$F = f_1 + \frac{1}{f_2 \log_{10} (f_3/V + 10)} \qquad (4)$$

where f_1, f_2, and f_3 are constants and V is the slip velocity. Note that (3) and (4) imply a similar type of time dependence of strength of the asperities that is consistent with the usual behavior of rock-forming minerals. The parameter f_3 has dimensions of microns per second. The specific form for (4) is as yet poorly defined, and it appears that a number of similar empirical relationships fit the data equally well. The following relationship for F agrees with the observations as well as (4) and has the slight advantage of making the relationship of C to F somewhat more transparent:

$$F = \frac{1}{f_1 + f_2 \log_{10} (f_3/V + 1)} \qquad (5)$$

The constants f_1, f_2, and f_3 do not have the same numerical values as (4). Note that the form of (5) is the inverse of (3) if f_3 has the dimensions of velocity. Equation (5) is used in preference to (4) for the remainder of this paper.

Because of fault displacement, the load-bearing contacts across the fault or between particles of gouge in the fault are continuously created and destroyed at a rate that is proportional to the rate of slip. For steady state slip at a constant velocity the average age of the contacts is

$$\theta = d_c/V \tag{6}$$

where d_c is proportional to the displacement required to change an existing population of asperities completely. The use of (6) in (3) along with (5) gives the steady state or residual strength as a function of velocity. Because a finite displacement proportional to d_c is required to change the contacts, a change of velocity does not result in an immediate change of θ. In order to use these relationships for computations to follow the change in strength following a change in velocity, it is necessary to follow the resulting change of θ with displacement. *Dieterich* [1979a, b] proposed the following function to follow the evolution of θ with displacement d for a step change to velocity V at displacement d:

$$\theta = (d_c/V) * (\theta_0 V/d_c)^{\exp[(d_0-d_c)/d_c]} \tag{7}$$

The time of contact at d_0 is θ_0. Note that as d increases from d_0 to $d \gg d_0$, θ goes from θ_0 to d_c/V. Detailed simulations of experiments in which the velocity changes continuously with displacement are possible using (7) by breaking the velocity into small constant velocity steps [*Dieterich*, 1979a, b]. Recently, *Rice* [1981], A. Ruina (unpublished manuscript, 1980), and *Kosloff and Liu* [1980] have investigated differential formulations based on (7) or similar functions that permit continuously varying velocity histories to be more accurately represented. A. Ruina (personal communication, 1980) has pointed out a potential problem with (7) that can arise for sudden decreases in velocity. As V goes to zero, $d\theta/dt$ also goes to zero. This is undesirable if the original concept that θ increases with time for stationary surfaces is correct. For the following equation, which is based on one proposed by Ruina, $d\theta/dt$ goes to 1 as V approaches zero:

$$\theta = (d_c/V) + (\theta_0 - d_c/V) \exp\left(\frac{d_0 - d}{d_c}\right) \tag{8}$$

The observations of time, velocity, and displacement dependence described above agree with the previous observations for friction of gouge-free surfaces [*Dieterich*, 1979a] and give added support for the validity of the general form of the model. Those observations have been used to derive specific values for the parameters in (3), (5), and (6) for the various combinations of experimental variables. The values are given in Table 1. Because the magnitudes of the time and velocity effects are small in comparison to the magnitude of the total strength of the

gouge, the approximate magnitude of μ for the fault is controlled by the relative values of c_1 and f_1. By arbitrarily setting the value of f_1 at 1.0, c_1 is then approximately the value of the μ in the absence of the time and velocity effects. In general the results summarized in Figures 7, 9, and 10 indicate that d_c and the time and velocity dependence are insensitive to gouge layer thickness. Consequently, the data for different layer thicknesses have been lumped within a single category for particle size and surface roughness. Note that parameter c_3 controls the lower limit that time-dependent effects can be seen and that f_3 similarly controls the maximum velocity for which velocity-dependent effects are seen. Because the present experiments were over a limited range of times and velocities, c_3 and f_3 are undefined, and the values given in Table 1 represent the approximate lower limits permitted by the observations. To obtain a fit of the data for c_2, it is necessary to know the value of d_c because the value of θ after a hold period is approximately the sum of the age of the contacts during the slip that immediately preceded the hold, θ_0, and the hold time, θ_h:

$$\theta = \theta_0 + \theta_h \tag{9}$$

where from (6),

$$\theta_0 = d_c/V \tag{10}$$

The values of d_c in Table 1 are those that were used to obtain c_2 and give the best fit of the time-dependent data using the numerical model described in the following section.

From (3), (5), and (6) for steady state slip at a constant velocity V (i.e., the displacement velocity at V is greater than d_c),

$$\mu = \frac{c_1 + c_2 \log_{10} (c_3 d_c/V + 1)}{f_1 + f_2 \log_{10} (f_3/V + 1)} \tag{11}$$

Hence parameter c_2, which controls the time dependence, also gives the inverse velocity-dependent effect. Parameter f_2 controls the direct velocity effect. The total residual velocity dependence B in Figure 7, observed when comparing the steady state values for μ at different velocities, is determined by both c_2 and f_2. In Table 1, note that c_2 shows a weak but systematic correlation with surface roughness and that f_2 shows a stronger correlation with surface roughness. As a result of those variations with roughness it is possible to identify which of the competing velocity effects is most important in controlling the observed variations of B with surface roughness. Table 1 shows that the inverse velocity effect is slightly enhanced by increasing the roughness of the surfaces (tends to decrease B slightly) and the direct velocity effect is greatly enhanced for the rough surfaces (tends to increase B greatly). The observed higher values for B for rough surfaces therefore are caused by enhancement of the direct velocity effect as suggested by the earlier observations on comminution and not by a weakening of the process con-

TABLE 1. Constitutive Parameters

Gouge Size, μm	Roughness	d_r, μm	d_c, μm	c_1	c_2	c_3, s^{-1}	f_1	f_2	f_3, μm/s
125–250	R	225	50	0.60	0.017	>0.5	1.0	0.025	>25
125–250	F	115	25	0.50	0.014	>1.0	1.0	0.012	>25
<250	R	140	37	0.63	0.019	>0.07	1.0	0.027	>25
<250	M	65	12	0.60	0.019	>2.0	1.0	0.023	>25
<250	F	50	7	0.57	0.019	>3.6	1.0	0.021	>25
<85	R	125	37	0.65	0.020	>0.07	1.0	0.027	>25
<85	M	75	17	0.64	0.019	>1.5	1.0	0.022	>25
<85	F	20	4	0.61	0.017	>6.3	1.0	0.018	>25

trolling the inverse effect. Figure 10 supports a similar conclusion for the variation of B with displacement illustrated in Figure 8. In Figure 10 the magnitude of the time dependence and hence c_2 do not change for displacements to 5–6 mm.

Numerical Model for Fault Interactions

It was noted above that this study was done from the perspective that slip phenomena observed for simulated faults and for natural active faults arise from the combined effects of the constitutive properties of the fault and the total mechanical system that loads the fault. Attempts to understand the observations without accounting for these interactions can be seriously incomplete. To simulate the specific experimental results of this study and to explore some of the implications of the interactions between the constitutive properties of a fault, a simple deterministic numerical model has been developed. The analysis is based on a zero-mass slider and spring system (Figure 14). The spring with stiffness K represents the combined elastic properties of the system that interacts with the fault. At this point it should be noted that it is surely an understatement to say that this model is a simple idealization of laboratory or natural faulting. Natural faults, especially, certainly involve three-dimensional interactions in which heterogeneity of properties on the fault and geometric deviations from perfectly planar fault surfaces probably have fundamental importance. However, previous investigations [*Dieterich*, 1978a, 1979a] indicate that the slider-spring model is sufficient to explore some of the first-order interactions that underly the dif-

ferent slip modes. The model presented here is intended to illustrate some of those interactions in more detail and to provide a detailed check of the applicability of the constitutive model to simulated gouge.

With this model the slider is externally loaded by specifying the displacement D, velocity V, or loads at the loading point P (Figure 14). For simulation of the present experiments the motion of the loading point P corresponds to the observed displacements because those displacements were controlled by the servo control system. Displacement and velocity of the slider are d and v, respectively. The area of the slider is unity. Normal stress is applied to the block independent of the conditions at the loading point. Shear stress τ is specified by the deformation of the spring:

$$\tau = K(D - d) \tag{12}$$

For the purposes of the computations it is convenient to nondimensionalize the equations for displacements and stresses. Below, the normalized quantities are indicated by primed symbols. Terms containing displacements are divided by the critical displacement d_c, and terms containing stress are divided by σ:

$$D' = D/d_c \tag{13}$$

$$V' = V/d_c \tag{14}$$

$$d' = d/d_c \tag{15}$$

$$v' = v/d_c \tag{16}$$

$$\tau' = \tau/\sigma = \mu \tag{17}$$

$$K' = (K)(d_c/\sigma) \tag{18}$$

Note that shear stress τ reduces to μ and that changes in the normalized stiffness of the spring can be brought about by changes in σ, d_c, or K.

The computational procedure employed for this model is deterministic and uses a displacement marching scheme in which the slider velocity v is assumed to be constant during each step. Displacement steps are generally $d_c/20$. For each displacement step the principal unknown of the computation is the slider velocity. An iterative procedure is employed to find v such that the friction force at the end of the step equals the spring force. At the beginning of

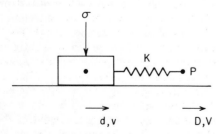

Fig. 14. Spring and slider model.

each displacement step a trial slip velocity v is used to define the duration T of the step:

$$T = d_c / (20v)$$

Using T the displacement at the loading point appropriate to the specified loading history is set. The evolution of θ is then calculated from (7) or (8), and the friction force is found from v and θ. If the absolute difference between μ and $(K)(d - D)$ is greater than 10^{-6}, then a correction is added to the trial velocity, and the procedure is repeated until μ and the spring force are equal. When v is determined, the displacement is then incremented, and the iteration for the next step is begun.

The computational procedure is modified slightly for simulation of the time dependence and creep tests where the velocity may become very small and consequently T becomes very large. In those situations the computer program reduces the displacement step from $d_c/20$ to a suitably small value such that the time required for a displacement step does not exceed the time required for a specified change in the applied loading conditions (for example, resumed action of the loading point following a hold period).

A slip instability can arise with this model if the decrease in μ with displacement exceeds the slope of the unloading curve which is fixed by K'. As a slip instability is approached, the system responds by increasing the slider velocity such that the direct velocity term in (5) is sufficient to keep the friction force in balance with the spring force. Because of the higher velocity, θ decreases with displacement, and to maintain equilibrium, a higher v is required for the next displacement step, and so on. However, as $v \gg f_3$, further velocity increases no longer directly affect μ, and the evolution of θ alone controls the change of μ with displacement. This causes μ to decrease rapidly with displacement. In this case the applied load exceeds the frictional resistance, and the assumption of quasi-static equilibrium employed for the computations is violated. During high-velocity slip, changes of μ are controlled initially by the decrease of θ with displacement. For the case $\theta \ll 1/c_3$, μ is insensitive to θ. A procedure that permits unstable slip in the present model is to assume an upper limit for the slider velocity of $v = 100f_3$, since higher velocities do not contribute to the direct velocity dependence. During the computations, if v exceeds $100f_3$, an instability will occur because μ will drop suddenly independent of v. When this occurs, the slider velocity is fixed at $100f_3$, and computations for μ proceed independent of the assumption of equilibrium. This assumption of instability continues until the displacement of the slider is sufficient to reduce the spring force to equal the frictional strength, whereupon the assumption of quasi-static equilibrium is reasserted. Clearly, a number of simple refinements could be added to this procedure to represent better the conditions during an instability. The purpose of the present procedure is not to follow the detail of interactions during an instability

but to allow the computations to continue through the instability without becoming trapped by a numerical condition that could not satisfy the assumptions of quasi-static equilibrium.

Simulation of Slip Modes

Figure 15 gives three examples of simulations using different stiffnesses K' that illustrate some features of the computations and show the details of the onset of a stick slip instability following a hold period of 100 s. Parameters for the constitutive law are $d_c = 20$ μm, $c_1 = 0.60$, $c_2 = 0.014$, $c_3 = 1.0$, $f_1 = 1.0$, $f_2 = 0.014$, and $f_3 = 1.0$. Effective stiffnesses are 0.01250, 0.00236, and 0.00125 for Figures 15a, 15b, and 15c, respectively. The motion of the loading point is initially 2.5 μm/s (points 1-2, Figure 15), followed by a 100-s hold (2-3), followed by resumed motion at 2.5 μm/s. The evolution function for θ given by (7) was used for each simulation. For the case in Figure 15a, sliding is stable at all times. Note the overall similarity of the stress versus displacement record for Figure 15a with that observed experimentally (Figure 10), including the stress relaxation due to slip during the hold interval. The increase in strength arises because θ increases during the hold when the slider moves at progressively slower velocity. The finite slope of the stress drop during the hold cycle of each simulation (2-3 in Figure 15) differs from the vertical slope of the experimental record. This difference results solely from the fact that the simulations show the motion of the block while the experiments give the motion of the loading point. The slopes of the stress drop equal the stiffness K'. The continued motion of the slider after the loading point is held stationary is a consequence of the direct velocity term in the constitutive law and finite stiffness of the spring. Additional slip results in a decrease in

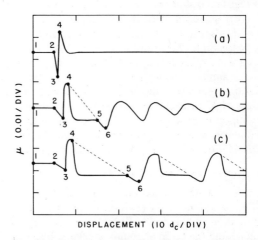

Fig. 15.   Simulations showing the effect of stiffness K' on a test in which the loading point is held stationary for 100 s. The dashed line shows the spring force acting on the slider during an unstable slip event. Stiffnesses are 0.0125, 0.00236, and 0.00125 for Figures 15a, 15b, and 15c, respectively.

stress with displacement that has a slope determined by K'. As a consequence of the direct velocity term F, slip can occur at reduced stress if v is sufficiently reduced also. This slip of the block decays with time and results in a relaxation of the stress. The rate of stress relaxation is controlled by the rate of slip and the stiffness of the spring.

For intermediate and low stiffnesses (Figures 15b and 15c), unstable slip follows the resumption of loading because the transient increase of μ caused by the hold cycle decays with displacement at a rate that exceeds K'. The dashed line (points 4–5) gives the stress acting on the slider by the spring and has a slope of $(-K')(1 - V/100f_3)$. The solid line gives the fault strength. The imbalance of spring and frictional stress provides the driving force for the slip instability. It is important to point out that to maintain equilibrium, slider velocity increased from $v \ll 2.5$ μm/s at point 3 to $v = 100$ μm/s at point 4 in Figures 15b and 15c. This type of accelerating preinstability slip is a common feature of laboratory stick slip [Byerlee, 1967; Logan et al., 1972; Scholz et al., 1972; Dieterich, 1978b]. Between points 4 and 5 in Figure 15 there is no slider velocity that permits quasi-static equilibrium. At point 5 the displacement of the slider is sufficient to reduce the stress applied to the slider by the spring to the fault strength, and the equilibrium condition is again asserted for the computations. The continued stress drop from points 5 to 6 results from interactions that are similar to the stress drop during the hold cycle. At point 5 the slider velocity is $100f_3$ (100 μm/s), and at point 6 the slider velocity and velocity of the loading point are equal (2.5 μm/s). Because of the direct velocity-dependent term, the deceleration of the slider results in a decrease of the frictional strength, which is followed by the spring.

For the simulations of Figure 15, note that for intermediate K' the slip instability is followed by stable but oscillating slip and the low-K' example has repeated unstable slip (stick slip). This change in slip mode for change in K' corresponds directly to the experimental observations of Dieterich [1978a] that show the transition from stable to unstable slip occurs when σ is increased, machine stiffness is decreased, or d_c is decreased. Recall from (18) that K' is similarly dependent upon K, σ, and d_c. For the simulations, oscillatory slip occurs only for a narrow range of values of K' near the transition from constant velocity stable slip and stick slip. This corresponds to the experimental observations of oscillatory slip near the transition from stable to unstable slip by Scholz et al. [1972].

Simulation of Time Dependence Tests

Figure 16 shows simulations of a time-dependent test that illustrate some of the characteristics of the two evolution functions (7) and (8). Figure 16a gives the experimental results for 1 mm of gouge, <250 μm in contact with rough surfaces. The simulation shown by Figure 16b uses (7) and has an overall form that resembles the experimental data, but the magnitudes of the peak strength following the hold are less than the magnitude observed in the

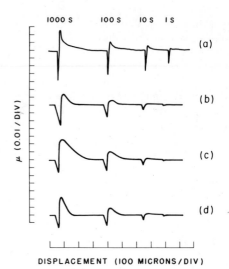

Fig. 16.   (a) Experimental results of a time dependence test using 1 mm of gouge, <250 μm in contact with rough surfaces. (b, c, d) Simulations with $c_1 = 0.63$, $c_2 = 0.019$, $c_3 = 1.0$ s^{-1} and $f_1 = 1.0$, $f_2 = 0.027$, $f_3 = 1.0$ μm/s.

experiments. That is because of the characteristic noted above that $d\theta/dt$ goes to zero as the velocity of slip goes to zero. As a result, θ at the end of the hold cycle is too small, resulting in a lower peak when loading velocity is reapplied. Figure 16c is an identical simulation that uses evolution equation (8) instead. In the case of Figure 16c the height of the peaks agrees with the observations. The decay of strength following peak strength for Figure 16c is much slower than that for Figure 16b. This is because the change of θ with displacement for $\theta > (d_c/v)$ using (8) is not as rapid as (7). Figure 16d employed a composite of (7) and (8) to compute the evolution of θ. In the case of Figure 16d, (7) was used for $\theta > (d_c/v)$ and (8) was used for $\theta < (d_c/v)$.

Velocity Simulations

Figure 17 gives a comparison of the simulated multiple-velocity tests with the experimental stress versus displacement curves. The example is for gouge of <85 μm, 2 mm thick, and in contact with a rough surface. The constitutive parameters used are those given in Table 1.

For Figure 17, as in the case of the simulations of the time-dependent tests of Figure 16, the curves in Figures 17b, 17c, and 17d use evolution equation (7), evolution equation (8), and the composite of (7) and (8), respectively. The simulations of the velocity tests in Figure 16, unlike those of Figure 15, show little sensitivity to choice of the evolution equation.

Figures 18 and 19 give experimental curves and simulations for the multiple-velocity tests using <85-μm gouge, 2 mm thick, in contact with M and F surfaces, respectively (i.e., Figures 15, 16, 17 give the series of velocity tests with only surface roughness varying between tests). The results using evolution equation (7) are shown. The other evolution models for θ give essentially similar results.

Simulation of Creep Tests

Figure 20 gives the results of a simulation of a constant stress creep experiment. Shear stress τ was held constant for an interval of 30 min and then increased to a higher level and held constant again for 30 min, and so on until a slip instability occurred. The initial value for θ was arbitrarily set at 10 s. The simulation attempts to approximate the conditions of Figure 12, which did not have an episode of slip preconditioning. The experiment used gouge of <250 μm in contact with rough surfaces giving $d_c = 37$ from Table 1. Because the gouge layer did not experience slip preconditioning, the data of Figure 8 indicate that B, the parameter for residual velocity dependence, had a positive value of approximately 0.009. Accordingly, constitutive parameters of $c_1 = 0.63$, $c_2 = 0.019$, $c_3 = 1.0$ s and $f_1 = 1.0$, $f_2 = 0.044$, $f_3 = 1.0$ μm/s were chosen for the model to give $B = 0.009$. In the absence of preconditioning, the results of this study show that velocity dependence, time dependence, and the gross strength level for constant velocity slip all change rapidly as a function of displacement in the early stages of a test. This simulation assumes all constitutive parameters to be independent of total displacement, and it is therefore only a rough approximation to the probable experimental conditions. It does, however, show the same general features as Figure 12. Creep with displacement rates that decay with time occurs at the lower stress levels. Because the stress and hence V are less than that required for steady state slip, $\theta > d_c/V$. This results in an increase of θ with time. That increase of θ tends to strengthen the gouge for constant velocity slip, but the direct velocity term F in the constitutive law permits the strength to remain constant if V decreases. The rate of velocity decrease in the simulations is less than that for

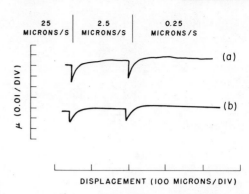

μ (0.01 / DIV)

(a)

(b)

DISPLACEMENT (100 MICRONS/DIV)

Fig. 18. Comparison of (a) experiment and (b) simulation for a multiple-velocity test with 2 mm of gouge, <85 μm in contact with medium surfaces.

the experiments, indicating that the evolution function for θ needs some refining. Total displacement for a 30-min constant stress interval and hence the average rate of slip increase with the applied shear stress. At the highest stress levels in the simulations and experiment, creep rapidly accelerates to unstable failure. Accelerating slip occurs because the stress increments eventually result in a slip velocity such that $\theta < d_c/V$. As a result, V must accelerate to keep the stress constant.

The simulation of Figure 21 corresponds approximately to the creep experiment of Figure 13. For that experiment the gouge had been fully preconditioned by an initial episode of constant velocity slip. Therefore constitutive parameters for this simulation were taken directly from Table 1 for the case of gouge of <250 μm in contact with rough surfaces. The constitutive parameters for the present model are identical to those of the previous model except that $f_2 = 0.044$ for the former and $f_2 = 0.27$ for the latter. Because of the preconditioning, the data of Figure 7 and indirectly the parameters from Table 1 indicate that the strength of gouge at steady state sliding is an inverse function of velocity. An interesting feature of the experiment and simulation is that the average velocities during the constant stress holds in the absence of other information would lead to the incorrect conclusion that gouge strength is a direct and not an inverse function of velocity. Note also that the results of Figure 21 compared to Figure 20 show that the effect of reducing f_2 from 0.044 to 0.027 is to reduce the amount of creep.

Discussion

These experiments display time-, velocity-, and displacement-dependent effects that qualitatively resemble effects reported previously for rock friction on initially clean surfaces [Dieterich, 1979a]. In addition, the constant stress creep experiments at successively higher stresses give results that appear to correlate with the high-pressure creep in simulated gouge reported by Solberg et al. [1978].

Following a change of velocity the displacement re-

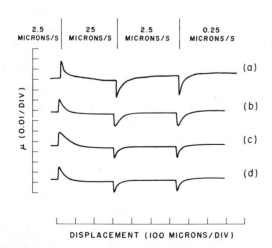

μ (0.01 / DIV)

(a)

(b)

(c)

(d)

DISPLACEMENT (100 MICRONS/DIV)

Fig. 17. Comparison of (a) experiment and (b, c, d) simulations for a multiple-velocity test with 2 mm of gouge, <85 μm in contact with rough surfaces.

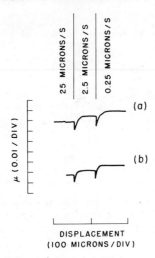

Fig. 19. Comparison of (a) experiment and (b) simulation for a multiple-velocity test with 2 mm of gouge, <85 μm in contact with a fine surface.

quired to reach a stable strength level d_r appears to reflect the displacement needed to change the population of intergrain contacts that govern the strength of the gouge. The parameter d_r is much larger for the gouge than it is for friction of surfaces free of gouge. Gouge particle size and roughness of the surface in contact with the gouge, but not the thickness of the gouge layer, control the magnitude of d_r. This result indicates that the deformation in the gouge is highly localized, probably at the gouge/rock contact. This conclusion is supported by previous studies of deformation structures from laboratory experiments

[*Byerlee et al.*, 1978; *Logan*, 1979]. The observed time dependence, velocity dependence, and related displacement-weakening properties of the gouge probably contribute to the localization process. Because increase of the age of the contacts, θ, tends to strengthen the gouge, perturbations of the deformation rate within the gouge will tend to grow. Zones that deform at relatively slower rates will on the average have interparticle contacts that are older than more rapidly deforming zones. As a result of the greater contact times, the more slowly deforming zones will become stronger and the more rapidly deforming zones will become weaker, resulting in greater concentration of the deformation in the weakening zone.

These experiments and the previous experiments of *Dieterich* [1979a] give clear evidence for two competing velocity-dependent processes. Because one process responds immediately to velocity changes and the other requires a finite displacement d_r to become fully effective, the two processes become separable when the sliding velocity is suddenly changed to a new value. The immediate process acts to give a direct dependence between strength and velocity. Hence increases in velocity of slip are marked by an immediate increase in strength. The process that requires displacement d_r to become fully effective produces an inverse dependence of strength on velocity. As a result of the competition of the two processes, an increase of velocity first causes an increase in strength followed by a decrease of strength with displacement. The final residual change in strength caused by a change of slip velocity may be either positive or negative depending upon the relative magnitudes of the two processes. Both positive and negative changes are seen for gouge. The results of these tests

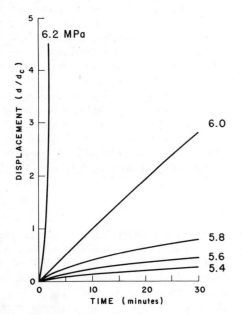

Fig. 20. Simulation of a creep test in which the constitutive parameters give the strength to be a positive function of steady state velocity ($B = 0.009$).

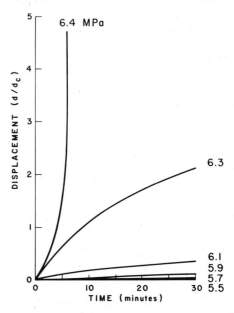

Fig. 21. Simulation of a creep test in which the constitutive parameters require the strength to be a negative function of steady state velocity ($B = -0.003$).

indicate that the process giving the negative effect is related to the time dependence and that it is relatively insensitive to experimental conditions. Experimental conditions more strongly influence the process giving direct positive velocity dependence. Apparently, the rate of comminution of the gouge directly affects the direct dependence on velocity. The magnitude of the direct velocity-dependent process and the strength of the gouge are greater for conditions that lead to maximum rates of comminution.

Numerical computations using the constitutive model with a simple spring and slider are in good agreement with the various experimental observations reported here. The significant aspect of the simulations is not that a single result can be reproduced in detail but that the model is capable of representing the full range of experimental effects. The results of simulations of the multiple-velocity tests are largely insensitive to the choice of evolution law for θ. Simulations of the time dependence tests all have the same general form as the experimental results, but there are some differences in the peak strength and decay of the peak with displacement that indicate further refinements in the evolution equations for θ may be required. An interesting aspect of the experiments and simulations of the time-dependent tests is that during the hold cycle, when the loading point is held stationary, the slider continues to move at rates that decay with time. This motion has the same underlying mechanism as that causing creep of the gouge at constant stresses. It arises because of the direct velocity term in the constitutive law and the finite stiffness of the system. In a system with infinite stiffness any motion of the slider would cause the stress to drop to zero. Hence continued slip would not be possible. Model and experimental creep tests show similar features. A specific result of the creep experiments and simulations is that although both the competing direct and inverse velocity-dependent processes operate during the deformation, the results taken alone could be erroneously interpreted to be evidence for only a direct, positive dependence on velocity. However, if an interpretation is adopted that only a positive dependence on velocity exists, then the acceleration of slip to an instability at the highest stress is unaccounted for, because instability requires some form of weakening with displacement.

Finally, the model is able to reproduce different modes of sliding including stable slip, oscillatory slip, and unstable slip by varying the stiffness of the system that loads the fault, the normal stress, or the gouge/surface property measured by d_r. Accelerating slip always precedes the onset of unstable slip. At the beginning of this report it was pointed out that these modes of slip are analogous to modes of slip seen for active faulting. In addition, the observation of earthquake afterslip, which shows a decay of the rate of slip with time, may be equivalent to the slip seen during the hold periods of the time tests and the decelerating slip of the creep tests. If this interpretation is correct, then the mechanism allowing afterslip might involve the following: During earthquake slip the velocity is high, and therefore θ, the average age of the stress bearing contacts in the slip zone, will be small. Because of the direct velocity effect, slip may continue at a reduced stress following an earthquake if the sliding velocity is low. At the lower velocity of slip the average age of the contacts will increase, resulting in a strengthening of the fault gouge and a further slowing of the afterslip.

References

Byerlee, J. D., Frictional characteristics of granite under high confining pressure, *J. Geophys. Res.*, *72*, 3639–3648, 1967.

Byerlee, J., and R. Summers, A note on the effect of fault gouge thickness on fault stability, *Int. J. Rock Mech. Mining Sci. Geomech. Abstr.*, *13*, 35–36, 1976.

Byerlee, J., V. Mjachkin, R. Summers, and D. Voevoda, Structures developed in fault gouge during stable sliding and stick-slip, *Tectonophysics*, *44*, 161, 1978.

Dengo, C. A., and J. M. Logan, Correlation of fracture patterns in natural and experimental shear zones (abstract), *Eos Trans. AGU*, *60*, 955, 1979.

Dieterich, J. H., Time-dependent friction in rocks, *J. Geophys. Res.*, *77*, 3690–3697, 1972.

Dieterich, J. H., Time-dependent friction and the mechanics of stick-slip, *Pure Appl. Geophys.*, *116*, 790–806, 1978a.

Dieterich, J. H., Preseismic fault slip and earthquake prediction, *J. Geophys. Res.*, *83*, 3940–3948, 1978b.

Dieterich, J. H., Modeling of rock friction, 1, Experimental results and constitutive equations, *J. Geophys. Res.*, *84*, 2161–2168, 1979a.

Dieterich, J. H., Modeling of rock friction, 2, Simulation of preseismic slip, *J. Geophys. Res.*, *84*, 2169–2175, 1979b.

Engelder, J. T., J. M. Logan, and J. Handin, The sliding characteristics of sandstone on quartz fault-gouge, *Pure Appl. Geophys.*, *113*, 69–86, 1975.

Kosloff, D. D., and H.-P. Liu, Reformulation and discussion of mechanical behavior of the velocity-dependent friction law proposed by Dieterich, *Geophys. Res. Lett.*, *7*, 913–916, 1980.

Logan, J. M., Laboratory and field investigations of fault gouge, contract report, grant No. 14-08-0001-17677, 89 pp., U.S. Geol. Surv., Menlo Park, Calif., 1979.

Logan, J. M., T. Iwasaki, M. Friedman, and S. Kling, Experimental investigation of sliding friction in multilithologic specimens, *Eng. Geol. Case Hist.*, *9*, 55–67, 1972.

Rice, J. R., The mechanics of earthquake rupture, in *Proceedings of the International School of Physics, 'Enrico Fermi,' Course LXXVIII on Physics of the Earth's Interior*, edited by E. Boschi, North-Holland, Amsterdam, in press, 1981.

Scholz, C. H., and J. T. Engelder, The role of asperity indentation and ploughing in rock friction, I, Asperity creep and stick-slip, *Int. J. Rock Mech. Mining Sci. Geomech. Abstr.*, *13*, 149–154, 1976.

Scholz, C. H., P. Molnar, and T. Johnson, Detailed studies of frictional sliding of granite and implications for the earthquake mechanism, *J. Geophys. Res., 77*, 6392-6406, 1972.

Solberg, P., and J. Byerlee, Strain-rate dependent strain hardening and experimental fault creep (abstract), *Eos Trans. AGU, 60*, 956, 1979.

Solberg, P. H., D. A. Lockner, R. S. Summers, J. D. Weeks, and J. D. Byerlee, Experimental fault creep under constant differential stress and high confining pressure, *Proc. U.S. Symp. Rock Mech. 19th*, 118-120, 1978.

Summers, R., and J. Byerlee, A note on the effect of fault gouge composition on the stability of frictional sliding, *Int. J. Rock Mech. Mining Sci. Geomech. Abstr., 14*, 155-160, 1977.

Tchalenko, J. S., Similarities between shear zones of different magnitude, *Geol. Soc. Am. Bull., 81*, 1625-1640, 1970.

Teufel, L. W., and J. M. Logan, Effect of shortening rate on the real area of contact and temperatures generated during frictional sliding, *Pure Appl. Geophys., 116*, 840-865, 1978.

Weeks, J., and J. Byerlee, Preliminary investigation of volume changes in crushed granite preceding stick-slip failure, *Geophys. Res. Lett., 5*, 832-834, 1978.

Laboratory Studies on Natural Gouge From the U.S. Geological Survey Dry Lake Valley No. 1 Well, San Andreas Fault Zone

John M. Logan, N. G. Higgs, and M. Friedman

Departments of Geology and Geophysics, Center for Tectonophysics
Texas A&M University, College Station, Texas 77843

Sidewall cores from depths up to 252 m taken from the U.S. Geological Survey Dry Lake Valley No. 1 well drilled into the San Andreas fault zone have been examined. Studies have investigated (1) composition, (2) texture and fabric, and (3) mechanical behavior. The composition, as determined by X ray diffraction techniques of four samples, is quite uniform with 26-33% quartz, 12-17% feldspar, 0-8% kaolinite, 4-6% illite, 9-16% chlorite, and 29-35% montmorillonite. Microscopic observations show laminated texture of shale and siltstone with irregular and discontinuous laminae up to 1 cm thick. Natural fractures cross and offset some textural boundaries but are terminated by others. Fractures induced by the coring process are very prominent. The lack of major deformational features suggests that either (1) the well is not on an active strand of the fault zone or (2) the shear strain is accommodated at a scale larger than the volume sampled. Disaggregating the gouge material and shearing it at a constant shear rate of 10^{-2} s^{-1} along a 35° saw cut within drained specimens of Tennessee sandstone at 150 MPa confining pressure, room dry, and temperatures to 300°C produce significant differences in behavior. At 25°C, major sliding occurs at σ_{12} of 115 MPa and is stable. At 300°C the comparable value of σ_{12} is raised to 180 MPa, and the displacement mode changes to stick slip. Studies on monomineralic gouges of quartz, bentonite, and chlorite show a strong control within the San Andreas gouge by chlorite on the strength properties and by montmorillonite on the sliding mode. Additionally, the increase of strength and transition from stable sliding to stick slip with increasing temperature is demonstrated to be a function of dehydration of the clays resulting in changes in the effective normal stress. The strength of the San Andreas gouge material precludes an inference of very low shear strength of the San Andreas fault zone just due to the presence of clays in the gouge zone.

The mechanical properties of rocks can be measured in the laboratory, and the processes involved in rock deformation can be observed. If the concomitant experimental research is seriously guided by theory at one hand and the facts of nature on the other, it can (1) aid in the selection of proper rheological models for analytical solutions of structural problems, and provide the physical constants of those models; (2) determine the mechanical properties of the prototype so that properly scaled models can be constructed; (3) furnish material deformed under controlled conditions for studies of deformation mechanisms.

John W. Handin (1961)

Introduction

A knowledge of the physical and mechanical properties of fault zones is desirable for many aspects of scientific research. Among others, (1) it is necessary for any intelligent program of earthquake prediction, (2) it is important for exploration and exploitation of hydrocarbons, (3) it is advantageous in thermal energy development programs, and (4) it is necessary in programs of nuclear waste disposal. Studies of surface exposures of faults offers the most obvious source of information of their geometrical and physical properties. Unfortunately, although relatively inexpensive and providing the opportunity of sampling extensive lateral surface outcrop, there are inherent disadvantages of this approach. There is at present a doubt as to the representative nature of the surface material with respect to that at depth. Has weathering and chemical activity altered the composition, texture, and fabric of the material once it is exposed at the surface? Does the fluid content accurately portray that which exists at depth? Does the process of changing the stress and temperature environment to bring the material to the surface change its properties? Thus it is desirable to examine the fault zone at depth. Unfortunately, this has seldom been done. Although some faults have been drilled during petroleum exploration, they are difficult to drill, hard to recover core material from, and generally pose technical problems that discourage their investigation. As a result, fault zones are generally eschewed. To fill this gap in our scientific knowledge, two wells have recently been drilled by the Office of Earthquake Studies of the U.S. Geological Survey into the San Andreas fault zone. The purpose of the drilling program is to (1) obtain material from some depth within the fault zone to determine composition, fabric, and physical properties, (2) allow in situ tests of the physical state of the fault zone, such as state of stress, pore pressure, permeability, temperature, etc., and (3) permit the installation of instruments at depth to measure changes within the fault zone.

In an attempt to drill to the depth of earthquake foci, the first well was drilled about 22 kilometers south of Hollister at the Dry Lake Valley site (Figure 1). The well reached a depth of 354 m, where a high pressure zone of gas was encountered and drilling was terminated. Although continuous core was not recovered, side wall cores were taken every 4.6 m from depths between 33.5 and 284.7 m. We investigated a number of these cores, which are about 2.5 cm in diameter and 4.5 cm long, over the depth interval from 120.3 to 252.9 m. Our studies have investigated (1) composition, (2) original texture and fabric, and (3) mechanical behavior. These will be discussed in this order.

Compositional Studies

Compositional determinations have been made by thin section observations and by quantitative X ray analysis. X ray analyses were done on four samples from well no. 1 (Table 1). All analyses were conducted on disaggregated core material which was homogeneously mixed. Thus the compositions represent averages of the total core material and are not made for specific portions or size fractions of the material.

Other analyses from the same well are reported by *Liechti and Zoback* [1979]. Our analysis technique is similar to that employed by them. There are, however, two significant differences in the analysis given here and that reported by *Liechti and Zoback* [1979]. First, they report that over 80% of the composition is clay, while our analysis

shows that the clay minerals compose only about 53-59% of the total sample. Second, they report significant amounts of kaolinite, where our analysis does not indicate any at three depths and only 8% at 252 m. *Liechti and Zoback* [1979] only looked at the fine-grained fraction. The recognition of kaolinite is interpretative, and additional work is in progress to try to resolve this difference.

From our analysis of the composition a number of points are worth noting. The material from the well is composed primarily of 40-57% quartz and feldspar. Clays make up the rest of the rock with montmorillonite the most abundant clay, ranging from about 30-35%. Chlorite is the next most prevalent clay, ranging from about 10-15%. Lesser amounts of illite are present, and kaolinite is absent except in one sample. Clinoptilolite, a zeolite, appears as filling in fractures. In addition to the minerals recognized by X ray analysis, microscopic observations show the presence of amorphous limonite.

The second feature of the analysis is that the material is quite homogeneous over the interval sampled. We would expect rather uniform mechanical properties and that the behavior should reflect an interaction of the quartz and feldspar with the montmorillonite and chlorite.

Studies of stability fields of clay minerals would suggest that the mineral assemblage found in either of the wells is not in equilibrium [*Deer et al.*, 1966; *Hower et al.*, 1976]. Additionally, studies of depositional shales in other areas suggest that similar mineral assemblages are common in the lower Paleozoic sequences of the eastern United States [*Parham*, 1966]. These sequences have been exposed to only moderate temperature and burial conditions. Thus there is no diagnostic information from this compositional data from the well to suggest equilibration at high temperatures or stress conditions.

Texture and Fabric Observations on Core Material

Although the sidewall core material is unoriented, it is felt that some benefit may be derived by characterizing the texture and fabric of the material. Accordingly, samples from each of the four depths were impregnated with epoxy and were thin sectioned. Microscopic observations were then made from which several relevant points can be made.

Texture

Figures 2 and 3 show characteristic textures from three depths. It is noticeable that there are grain size changes from shale to siltstone, which produce irregular banding with widths of about 1 cm. Whether these textural changes are also compositional variations is undetermined.

Grains of quartz and feldspar, having an average grain size of about 0.25 mm, are in a matrix of clay (Figure 4). Grain-to-grain contacts are present but not universal, with many clastic grains completely embedded in the clay matrix. Many quartz grains appear to be composite grains composed of smaller domains that go to extinction at dif-

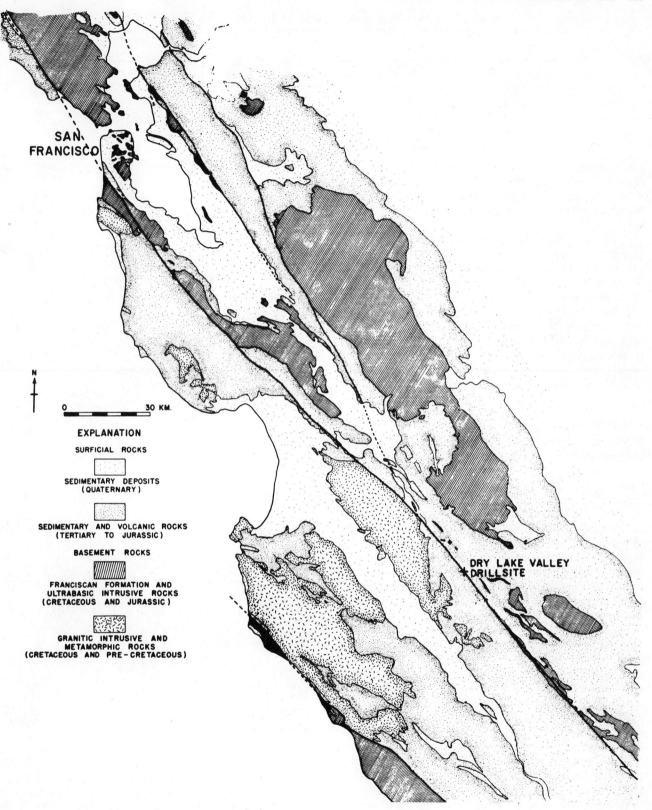

Fig. 1.   Generalized map of the San Andreas fault region around San Francisco and the Monterey peninsula. The site of the Dry Lake Valley no. 1 well is shown.

TABLE 1. X Ray Compositional Analysis of Dry Lake Valley No. 1 Well

Depth, m	Quartz, %	Feldspar, %	Chlorite, %	Kaolinite, %	Illite, %	Montmorillonite, %	Clinoptilolite, %
120.3	26	15	16	⋯	8	31	4
143.2	33	14	13	⋯	5	33	2
193.5	33	12	16	⋯	4	35	Tr
252.9	30	17	9	8	4	29	3

ferent positions under crossed polarizers. The grains are equant in shape, showing little dimensional elongation. Few intergranular microfractures or other deformation features are present except for the fractures discussed below.

From the compositional analyses we have suggested that the mechanical behavior should be an interaction of the quartz and feldspar grains and the dominant clays: montmorillonite and chlorite. These textural observations allow some refinement of this statement. As the quartz and feldspar grains appear to be largely 'floating' in the clay matrix with few grain-to-grain contacts, it would appear that the physical and mechanical behavior should be largely controlled by the response of the clay matrix. Specifically, an interaction of the montmorillonite and chlorite might be expected to control the mechanical properties.

Fractures

Natural fractures of two types are found: (1) those that are 'contained' at a shale-siltstone boundary (Figure 2a) and (2) those that cross and frequently offset the textural boundaries (Figures 2a and 3b). The first of these fractures is particularly interesting. As is observed in Figure 2a, these fractures extend at high angles to a shale-siltstone interface, where they appear to change direction abruptly, and continue along the interface. This geometry is remarkably similar to that found by *Teufel* [1979] in his study of propagating hydrofractures. He demonstrated that depending upon the normal stress across the interface and therefore the frictional resistance to shear, the fracture would either propagate along the interface or continue into the adjacent block. This 'containment' of the fracture at an interface appears similar to the geometry seen from the fault material (Figure 2a). It is of interest that Teufel demonstrated that the magnitude of the normal stress necessary to modify the interface to allow propagation of the fracture is related to the strength of the material. The containment of the fractures at interfaces of relatively weak material, i.e., the clays, suggests that high normal stresses were not operational at the time of fracture propagation. If they had been, we would expect all of the fractures to cross the interfaces. Although these fractures are contained in a manner similar to those of hydraulic fractures produced in the laboratory, there is no independent evidence to indicate that these are natural hydraulic fractures or that high pore pressures were present.

This possibility should be entertained pending additional information.

The second type of fractures are found to offset the shale-siltstone interface (Figures 2a and 3a). One of these

Fig. 2. Photographs of cut sections of sidewall cores. (a) Depth 120m. Note fractures which intersect layer boundary and are contained by it, extending along the interface. Also note offset at upper contact between shale and siltstone by fractures. (b) Depth 120 m. Scale is the same for both photographs.

features is shown at higher magnification in Figure 5a, as it is manifested at the lithologic interface. A similar fracture within the siltstone is shown in Figure 5b. These clearly show the grain size reduction associated with displacement and even some secondary fractures, such as 'pull apart' structures (arrow number 1, Figure 5b) or en echelon fractures (arrow number 2, Figure 5b), which can be used to infer the sense of displacement. As the displacements are very small and the core material unoriented, there is no evidence in the observations as to the origin of these fractures nor as to their tectonic significance.

A third type of fracture is found that is quite different but important in any physical property measurements at-

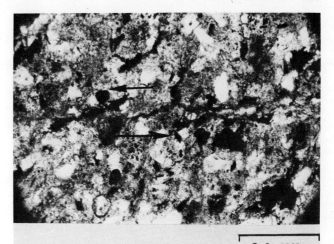

Fig. 4. Photomicrograph of sidewall cores showing quartz and feldspar grains in clay matrix. Note general lack of grain to grain contacts and equant shape of grains.

tempted on these cores. These are the fractures induced during the coring process. They are clearly discernible (Figure 6), as they extend from the edge of the cores and die out toward the center. Physical property measurements of strength, permeability, resistivity, etc. will undoubtedly be affected not only in their absolute values but also in affecting the reproducibility of the measurements.

Discussion

From both the compositional data and the textural and fabric studies, one significant observation concerning the site of the well seems warranted. Although the core material does contain natural fractures, there is not any evidence at this scale of extreme deformation analogous to that found in the laboratory-deformed specimens, where large shearing strains are produced (see the later discussion). It appears that the well is either not on an active strand of the fault zone or that the shear strain has been accommodated at a scale larger than that of the well cores. In either case, it suggests that the deformation is quite heterogeneous within the fault zone.

Mechanical Behavior

In an effort to assess the mechanical behavior of the natural gouge material and compare it with laboratory simulated gouges [*Engelder et al.*, 1975; *Logan et al.*, 1979] we have investigated a sample of the core material. The particular sample used is from a depth of 143 m and was chosen because of its fine grain size and homogeneity. The composition of this sample is primarily quartz (33%) and montmorillonite (33%), with significant quantities of chlorite (13%) and feldspar (14%), as determined from X ray diffraction analyses (see Table 1).

Although extensive experimental studies have been undertaken in the last 40 years to determine the mechanical behavior of rocks, most of this work has been concerned

Fig. 3. Photographs of cut sections of sidewall cores. (a) Depth 252 m. (b) Depth 193 m. Note fractures offsetting siltstone-shale contact. Scale same for both photographs.

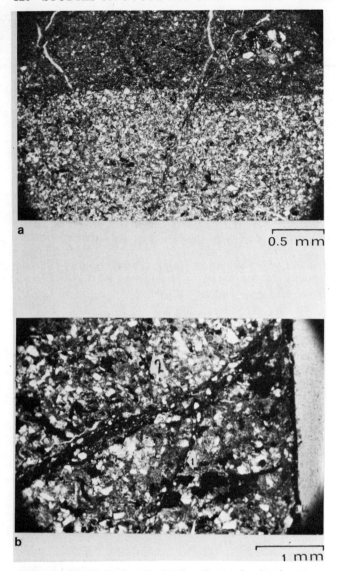

Fig. 5. Photomicrographs of sidewall cores showing fractures. (a) Depth 120 m, showing offset of interface between shale and siltstone. (b) Depth 252 m, showing fracture. Note secondary fractures indicated by white arrows. A 'pull apart' feature is shown by arrow 2.

with monomineralic rocks and a few specific granites. As a result, at present we have few insights that allow us to predict the behavior of polymineralic rocks. How do the mineral species interact? How do they contribute to a specific phenomenon? What amounts of a particular species are necessary to influence the mechanical behavior? Some insights into the last question have been provided by the studies of *Shimamoto* [1977], who looked systematically at mixtures of quartz and calcite in simulated gouge. He found that as little as 15% calcite could change the sliding behavior from stable sliding to stick slip and that the stress drops increased monotonically as the percentage increased.

Experimental Procedure

Attempts were made to cut thin slices of the gouge material to be sheared in the 35° saw cut configuration. However, the friable nature of the gouge precluded this approach, and the material was subsequently disaggregated by ball milling for 15 min. Thin layers (0.4 mm) of this powder are introduced along a 35° saw cut in right circular cylinders of Tennessee sandstone. This rock is used because it is relatively rigid under the experimental conditions and contributes very little to the inelastic deformation. It has a permeability of about 10^{-6} darcies at 150 MPa effective pressure. The specimens are about 1.9 cm in diameter and 4.2 cm long. Specimens are assembled in annealed copper jackets of wall thickness of 230 μm and deformed in a high temperature triaxial apparatus [*Friedman et al.*, 1979]. The specimen ends are lubricated with MoS_2 prior to assembly. All tests are conducted at a confining pressure of 150 MPa, corresponding to a depth of 6 km in dry crust. Temperatures are either 25° or 300°C, and the shear strain rate is constant at $10^{-2}\,s^{-1}$. All specimens are initially room dry. The lower piston contains a hole which vents the specimens to the atmosphere. For the time involved in these experiments and the permeability of the Tennessee sandstone any water driven off from the gouge layer should be dissipated through this hole in the piston.

Experimental Results

The mechanical behavior of the sheared San Andreas gouge material at 25° and 300°C is illustrated in Figure 7. The room temperature yield strength of the gouge material occurs at shear stress, σ_{12} of 115 MPa, and subsequent deformation is stable, with slight work hardening. At 300°C the yield point is higher at 180 MPa, and subsequent

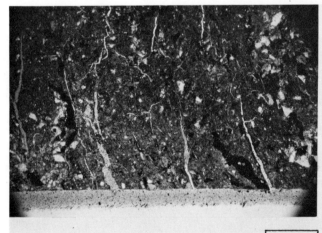

Fig. 6. Photomicrograph of sidewall core showing fractures induced by coring processes. Edge of core is at lower edge of photograph. Note fractures decrease in width toward center (upper) part of core.

deformation is unstable, with stick slip stress drops of 11 MPa. Premonitory slip is observed.

Discussion

The increase of yield and ultimate strength with increasing temperature is directly opposite to that found for most rocks [*Handin and Hager*, 1958]. Additionally, the transition from stable sliding at room temperature to stick slip at elevated temperatures is puzzling. *Brace and Byerlee* [1970] found that in granite and dunite the sliding mode changes in the opposite direction. That is, with increasing temperature stick slip is replaced by stable sliding. This is also found in sandstone [*Friedman et al.*, 1974] and in simulated gouge of calcite [*Logan et al.*, 1979].

In order to investigate this apparently anomalous behavior, three of the main constituents: quartz, montmorillonite (represented by Na-bentonite), and chlorite were studied. Monomineralic simulated gouges were tested under the same conditions as the San Andreas material. The quartz was obtained by ball milling pure quartz sand. The grain size of the quartz powder ranges up to 100 μm, with most <10 μm. The chlorite and bentonite are powders obtained from Wards Scientific Supplies.

The simulated gouge of pure quartz shows yield and ultimate strengths almost twice as high as that of the San Andreas material at 25°C (Figure 8), although both show stable sliding. At 300°C there is little change in the

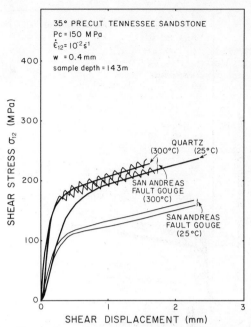

Fig. 8. Shear stress-shear displacement curves for simulated gouge of dry quartz compared with gouge from San Andreas fault zone.

strength, the slight strengthening falling within the experimental error. Under these conditions the knee on the curve is comparable with the San Andreas gouge. Additionally, at 300°C, quartz continues to show stable sliding.

A comparison of simulated gouges of pure bentonite with the San Andreas material is shown in Figure 9. At 25°C the bentonite is noticeably weaker than the San Andreas gouge material, with the knee at about 30 MPa. However, the sliding is stable, as is the San Andreas gouge. Upon increasing the temperature to 300°C, the bentonite shows an increase in strength similar to that found for the San Andreas gouge, although it is still weaker, with knee on the stress-strain curve about 50% less than that of the San Andreas material. Of equal interest is the transition from stable sliding at 25°C to stick slip at 300°C. This behavior is similar to that found for the San Andreas core material.

The behavior of the third major component of the San Andreas gouge, chlorite, is compared next (Figure 10). Here we see that at both 25°C and at 300°C the strength of chlorite is very similar to that of the San Andreas gouge. Under both conditions, however, chlorite shows only stable sliding.

It appears then, as is predicted, that the quartz (and feldspar) probably do not strongly influence either the strength or sliding mode of the San Andreas gouge. The mechanical behavior appears to be a complex interaction of the montmorillonite and the chlorite. The strength is apparently controlled by the properties of the chlorite, while the montmorillonite dictates the sliding mode. This

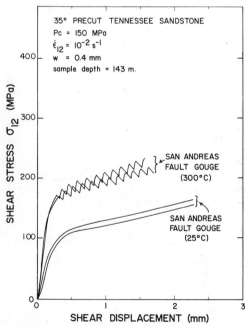

Fig. 7. Shear stress-shear displacement curves for gouge from the San Andreas fault, deformed at the conditions indicated. Two tests were conducted for each condition, the differences representing a measure of reproducibility. P_c is confining pressure, $\dot{\epsilon}_{12}$ is shear strain rate, and w is shear zone thickness.

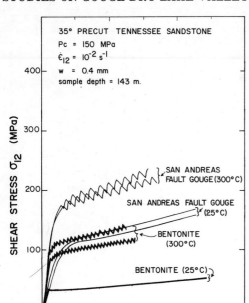

Fig. 9.   Shear stress-shear displacement curves for simulated gouge of bentonite compared with gouge from San Andreas fault zone.

interaction may be explained by the long-term mechanical properties such as strength responding to one set of compositional and physical properties, while the sliding mode reflects physical characteristics that are important on the short time scale. Such an interaction of two or more components has been reported for dilithologic specimens of sandstone and dolomite in frictional sliding [*Logan et al.*, 1972] and the mechanical behavior of granite in a partial melt condition [*van der Molen and Paterson*, 1979].

The increase in strength with increasing temperature in the San Andreas bentonite and chlorite gouges is believed to be closely related to the dehydration properties of clays as a function of temperature [*Mackenzie*, 1957; *Burst*, 1969]. Differential thermal analysis indicates that the interlayer water in smectites is lost reversibly mostly between 100° and 250°C, but some remains to about 300°C, at which point slow loss of constitutional (OH) water begins. *Summers and Byerlee* [1977] explain the low shear strength of hydrated clays at room temperature by the development of a 'pseudo pore pressure' capable of lowering the effective stress. This idea is supported by the anomalously low measured value (0.12) for the ratio of shear stress to normal stress ('coefficient of friction') for the room temperature test on bentonite (Figure 9). Adopting this as a working hypothesis, further tests were conducted on bentonite gouge as a function of temperature (Figure 11). The increased shear strength at 150°C may be explained as a partial dehydration phenomenon associated with an increased effective normal stress, the water released being vented to atmosphere via the piston port in

the apparatus. The continued increase in shear strength to 300°C may be similarly explained. Evidence that water is being driven off with increasing temperature is supported by the repeated presence of a patch of rust at the hole in the piston for tests at elevated temperatures. There is no rusty patch in the room temperature tests.

The transition from stable sliding to stick slip may also be associated with the dehydration of the clays. If a pseudo pore pressure is developed at low temperatures, as suggested by *Summers and Byerlee* [1977], this should be reduced as the temperature is increased, driving the water from the clay zone. The loss of water should reduce the pore pressure, raise the effective normal stress, and facilitate the transition to stick slip [*Humston and Logan*, 1972].

The behavior of bentonite at further increases of temperature is shown in Figure 11, where it is observed that a second transition in sliding mode occurs, this time a return to stable sliding. An explanation for this transition may lie in some recent experiments conducted by J. H. Dieterich (personal communication, 1978). He has demonstrated that by entirely removing adsorbed water by argon drying, during frictional sliding of two blocks of rock, stick slip is completely suppressed. Conversely, the presence of trace amounts of water or water vapor will permit unstable behavior, given that the conditions are otherwise favorable. Taking this as a working hypothesis, one might explain this transition by suggesting that at 450°C all of the water is removed from the bentonite gouge producing the transition to stable sliding. We would hypothesize that

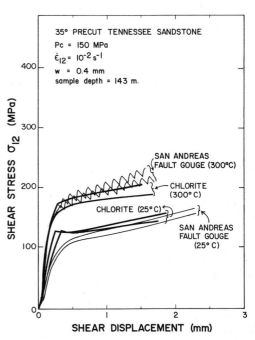

Fig. 10.   Shear stress-shear displacement curves for simulated gouge of chlorite compared with gouge from San Andreas fault zone.

a similar change in sliding mode should be observed in the San Andreas gouge material with increasing temperature. Support for the dewatering hypothesis is found by varying the heating time at a constant temperature, 300°C (Figure 12). The heating time after the application of confining pressure but prior to the superposition of the axial load was varied from 5×10^3 (<2 hours) to 2×10^5 s (>2 days). Two effects are clear: (1) the strength of the gouge increases with increased heating time and (2) there is a transition from stick slip to stable sliding. This strongly supports the contention that both properties are influenced by the loss of water. It is possible that a transition from montmorillonite to illite is also occurring, although X ray analysis has not confirmed this hypothesis. The time-dependent nature of the changes may result from (1) a time-dependent removal of the water from the clay structure, (2) the time necessary for the water to move out of the gouge zone, or (3) a combination of both effects.

One final point should be made concerning the mechanical behavior of the San Andreas gouge material. Neither at room temperature, where a low effective pressure is predicted because of the creation of a pseudo pore pressure, nor at temperatures to 300°C is the material noticeably weak. Significant displacement does not take place along the shear zone until shear stresses of about 115 MPa are reached. If the strength properties are largely controlled by the mechanical behavior of the chlorite, we would not expect to see weakening until temperatures of about 450°C (Figure 13), where presumably intra-

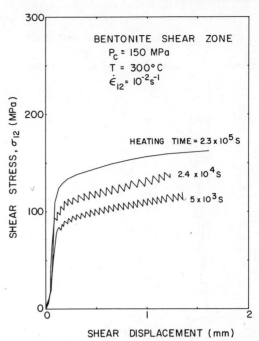

Fig. 12. Shear stress-shear displacement curves for simulated gouge of Na-bentonite, deformed at 300°C. Heating time is that after application of confining pressure and before differential axial stress.

crystalline flow becomes important. The possibility of clays acting as a low strength gouge material along the San Andreas fault has been suggested [*Wu*, 1978; *Wang et al.*, 1980]. This appears as an explanation for the postulated condition of low shear stress [*Brune et al.*, 1969; *Zoback and Roller*, 1979; *Raleigh and Evernden*, this volume]. Our data do not support the contention that clays, even if they persist to greater depth, offer a mechanical explanation for low shear stress of the order of tens of bars. An alternate method of obtaining low strengths such as high pore pressure seems to be necessitated by the laboratory data at this time.

Even appealing to lower displacement rates does not a priori provide an explanation. Frictional sliding strengths, in contrast to those of intact rock, appear to increase with decreasing displacement rate [*Teufel and Logan*, 1975] as long as the real area of contact is some fraction of the apparent area of contact. Thus as long as the displacement occurs along discrete surfaces, as argued later, and the time-dependent growth of asperities dominates the behavior of these surfaces, the zone should be expected to be stronger with decreasing displacement rates.

Microscopic Observations

Observations of thin sections of the experimentally deformed San Andreas fault gouge have revealed some interesting deformation fabrics and some of the first differences that we have been able to discern between stable sliding and stick slip in gouge material.

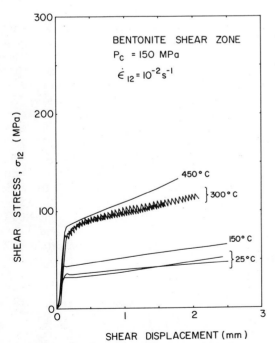

Fig. 11. Shear stress-shear displacement curves for simulated gouge of Na-bentonite deformed at the conditions indicated.

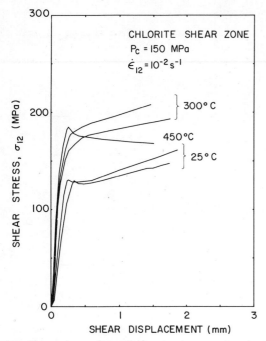

Fig. 13. Shear stress–shear displacement curves for simulated gouge of chlorite deformed at the conditions indicated.

The room temperature tests, where stable sliding is present, are characterized by a quasi-homogeneously deformed shear zone, with pervasive development of microscopic Riedel shear orientations [*Logan et al.*, 1979] manifest by the birefringence of aligned clay particles. The majority of montmorillonite particles have clearly been reoriented parallel to the R1 and R2 Riedel shear directions (Figure 14a). There are not any fractures (i.e., loss of cohesion) in the orientation of Riedel shears, these usually being the most common structures associated with experimentally deformed shear zones. Quartz and feldspar clasts appear randomly oriented and are largely undeformed except for brittle fracturing and strain shadows known to be present prior to deformation. Offsets of quartz clasts on R2 Riedel shears measure as much as 25 μm (Figure 14a). This value is large in relation to the thickness of the layer (400 μm), and it is thought that combined operation of differential shearing on both R1 and R2 Riedel shears at the scale of the clay matrix particles can overcome the displacement incompatibilities apparently imposed by the rigid nature of the shear zone boundary. The more ductile minerals for the conditions of the test (i.e., chlorite and oxides) undergo large strains within the shear zone, forming wavy stringers, oriented 150–180° to the direction of shear. Detailed observation of these stringers reveals that their waviness is imparted by heterogeneities in the displacement field of the clay matrix. The texture may be due to differential shearing on the microscopic Riedel shears discussed above (see Figure 15).

At 300°C, where stick slip is observed, the fabric displays many of the features observed at room temperature and some additional ones. The clay matrix still deforms quasi-homogeneously by shearing displacements on microscopic Riedel shears. Ductile stringers of chlorite and oxides are very prominent, attaining aspect ratios of 75. Two outstanding features exist which were not observed in the room temperature tests: (1) highly elongated ductile stringers of quartz and feldspar and (2) abundant discrete shear surfaces in anastomosing R1 and layer parallel orientations lined with an optically isotropic material.

The elongated stringers of quartz and feldspar occur in orientations 150°–180° to the shear direction and display the characteristic waviness imparted by differential movements in the clay matrix (Figure 16). They often highlight regions of locally high shear strain, such as at the gouge-rock interface and along the discrete shear surfaces within the shear zone (Figure 16). Their extremely large elongations, with aspect ratios as high as 85, reflect a cataclastic ductility of the material not observed in pure quartz shear zones under similar conditions. It is postulated that the dehydration of montmorillonite produces a wet environment during the experiment. This water apparently enhances the development of microfractures. Thus the ductile clay material appears to enhance the cataclastic ductility of quartz. The shear surfaces parallel to the shear zone interface observed at 300°C are extremely discrete, very linear, forming anastomosing networks, and are lined by a material that is isotropic (Figure 14b). Whether it is glass due to melting [*Friedman et al.*, 1974] or ultrafine material created by comminution is undetermined at present.

It is suggested that stick slip displacements during the 300°C tests take place along these discrete, planar zones, both within the shear zone and at the gouge-rock interface [*Logan and Shimamoto*, 1976]. In light of this additional work, modification of this concept appears necessary. During small displacements or shear strains it does appear that most of the displacement takes place at or close to the gouge-host rock interface. As a result, the mechanical behavior is an interaction between the two materials [*Logan and Teufel*, 1978]. However, as the displacement and/or the shearing strain increase, the major displacement appears to shift to the fractures parallel to the shear zone boundary. The displacement is not continuously along the same fracture but may shift from one to another and back to the first, even possibly to the host rock interface for brief periods. Thus it is now suggested that the properties of host rock-gouge interface and the properties of the gouge material are both important in controlling the mechanical behavior.

Conclusions

At this time a number of conclusions from this study seem warranted:

1. The composition of the sidewall cores is about 41–47% quartz and feldspar in the silt to shale size range, with clays composing the remaining volume of the rock. Of the

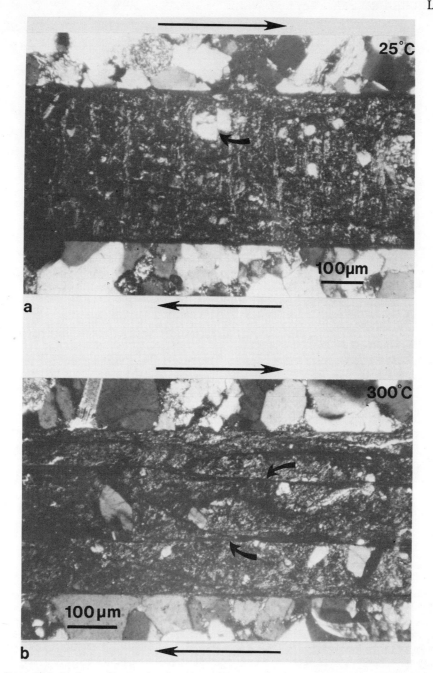

Fig. 14. Crossed nicols photomicrographs of sections of San Andreas fault gouge, sheared at 150 MPa confining pressure, 25° and 300°C, and 10^{-2} s^{-1} shear strain rate. (a) At 25°C, note alignment of clay particles along R1 and R2 Riedel shears, offset of quartz clast along R2 shear (curved arrow), and absence of discrete, through-going R1 shear fractures. (b) At 300°C, note discrete through-going shear surfaces parallel to R1 Riedel shear orientation and the shear zone boundary. These surfaces are isotropic under crossed nicols. Straight arrows indicate overall sense of shear.

latter minerals, montmorillonite (29–35%) and chlorite (9–16%) appear to affect the mechanical behavior.

2. Texture and fabric studies, although limited by the lack of oriented cores, do show natural fractures, which are both contained by and offset by the textural boundaries.

3. There is no other evidence at the scale of the samples (2.5 cm in diameter and 4.5 cm long) of deformation.

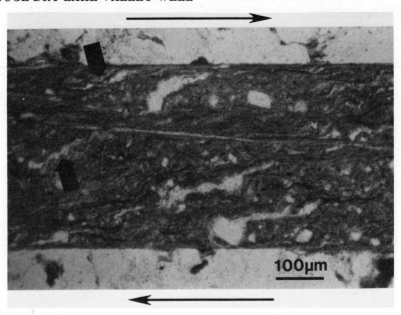

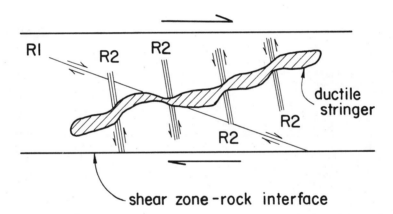

Fig. 15. Photomicrograph and diagram illustrating formation of wavy, ductile stringers in deforming shear zone. The photomicrograph, imaged in plane polarized light, is of San Andreas fault gouge sheared at 150 MPa confining pressure, 300°C, and 10^{-2} s^{-1} shear strain rate. Note ductile stringers (large arrows, for example), oriented 150°–180° to shear direction. Combined displacements on R1 and R2 Riedel shear sets in the clay matrix are thought to account for their orientation, and heterogeneities in the displacement field account for their waviness. Large arrows indicate overall sense of shear. The diagram illustrates the mechanism more clearly.

Noticeably absent are fractures or small faults analogous to those found in laboratory-deformed specimens that have undergone large shearing strain. Also absent are elongated grains of quartz and feldspar which have been found at other locations along the San Andreas and produced in laboratory experiments on the same material. This absence of diagnostic features may be a function of the strain being accommodated at another site within the fault zone or at a larger scale. Clearly the shear strain is not pervasive.

4. Triaxial shearing tests of the natural gouge, disaggregated and spread along a precut, show stable sliding at 25°C and a confining pressure of 150 MPa, with σ_{12} of 115 MPa at the onset of significant sliding. At 300°C, σ_{12} for a similar point on the stress-strain curve increases to 180 MPa, and the sliding mode changes to stick slip. A comparison with studies of simulated gouge shows marked similarities in that the natural material apparently reflects a subtle combination of the short-term frictional properties of montmorillonite to control the sliding mode

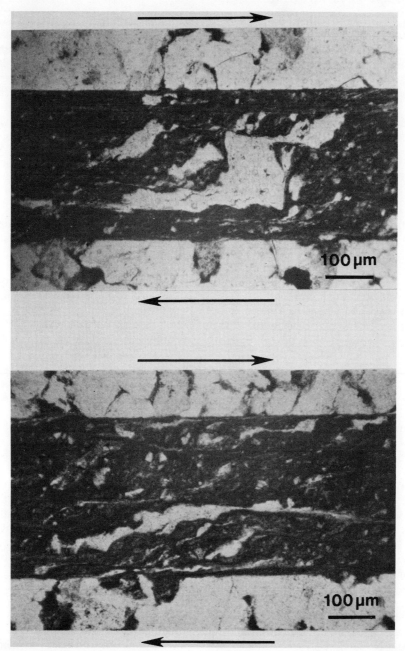

Fig. 16. Photomicrographs imaged in plane polarized light, further illustrating the cataclastic ductility of quartz, feldspar, and oxide clasts within a deforming clay matrix at a confining pressure of 150 MPa, 300°C, and 10^{-2} s^{-1} shear strain rate. Note accentuation of regions of locally high shear strain by shape of deformed clasts along R1 shears and near the shear zone boundary. Arrows indicate sense of shear.

and the long-term strength properties of quartz and chlorite to dominate the sliding stress.

5. An increase of strength and a transition from stable sliding to stick slip are found with an increase in temperature to 300°C. Experimental studies on chlorite and bentonite under similar temperatures and different heating times suggest that these changes in properties are associated with the release of water during the heating process.

This decreases the effect of a pseudo pore pressure within the clay structure and increases the effective normal stress resulting in the expected changes in strength and sliding mode. This dehydration of the clays appears to be time-dependent.

6. The shear stress necessary to cause deformation of the shear zones in the laboratory preclude clays by themselves as offering a mechanical explanation for low shear

stress conditions along fault zones. Other mechanisms, such as high pore pressure, appear necessary.

7.   Stable sliding in this gouge is characterized by more homogeneous behavior, while stick slip is marked by displacements along discrete boundary-parallel zones. Ductile deformation of the quartz clasts by cataclastic flow shows aspect ratios up to 85, suggesting very large shearing strains. Such ductility is thought to be controlled by the behavior of the clays which are postulated to release water at elevated temperatures.

Acknowledgments. We would like to thank John Handin for his profound influence on our outlook on scientific research. It is probably no understatement that none of us would be contributing to this work that has been advocated for many years by John without this remarkable association. We would like to thank Barry Raleigh, who led the effort to drill the San Andreas and to Mark Zoback who provided the core material and critically read the manuscript. Mineralogy, Incorporated provided us with the X ray analysis. Jack Nash, especially, has spent long hours trying to solve the compositional differences. This work was generously supported by the U.S.G.S. grant 14-08-0001-17677.

References

Brace, W. F., and J. D. Byerlee, California earthquakes: Why only shallow focus, *Science, 168,* 1573-1575, 1970.

Brune, J. N., T. L. Henyey, and R. F. Roy, Heat flow and rate of slip along the San Andreas fault, California, *J. Geophys. Res., 74,* 3821-3827, 1969.

Burst, J. F., Diagenesis of Gulf Coast clayey sediments and its possible relation to petroleum migration, *Am. Assoc. Pet. Geol. Bull., 53,* 73-93, 1969.

Deer, W. A., R. A. Howie, and J. Zussman, *An Introduction to the Rock Forming Minerals,* 528 pp., Longman Group Limited, London, 1966.

Engelder, J. T., J. M. Logan, and J. Handin, The sliding characteristics of sandstone on quartz fault-gouge, *Pure Appl. Geophys., 113,* 69-86, 1975.

Friedman, M., J. M. Logan, and J. A. Rigert, Glass-indurated quartz gouge in sliding-friction experiments on sandstone, *Geol. Soc. Am. Bull., 85,* 937-942, 1974.

Friedman, M., J. Handin, N. G. Higgs, and J. R. Lantz, Strength and ductility of four dry igneous rocks at low pressures and temperatures to partial melting, *Proc. U.S. Symp. Rock Mech., 20th,* 35-50, 1979.

Handin, J. W., and R. V. Hager, Jr., Experimental deformation of sedimentary rocks under confining pressure: Tests at high temperature, *Am. Assoc. Pet. Geol. Bull., 42,* 2892-2934, 1958.

Hower, J., E. V. Eslinger, M. E. Howers, and E. A. Perry, Mechanism of burial metamorphism of argillaceous sediment, 1, Mineralogical and chemical evidence, *Geol. Soc. Am. Bull., 87,* 725-737, 1976.

Humston, J., and J. M. Logan, Stick slip in Tennessee sandstone (abstract), *Eos Trans. AGU, 53,* 512, 1972.

Liechti, R., and M. D. Zoback, Preliminary analysis of clay gouge from a well in the San Andreas fault zone in central California, in Proceedings of Conference VIII: Analysis of Actual Fault Zones in Bedrock, *Geol. Surv. Open File Rep. U.S., 79-1239,* 268-275, 1979.

Logan, J. M., and T. Shimamoto, The influence of calcite gouge on the frictional sliding of Tennessee sandstone (abstract), *Eos Trans. AGU, 57,* 1011, 1976.

Logan, J. M., and L. W. Teufel, The influence of rock type on the sliding behavior of rock gouge systems with quartz gouge (abstract), *Eos Trans. AGU, 59,* 1208, 1978.

Logan, J. M., T. Iwaski, M. Friedman, and S. Kling, Experimental investigation of sliding friction in multi-lithologic specimens, *Eng. Geol. Case Hist., 9,* 55-67, 1972.

Logan, J. M., M. Friedman, N. Higgs, C. Dengo, and T. Shimamoto, Experimental studies of simulated gouge and their application to studies of natural fault zones, in Proceedings of Conference VIII: Analysis of Actual Fault Zones in Bedrock, *Geol. Surv. Open File Rep. U.S., 79-1239,* 305-343, 1979.

Mackenzie, R. C. (Ed.), *The Differential Thermal Investigation of Clays,* Mineralogical Society, London, 1957.

Parham, W. E., Lateral variations of clay mineral assemblages in modern and ancient sediments, *Proc. Int. Clay Conf., 2,* 135-146, 1966.

Raleigh, B., and J. Evernden, The case for low deviatoric stress in the lithosphere, this volume.

Shimamoto, T., Effects of fault gouge on frictional properties of rocks: An experimental study, Ph.D. dissertation, 198 pp., Tex. A&M Univ., College Station, 1977.

Summers, R., and J. Byerlee, A note on the effect of fault gouge composition on the stability of frictional sliding, *Int. J. Rock Mech. Mining Sci., 14,* 155-160, 1977.

Teufel, L. W., An experimental study of hydraulic fracture propagation in layered rock, Ph.D. dissertation, 99 pp., Tex. A&M Univ., College Station, 1979.

Teufel, L. W., and J. M. Logan, Time-dependent friction in Tennessee sandstone (abstract), *Eos Trans. AGU, 56,* 1061, 1975.

van der Molen, I., and M. S. Paterson, Experimental deformation of partially-melted granite, *Contrib. Mineral. Petrol., 70,* 299-318, 1979.

Wang, C., N. Mao, and F. T. Wu, Mechanical properties of clays at high pressure, *J. Geophys. Res., 85,* 1462-1468, 1980.

Wu, F. T., Mineralogy and physical nature of clay gouge, *Pure Appl. Geophys., 116,* 655-689, 1978.

Zoback, M. D., and J. C. Roller, Magnitude of shear stress on the San Andreas fault: Implications of a stress measurement profile at shallow depth, *Science, 206,* 445, 1979.

Pore Volume Changes During Frictional Sliding of Simulated Faults

L. W. TEUFEL

Geomechanics Division, Sandia National Laboratories, Albuquerque, New Mexico 87165

Dilatancy is well established as preceding and accompanying the fracturing of intact rock in the laboratory, but few detailed experiments have been done to determine if dilatancy precedes slip along faults and/or occurs during frictional sliding. The general problem of dilatancy along simulated faults has been investigated by monitoring pore volume changes in drained, triaxial compression, pore pressure experiments on 35° precuts of Coconino sandstone. Tests were conducted at effective confining pressures to 90 MPa, an axial displacement rate of 5×10^{-5} cm/s, and fault displacements up to 8 mm. All of the tests show an initial compaction of the specimens upon application of the axial load. As the peak stress for the onset of slip is reached and sliding begins, compaction decreases to zero, and dilatancy occurs. The amount of dilatancy increases with increasing fault displacement. With increasing normal stress there is an increase in the amount of dilatancy for equivalent fault displacements. Petrographic studies of the deformed specimens show that microfracture development is localized within a narrow zone along the fault. These fractures occur only after slip has occurred and are considered to be contact-induced extension fractures, developing directly from tensile stress components which are generated from high stress concentrations at grain to grain (asperity) contacts. The density of these microfractures increases with increasing fault displacement and increasing normal stress. Also, the average length of these microfractures increases with increasing fault displacement and increasing normal stress. Calculations of the total microfracture volume along the fault for different displacement and normal stress conditions correlate with the measured increase in dilatant pore volume. It is concluded that for simulated faults deformed in triaxial compression, dilatancy can occur only during frictional sliding and that prior to slip, compaction occurs. In addition, dilatancy during frictional sliding is a direct consequence of an increase in the fracture porosity resulting from the creation, opening, and extension of microfractures along the fault.

Introduction

A period of optimism in earthquake prediction was created about 5-10 years ago by attempts to correlate geophysical precursors with a dilatancy diffusion model [*Scholz et al.*, 1973; *Whitcomb et al.*, 1973]. Aspects of this model have received considerable discussion and attention. A 'dry' dilatancy model also has been proposed [*Mjackin et al.*, 1975] to provide an alternative hypothesis for some of the postulated precursors. Both of these models are based upon the concept of strain accumulation resulting in preseismic deformation (e.g., dilatancy) and corresponding changes in geophysical properties in the focal region in preparation for the earthquake. The expected changes in geophysical properties are based primarily on laboratory observations of phenomena associated with dilatancy preceding failure of intact, stressed, dry or fluid-saturated rock specimens [*Brace*, 1978]. However, since most shallow focus earthquakes take place along existing faults and since friction is probably the major controlling factor [*Brace and Byerlee*, 1966], the conclusions derived from laboratory experiments on intact rocks alone may be questioned. In fact, the presence of precursors, such as V_p/V_s ratios, has been questioned [*Lockner and Byerlee*, 1978], and the general validity of either model is now in doubt.

Although the presence and position of dilatant zones along faults have been postulated, to my knowledge no area of fractured rock has actually been recognized in the field as acting in a dilatant fashion. Laboratory studies have also failed to substantiate the importance of dilatant zones in producing significant changes in geophysical measurements prior to or during displacement along existing fractures or simulated joints [*Wang et al.*, 1975*a*, *b*; *Logan*, 1978]. Thus it appears that both field and laboratory studies to date have failed to reach the levels of confidence predicted by the early wave of optimism in the dilatancy models. Therefore it is necessary to ask the fundamental question, does dilatancy exist prior to and/or during movement along existing faults under high effective confining pressure? If dilatancy does exist, then what is the size and extent of the dilatant zone?

As dilatancy was first recognized in the laboratory, it appears reasonable to continue laboratory studies as a first approach to the problem. In the present study, a series of drained, triaxial compression, pore pressure experiments have been performed on 35° precut specimens of Coconino sandstone in order to determine the influence of confining pressure and normal stress across the sliding surface upon the dilatant characteristics of simulated faults. The design of the pore pressure monitoring system allows pore volume measurements to be made during slip along the faults. In addition, detailed thin section examination of selected specimens provides direct implications as to the relationship between volume changes and deformation in the precut specimens.

Experimental Procedure

In this work the general problem of dilatancy along sliding surfaces has been investigated for drained, triaxial compression, pore pressure experiments. Volume changes were measured by monitoring the adjustment in pore fluid volume required to maintain a constant pore pressure during the course of the test. This technique is essentially identical to that used by *Handin et al.* [1963], and it is also widely used in soil mechanics. The technique is applicable for measuring volume changes for intact specimens undergoing large deformations as well as pore volume measurements associated with slip along a single discontinuity. However, it is limited in application by the requirement of adequate permeability and interconnection of pores. In using this technique to measure pore volume changes associated with movement along a simulated fault, two conditions are necessary: (1) communication of the pore fluid (distilled water) between the simulated fault and the pressure transducer and (2) an ability to determine how much of the volume change is produced by deformation associated with sliding and how much is produced by elastic effects of the intact rock. In order to achieve these conditions right-circular cylindrical specimens with a precut 35° to the cylinder (load) axis of Coconino sandstone were deformed at effective confining pressures to 90 MPa, an axial displacement rate of 5×10^{-5} cm/s, and fault displacements up to 8 mm. In all of the tests the pore pressure was a constant 10 MPa. The porosity and permeability of this particular block of Coconino sandstone (8% and 84 md, respectively) is such as to insure equilibrium of pore pressure in the system under the displacement rate utilized.

The apparatus used in these experiments is fully described by *Handin et al.* [1972]. Force measurements were recorded on the upper piston with an accuracy of ±5 × 10⁷ dyn. The displacement measurements were made with a linear variable differential transducer mounted parallel to the long axis of the precut specimens, inside the pressure vessel, with an accuracy of ±1 μm. Shear displacement along the fault was limited to 8 mm to reduce alignment problems. Interstitial pore fluid (distilled water) was pumped into the specimens through the hollow loading piston, which was connected to an external pressure cell. This system is sensitive to pore volume measurements of 2 mm³.

The 35° precut specimens were 10.0 cm long and 4.76 cm in diameter. The precut angle was accurate to within 0.5° for each half of the specimen. The ends of the assembled specimen were within 0.1° of parallelism. The sliding surface and ends of the specimens were surface ground with an 80-grit wheel. Four polyolefin jackets were used to isolate each of the specimens from the confining fluid. Lubrication in the form of a thin layer of Molykote (MoS_2) was applied to the specimen-piston interfaces.

In order to ascertain how the deformation is distributed within the specimen and provide some implications as to the relationship between dilatancy and the internal mechanisms of deformation, observational studies were done on thin sections of the deformed specimens. After each experiment, specimens were impregnated with epoxy resin. Thin sections were then made parallel to the long axis of the specimen and normal to the plane of the fault zone and examined with a petrographic microscope.

Results

Deformational and Dilatant Character of Precut Specimens

Specimens containing 35° precuts were deformed at an axial displacement rate of 5×10^{-5} cm/s and under effective confining pressures from 15 to 90 MPa (Figure 1). At all of these effective confining pressures the axial shortening of the specimen was achieved by stable sliding along the precut. With increasing confining pressure there is a linear increase in the yield stress required to initiate slip along the precut, indicating that the coefficient of friction at the initiation of slip, μ_s, is essentially a constant independent of confining pressure and the normal stress across the sliding surface. These results are consistent with *Byerlee*'s [1978] summary of the rock friction literature. For water-saturated Coconino sandstone at these test conditions, μ_s is 0.78.

Plots of pore volume change as a function of axial displacement (Figure 2) indicate that an initial compaction of the precut specimen occurs upon application of the axial load at all effective confining pressures which were used. In this study, compaction is shown as negative volumetric strain. At each effective confining pressure, as the yield stress for the onset of slip is reached and sliding begins, the pore volume change/displacement curve becomes non-

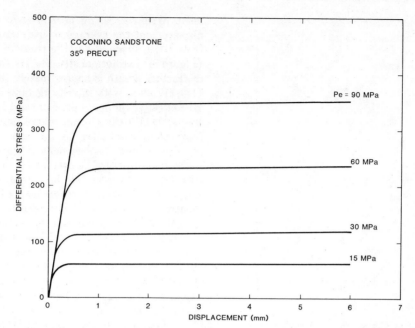

Fig. 1.   Differential stress versus axial displacement curves as a function of effective confining pressure.

linear, corresponding to the nonlinear shape of the differential stress/displacement curve. As sliding begins, compaction of the specimens gradually decreases to zero, and dilatancy dominates, as is indicated by the subsequent reversal in the slope of the pore volume change curve. With increasing effective confining pressure there is an increase in the yield stress required for the onset of sliding and a corresponding increase in the compaction of the specimen.

After the onset of sliding, increasing displacement along the precut results in an increase in the amount of

dilatancy. Increasing the normal stress across the sliding surface (by increasing the confining pressure) results in an increase in the amount of dilatancy for equivalent fault displacements. After 6-mm total displacement along the precut the dilatant pore volume change associated with movement along the precut increases from 8 mm^3 at 31-MPa normal stress to 512 mm^3 at 208-MPa normal stress (Figure 3). The normal stress magnitudes plotted in Figure 3 are the maximum normal stress across the precut calculated from the differential stress/displacement curves.

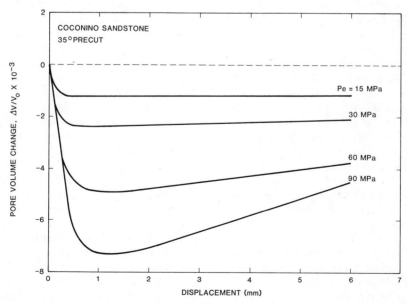

Fig. 2.   Pore volume change versus axial displacement curves as a function of effective confining pressure.

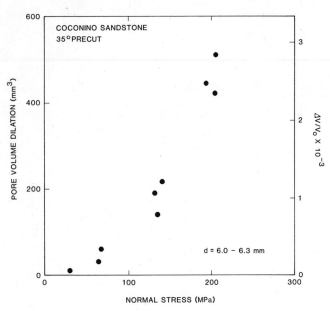

Fig. 3. Dilatant pore volume change versus normal stress for 35° precuts after 6- to 6.3-mm fault displacement.

Observational Studies

Microscopic examination of thin sections taken from the deformed precut specimens clearly indicates that microfracture development is localized within a narrow zone along the simulated fault and that a thin layer of fine grain fault gouge is present at the higher normal stress and fault displacement conditions (Figure 4). The fault gouge has been generated from the fractured sliding surface. At 90-MPa confining pressure and normal stresses from 186 to 203 MPa the width of the fault gouge increases with increasing fault displacement to an average width of 262 ± 70 μm after 7.3-mm displacement (Figure 5). Fault gouge development decreases rapidly with decreasing normal stress (Figure 6).

Microfractures along the simulated fault are often wedge shaped and generally die out within a few grain diameters of the sliding surface. Most of the microfractures run through quartz grains and rarely along grain boundaries. Careful measurements and plotting of the direction of the microfractures indicate that they are oriented at an acute angle of about 20°–40° to the fault surface, dipping toward the shear direction of the overlying fault block at the smaller fault displacement conditions (Figure 7a). However, with increasing fault displacement at the higher normal stresses the orientation becomes slightly more diverse with the occurrence of nearly vertical fractures (Figure 7b).

The average width of the microfractures is about 3 μm and rarely exceeds 6 μm in their present unstressed state. The average length of a continuous microfracture is largely dependent on the normal stress across the fault and to a lesser extent on the amount of fault displace-

ment. With increasing normal stress after about 6-mm displacement the average microfracture length increases from 122 ± 40 μm at 35 MPa to 264 ± 108 μm at 203 MPa (Figure 8). For normal stresses less than 138 MPa the microfracture length is approximately that of the grain size (140–210 μm). With increasing fault displacement at 80-MPa effective confining pressure and normal stresses from 186 to 203 MPa the average microfracture length increases from 180 ± 52 μm after 1.8-mm displacement to 272 ± 105 μm after 7.3-mm displacement (Figure 9).

The microfracture density along the simulated fault is also largely dependent on the normal stress and fault displacement. The density of microfractures was obtained by counting the number of microfractures within 1 mm of both sides of the precut sliding surface and dividing by the observed area. It should be noted that no microfractures were observed outside the area examined. After about 6-mm displacement and with increasing normal stress from 31 to 203 MPa the average microfracture density along the simulated fault increases from 1.2 ± 0.9 to 14.1 ± 4.2 fractures/mm^2, respectively (Figure 10). With increasing fault displacement at 90-MPa confining pressure and normal stress from 186 to 203 MPa the microfracture density increases from 3.7 ± 2.1 fractures/mm^2 after 1.8-mm displacement to 15.2 ± 4.3 fractures/mm^2 after 7.3-mm displacement (Figure 11).

Discussion
General

Although dilatancy is well established as preceding and accompanying the failure of intact rock in the laboratory [*Brace*, 1978], few detailed experiments have been done to determine if dilatancy precedes slip along faults and/or occurs during frictional sliding. In this paper the general problem of fault dilation has been investigated by monitoring pore volume changes in drained, triaxial compression, pore pressure experiments on 35° precuts of Coconino sandstone. The results of these experiments clearly indicate that prior to slip, only compaction occurs but that dilatant pore volume changes can occur during frictional sliding. With increasing fault displacement there is an increase in the amount of dilatancy. With increasing normal stress there is an increase in the amount of dilatancy for equivalent fault displacements.

For intact rock, dilatancy is presumed to be a consequence of the development and opening of microcracks in the rock [e.g., *Brace et al.*, 1966; *Bieniawski*, 1967; *Hallbauer et al.*, 1973]. Some of these microcracks eventually are thought to coalesce and form the throughgoing fracture associated with macroscopic failure of rock in the brittle regime. Following similar reasoning, it may be assumed that the mechanisms producing dilatancy during the frictional sliding of the precut specimens are the development and opening of microfractures along the sliding surface. Detailed microscopic examination of thin sections of the deformed precut specimens supports this

Fig. 4.   Photomicrograph of a 35° precut sliding surface of Coconino sandstone. Large arrows indicate direction of slip, and small arrows point to microscopic feather fractures. The scale is 300 μm. Test conditions were 203-MPa normal stress and 6.0-mm displacement. Note the fine grain fault gouge separating fault surfaces.

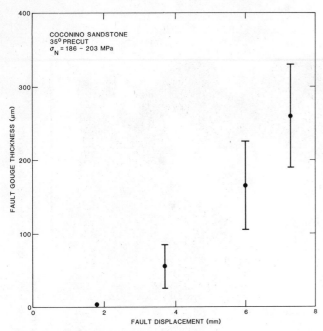

Fig. 5. Fault gouge thickness versus fault displacement for 35° precuts subjected to maximum normal stress of 186-203 MPa. Gouge thickness is given as an average plus or minus one standard deviation.

hypothesis, showing only localized microfracture development along the sliding surface.

Nature of Microfractures

The nature and orientation of most of the microfractures adjacent to the fault are consistent with the observa-

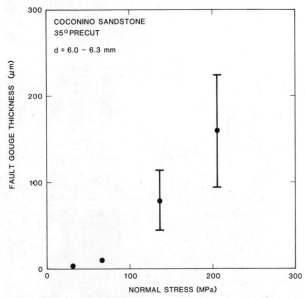

Fig. 6. Fault gouge thickness versus normal stress for 35° precuts after fault displacements of 6-6.3 mm. Gouge thickness is given as an average plus or minus one standard deviation.

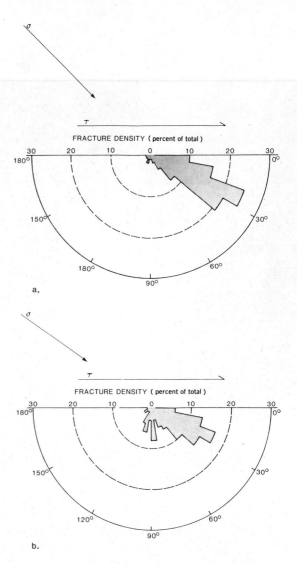

Fig. 7. Histograms showing fracture orientation along sliding surface of 35° precuts of Coconino sandstone. The τ arrow indicates the slip direction of the overlying fault block, and the σ arrow indicates the axial load direction. (*a*) Test conditions were effective confining pressure equal to 90 MPa, normal stress equal to 186 MPa, and fault displacement equal to 1.8 mm. (*b*) Test conditions were effective confining pressure equal to 90 MPa, normal stress equal to 203, and fault displacement equal to 6.0 mm.

tions of *Conrad and Friedman* [1976], who have identified this type of fracture development, immediately adjacent to the sliding surface, as microscopic feather fractures. They found that these fractures occur only during frictional sliding. These fractures are considered to be contact-induced extension fractures, developing directly from tensile stress components which are generated from high stress concentrations at grain to grain (asperity) contacts and the resulting shear tractions at the contact interfaces. Elastic analyses by *Johnson* [1975] have shown that as slip occurs, local frictional heterogeneities can occur at the

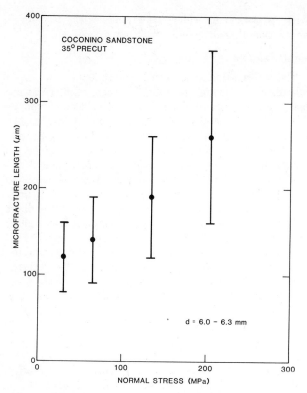

Fig. 8. Microfracture length versus normal stress for 35° precuts after fault displacements of 6-6.3 mm. The microfracture length is given as an average plus or minus one standard deviation.

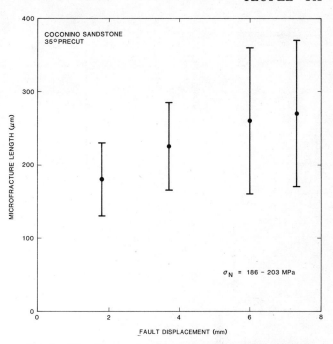

Fig. 9. Microfracture length versus fault displacement for 35° precuts subjected to maximum normal stress of 186-203 MPa. The microfracture length is given as an average plus or minus one standard deviation.

contact interfaces, resulting in the development of tensile stresses in the immediate vicinity of the edge of the contact. Extending downward from the contact interface, the tensile stresses decay rapidly. According to Johnson this tensile stress region appears to play the dominant role in both the initiation and the growth of contact-induced fractures. Wedge-shaped fractures can develop once the tensile strength of the material is exceeded. The morphology and size of these fractures would depend on the magnitude and distribution of the stresses at individual asperity contacts. An increase in the magnitude of the applied stresses on the asperities results in a larger tensile stress region beneath the contact interface and a corresponding increase in the potential fracture length. Similarly, observations of the average microfracture length adjacent to the sliding surface of Coconino sandstone clearly show an increased fracture length with increasing normal stress (Figure 8).

As a consequence of the discrete nature of the real area of contact on a sliding surface, different asperity contacts are formed during sliding at successive intervals of time. Accordingly, the density of contact-induced fractures is directly related to the displacement history of the real contact area on the sliding surface as reflected by the size and density of asperity contacts, provided that the stresses at individual contacts are sufficient to cause frac-

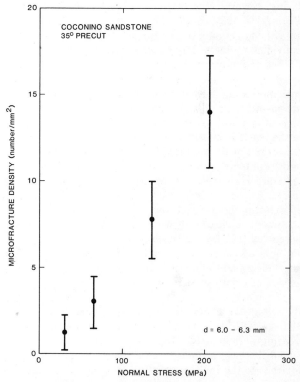

Fig. 10. Microfracture density versus normal stress for 35° precuts after fault displacements of 6-6.3 mm. The microfracture density is given as an average plus or minus one standard deviation.

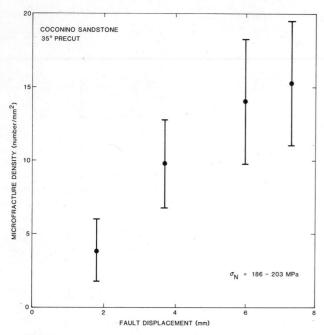

Fig. 11. Microfracture density versus fault displacement for 35° precuts subjected to maximum normal stress of 186–203 MPa. The microfracture density is given as an average plus or minus one standard deviation.

turing. Several studies [e.g., *Bowden and Tabor*, 1950; *Kragelskii*, 1965; *Teufel*, 1976] have shown that the real area of contact on a sliding surface and the stresses at the individual contacts increase with increasing normal stress. Correspondingly, the measured density of microfractures along the Coconino sandstone sliding surface also increase with increasing normal stress (Figure 10). Moreover, with increasing fault displacement more asperity contacts are formed and broken, and consequently there is an increase in the fracture density with increased fault displacement. As a result of the repeated formation and breakage of asperity contacts with increasing fault displacement, existing fractures resulting from previous asperity contact interactions can also be opened and extended. The extension of existing microfractures with increasing fault displacement is clearly shown in Figure 9. Therefore it is suggested that fault dilation in these experiments occurs not only because of the creation of microfractures along the sliding surface but also as a result of opening and extension of existing microfractures.

It should also be noted that, in general, a small amount of fault gouge begins to accumulate between the sliding surfaces with increasing fault displacement and with increasing normal stress. This trend corresponds to the observed increase in density and length of microfractures adjacent to the fault, suggesting that the gouge develops primarily from the destruction of grains containing microfractures. As the microfractures continue to form, there is an increase in the gouge zone width to some critical width when microfracturing ceases. This conclusion is consistent with the work of *Engelder* [1974] and *Teufel* [1978] on the generation of fault gouge.

Fault Dilation

The information about the density, shape, and size of microfractures along the sliding surfaces at different normal stresses and fault displacements, as obtained from detailed microscopic observations, enables investigation of the validity of the above hypothesis by calculating the total microfracture volume along the sliding surface and comparing it to the measured dilatant pore volume associated with frictional sliding. Following *Hallbauer et al.* [1973], the average volume V_M of a single microfracture is calculated from the average fracture length at the different normal stresses and fault displacements, from the average fracture width (which is 3 μm and independent of normal stress and fault displacement), and by assuming a thin section thickness of 30 μm. Since all of the deformation is localized within 1 mm of either side of the simulated fault, the total microfracture volume V_{TM} can then be estimated as

$$V_{TM} = V_M \cdot N \cdot (V_F/V_S)$$

where

V_M volume of an average microfracture;
N number of microfractures per 1-mm^2 section with 30-μm thickness;
V_F volume of simulated fault zone (2-mm-thick section multiplied by area of sliding surface);
V_S volume of 1-mm^2 section with 30-μm thickness.

Using the above relationship, the total microfracture volume has been calculated for precut specimens which have been deformed at different normal stresses and fault displacements (Table 1). Both the calculated total microfracture volume and the measured dilatant pore volume show an increase with increasing normal stress and increasing fault displacement. At all of the test conditions the microfractures accounted for less than 25% of the dilatant pore volume. However, it should be noted that the observed average width of the microfracture (about 3μm) is that of a fully unstressed fracture. It is more than likely that the width of a microfracture created and opened at a highly stressed asperity contact on the sliding surface would be greater than the present 3 μm, resulting in a larger microfracture volume. In addition, since fault gouge is generated during sliding and the gouge is presumably a direct consequence of the destruction of grains containing microfractures, the observed fracture density is probably lower than the actual total microfracture density that contributed to the measured dilatant pore volume. Moreover, the ratio of the calculated total microfracture volume to the measured dilatant pore volume tends to decrease with increasing fault displacement and increasing normal stress, corresponding to the trend of increasing fault gouge. When these two factors are taken into account, the total microfracture volume calculations

TABLE 1. Comparison of Calculated Total Microfracture Volume and Measured Pore Volume Dilation as a Function of Normal Stress and Fault Displacement

Normal Stress, MPa	Fault Displacement, mm	Total Microfracture Volume, mm³	Pore Volume Dilation, mm³	Ratio of Total Microfracture Volume to Pore Volume Dilation
31	6.2	2	8	0.25
64	6.3	7	29	0.24
138	6.1	28	142	0.20
203	6.0	65	426	0.15
186	1.8	11	39	0.28
191	3.7	39	187	0.21
198	7.3	72	583	0.12

support the suggestion that the measured pore volume dilations for these normal stress and displacement conditions are a direct consequence of the creation, opening, and extension of microfractures along the fault.

Since the measured dilatant pore volume is a direct result of microfracturing within a narrow zone adjacent to the fault, then the dilatant volumetric strain should only reflect the fault zone dilation and not the dilation of the entire precut specimen. Accordingly, if the fault zone is considered to be identical to the volume used in the total microfracture calculations (2-mm-thick zone times the area of the sliding surface), then the dilatant volumetric strain associated with deformation after fault movement of about 6 mm increases up to 8.2% as the normal stress increases to 208 MPa (Figure 12). This fault zone dilatant volumetric strain is about 30 times larger than the volumetric strain calculated for the entire precut specimen (Figure 3), and more important, it reflects the actual intensity of the deformation along the fault. The localized increase in the fracture porosity along the fault with increasing shear deformation indicates that the fault changes from a discrete slip plane to a zone of granular material. *Jamison and Teufel* [1979] previously recognized localized deformation and large dilatant volumetric strain along induced shear fractures in intact sandstone.

Implications for Earthquake Prediction

Earthquakes are thought to occur after a period of increasing stress and strain accumulation along existing fault zones. One approach to earthquake prediction is the monitoring of geophysical properties considered to be stress-sensitive, such as seismic velocities and electrical resistivity. On the basis of laboratory studies these properties have been directly related to dilatancy.

Although large changes in seimsic velocities and electrical resistivity occur prior to and during fracturing of intact rock in the laboratory [*Brace*, 1978], variations in these same properties are very small or undetectable prior to and during frictional sliding on simulated faults or

fractures [*Wang et al.*, 1975a, b; *Logan*, 1978]. This study has shown that dilatancy can occur during frictional sliding of simulated faults; but only compaction occurs prior to fault movement. The deformation responsible for the increased dilatant pore volume during fault movement is localized within a narrow zone immediately adjacent to the fault. Consequently, laboratory measurements of geophysical properties which depend on large pervasive dilatant strain within the rock, such as seismic velocities, are probably not affected during frictional sliding, since the relative distance of travel of seismic waves through the fault zone is small and the corresponding travel time changes due to dilatancy along the fault are negligible. If these results are applicable to the natural situation, then dilatancy and possibly related changes in geophysical properties may occur locally within the fault zone during slip but not in the country rock. The magnitude of change in geophysical properties will depend on the size and extent of the dilatant fault zone and the intensity of deformation within the zone.

The lack of any dilatancy prior to fault movement in these experiments, regardless of the magnitude of the normal stress across the fault and the relative stress changes prior to slip, suggests that for preseismic dilatancy to occur within a fault zone, preseismic slip may be necessary. *Dieterich* [1978] has proposed a model which specifically identifies preseismic slip as the underlying cause of earthquake precursors attributed to dilatancy. His model is based on evidence that preseismic fault slip is known to precede some earthquakes and, by analogy, that premonitory slip almost universally precedes the major slip event during laboratory stick slip sliding. In addition, recent observations of pore pressure variations during stick slip sliding on simulated faults [*Teufel*, 1980] has

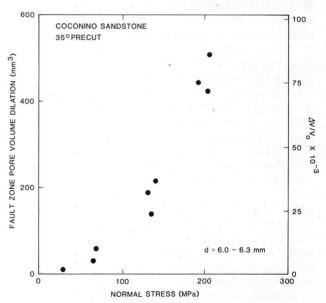

Fig. 12. Fault zone pore volume dilation versus normal stress for 35° precuts after fault displacements of 6–6.3 mm.

shown a direct correlation between the magnitude of pre-seismic dilatancy and the amount of premonitory slip preceding individual stick slip events. However, the observed dependence of fault dilation on the onset and amount of premonitory slip preceding individual stick slip events is essentially identical to that occurring during continued stable sliding which does not lead to stick slip. Therefore if the onset and magnitude of dilation and associated changes in geophysical phenomena along faults is dependent on fault displacement, then the ability to discriminate between stable sliding or aseismic fault creep and preseismic slip prior to an impending earthquake may be severely limited. Clearly, more work on preseismic slip, aseismic creep, and fault dilation is warranted.

In applying laboratory observations directly to earthquake faulting there is some degree of uncertainty, particularly the question of scaling. Dilatant behavior in the laboratory precut specimens is dominated by changes in the size, shape, and density of microfractures along the sliding surface, whereas in nature the response of joints or other large-scale discontinuities would most likely be the dominant factor influencing any dilatancy in the vicinity of a fault zone. Therefore it is suggested that detailed examination of joints and fracture systems along exhumed, ancient faults may provide useful information as to the potential size of dilatant zones along active faults and suggest whether or not these zones are sufficiently large to affect any of the geophysical properties currently being used for earthquake prediction.

Conclusions

Pore volume measurements during drained, triaxial compression, pore pressure experiments on 35° precut specimens of Coconino sandstone and detailed petrographic examination of thin sections of the deformed specimens seem to warrant the following conclusions.

1. Compaction always precedes the onset of slip along simulated faults in the triaxial test where the normal stress is increasing with increasing differential stress.

2. The amount of compaction increases with increasing confining pressure and normal stress across the fault.

3. Dilatant pore volume changes can occur during frictional sliding of simulated sandstone faults.

4. The amount of dilatancy increases with increasing fault displacement and increasing normal stress.

5. During frictional sliding, deformation is localized along the fault, and the resulting microfracture density and microfracture length increase with increasing fault displacement and increasing normal stress.

6. Calculations of the total microfracture volume along the fault correlate with the measured dilatant pore volume, suggesting that dilatancy during frictional sliding is a direct consequence of an increase in the fracture porosity resulting from the creation, opening, and extension of microfractures along the fault.

Acknowledgments. Stimulating discussions with John Handin and John Logan instigated this work. The experimental work was conducted at the Center for Tectonophysics, Texas A&M University, and financial support was contributed from the U.S. Geological Survey grant 14-08-0001-6-460. I am thankful to W. A. Olsson and D. J. Holcomb for their careful and critical reading of the manuscript. This work was supported by the U.S. Department of Energy (DOE) under contract DE-AC04-76-DP00789. Sandia National Laboratories is a DOE facility.

References

Bieniawski, Z. T., Mechanisms of brittle fracture of crystalline rocks, *Int. J. Rock Mech. Min. Sci.*, *4*, 407–423, 1967.

Bowden, F. P., and D. Tabor, *The Friction and Lubrication of Solids*, Oxford University Press, New York, 1950.

Brace, W. F., Volume changes during fracture and frictional sliding, *Pure Appl. Geophys.*, *116*, 603–614, 1978.

Brace, W. F., and J. D. Byerlee, Stick-slip as a mechanism for earthquakes, *Science*, *153*, 990, 1966.

Brace, W. F., B. W. Paulding, and C. H. Scholz, Dilatancy in fracture of crystalline rocks, *J. Geophys. Res.*, *71*, 3939–3953, 1966.

Byerlee, J., Friction of rocks, *Pure Appl. Geophys.*, *83*, 615–626, 1978.

Conrad, R. E., and M. Friedman, Microscopic feather fractures in the faulting process, *Tectonophysics*, *33*, 187–198, 1976.

Dieterich, J. H., Preseismic fault slip and earthquake prediction, *J. Geophys. Res.*, *83*, 3940–3948, 1978.

Engelder, J. T., Cataclasis and the generation of fault gouge, *Geol. Soc. Am. Bull.*, *85*, 1515–1522, 1974.

Hallbauer, D. K., H. Wagner, and N. G. Cook, Some observations concerning the microscopic and mechanical behavior of quartzite specimens in stiff, triaxial compression tests, *Int. J. Rock Mech. Min. Sci.*, *10*, 713–726, 1973.

Handin, J., R. V. Hager, M. Friedman, and J. N. Feather, Experimental deformation of sedimentary rocks under confining pressure: Pore pressure tests, *Am. Assoc. Pet. Geol. Bull.*, *47*, 717–755, 1963.

Handin, J., M. Friedman, J. M. Logan, L. J. Pattison, and H. S. Swolfs, Experimental folding of rocks under confining pressure: Buckling of single-layer rock beams, in *Flow and Fracture of Rocks, Geophys. Monogr. Ser.*, vol. 16, edited by H. C. Heard, I. Y. Borg, N. L. Carter, and C. B. Raleigh, pp. 1–28, AGU, Washington, D. C., 1972.

Jamison, W. R., and L. W. Teufel, Pore volume changes associated with failure and frictional sliding of a porous sandstone, *U.S. Symp. Rock Mech., 20th*, 163–170, 1979.

Johnson, C. B., Characteristics and mechanics of formation of glacial arcuate abrasion cracks, Ph.D. dissertation, Pa. State Univ., University Park, 1975.

Kragelskii, I. V., *Friction and Wear*, Butterworths, London, 1965.

Lockner, D. A., and J. D. Byerlee, Velocity anomalies: An

alternative explanation based on data from laboratory experiments, *Pure Appl Geophys.*, *116*, 765-772, 1978.

Logan, J. M., Creep, stable sliding, and premonitory slip, *Pure Appl. Geophys.*, *116*, 773-789, 1978.

Mjackin, V. I., W. F. Brace, G. A. Sobolev, and J. H. Dieterich, Two models for earthquake forerunners, *Pure Appl. Geophys.*, *113*, 169-181, 1975.

Scholz, C. H., L. R. Sykes, and Y. P. Aggarwal, Earthquake prediction: A physical basis, *Science*, *181*, 803-810, 1973.

Teufel, L. W., The measurement of contact areas and temperature during frictional sliding of Tennessee sandstone, M.S. thesis, Dep. of Geol., Texas A&M Univ., College Station, 1976.

Teufel, L. W., Generation of fault gouge in sandstone: An experimental study, *Geol. Soc. Am. Abstr. Programs*, *10*, 503-504, 1978.

Teufel, L. W., Precursive pore pressure changes associated with premonitory slip during stick-slip sliding, *Tectonophysics*, *69*, 189-199, 1980.

Wang, C.-Y., R. E. Goodman, and P. N. Sundaram, Variations of V_p and V_s in granite premonitory to shear rupture and stick-slip sliding: Application to earthquake prediction, *Geophys. Res. Lett.*, *2*, 309-311, 1975a.

Wang, C.-Y., R. E. Goodman, P. N. Sundaram, and H. F. Morrison, Electrical resistivity of granite in frictional sliding: Application to earthquake prediction, *Geophys. Res. Lett.*, *2*, 525-528, 1975b.

Whitcomb, J. H., J. D. Garmany, and D. L. Anderson, Earthquake prediction: Variation of seismic velocities before the San Fernando earthquake, *Science*, *80*, 632, 1973.

Fluid Flow Along Very Smooth Joints at Effective Pressures up to 200 Megapascals

TERRY ENGELDER AND CHRISTOPHER H. SCHOLZ

Lamont-Doherty Geological Observatory and Department of Geological Sciences
Columbia University, Palisades, New York 10964

Flow rates and thus permeability were measured within very smooth joints in Cheshire quartzite. Measurement of the change in aperture with effective pressure shows that at effective pressures of less than 20 MPa, changes in confining pressure have a larger influence on the permeability than changes in pore pressure. Although a 'cubic law' model for flow within a joint gives a rough estimate of joint permeability, the use of joint aperture as the only variable in the cubic law model for flow is inadequate for calculating joint permeability. We suggest that the effective cross section available for flow changes with effective pressure in a nonlinear manner and we present a modified cubic law to account for this behavior.

Introduction

Flow of fluid in rocks of very low permeability is primarily along joints. At midcrustal depths the degree to which joints close under pressure ultimately influences the bulk permeability of rocks with low intrinsic permeability. Here we examine some details concerning the effect of joint closure on joint permeability.

The rate of flow of a fluid along a nominally closed joint is related to the aperture of the joint. For parallel plates there is a direct relationship between aperture and fluid flow based on the Hele-Shaw model [*Harr*, 1962]. This relationship can be developed from Darcy's law and simplified as

$$Q/\Delta h = Cd^3 \qquad (1)$$

where Q is the flow rate, Δh is the drop in head, d is the joint aperture, and C is a function of fluid properties as well as geometry of the flow path [*Witherspoon et al.*, 1980]. We shall refer to this relationship as the cubic law, and its validity has been discussed in a series of papers including those by *Lomize* [1951], *Sharp* [1970], *Sharp and Maini* [1972], and *Gale* [1975]. *Gangi* [1978], using the experimental data of *Nelson* [1975], argues that while the permeability of a joint may be stress history dependent, it is uniquely defined by the joint aperture. The stress history determines the asperity distribution function of the joint and, consequently, its aperture for a given stress. The experimental data of *Witherspoon et al.* [1980] confirm the validity of the cubic law for laminar flow along tensile joints with apertures down to 4 μm and effective pressures of 20 MPa. They also argue that permeability is uniquely defined by joint aperture. *Kranze et al.* [1979]

present experimental data on very smooth joints which indicate that parameters must be added to the simple cubic law to account for stress history. The purpose of this paper is to present data suggesting that for very smooth joints the rate of change of a joint's effective cross section with pressure is higher than the rate of change of a joint's closure with pressure. Hence one additional parameter accounting for effective cross section must be included in the cubic law to make it a more general equation. By effective cross section we mean the area in an imaginary plane intersecting a joint through which fluid can flow unobstructed by asperities.

Experimental Procedure

Our experimental procedure for fluid flow tests was much the same as described by *Kranz et al.* [1979]. Experiments were conducted in a triaxial, servocontrolled, hydraulic press equipped with a 5.08-cm bore pressure vessel. Kerosene was used as the pressure medium as well as the fluid pumped through the rock. It is chemically inert with respect to the rock, so that only mechanical effects of pressure were investigated. The confining pressure system was independently controlled and separate from the internal fluid system. All samples were covered with a polyolefin jacket. Steel end caps with center holes and radial grooves were affixed to the samples.

Split cylinders were used to model joints. A block of Cheshire quartzite was saw cut into large prismatic sections with ground sides. Two sections with parallel sides were clamped together and cores, also approximately 3.5 cm in diameter, were taken centered on the joint between the sections. These split, cylindrical samples were then

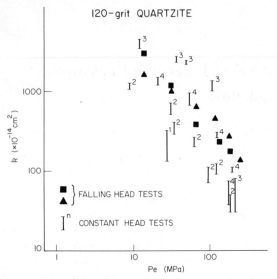

Fig. 1. Whole rock plus joint permeability k at various effective pressures P_e for joints in Cheshire quartzite prepared with 120-grit polishing compound. Constant head data come from four different samples (1–4), whereas falling head data come from two different samples. The error bar for permeability shown for constant head tests indicates the variability of k for repeated tests during which confining pressure is held constant and hydraulic head is varied.

reclamped and saw cut to be approximately 9 cm in length, and the ends were ground parallel. The split samples were then unclamped and the interior, opposing surfaces were ground with either number 80 grit or 120 grit.

Changes in joint aperture were measured with a four-armed dilatometer, consisting of four thin beryllium-copper beams connected in a radially symmetric pattern to an aluminum ring which was slipped over the sample. The output from two arms measuring rock compression was subtracted from the output of the other two arms measuring the sum of joint closure and rock compression so that the combined output was proportional to joint closure. Calibration included subtracting the effect of pressure on the dilatometer. Measurements made with the dilatometer do not give an absolute value for the joint aperture but rather a change in aperture. To get an absolute value, the joint must be closed down to the point where no further changes in aperture are discernable, and that point taken as $d = 0$. We did not reach such a point in any of our experiments. From the asymptotic approach to complete closure we can, with some uncertainty, get absolute values of d as a function of $P_c - P_f$. At low pressures the estimated uncertainty in the aperture may be as high as 200%, but it decreases rapidly with pressure, so that above 10 MPa the estimated uncertainty in the aperture of the joint may be considered as ±25%.

For the experiments reported here, the fluid flow apparatus was modified to measure flow rates during constant head tests, whereas *Kranz et al.* [1979] reported data

from falling head tests. For the constant head test the pore fluid systems at the top and bottom of the sample were isolated. The pore pressure on the high-pressure end of the sample was generated by servocontrolling a double-acting Aminco pump. Feedback for servocontrolling the pump was from a BLH pressure cell. Pore pressure on the low-pressure end of the sample was maintained with a servocontrolled piston moving within a small pressure vessel serving as a pore fluid collector. Here again the feedback element was another BLH pressure cell. The signals from the two BLH pressure cells were compared to maintain a constant pore-pressure gradient with the signal driving the Aminco pump as the master maintaining a constant absolute pore pressure. The BLH pressure cells were capable of detecting pressure differences of 0.01 MPa at 200 MPa ambient pressures. Using a linear variable displacement transducer (LVDT), flow rate was measured by monitoring the withdrawal of the piston from the pressure vessel collecting pore fluid. Strokes of the Aminco pump were not used to measure flow because of back leaks through ball-check valves.

Pulse Decay Versus Constant Head Tests

The purpose in using a constant head test was to check data using the falling head tests, as reported by *Kranz et al.* [1979]. The constant head test assumes that Darcy's law holds and there is a linear pressure gradient along the length of the sample. That is,

$$Q = \frac{kA}{\mu} \frac{dP}{dL} \qquad (2)$$

where Q is the volume flow rate, k is the intrinsic permeability, A is the cross section through which the fluid flows, μ is the dynamic viscosity, L is the length of flow path, and P is the fluid pressure. For the constant head test, Q is measured as the rate of flow into the small pres-

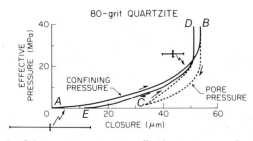

Fig. 2. Joint aperture versus effective pressure where the solid curve represents closure during adjustment of confining pressure and the dashed curve represents closure during adjustment of pore pressure. The joint in Cheshire quartzite was prepared with 80-grit polishing compound. By the nature of the experiment, confining pressure is applied before pore pressure and removed after pore pressure. Pore P_p and confining P_c pressure conditions are as follows: point A, $P_c = P_p = 0.1$ MPa; point B, $P_c = 40$ MPa and $P_p = 0.1$ MPa; point C, $P_c = 40$ MPa and $P_p = 39.4$ MPa; point D, $P_c = 40$ MPa and $P_p = 0.1$ MPa; and point E, $P_c = P_p = 0.1$ MPa.

sure vessel from the sample. Permeability may be calculated using either the cross section of the sample (9.62 cm²) or the cross section of the joint whose aperture is estimated using the four-armed dilatometer. If the cross section of the sample is used for A, k is then 'rock permeability' in which the major flow capacity is governed by the joints. This is to be distinguished from 'joint permeability' which is calculated using the effective cross section of the joint. The ratio of rock permeability for rocks in which the major flow capacity is contributed by joints to joint permeability is equal to rock porosity due to joints [*Jones*, 1975].

The procedure for both the constant head and falling head tests was to prepare a sample for tests at a variety of effective pressures. For data shown in Figure 1 the tests started at low effective pressures with changes in effective pressure accomplished by increasing the confining pressure. For the falling head tests, one pulse-decay curve was recorded at each effective pressure, whereas several flow tests were recorded for each effective pressure during the constant head tests. In the latter case each flow test was accomplished at a different pore pressure gradient with a maximum range from 0.6 to 0.006 MPa/cm. Error bars in Figure 1 represent the range of permeabilities measured using a range of pore pressure gradients. The data in Figure 1 represent 220 flow tests.

Constant head tests show the same range in permeability as falling head tests for a joint prepared with 120 grit in Cheshire quartzite (Figure 1) where rock permeabilities are calculated. For comparison, intact Cheshire quartzite has a permeability of less than 10^{-14} cm² at ef-

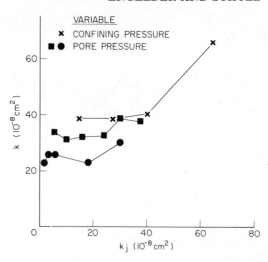

Fig. 4. Plot of joint permeability using the cubic law for fluid flow versus joint permeability using experimental data. Effective pressure was changed by changing either pore pressure or confining pressure with the other held constant.

fective pressures of less than 10 MPa. The variability in permeability at any effective pressure is in part associated with the preparation of six different samples for which data are shown. The grinding of surfaces by hand rarely produces a uniform surface finish, and so the initial surface finishes were probably different from sample to sample. The data converge at high effective pressures, suggesting that initial sample mismatches contribute larger variability to the permeability than would be the case if each sample assembly were exactly like its prede-

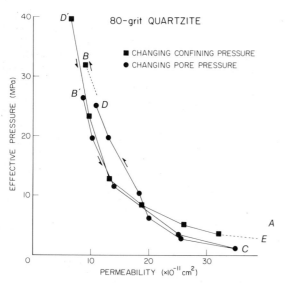

Fig. 3. Effective pressure versus permeability curves for a joint in Cheshire quartzite prepared with 80-grit polishing compound. Data come from constant head flow tests. Arrows indicate direction of change of either confining or pore pressure. Only one point was taken for permeability during initial application of confining pressure.

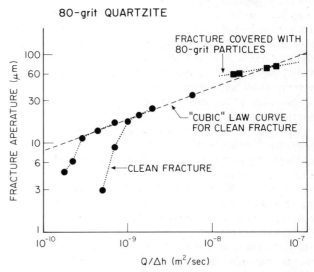

Fig. 5. Fluid flow versus aperature for clean joints and a joint lined with a layer of 80-grit particles. For the covered joint $Q/\Delta h$ decreases faster with d than as d^3, whereas for the clean joint at high pressure the opposite is true.

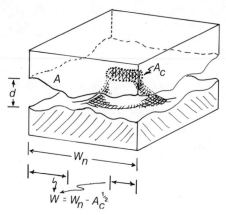

Fig. 6. Schematic of a joint where A is the nominal surface area, A_c is the real area of contact, d is mean aperture, W_n is nominal width of joint, and W is the effective cross-sectional width.

cessor. During constant head tests at a single effective confining pressure the use of different hydraulic heads gives different values for permeability. This may be attributed to different gradients in effective pressure down the length of the sample. Further variability may also arise from a slight misalignment of the saw cut sample and shear stresses at the piston-sample interface, causing part of the saw cut to pop open and thus have an excessively large aperture. From the comparison of constant head and falling head tests we conclude that both types of tests showed similar flow rates and flow rates varied in a similar manner with similar changes in effective pressure.

Change in Aperture With Pressure

Permeability of soil, intact rocks, and jointed rock follows a hysteresis loop when either confining pressure or pore pressure is cycled up and down [Freeze and Cherry, 1979; Kranz et al., 1979; Witherspoon et al., 1980]. Direct measurement of the aperture in jointed rock shows that the hysteresis in permeability is a direct consequence of the hysteresis in the response of the aperture to changing pore pressure and confining pressure [Kranz et al., 1979; Witherspoon et al., 1980]. The flow rates along joints are controlled by the cross-sectional area of the flow channel.

The effect of changing confining and pore pressure on aperture is shown in Figure 2. This curve is representative of four samples prepared with 80-grit polishing compound. Eighty grit was used to give joint surfaces capable of closing 40–60 μm. We found that the 15-μm total closure for 120-grit surfaces was too small to measure accurately aperture changes accompanying small changes in effective pressure. The error bars shown in Figure 2 represent the variation in position of the curves for the four experiments. Variation in position does not affect the shape or relative positions of the confining and pore pressure curves for each test. The loading path consisted of increasing the confining pressure to 40 MPa (from A to B) then increasing pore pressure to about 39.4 MPa (B to C).

Pore pressure was then decreased to 0.1 MPa (C to D) followed by the release of confining pressure (D to E). Hysteresis loops are evident for both pore and confining pressure cycles, and the top of the confining pressure cycle is offset by the residual amount of the pore pressure hysteresis loop. Likewise, the bottom of the confining pressure cycle does not return to 0 μm. During the pore pressure cycle, work was lost. This behavior is displayed in the conventional sense for the confining pressure cycle where lost work is represented by the area between the loading and unloading curves.

Kranz et al. [1979] present permeability as a function of pressure with the equation

$$dk = -a \, dP_c + b \, dP_f \qquad (3)$$

where $b/a < 1$ for jointed Barre granite. The physical reason for $b/a < 1$ is apparent from Figure 2. At effective pressures of less than 20 MPa, changes in confining pressure have a greater effect on changes in aperture than equivalent changes in pore pressure. A confining pressure of 40 MPa closed the joint in quartzite about 50 μm, whereas the injection of about the same pore pressure reopened the joint a little more than 20 μm. During the first cycle, such behavior may be attributed to changes in the asperity height distribution which changes less dramatically in the second cycle. However, for the second cycle in closing and opening the joint the change of confining pressures still has a larger effect despite the reverse order in which the changes took place. This result is unusual in light of several studies showing that permeability is governed by effective confining pressure and not greatly dependent on the absolute values of either pore pressure or confining pressure [Jones, 1975].

Both the confining and pore pressure curves exhibit marked changes in slope when the effective pressure is less than 25 MPa. However, close inspection shows that the change in slope for the confining pressure curve occurs

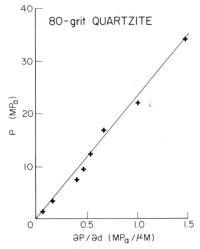

Fig. 7. Plot of joint stiffness $\partial P/\partial d$ versus confining pressure P. Data from loading curve in Figure 2.

at about 23-24 MPa, whereas the change in slope for the pore pressure curve is somewhat lower (~21 MPa). Again, this demonstrates the greater overall capacity for changes in confining pressure to affect joint aperture.

Joint Permeability Versus Effective Pressure

Change in flow rate was measured in conjunction with change in joint aperture for the split samples of Cheshire quartzite prepared with 80 grit. To illustrate details associated with change in effective pressure, the rock permeability was calculated using the cross-sectional area of the sample (9.62 cm^2). Figure 3 shows that the permeability versus effective pressure curves are similar to the joint aperture curves for the same experiment. The letters A through E correspond to the same steps shown in Figure 2. Although the absolute value of the permeability was different for each of four experiments, the relative shape and position of the curves for each experiment are consistent. During the application of confining pressure (A to B) only one permeability test was attempted for the experiment shown in Figure 3. However, consider the pore pressure cycle with confining pressure held constant (dots) during which the effective pressure was first decreased (B to C) and then increased in a series of steps (C to D). The permeability was lower during the initial decrease in effective pressure than upon the return to the same effective pressure. This is equivalent to the slight opening of the joint aperture seen during the pore pressure cycle in Figure 2. Likewise, during the confining pressure cycle with constant pore pressure (squares), permeability should have been higher during confining pressure changes that lowered the effective pressure (D to E) relative to the pore pressure cycle (C to D). This effect actually appears at effective pressures of less than 10 MPa where there is a distinct difference in the closure curves D to E and C to D (Figure 2). At low effective pressures, permeability changed more rapidly with changing confining pressure (Figure 3). At low confining pressures the permeability was higher than for high confining pressures at the same effective stress. Hence permeability varies in a complicated manner with effective pressure and is highly dependent on stress history.

Joint permeability may be calculated using the effective cross-sectional area of the joint, which is joint aperture times the diameter of the sample. Assuming that the effective cross-sectional area of the joint changes linearly with the joint aperture, this calculation requires the absolute aperture which can only be estimated [Witherspoon et al., 1980]. For our calculation we assume that at the lowest effective stress, flow follows the cubic law where $k_j = d^2/12$. Knowing the flow rate and drop in head for permeability tests at low effective stress, we find the aperture d_0 which sets the k_j equal to $k = (\mu Q/A)(dL/dP)$. Once d_0 is found, we calculate k for higher effective pressures using an aperture $d = d_0 - \Delta d$, where the change in aperture Δd is measured experimentally. A plot of k_j versus k shows that measured fluid flow does not follow the d^3 law at high

pressure (Figure 4). At low effective pressures and wide apertures the d^3 law is confirmed, as shown by the one-to-one change of k versus k_j for the confining pressure curve. However, for the two pore pressure curves with narrow apertures there is just a slight increase in joint permeability k with increase in aperture versus an apparent dramatic increase in k_j.

Failure of the plot of k versus k_j to follow a straight line suggests that the cubic law for fluid flow requires an additional parameter other than joint aperture. This is also seen in a plot of flow rate per hydraulic head Q/h versus aperture d (Figure 5). At high flow rates and wider apertures the curves appear to parallel a d^3 law, but at low flow rates and narrower apertures the d^3 law is inadequate.

Discussion

Our significant results include confirmation of the d^3 law at low pressure for clean joints, as did Witherspoon et al. [1980]. However, for filled joints permeability decreases faster with d than as d^3 and for clean joints at high pressure the opposite is true. We also reaffirmed the result of Kranz et al. [1979] that pore pressure has a weaker effect on fracture closure (permeability) than confining pressure.

One can begin to understand these results by considering the theory of closure of random surfaces of Greenwood and Williamson [1966] which has recently been partially confirmed for rock by Walsh and Grosenbaugh [1979]. This theory has the result that if two surfaces of random topography are in contact under a pressure P and if the deformation is elastic, the mean aperture d will be

$$d = d_0 - \Lambda \log (P/P_0) \qquad (4)$$

where Λ is the standard deviation of the topography of the surface and d_0 the aperture at P_0, a reference pressure. An additional result, originally from Archard [1957], is that the real area of contact, A_c, is dependent only on the load and not on the nominal pressure P; thus

$$A_c = \epsilon P A \qquad (5)$$

where ϵ is an elastic constant. We equate W, the effective cross-sectional width available for flow, and the real area of contact:

$$W = W_n - (\epsilon P A)^{1/2} \qquad (6)$$

where W_n is the nominal width, and from (4),

$$W = W_n - \left[w_0 + (\epsilon A P_0)^{1/2} \exp \left(\frac{d_0 - d}{2\Lambda} \right) \right] \qquad (7)$$

where w_0 is the contact cross section at zero pressure (Figure 6). Substituting this into equations (2) and (4) of Witherspoon et al. [1980] for linear flow through a crack of separation d, we obtain

$$\frac{Q}{\Delta h} = \left[\frac{W_n - [w_0 + (\epsilon A P_0)^{1/2} \exp ((d_0 - d)/2\Lambda)]}{2} \right] \frac{\rho g}{12 \mu L} d^3$$

$$(8)$$

Equation (4) fits the closure data (Figure 7) within the experimental error when pore pressure is fixed. Equation (8) shows why the permeability decreases faster with d than d^3 (Figure 5). *Witherspoon et al.* [1980] assumed that w was a constant and introduced a factor f to account for it. We show here (equation (6)) that w is a function of P and hence d.

If A_c is only a function of load and then if a pore pressure P_p is present,

$$A_c = \epsilon PA - \epsilon P_p(A - A_c) \qquad (9)$$

and we see why P_p plays a smaller role than P in changing the aperture and permeability. The above result was discussed by *Kranz et al.* [1979].

Acknowledgments. This work was supported by Department of Energy contract DE-AS02-76ER04054. Lamont-Doherty Geological Observatory contribution #3104.

References

Archard, J. F., Elastic deformation and the laws of friction, *Proc. R. Soc. London, Ser. A, 243,* 190–205, 1957.

Freeze, R. A., and J. A. Cherry, *Groundwater,* 604 pp., Prentice-Hall, Englewood Cliffs, N. J., 1979.

Gale, J. E., A numerical field and laboratory study of flow in rocks with deformable fractures, Ph.D. thesis, 255 pp., Univ. of Calif., Berkeley, 1975.

Gangi, A. F., Variation of whole and fractured porous rock permeability with confining pressure, *Int. J. Rock Mech. Mining Sci., 15,* 249–257, 1978.

Greenwood, J. A., and J. B. P. Williamson, Contact of nominally flat surfaces, *Proc. R. Soc. London, Ser. A, 295,* 300–319, 1966.

Harr, M. E., *Groundwater and Seepage,* 315 pp., McGraw-Hill, New York, 1962.

Jones, F. O., A laboratory study of the effects of confining pressure on fracture flow and storage capacity in carbonate rocks, *J. Petrol. Tech., 27,* 21–27, 1975.

Kranz, R. L., A. D. Frankel, T. Engelder, and C. H. Scholz, The permeability of whole and jointed Barre granite, *Int. J. Rock Mech. Mining Sci., 16,* 225–234, 1979.

Lomize, G. M., *Filtratsiya v Treshchinovatyk Porodakh,* 127 pp., Gosenergoizdat, Moscow, 1951.

Nelson, R., Fracture permeability in porous reservoirs: Experimental and field approach, Ph.D. thesis, Tex. A&M Univ., College Station, 1975.

Sharp, J. C., Fluid flow through fissured media, Ph.D. thesis, 181 pp., Univ. of London, Imperial College, London, 1970.

Sharp, J. C., and Y. N. T. Maini, Fundamental considerations on the hydraulic characteristics of joints in rock, paper presented at the International Society for Rock Mechanics Symposium on Percolation Through Fissured Rock, Stuttgart, 1972.

Walsh, J. B., and M. A. Grosenbaugh, A new model for analyzing the effect of fractures on compressibility, *J. Geophys. Res., 84,* 3532–3536, 1979.

Witherspoon, P. A., J. S. Y. Wang, K. Iwai, and J. E. Gale, Validity of cubic law for fluid flow in a deformable rock fracture, *Water Resour. Res., 16,* 1016–1024, 1980.

Thermomechanical Properties of Galesville Sandstone

J. D. BLACIC, P. H. HALLECK, P. D'ONFRO, AND R. E. RIECKER

Geosciences Division, Los Alamos Scientific Laboratory, Los Alamos, New Mexico

Elastic and strength properties of Galesville sandstone were experimentally determined at temperatures over the range 37°-204°C and effective confining pressures of 0-31 MPa. Young's modulus increases from 30 to 60 GPa with increasing temperature and effective pressure. Poisson's ratio decreases from 0.3 to 0.1 with increasing temperature and effective pressure. Pore space reduction apparent in hydrostatic compression and thermal expansion tests suggests that these changes are a direct result of densification of the sandstone with either increasing temperature or pressure. Ultimate brittle failure is associated with a critical inelastic volume strain of about 4.5×10^{-3}, which decreases with increasing temperature and increases with increasing effective pressure. Mohr failure envelopes are nonlinear owing to plastic flow of the dolomite cement. Low stress cyclic loading has little effect on elastic constants until 1000 cycles at 37° and 120°C and 10,000 cycles at all temperatures. Young's modulus increases and Poisson's ratio decreases with either increasing temperature or cycle number.

Introduction

Interest in mechanical properties of sandstones for structural geology and geoengineering applications led to the pioneering laboratory studies of *Handin and Hager* [1957, 1958] and *Handin et al.* [1963]. Although much additional work has been done since these early works, there is still little known about, for example, temperature effects on fundamental elastic and strength properties. Our work is in response to an examination of the problem of storing hot, compressed air in aquifers for power plant load leveling. This application requires knowledge of the temperature and effective pressure effects on elastic constants and strength as well as low stress cyclic fatigue properties of sandstone. We report here the results of laboratory experiments aimed at providing this information for a particular sandstone aquifer.

Experimental Method

Sample Material

Sample material is the Upper Cambrian Galesville sandstone obtained from subsurface core taken by Illinois Power Company in Warren County, Illinois. Galesville is a carbonate cemented sandstone with average grain size of approximately 0.1 mm. All carbonate is dolomite [*Emrich*, 1966] and in our samples averages 10-15%. However, the amount of dolomite is very inhomogeneous and can range from 0-30% within samples. Reservoir analysis by Geo-Engineering Laboratories, Inc., indicates that our samples have horizontal gas permeability of 100-600 mdarcies and 15-24% porosity. Sand grains exhibit abundant evidence of primary deformation (undulatory extinction, kink bands); dolomite is undeformed.

Test Procedure

Test samples are right circular cylinders 2.5 cm in diameter by 6.3 cm long. The samples are jacketed by coating with an RTV silastic. Jacketed samples are tested in a triaxial, externally heated pressure vessel, using silicone oil as a pressure-transmitting medium. A two-zone heater is used, resulting in temperature gradients of less than 1°C over the sample. Axial load is applied at a constant displacement rate with a servocontrolled hydraulic testing machine. Sample strains are determined with an internal system consisting of three axial and three radial LVDT displacement gages. The axial gages are contained in a ring that is attached to steel buttons epoxied to the ends of the sample. In this way, the gage length does not span the sample-piston interface. Radial LVDT cores impinge on small steel buttons epoxied to the midline of the sample. Volume strain is calculated from the average axial and radial strains.

Normal test procedure is as follows. The jacketed sample is first nominally saturated by pumping water through it at low pressure. Confining pressure is then applied, followed by argon gas pore pressure, and finally the sample is heated. Gas pore pressure is applied through a small separate reservoir. At temperatures above boiling, water vapor may pass into this separate vessel from the sample and condense. In this way, a relatively constant-humidity pore pressure is maintained during the test.

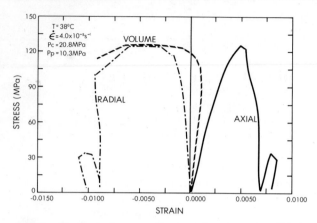

Fig. 1.   Stress-strain curves for experiment number C-5. $\dot{\epsilon}$ = strain rate, P_c = confining pressure, and P_p = pore pressure. Note that after failure, samples are reloaded to determine the residual strength.

Elastic Properties

Displacement versus load records are digitized, and stress-strain values are calculated and plotted. The stress-strain curves shown in Figure 1 are typical of the mechanical response of Galesville sandstone. Tangent Young's modulus is determined by linearizing small segments of the axial stress strain curve in a running least squares scheme and plotted versus stress (Figure 2). Poisson's ratio is determined in a similar way (Figure 3). Typically, modulus increases from a low value and stabilizes when about 25% of the failure stress is attained. As load increases toward failure, Young's modulus continually decreases. Poisson's ratio is usually characterized by unstable values at low load followed by a broad, fairly constant value. At still higher loads, Poisson's ratio increases rapidly as failure is approached, eventually attaining a value of 0.5. The reason for the latter behavior can be seen in the volume stress-strain curve (Figure 1). As failure is approached, the samples become dilatant. Even-

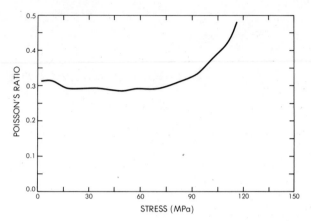

Fig. 3.   Poisson's ratio versus axial differential stress for experiment number C-5.

tually, the volume strain becomes equal to zero again. At this point the sample appears as though it were incompressible and, therefore, Poisson's ratio must equal one half. As volume strain then becomes negative, Poisson's ratio exceeds one half, though the strain ratio cannot be characterized as an 'elastic' constant since much of the strain is now not recoverable.

While the behavior described above is that most often observed, many samples exhibit little or no stable region for the elastic constants. For example, Poisson's ratio is often observed to increase continually from a low value to 0.5 at failure (Figure 4). Overall, the elastic 'constants' of this sandstone are nonlinear as a function of stress, characterized by the general features described above. Furthermore, these elastic constants are not constant as a function of temperature or effective confining pressure.

Variations of Young's modulus and Poisson's ratio with temperature and effective pressure (confining pressure minus pore pressure) are listed in Table 1 and are shown in Figures 5 and 6. The values plotted are the stable values (for example, between 15 and 45 MPa in Figures 2 and 3)

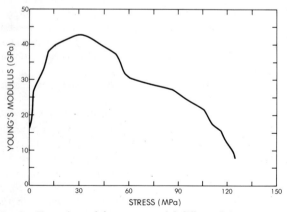

Fig. 2.   Young's modulus versus axial differential stress for experiment number C-5.

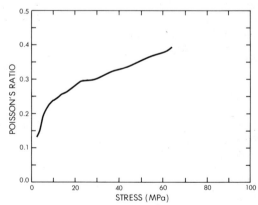

Fig. 4.   Poisson's ratio versus axial differential stress for experiment number C-15. Confining pressure, 0.5 MPa; pore pressure, 0; temperature, 37°C.

TABLE 1. Galesville Sandstone Experiments

Temperature (°C)	Confining Pressure (MPa)	Pore Pressure (MPa)	E (GPa)	ν	σ_y (MPa)	σ_u (MPa)	σ_y/σ_u	ε*_v (×10^-3)	N (cycles)	Cyclic Stress (MPa)	Δ E %	Δ ν %	σ_r (MPa)	θ (deg)	μ	n	Experiment Number
37	0	0	23	.33	38	69	.55	4.4	1	-	-	-	-	33	1.5	-	C-3
37	0.5	0	32	.15	32	79.5	.41	3.7	1000	10.4	+4	+32	3.5	26	2.0	1.2	C-16
37	0.5	0	33	.35	51	87	.59	4.0	1000	5.2	+8	-26	9.3	20	2.6	1.9	C-15
37	17.8	10.3	41	.27	74	126	.59	4.0	1	-	-	-	28.7	20	1.6	0.85	C-5
37	20.8	10.2	32	.19	79	160.5	.49	5.0	100	11.0	+7	-5	37.1	21	1.7	0.81	C-11
37	17.6	10.1	32	.28	80	124.5	.64	9.0	1000	15.7	+7	-14	31.5	25	1.6	0.92	C-12
37	17.6	10.3	58	.34	74	135	.55	5.8	10,000	10.4	-10	-26	22.8	26	1.6	0.77	C-13
37	20.7	10.3	39	.23	102	160.5	.64	6.5	1000	5.2	+2	-9	29.3	22	1.7	0.70	C-14
37	36.6	10.3	47	.22	162	211	.77	2.5	1	-	-	-	132.5	27	1.1	1.0	C-6
37	40.1	10.2	32	.14	128	211	.60	8.5	1000	10.4	+4	+14	77.6	25	1.2	0.68	C-17
120	0	0	36	.23	41	58	.71	1.7	1	-	-	-	1.4	25	2.1	2.1	C-4
120	0.5	0	32	.13	33	64.5	.51	4.0	1000	5.2	+15	+8	8.0	19	2.7	1.8	C-21
120	0.5	0	36	.13	43	76.5	.56	3.4	1000	15.5	+8	0	11.6	25	2.1	1.7	C-22
120	0.5	0	35	.18	19	67.5	.28	3.6	100	10.4	+4	-11	6.0	23	2.2	1.5	C-25
120	0.5	0	43	.23	45	81	.56	2.1	10,000	10.4	+19	+22	7.7	-	-	-	C-27
120	17.6	10.4	43	.13	91	154	.59	6.3	1	-	-	-	46.5	20	2.0	1.2	C-8
120	20.7	10.2	50	.21	92	153	.60	4.3	100	15.5	+7	+10	37.1	24	1.6	0.83	C-18
120	20.7	10.1	44	.13	120	134	.90	1.1	1000	15.5	-2	-8	29.3	27	1.4	0.71	C-19
120	20.7	10.1	50	.21	112	150	.75	4.9	1000	10.4	+3	-2	31.2	25	1.5	0.74	C-20
120	20.7	10.1	64	.33	90	142.5	.63	5.7	10,000	15.5	+28	-6	43.1	21	1.7	0.89	C-23
120	20.7	10.1	53	.32	75	121.5	.52	3.3	1000	10.4	-5	-6	30.2	26	1.4	0.73	C-24
120	20.7	10.3	48	.16	81	140	.58	2.3	100	5.2	+14	+19	57.7	27	1.4	1.0	C-26
120	38.6	10.6	45	.08	104	204	.51	-	1	-	-	-	-	35	-	1.0	C-9
121	41.4	10.2	35	.11	-	-	-	-	1	-	-	-	4.7	28	.06	-	C-34
204	0.5	0	38	.15	32	76	.42	3.3	1	-	-	-	5.6	-	-	-	C-31
204	0.5	0	47	.18	53	71	.74	1.7	1	-	-	-	4.1	25	1.3	2.1	C-29
204	0.5	0	47	.24	-	82.5	-	-	1	-	-	-	3.0	20	1.1	2.6	C-30
204	0.5	0	45	.12	41	69	.59	2.2	1000	10.3	+4	-17	17.7	-	-	-	C-39
204	20.7	10.2	50	.14	90	142.5	.63	3.0	1	-	-	-	-	30	-	1.3	C-33
204	20.7	10.2	50	.17	98	172	.57	3.9	1000	15.5	17	-59	49.1	17	.93	1.9	C-36
204	20.7	10.3	48	.15	128	148	.86	2.9	1	-	-	-	3.0	-	-	-	C-32
204	20.7	10.1	57	.08	115	143	.80	1.4	100	10.3	0	-12	87.0	-	-	-	C-37
204	20.7	10.1	52	.13	97	168	.58	6.4	10,000	10.3	+12	-54	66.0	-	-	-	C-38
204	41.4	10.4	60	.08	30	-	-	8.1	1	-	-	-	6.9	20	0.7	0.51	C-35

E = Young's modulus, ν = Poisson's ratio, σ_y = yield stress, σ_u = ultimate stress, σ_r = residual stress, θ = fault angle to compression axis, μ = coefficient of internal friction, n = coefficient of sliding friction, ε*_v = inelastic volume strain at ultimate stress. Strain rate equals 4×10^{-5} in all experiments loaded to failure. Cycle period equals 5 seconds in all cyclic tests.

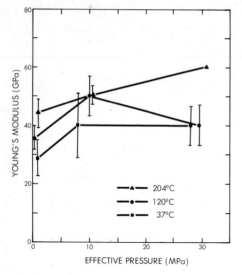

Fig. 5. Young's modulus versus effective confining pressure for 37°, 120°, and 204°C. Bars are 1 standard deviation.

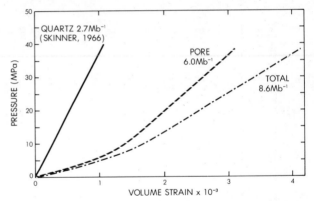

Fig. 7. Volume strain versus pressure for hydrostatic compression, experiment number C-35. Curve labeled 'pore' is the difference between the total observed strain and that expected for quartz.

or the value at about 50 MPa if there is no stable region. Data scatter is large, which we believe to be the result of the inhomogeneity in cementation and porosity of this sandstone as described above. However, the general trends are an increase in Young's modulus and a decrease in Poisson's ratio with increasing effective pressure and temperature. The variations with increasing temperature are particularly marked and are, to our knowledge, effects not previously noted for sandstone.

To identify the mechanism of elastic constant changes, we performed a few hydrostatic compression and thermal expansion tests on Galesville sandstone. A 20°C hydrostat is shown in Figure 7. A handbook compressibility value for quartz [*Skinner*, 1966] is subtracted from the total volume strain to demonstrate, as expected, that most of the compression is taken up by the pore space. Similarly, thermal expansion, compared to that expected for a zero poro-

sity quartz aggregate (Figures 8 and 9), indicates that grains are expanding into the pore space. Also, note the strong anisotropy of linear thermal expansion for this sandstone. Hence, in both cases, increasing effective pressure and increasing temperature have the effect of densifying the sample. This densification appears to be the reason why Young's modulus increases with increasing pressure and temperature since a zero porosity quartz aggregate should have a modulus of about 100 GPa [*Birch*, 1966]. In addition, a zero porosity quartz aggregate has a calculated Poisson's ratio of 0.07 [*Voight*, 1928]. At the highest pressures and temperatures of our tests, we observe values of Poisson's ratio below 0.1, indicating that densification is also the likely cause of the variation of this elastic constant. Thus, changes in pore space, whether caused by increasing effective confining pressure, differential thermal expansion, or dilatancy, can explain the large changes in elastic constants of this sandstone.

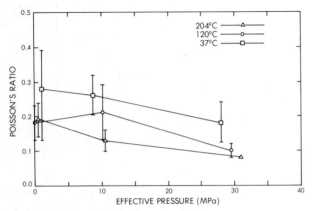

Fig. 6. Poisson's ratio versus effective confining pressure for 37°, 120°, and 204°C. Bars are 1 standard deviation.

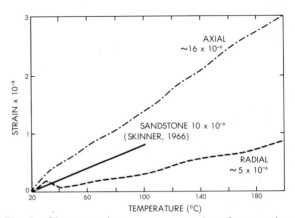

Fig. 8. Linear strain versus temperature for experiment number C-35. Axial strain is perpendicular to bedding; radial strain is parallel to bedding. Linear coefficients of thermal expansion are given compared to an average value for sandstone.

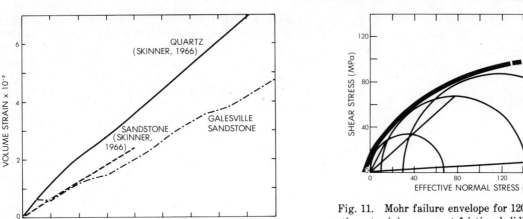

Fig. 9. Volume strain versus temperature for experiment number C-35 compared to quartz and an average for sandstone.

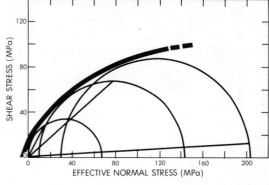

Fig. 11. Mohr failure envelope for 120°C. Sliding lines intersecting at origin represent frictional sliding on fault planes produce during the experiments.

Strength Properties

Failure of Galesville sandstone is reflected in the onset of nonlinear stress-strain behavior, as shown in Figure 1. We define yield stress as the stress at the first deviation from the linear portion of the volume strain curve. Hence, yielding is identified with the onset of dilatancy. Yield and ultimate stresses are given in Table 1.

It has been suggested that brittle failure is associated with a critical amount of inelastic volume strain [*Kranz and Scholz*, 1977]. Our experiments on Galesville sandstone are consistent with this idea in a general way. For example, in Figure 10 we show the inelastic volume strain (difference between total and elastic volume strain) at ultimate stress as a function of temperature for 0.5 MPa effective confining pressure. Although the scatter is large and increases at higher effective pressures, the following trends are apparent:

1. Volume strain at failure decreases with increasing temperature.

2. Volume strain at failure increases with increasing effective confining pressure. This agrees with the results of *Wawersik and Brown* [1973] but is contrary to those of *Handin et al.* [1963] who found an increasing reduction in porosity at failure with increasing confining pressure.

3. Over the whole range of temperature (37°-204°C) and effective pressure (0-31 MPa) of our tests, the critical inelastic volume strain is approximately $4.5 \pm 2 \times 10^{-3}$.

At all temperatures and pressures of our tests, failure occurs by faulting on one or more inclined shear zones accompanied by considerable crushing and disaggregation. Failure is brittle in all cases, although there is a small amount of permanent strain before failure at the highest temperatures and pressures. This is in contrast to the behavior of Berea sandstone reported by *Handin et al.* [1963] for which considerable permanent strain before failure is demonstrated. Axial strain at ultimate stress is approximately 2% in Berea, whereas in Galesville it is seldom more than 0.5%.

Effective principal stresses at the ultimate strength are represented in Mohr circle plots shown in Figure 11 for 120°C and summarized in Figure 12 for all temperatures. Shear strength is largely independent of temperature over

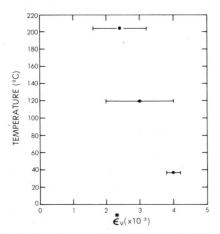

Fig. 10. Inelastic volume strain at ultimate uniaxial stress (ϵ_v^*) versus temperature. Bars are 1 standard deviation.

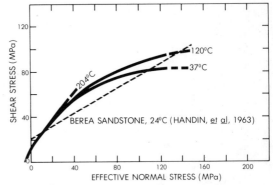

Fig. 12. Mohr failure envelopes at 37°, 120°, and 204°C compared to those for Berea sandstone at 24°C.

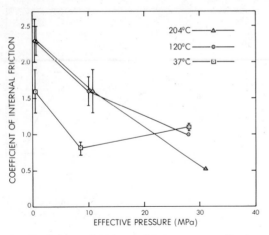

Fig. 13. Coefficient of internal friction versus effective confining pressure for 37°, 120°, and 204°C.

the range of our tests. The strong curvature of the Mohr envelopes relative to that of Berea sandstone is believed to be due to the extensive plastic deformation of the dolomite cement in Galesville sandstone evident in thin sections of the deformed samples. Frictional sliding lines are also shown in Figure 11, based on measured residual strength and fault angles. Coefficients of internal and sliding friction are given in Table 1 and are summarized in Figures 13 and 14, respectively. Both coefficients appear to decrease with increasing effective pressure and are largely independent of temperature over the range of our tests. Tensile strength of Galesville sandstone was also determined at room temperature by using the Brazilian test and was found to be 4 ± 0.5 MPa.

Low Stress Cyclic Loading Effects

Experiments to evaluate effects of cyclic stress on elastic constants and strength were performed in the following way. After attaining the planned temperature and pressure for a particular test, samples were loaded in compression to about 50% of the failure stress to determine

the initial elastic constants. After subsequent unloading, the samples were repeatedly loaded at stresses ranging from 5 to 15 MPa for up to 10,000 cycles, with a 5-s cycle period. The samples were then loaded to failure to determine the final elastic constant values and ultimate strength. Some of the results of these tests are listed in Table 2, extracted from Table 1 for clarity. It should be noted that changes in Young's modulus (ΔE) of ±10% and in Poisson's ratio ($\Delta \nu$) of ±15% are judged not to be significant because of inaccuracies in the method.

Results indicate that for up to 1000 cycles there is generally no significant change in elastic constants of Galesville sandstone at 37°C and 120°C. At 204°C and 1000 cycles, significant changes in both constants occur. At 10,000 cycles there are significant changes at all temperatures in our test range. It should be noted that in our tests the cyclic stresses are much lower than those commonly employed, 5-20% of failure stress instead of 75-95%, for example [Dunn and LaFountain, 1976]. The changes we see are, with some exceptions, consistent with the densification interpretation given above, that is, Young's modulus increases and Poisson's ratio decreases, probably as a result of pore space reduction. We could discern no effect of low stress cyclic loading on the ultimate strength within the rather large scatter of this property.

Summary

Elastic constants of Galesville sandstone are temperature and pressure dependent over the range 37°-204°C and 0-31 MPa, respectively. Generally, Young's modulus increases with increasing temperature and effective confining pressure. Poisson's ratio decreases with increasing temperature and effective confining pressure. Pore space reduction apparent in hydrostatic compression and thermal expansion tests suggest that these changes are a direct result of densification of the sandstone with either increasing pressure or temperature.

Failure occurs over the range of temperature and pressure noted above by brittle faulting in all cases. Ultimate failure is associated with a critical inelastic volume strain

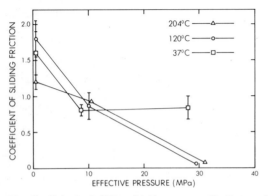

Fig. 14. Coefficient of sliding friction versus effective confining pressure for 37°, 120°, and 204°C.

TABLE 2. Changes in Elastic Constants Owing to Cyclic Loading

Temperature, °C	Effective Pressure, MPa	Cycles, N	ΔE, percent	$\Delta \nu$, percent
37	10.6	100	+7	−5
37	9.0	1000	+5	−11
37	7.3	10,000	−10	−26
120	0.5	100	+4	−11
120	0.5	1000	+8	0
120	0.5	10,000	+19	+22
120	10.5	100	+7	+10
120	10.6	1000	−1	−5
120	10.6	10,000	+28	−6
204	10.6	100	0	−12
204	10.5	1000	+17	−59
204	10.6	10,000	+12	−54

of about $4.5 \pm 2 \times 10^{-3}$, which decreases with increasing temperature and increases with increasing effective pressure. Mohr failure envelopes are nonlinear owing to plastic flow of the dolomite cement.

Low stress cyclic loading has little effect on elastic constants until 1000 cycles at 37°C and 120°C and 10,000 cycles at all temperatures. Changes are somewhat inconsistent but generally agree with the mechanism of densification, namely, Young's modulus increases and Poisson's ratio decreases with either increasing temperature or cycle number.

Acknowledgments. This research was performed in connection with subcontract B-63737-A-H, supported by the U.S. Department of Energy through Pacific Northwest Laboratory, operated for DOE by Battelle Memorial Institute under prime contract EY-76-C-06-1830. Sandstone samples were provided by Illinois Power Company.

References

Birch, F., Compressibility: Elastic constants, in Handbook of Physical Constants, edited by S. P. Clark, *Mem. Geol. Soc. Am., 97,* pp. 97-173, 1966.

Dunn, D. E. and L. J. LaFountain, Porosity-dependent strength reduction by cyclic loading, paper presented at the Proceedings of the 17th U.S. Symposium on Rock Mechanics, Snowbird, Utah, 1976.

Emrich, G. H., Ironton and Galesville (Cambrian) sandstone in Illinois and adjacent areas, *Circ. 403,* Ill. State Geol. Surv., Urbana, 1966.

Handin, J., and R. V. Hager, Experimental deformation of sedimentary rocks under confining pressure: Tests at room temperature on dry samples, *Am. Assoc. Petrol. Geol. Bull., 41,* 1-50, 1957.

Handin, J., and R. B. Hager, Experimental deformation of sedimentary rocks under confining pressure: Tests at high temperature, *Am. Assoc. Petrol. Geol. Bull., 42,* 2892-2934, 1958.

Handin, J., R. V. Hager, M. Friedman, and J. N. Feather, Experimental deformation of sedimentary rocks under confining pressure: Pore pressure tests, *Am. Assoc. Petrol. Geol. Bull., 47,* 717-755, 1963.

Kranz, R. H., and C. H. Scholz, Critical dilatant volume of rocks at the onset of Tertiary creep, *J. Geophys. Res., 82,* 4893-4898, 1977.

Skinner, B. J., Thermal expansion, in Handbook of Physical Constants, edited by S. D. Clark, *Mem. Geol. Soc. Am. 97,* pp. 75-96, 1966.

Voight, W., *Lehrbuch der Kristallphysik,* G. B. Teubner, Leipzig, 1928.

Wawersik, W. R., and W. S. Brown, Creep fracture of rock, *Tech. Rep. UTEC ME 73-197,* Univ. of Utah, Salt Lake City, 1973.

Field Mechanical Properties of a Jointed Sandstone

H. S. SWOLFS,[1] C. E. BRECHTEL,[2] W. F. BRACE,[3] AND H. R. PRATT[4]

The in situ mechanical and transport properties of a 2-m³ block of sandstone containing a near-vertical joint are measured at a field site in the Castlegate sandstone near Rangely, Colorado, as a function of compressive stress to 3 MPa. Prior to the static tests, the ambient (in situ) stress and dynamic elasticity of the undisturbed rock mass are determined; the average ambient horizontal stress in jointed but otherwise homogeneous and transversely isotropic sandstone is about 1 MPa. During the static loading tests, either permanent compaction or dilation and shear displacement of the through-going joint begins at applied stresses only slightly higher than 1 MPa. In comparison with our previous block test in jointed granite, jointing in the sandstone plays a greatly subordinate role. Seismic velocity is only slightly affected by the joints and fracture permeability actually appears to be less than that of the rock matrix. Subsequent excavation of the sandstone block alters the elastic and, presumably, the transport properties by an amount that is restored once the in situ value of horizontal compression is reapplied. Ambient stress should be taken into account in in situ tests to obtain accurate and meaningful values of mechanical properties.

Introduction

The role of joints and other natural discontinuities in the mechanical and transport behavior of rock is still poorly understood in spite of a number of recent studies devoted to this question [e.g., *Pratt et al.*, 1977]. Field testing of large jointed blocks of rock is the most straightforward method of studying the effects of joints on rock mass behavior, and we report here a continuation of our block testing program. We previously described the behavior of a 3-m cube of granite near Laramie, Wyoming [*Pratt et al.*, 1977]; here we report the mechanical and hydraulic behavior of a large block of jointed sandstone, but more attention is given to changes that occur when the block of sandstone is excavated. Following the description of the site and excavation procedures, we report measurements of displacements and sonic velocities under uniaxial and biaxial loading and attempts to measure shear strength and permeability of a joint.

Site and Lithology

The test area (T. 2 N.; R. 103 W.; Sec. 4; SW) is located on the nothern flank of the Rangely Anticline in northwestern Colorado. This west-north-west trending structure is a simple, elongate, plunging monoclinal flexure with surface dips varying from 2° to 4° on the northern flank to 15° to 40° on the southern flank. In this area the exposed Castlegate sandstone is the lowest member of the Price River formation (Mesaverde group) and overlies the Mancos shale [*Hale*, 1959].

Cullins [1969] divided the Castlegate sandstone into two units: an upper unit and a lower unit. The upper unit is a very light gray to gray, massive, limy, and fine-grained lagoonal deposit, and the lower unit is a gray-orange, yellowish-brown weathering, limy, friable, and very fine-grained littoral marine sediment. The test block is located in the lower unit of the Castlegate sandstone about 9 m north of a shallow, south-facing cliff (Figure 1). The physical and mechanical properties of this sandstone unit, measured on selected cores in the laboratory, are shown in Table 1.

The near-vertical joints in the otherwise massive lower unit are divided into three dominant sets: N 35°W, N 55°E, and N 85°E (Figure 2). The first two sets contain rather inconspicuous, hairline fractures spaced about 60 to 90 cm apart. The third and most conspicuous set, striking east-west, is composed of long, filled joints (up to 0.3 cm thick) that are spaced about 180 cm apart; their spacing, however, decreases to about 30 cm toward the edge of the south-facing cliff. This set of joints appears to be a stress-relief feature.

The test block is located adjacent to a joint that strikes N 83°E and dips 88°S (Figure 1). The width of the joint as seen on the surface and in cores averages about 0.15 cm. The filling material is almost entirely calcite; the average grain size is 0.02 mm.

Five seismic-refraction profiles are shot in the test area

[1] U. S. Geological Survey, Denver, Colorado 80225.
[2] Agapito and Associates, Inc., Grand Junction, Colorado 81507.
[3] Massachusetts Institute of Technology, Cambridge, Massachusetts 02139.
[4] Terra Tek, Inc., Salt Lake City, Utah 84108.

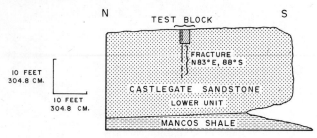

Fig. 1.   Cross section through test site showing lithologies, test block, attitude of near-vertical joint, and south-facing cliff.

to determine the compressional (P) wave velocity in the sandstone as a function of azimuth. True velocities are obtained over travel distances of about 23 m by reversing the profiles. The average seismic P wave velocity in the lower sandstone unit is 1.5 km/s and is clearly independent of joint direction, spacing, and frequency (Figure 2). The jointed sandstone appears to be seismically homogeneous.

Procedure

A site is selected adjacent to an east-west trending joint (Figure 1). Within a 2-m² area, but prior to the excavation of the block, surface locations are prepared for the application of strain gage rosettes (S_1, S_2, and S_3; see Figure 3), and shallow holes are core drilled for the emplacement of a stress-monitoring gage (SG), a displacement gage (DG), and an ultrasonic velocity transmitter (VT) and receiver (VR). Three additional holes (P_1, P_2, and P_3) are drilled at an angle to the surface to intersect the joint at 60-cm depth and allow the emplacement of pressure transducers for joint permeability tests. At some distance from the block location an additional strain gage station (S_4) is prepared to be overcored at some later time.

The block is excavated by line drilling and broaching three vertical slots 1.2 m deep and 3 cm wide through the use of pneumatic equipment. The two side slots are 90 cm long and extend 15 cm beyond the joint (Figure 3). The

third slot is 2.1 m long and parallel to the joint. The resulting block, bounded on the north side by the near-vertical joint, essentially is free along its sides but remains attached at the bottom. This particular slot pattern is determined in advance by a finite element study to ascertain the most appropriate block geometry that would insure uniform loading of the isolated joint. The excavation of the block is accomplished in five days (July 28–August 1, 1976).

Physical Changes During Excavation

The primary aim of field testing is to perform geotechnical experiments on rock samples that are undisturbed and therefore representative of the host rock. Consequently, it is important to know to what degree the rock sample is disturbed, if at all, during its excavation. The principal causes of sample disturbance are (1) change in stress conditions due to unloading and (2) change in rock fabric or structure resulting from the excavation process itself. While the latter effect is usually negligible in large rock samples [Pratt et al., 1972], the former effect is unavoidable. The material changes that take place due to unloading will be evident from measurements of stress, strain, displacement and sonic velocity.

We attempt to measure stress changes directly during excavation of the block by using a modified version of a recently developed stress meter [Swolfs and Brechtel, 1977]. This meter embodies flatjacks that, when emplaced in a borehole (SG in Figure 3), give the normal stress change along three different azimuths. Malfunction of one of the jack systems prevents a full stress determination during excavation. Nevertheless, we observe at 60-cm depth a horizontal decompression of 0.55 MPa at N 7°W and of 1.08 MPa at N 53°E.

A U. S. Bureau of Mines displacement gage (DG) [Hooker et al., 1974] at a 60-cm depth registers strain changes in all three horizontal directions (Figure 4). The associated stress changes are calculated by using 2.3 GPa for Young's modulus and 0.28 for Poisson's ratio [Ageton, 1967]: 0.55 MPa (N 7°W), 1.06 MPa (N 53°E), and 1.91

TABLE 1.   Properties of Castlegate Sandstone

Parameter	Value
Physical properties	
Grain size	0.05 mm
Grain density	2.63 g/cm³
Bulk density	1.97 g/cm³
Porosity	25%
Permeability	190 ± 80 mdarcies
Mechanical properties	
Unconfined compressive strength	11.4 MPa
Young's modulus	2.3 GPa
Poisson's ratio	0.28
Compressional wave velocity	
Dry	2.1 km/s
Moist	1.7 km/s
Shear wave velocity	
Dry	1.4 km/s
Moist	1.1 km/s

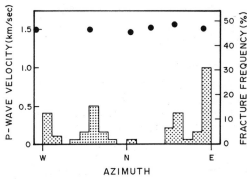

Fig. 2.   Relationship between frequency of vertical joint sets (65 joints measured) and azimuthal distribution of compressional (P) wave velocity measured by the seismic refraction method. (represented by the closed circles).

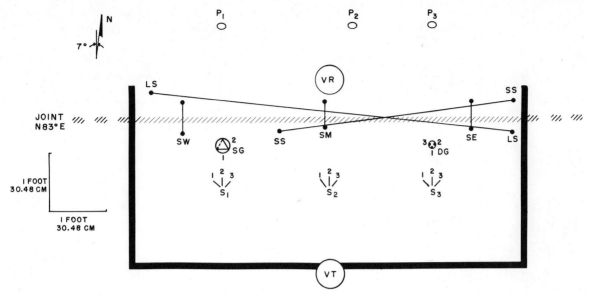

TEST BLOCK, CASTLEGATE SS, RANGELY ANTICLINE, COLORADO

Fig. 3. Schematic of the top surface of the test block excavated to include portion of the east-west trending joint. Heavy black lines indicate the position of the vertical slots and loading flatjacks. Letters refer to surface and bore hole instrumentation identified in the text.

MPa (N 67°W). The secondary principal stress magnitudes and directions are 1.57 MPa (N 79°W) and 0.77 MPa (N 11°E).

Strain changes are monitored by the surface strain gage rosettes (S_1, S_2, and S_3), but they show considerable scatter and probably are unreliable; one typical result is shown in Figure 5. Their ineffectiveness is due primarily to the high porosity of the host rock, which makes it difficult to cement properly the strain gages to the surface of the rock. Rosette S_4, located some 3-m east of the block, was overcored on August 13, 1976, using a 15-cm diameter core bit (Figure 6). The secondary principal stresses, calculated by using 2.3 GPa for Young's modulus, are 0.9 MPa (N 88°W) and 0.4 MPa (N 2°E).

To summarize the ambient or in situ state of stress in the sandstone, the maximum horizontal compression strikes about east-west and is about twice the magnitude of the minimum compressive stress. It is of interest to note that previous stress measurements [de la Cruz and Raleigh, 1972] at various sites around the Rangely Anticline have yielded comparable results. Using a variety of techniques, the surface stress magnitudes averaged about 1 MPa. The principal directions of stress determined by surface measurements, earthquake focal-plane solutions, and hydraulic fracturing [Raleigh, 1972] also are reasonably consistent; the maximum horizontal stress component strikes about west-north-west, subparallel to the fold axis of the monocline, and the minimum horizontal stress component strikes north-north-east.

The consequences of relieving about 1 MPa of in situ stress on the mechanical properties of the rock sample are determined by ultrasonic measurements across the block.

Using shear wave transducers (70 kHz) both the compressional (P) wave and shear (S) wave velocities are measured along a path slightly less than a meter and at various depths, from hole VT to hole VR (Figure 3) before and after the excavation of the block. Prior to the isolation of the block, the P and S wave velocities are 1.5 km/s and 0.8 km/s, respectively (Figure 7, closed symbols). The P wave velocity of the intact block is thus the same as the velocity measured by seismic refraction across longer travel distances by using 45-g explosives (Figure 2) and close to the value determined in the laboratory on moist samples (Table 1).

After the block is isolated, the P and S wave velocities are remeasured (Figure 7, open symbols). Both quantities are reduced by about 40% to 0.9 km/s (P) and 0.5 km/s (S). The dynamic Young's modulus, calculated from the measured velocities and using a rock density of 1.97 gm/cm^3, decreases from 3.3 GPa to 1.2 GPa (60%). It seems reasonable to assume that the static modulus of the block decreases as well and perhaps by a similar amount, although there is no method available to measure directly this modulus. The static modulus of the isolated block may, at most, be lower by 10-20% than that of the host rock, owing to stress relaxation. It follows then that considerable care is needed to interpret properly the results of the subsequent loading tests that will be described in the next section.

Static Loading Tests

In preparation for the static loading tests, flatjacks are cemented with high strength Hydrostone in the open, vertical slots. The flatjacks are made by welding two thin,

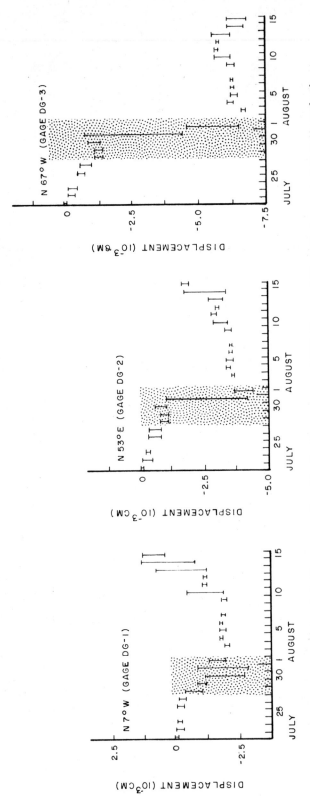

Fig. 4.   Change in bore hole diameter monitored by the three-component U. S. Bureau of Mines displacement gage at 60-cm depth (DG in Figure 3) during excavation of the test block (shaded region). Vertical bars indicate maximum change registered per day. Positive displacement is compression.

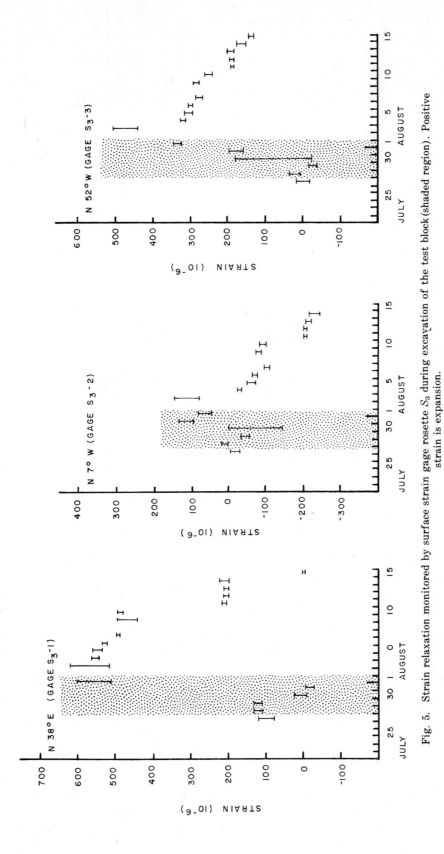

Fig. 5. Strain relaxation monitored by surface strain gage rosette S_3 during excavation of the test block (shaded region). Positive strain is expansion.

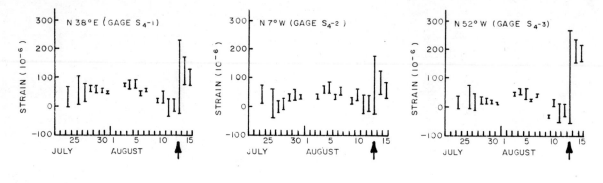

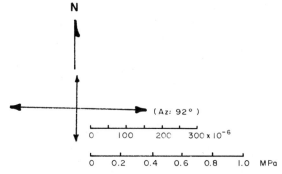

Fig. 6. Strain relief monitored by surface strain gage rosette S_4 after overcoring on August 13, 1976 (arrow). S_4 was located 3 m east of test block.

stainless steel plates (90 × 120 cm) together along the edges. A thin walled steel tube conducts water under pressure to the flatjacks. Pressurized accumulators are used to increase the water pressure at a constant rate. Three flatjacks are installed in each of the two side slots and four flatjacks similarly are placed in the long slot, two flatjacks on either side of hole VT (Figure 3).

Five direct current displacement transducers (DCDT's) are used to monitor the closure and opening of the joint during the loading tests. Three of these (SW, SM, and SE; see Figure 3) are to measure the normal displacement across the joint, whereas two (LS and SS) with long gagelength record the shear displacement along the joint. All these devices are calibrated before the start of each loading test.

Measurements of joint permeability follow the methods used in our previous study [Pratt et al., 1977]. Water is injected under constant pressure into the central hole P_2 (Figure 3) and either the volume flowing along the joint in a given time or the increase in pressure is measured at holes P_1 and P_3. Laboratory measurements of permeability of intact sandstone (Table 1) range from 0.1 to 0.3 μm^2 (100 to 300 mdarcies).

Observations

Five successful loading tests are run on the sandstone block: one uniaxial test, two biaxial tests, and two shear tests. The results of each group of tests will be discussed separately.

Uniaxial Test

This test consists of loading the block in only one direction; namely, perpendicular to the joint. The four flatjacks in the long slot are pressurized at a constant rate to load the block to a peak stress of about 3.3 MPa. The recorded displacements during loading and unloading are shown in Figure 8. Only two of the normal displacement transducers functioned during the test (SW and SM; Figures 8a

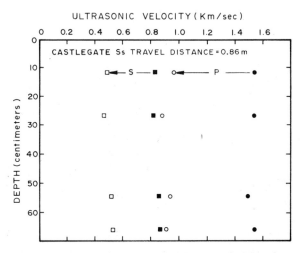

Fig. 7. Change in ultrasonic P and S wave velocities after excavation of the test block (open symbols). Closed symbols indicate preexcavation values at various depths.

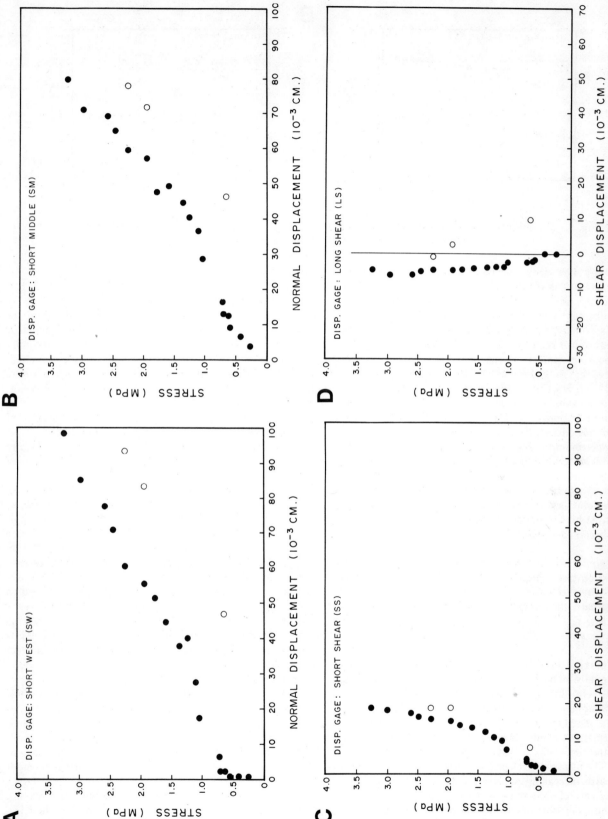

Fig. 8. Uniaxial stress displacement curves. Positive displacement is closure or shortening. (a, b) Normal displacement across the joint. (c, d) Shear displacement along the joint. The loading cycle is represented by the solid circles and the unloading cycle is represented by the open circles.

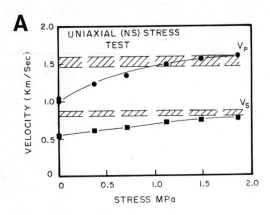
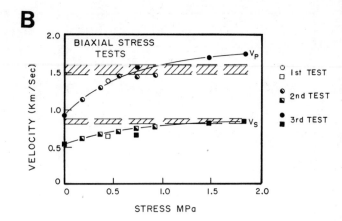

Fig. 9. Velocity stress curves during the (a) uniaxial and (b) biaxial tests. Horizontal broken bars indicate the magnitude of velocities measured prior to block excavation.

and b); they show closure of the joint at a rate of about 3×10^{-2} cm/MPa. The displacements parallel to the joint recorded on SS and LS (Figures 8c and d) show what appears to be contrary behavior; the short shear transducer indicates shortening along the joint, whereas the long shear transducer records a slight expansion. The LS transducer monitors almost the entire length of the joint and, therefore, registers the Poisson effect in the block under uniaxial load. The SS transducer monitors a little over half the joint length, but its behavior is not understood. It should be mentioned at this point that the sides of the block are not entirely free to expand.

All displacement curves show a slight inflection between 0.5 and 1.0 MPa. Recalling that this stress is about equal to the magnitude of the preexisting component normal to the joint, it appears that the behavior of the material in or near the joint is different below the prestress level than above it. At higher stress levels the material becomes permanently compacted by about 3×10^{-2} cm (Figures 8a and b).

Ultrasonic measurements made during the uniaxial compression test (Figure 9) show an increase in both the P and S wave velocities. Both quantities, however, level off and approach the preexcavation velocities at higher applied stresses.

Biaxial Tests

Two biaxial tests are made by increasing the applied stress equally on both sides of the block to 1.6 and 3.3 MPa, consecutively (Figure 10). The joint compacts uniformly in all directions until, at about 3.3 MPa of applied stress, the joint begins to creep and lose its load-bearing capacity. The permanent compaction in the joint material at the end of the tests is in excess of 6×10^{-2} cm (Figures 10a and b). The ultrasonic velocities across the entire block increase with increasing applied stress (Figure 9b). However, the rate of change in velocity is greater for the biaxial than for the uniaxial condition. Similarly, the velocities reach the preexcavation values at lower applied stress levels.

Shear Tests

These tests are run as follows: first, the joint is loaded to a constant normal stress by inflating the flatjacks in the long slot; second, the pressure in the east slot is increased to apply a uniformly increasing shear stress on the joint. The normal stress in the first test is held constant at about 0.7 MPa, and the shear stress is increased to about 1.4 MPa and then reduced. In the second test the normal stress is held constant at 1.4 MPa, while the shear stress is increased to 3.1 MPa, at which point the eastern portion of the block surface spalls and terminates the test.

The results of the two shear tests are shown in Figure 11. The normal displacement across the joint remains essentially at zero (Figure 11a) at the beginning of the tests until the shear stress exceeds the normal stress on the joint (0.7 MPa or 1.4 MPa) and the joint begins to slide. At higher levels of shear stress the joint begins to open up or dilate (Figure 11a) and the shear displacement (Figure 11b, SS) reverses and begins to increase. The inflection points C' and C'' may represent the shear stress levels required to initiate sliding on the joint at two different normal stresses. Figure 11c shows the shortening recorded on LS as the block is compressed during the two shear tests. A slight change in the rate of displacement, however, is evident in the curves at higher shear stress levels. The difference in behavior between the SS and LS transducers can be analyzed (Figure 3) by noting their respective anchoring points near the eastern flatjacks.

It should be noted at this point that the block as a whole does not move because it is still firmly attached at the bottom. The data, however, do indicate that the stress conditions for sliding are momentarily reached in the upper part of the block. The fact that the stress displacement curves continue to increase also argues against movement of the entire block.

The change in ultrasonic velocities with increasing shear stress is, as expected, not very large after their initial increase due to the application of normal stress (Figure 12).

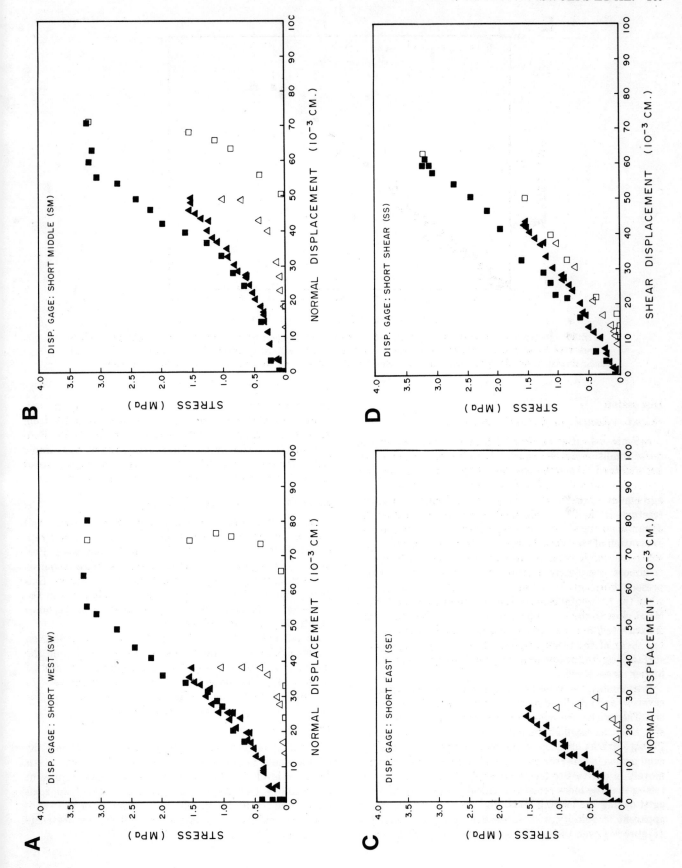

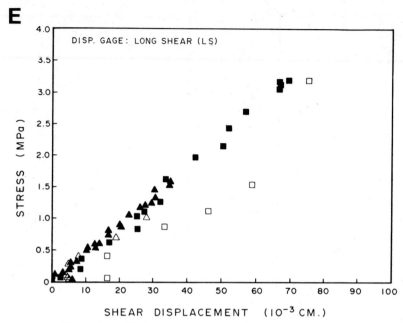

E

DISP. GAGE: LONG SHEAR (LS)

SHEAR DISPLACEMENT (10⁻³ CM.)

Fig. 10. (opposite) Biaxial stress displacement curves (the first cycle is represented by triangles; second cycle is represented by squares). Positive displacement is closure. (a, b, c) Normal displacement across the joint. (d, e) Shear displacement along the joint loading cycles (solid symbols) and the unloading cycles (open symbols).

Discussion

Seismic Velocities in the Castlegate Sandstone

Seismic velocities in small laboratory samples in the moist condition agree remarkably well with the sonic values obtained across the unrelieved block and the seismic refraction values in the area around the test block. This agreement suggests that the joints and smaller natural fractures in the Castlegate sandstone have no effect on the dynamic elasticity of this rock. Velocities are affected by excavation of the block, however. It is interesting that the 40% reduction in velocity caused by excavation (Figure 7), is almost completely restored when the block is compressed horizontally (Figure 9). Recovery is achieved when the horizontal compression reaches a value (0.7 to 1.5 MPa) close to the in situ prestress relieved when the block is excavated. Seismic velocity changes are very small during shear of the block (Figure 12) along the major joint surface despite the apparent dilation across this surface at higher stress levels.

Presumably, the velocity changes we observe during excavation and later loading reflect changes in crack geometry. The cracks are apparently on the scale of the grain size, since, as noted above, joints seem to play no role here. Perhaps cracks open up in the material that cements the sand grains in this rock when the in situ compression is removed. In contrast to this microscopic 'damage' it is interesting that the block remains elastically relatively undamaged during the loading tests. For example, no change is apparent in velocity with subsequent biaxial load cycles (Figure 9b) even though the applied stress reaches two to three times the prestress level. This insensitivity to moderate stress may be typical of porous rocks, to judge from studies of Kayenta sandstone [*Swolfs et al.*, 1976], Bedford limestone [*Brace and Riley*, 1972], and many volcanic tuffs [*Morrow et al.*, 1977]. On the same basis, we would expect to see large permanent changes at high stress.

Local Failure of the Test Block

Slight permanent compaction under normal stress and dilation under shear stress (Figure 11, for example) represent the only evidence of failure in the test block. Perhaps it is significant that this local failure requires stresses only a little higher than the in situ stress. Evidently, differential movement on joints could occur in the Castlegate sandstone once the applied stress becomes slightly higher than ambient.

An estimate of the coefficient of friction for sliding on the filled joint in the test block can be obtained from the shear and normal stresses at the inflection points C' and C'' (Figure 11b). The value is close to 1.0, which agrees with average values of the coefficient of sliding friction for rocks at these low pressures [*Byerlee*, 1978].

Permeability

Pumping tests across pairs of holes intersecting the joint at depth failed to produce flow along the joint at our detection level. A very rough estimate of maximum joint permeability (see Appendix) gives a value of approximately one-tenth that in laboratory samples of the sandstone. The calcite filling in the joint effectively obstructs flow along the entire joint surface.

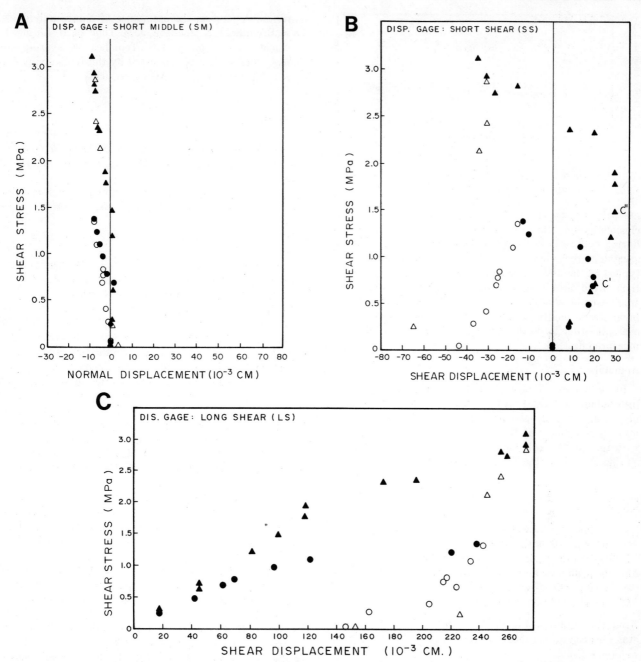

Fig. 11. Shear stress displacement curves (the first test is indicated by the circles; second test is indicated by the triangles). Positive displacement is shortening. (a) Normal displacement across the joint. (b, c) Shear displacement on SS and LS transducers, respectively, along the joint. Loading cycles are represented by the solid symbols and the unloading cycles are represented by open symbols. Inflection points C' and C'' in (b) indicate reversals in shear displacements and initiation of sliding along the joint.

Significance of the Prestress Measurements

Both the present results and observations made in similar field tests in Kayenta sandstone [*Swolfs et al.*, 1976] point to significant changes in properties of rocks in situ as stresses are altered by even partial excavation. This probably also is true for crystalline rocks, although pre-excavation stress values are not available for our earlier block experiment [*Pratt et al.*, 1977]. Thus, for example, the very low stiffness of the Sherman granite block below 1-MPa stress (*Pratt et al.* [1977], Figure 13) may not be characteristic of the granite at ambient stress levels pres-

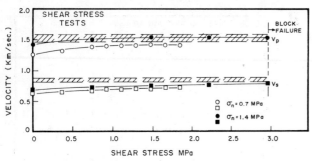

Fig. 12.   Velocity stress curves during shear tests for two different normal stresses (see Figure 9).

ent prior to excavation. This suggests an important conclusion; namely, that the in situ stress must be reapplied during field tests in porous sedimentary as well as dense crystalline rocks in order to obtain properties that will be typical of the undisturbed rock mass. In other words, although the block test is an important improvement over small sample laboratory tests, it needs to be done at ambient stress conditions to be completely meaningful. Obviously, in order to do this, in situ stresses will need to be measured first.

Appendix

In an attempt to measure joint permeability, water is injected under pressure at hole P_2 (Figure 3). Although no flow is detected at holes P_1 and P_3, an estimate of maximum permeability of the joint is made for comparison with that of the rock matrix measured in the laboratory (Table 1). We use the Thiem equation [*Todd*, 1959]:

$$Q = 2\pi k b \, (\Delta h) / \ln \, (r_e / r_w)$$

where

r_e the distance between the holes P_2 and P_3 (50 cm);
r_w the hole radius (2 cm);
b the length of joint intersected obliquely (10 cm);
Δh the head drop between the holes (7×10^3 cm);
Q the flow rate (5 cm³/s); and
k the permeability in cm/s.

From this equation, k is 3.7×10^{-5} cm/s. For water under standard conditions this is equivalent to 3.7×10^{-2} or 37 mdarcies.

Acknowledgments.   This paper is dedicated to John Handin. We owe John a debt of gratitude for his constant and continuing interest and participation in this line of research. If, out of the rock dust and pungent odor of hydraulic oil, new truths emerge about the constitution of the earth's crust, a large measure of the credit belongs to him. Field experiments of this magnitude and scope require and depend on the assistance of many individuals. To F. Bronner, J. N. Johnson, J. M. Thomas, R. Lingle, L. Buchholdt, C. Nelson, and G. Toombes we express our thanks. This work was supported by the U. S. Army Research Office, contract DAAG29-75-C-0003.

References

Ageton, R. W., Deep mine stress determinations using flatjack and borehole deformation methods, *Rep. Invest. U. S. Bureau Mines 6887*, 25 pp., Washington, D. C., 1967.

Brace, W. F., and D. K. Riley, Static uniaxial deformation of 15 rocks to 30 kb, *Int. J. Rock Mech. Mining Sci., 9*, 271–288, 1972.

Byerlee, J. D., Friction in rocks, *Pure Appl. Geophys., 116*, 615–626, 1978.

Cullins, H. L., Geologic map of the Mellen Hill Quadrangle, Rio Blanco Moffat Counties, Colorado, *Geol. Map GQ 835*, U. S. Geol. Surv., Washington, D. C., 1969.

de la Cruz, R. V., and C. B. Raleigh, Absolute stress measurements at the Rangely Anticline, northwestern Colorado, *Int. J. Rock Mech. Mining Sci., 9*, 625–634, 1972.

Hale, L. A., Intertonguing Upper Cretaceous sediments of northeastern Utah–northwestern Colorado, paper presented at Rocky Mountain Association of Geology 11th Annual field conference, Denver, Colo., October 1959.

Hooker, V. E., J. R. Aggson, and D. L. Bickel, Improvements in the three-component borehole deformation gage and overcoring techniques, *Rep. Invest. U. S. Bureau Mines. 7894*, 29 pp., Washington, D. C., 1974.

Morrow, C. A., W. F. Brace, and E. Carter, Dramatic electrical changes in porous rocks at low stress (abstract), *Eos Trans. AGU, 58*, 1235, 1977.

Pratt, H. R., A. D. Black, W. W. Brown, and W. F. Brace, The effect of specimen size on the mechanical properties of unjointed diorite, *Int. J. Rock Mech. Mining Sci., 9*, 513–529, 1972.

Pratt, H. R., H. S. Swolfs, W. F. Brace, A. D. Black, and J. W. Handin, Elastic and transport properties of an in situ jointed granite, *Int. J. Rock Mech. Mining Sci., 14*, 35–45, 1977.

Raleigh, C. B., Earthquakes and fluid injection, *Mem. Am. Assoc. Pet. Geol., 18*, 273–279, 1972.

Swolfs, H. S., and C. E. Brechtel, The direct measurement of long-term stress variations in rock, in *Proceedings of the 18th U. S. Rock Mechanics Symposium*, pp. 4C51-3, Colorado School of Mines, Keystone, Colo., 1977.

Swolfs, H. S., C. E. Brechtel, and J. Handin, Stress-relief measurements in large rock specimens, in *Proceedings of the 17th U. S. Rock Mechanics Symposium*, pp. 4B51-4, University of Utah, Salt Lake City, 1976.

Todd, G. W., *Geohydrology*, John Wiley, New York, 1959.

Case for Low Deviatoric Stress in the Lithosphere

BARRY RALEIGH AND JACK EVERNDEN

United States Geological Survey, Menlo Park, California 94025

The case for low (i.e., 100 bars), average ambient shear stress on major plate boundary faults depends on several fundamental observations. The absence of anomalous heat flow over the San Andreas fault zone implies that most of its displacement has occurred at low shear stress. No plausible mechanisms other than conduction exist for removal of the excess seismogenic heat. Stress drops in earthquakes are rarely greater than a hundred bars. If water is present in the fault zone, seismogenic heating will raise the water pressure to equal the normal stress on the fault provided the shear strain in the slip zone during the earthquake is 2 or greater. Therefore one might expect nearly total stress drop in earthquakes because of the effective removal of the frictional resistance to slip. If the stress drops are nearly total, then the average ambient shear stress should be 100 bars or so. Asperities carrying higher stress may yield stress drops of several hundred bars, but such stress drops are uncommon and do not represent the average resisting stress along plate boundaries. There are other more ambiguous but still suggestive observations such as steady fault creep on the Central San Andreas, where low P wave velocities may be related to high fluid pressure in the fault. Intraplate earthquakes in the central and eastern United States have moments 2 to 3 orders of magnitude greater for a given length of fault break than earthquakes in the western United States. Our explanation calls for the fault zones in the east to be soft inclusions in a rigid crust, whereas the crust in the west is relatively less rigid. Thus the fault zones in the east and in the west are not appreciably different, only the surrounding crust. Faulting in soft inclusions at low stress in the east occurs in an environment of highly stressed but strong crust.

Introduction

We undertake to marshall evidence demonstrating that large interplate and intraplate earthquakes occur under conditions of low ambient shear stress, i.e., in the neighborhood of 100 bars. Until the demonstration by *Lachenbruch and Sass* [1973] that the large, late Cenozoic and Quaternary displacements on the San Andreas fault have not generated a measurable thermal anomaly, little compelling evidence had been offered to indicate that the shear stress responsible for earthquakes was less than the kilobar levels called for by laboratory-determined values of the strength of rocks. One of us [*Raleigh and Paterson,* 1965] had proposed that intermediate and deep focus earthquakes must occur in an environment of high pore fluid pressure to promote brittle failure; otherwise the rocks would surely flow at the high shear stresses required by their dry frictional strength at such elevated normal stresses. In recent years, however, the accumulated evidence has led us to the conclusions that most, if not all, earthquakes occurring in the upper crust, both along plate boundaries and internal to plates, are driven by low shear stress. If correct, this conclusion has far-reaching implications. The dynamics of plate motion, the mechanics of faulting, and the nature of the failure process and its predictability are critically dependent on the magnitudes of the stresses at which faulting takes place.

Our hypothesis is that the environment in which nearly all earthquakes occur includes low mean shear stress but with stronger asperities of varying sizes and density of distribution along the fault surface. We consider that localized evidence of shear stresses of several hundred bars is convincing but inadequate to refute the data implying low mean shear stress. This paper constitutes a presentation of arguments demonstrating the prevalence of low mean shear stress driving fault motion.

The Absence of a Thermal Anomaly Over the San Andreas Fault

Brune et al. [1969] first noted the absence of any anomaly in heat flow over the San Andreas fault and drew the conclusion that low driving stress was a likely cause. *Lachenbruch and Sass* [1973] developed the argument with the aid of much more data in central California. They [*Lachenbruch and Sass*, 1980] have since extended their measurements to the Mojave Desert region, finding that over a segment of the San Andreas fault that is not creeping and produced a great earthquake in 1857, no measurable perturbation of the regional value of 1.6 heat flow units (1 HFU = 10^{-2} cal/m^2 s) occurs.

There appear to be only two avenues of escape from the conclusion that the ambient shear stress on the San Andreas fault is less than 150 bars. *Lachenbruch and Sass* [1973] noted that heat generated by long-term fault motion, if driven at shear stress of a kilobar or so, might be dissipated by convective circulation of water. In this case, thermal springs would provide a localized heat sink. The excess heat to be so dissipated is readily calculable. Assuming that the average shear stress τ in excess of that called for by the absence of a conductive anomaly is about 1000 bars, the yearly excess heat production per centimeter of fault surface is just

$$\Delta W = \tau \dot{u} \qquad (1)$$

where $\dot{u}$ is the long-term average displacement rate of about 3 cm/yr and all the work done goes into heat. Using a fault depth of 10 km (e.g., from 5 to 15 km) over which the heat is generated, the excess heat is 8×10^{12} cal/yr for each kilometer of the fault length. If the heat escapes in water heated to 100°C, springs flowing at 10,000 l/hr at 100°C would be required per kilometer of fault length. Except for the Imperial Valley region, where young, volcanic heat sources are available, the San Andreas fault system is notable for the absence of thermal springs [*Waring*, 1915]. If the heat were dissipated by convection with transport to distances of tens of kilometers from the fault, a broad heat flow anomaly would result. The central California Coast Ranges do show such a high average heat flow, about 2.1 heat flow units. However, there is no peak in heat flow over the fault. Rather special pleading would be required to call for heterogeneous permeabilities so distributed in the crust as to allow the water to carry the same amount of heat to the surface irrespective of distance from the fault. Furthermore, the regional heat flow of 1.6 heat flow units [*Lachenbruch and Sass*, 1980] in the Mojave Desert is not anomalously high, and there is likewise no peak over the fault. Convection, therefore, is an unlikely mechanism for dissipation of seismogenic heat.

In principle, energy can be dissipated by mechanisms other than generation of heat. However, we have found no acceptable candidates. Creation of new surface area of silicate crystals requires 10^2–10^3 ergs/cm^2 [*Brace and Walsh*, 1962]. Taking the higher estimate, we find that comminution of the entire fault zone to a grain size of 1 μm over a width of 5 km would require only 10^4 yr at 1-kbar shear stress and 3-cm/yr displacement rate. To put it another way, given 300 km or so displacement on the San Andreas fault at 1 kbar shear stress and assuming all energy released during slippage is expended on creation of new surface area, there would result a zone 5 km wide in which the grains were milled down to the dimensions of a unit cell, about 10 Å.

The available energy from the total slip on the San Andreas fault is very large, about 3×10^{16} ergs/cm^2 when assuming that a kilobar of shear stress is acting during the slip. As a result, other energy-absorbing processes such as phase changes are totally inadequate to explain the absence of a heat flow anomaly. For example, dehydration requires about 100 cal/g for the enthalpy of reaction, so a fault zone 10 km wide could be completely dehydrated by the fault displacement.

There is unequivocal evidence for the absence of such widespread phase changes as dehydration or melting, which could absorb the required energy, for the absence of the intensive comminution that we calculate would be needed, and for the absence of thermal springs. Consequently, we agree with *Lachenbruch and Sass'* [1973] conclusion that shear stress of less than about 150 bars must drive the San Andreas fault.

Measurements of Absolute Stress in the Crust

Although there are presently no direct measurements of in situ stress from seismogenic regions of plate boundary faults, the available data suggest shear stress is likely to be low. *McGarr and Gay* [1978] have compiled measurements of in situ stresses and find that the average maximum shear stress in the crust between 500 and 4000 m depth to be about 150 bars. The maximum value (one measurement) is 320 bars, and all others are less than 250 bars. These measurements are from mines, generally in older, crystalline rock terrains where fluid pressures are hydrostatic or even subnormal. If, as *Lachenbruch and Sass* [1973] propose, the San Andreas and other plate-boundary faults are brittle relative to the underlying material, the shear stress in the adjacent crustal blocks is likely to be higher than that along the fault itself. The stresses measured within the plate would thus represent high values relative to the plate boundary faults.

Measurements of absolute stress in holes as deep as 900 m have been made by *Zoback et al.* [1980] near the San Andreas fault. The results leave the question of the state of stress at hypocentral depths unresolved because of the dubious assumptions that must be made to permit any extrapolation from shallow observations to those greater depths. Resolution of the problem by direct measurements will probably require deeper drilling.

Arguments in Favor of Total Stress Drop

The large displacements at slip velocities of about 1 m/s in great earthquakes have prompted various analyses of the seismogenic heat and its effect on the properties of the fault surface [*McKenzie and Brune*, 1972; *Sibson*, 1973; *Raleigh*, 1977]. *McKenzie and Brune* [1972] proposed that melting could take place. *Raleigh* [1977] showed that explosively rapid dehydration of silicates in the fault zone

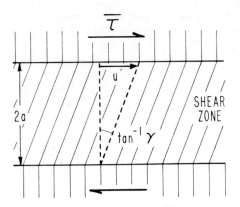

Fig. 1. Diagram of fault zone showing displacement μ during an earthquake distributed so that shear strain γ over the zone of width $2a$ is uniform; $\bar{\tau}$ is the average shear stress, assumed to fall linearly with displacement.

during large earthquakes could be expected given high shear stresses (1 kbar) or with shear strains around 100 for faulting at shear stress of 100 bars. Either melting or dehydration should serve to reduce the local effective normal stress to a value near zero. *Sibson* [1973] proposed that total stress drop might occur as the result of the rise in pressure in occluded water along the fault surface caused by seismogenic heating. This suggestion is worthy of closer examination, since water seems likely to be present in major fault zones and the effect is large.

To calculate the pressure increase in a volume of water V that fills the pore space in rock upon a rise in temperature dT, we let the volume expansion dV of the water and the pore be equal and

$$dV = V\alpha_w \, dT - V\beta_w \, dP \qquad (2)$$

$$dV = V\alpha_r \, dT + V\beta_r \, dP \qquad (3)$$

where α_w and α_r are thermal expansion coefficients of water and rock, and β_w and β_r are their compressibilities. Then

$$\frac{dP}{dT} = \frac{\alpha_w - \alpha_r}{\beta_w + \beta_r} \qquad (4)$$

and since $\alpha_w \gg \alpha_r$ and $\beta_w \gg \beta_r$, a reasonable assumption at moderate effective confining pressure, then

$$\frac{dP}{dT} = \frac{\alpha_w}{\beta_w} \qquad (5)$$

For reasonable ranges of crustal temperatures and pressures, α_w/β_w falls between 15 and 20 bar/°C [*Kennedy and Holser*, 1966]. To raise the pore fluid pressure of water from hydrostatic to lithostatic at a depth of 10 km, for example, would therefore require a temperature rise of only about 100°C.

We calculate the temperature increase during fault slip from the following simplifying assumptions: Let the fault displacement u take place at constant slip velocity over

time τ at an average shear stress $\bar{\tau}$ (Figure 1). The work done is assumed to be converted entirely to heat, thus neglecting radiated elastic energy or other dissipative processes. Since we are concerned with only the maximum temperature reached, the heat is taken to be liberated uniformly over a fault zone of constant width $2a$, and $2a$ is large enough so that faulting ceases before heat leaves the center of the zone by conduction. The condition that satisfies this requirement is

$$a \geq (4ht)^{\frac{1}{2}} \qquad (6)$$

where h is the thermal diffusivity and t the time. Given the short times involved in the slip at some point along a fault during an earthquake, i.e., 10 s or less, the width of the shear zone need not be more than 1 cm where $h = 0.01$ cm^2 s^{-1}. The temperature increase at the center of the shearing slab can now be calculated from

$$\Delta T = \frac{\bar{\tau}u}{2a\rho C} \qquad (7)$$

$$\gamma = \frac{u}{2a}$$

$$\Delta T = \frac{\bar{\tau}\gamma}{\rho C} \qquad (8)$$

where γ is the shear strain, ρ is density, and C the heat capacity. Taking ρ as 2.7 g cm^{-3} and $C = 0.25$ cal g^{-1} deg^{-1}, then $\Delta T = 0.0358$ in degrees centigrade per bar of shear stress.

The change in fluid pressure due to seismogenic heating is, then,

$$\frac{dP}{d\bar{\tau}} = 0.6\,\gamma \qquad (9)$$

For example, 10 cm of displacement distributed over a fault zone 5 cm wide, a shear strain of 2, would generate a fluid pressure rise of about 1.2 bar per bar of shear stress. As the shear stress is about 0.6 times the effective normal stress at failure, then the fluid pressure increase would cause the effective normal stress to vanish when the strain exceeds 2.5, and near total stress drop will occur. This point arises also from similar analyses by *Lachenbruch* [1980] and *Sibson* [1980].

Now, the fluid pressure increase due to the seismogenic rise in temperature depends not only upon the thermal expansion of water in pores but on the permeability of the shear zone being heated. If the hydraulic diffusivity is very much larger than the thermal diffusivity, the pressure increase with temperature will be less than that given by (5) due to escape of the heated water from a narrow shear zone. The permeabilities of deeply buried crustal rocks are so poorly known that much attention paid to exact calculation of the coupled thermal/fluid pressure problem would be unrewarding. However, the same relationship used to calculate a minimum width of the shear

TABLE 1.   Maximum Increase in Pressure of Water in Fault Due to Seismogenic Heating at Slip
Velocity in Meters per Second, Where the Width $2a$ of the Fault Undergoing Uniform Shear
Strain Is Greater Than $(4h_wt)^{1/2}$ and $h_w = k/(\phi B_w \mu) = 10^{15}k$

Average Shear Stress $\bar{\tau}$, bars	Displacement u, cm	Width $2a$, cm	Shear Strain	Time, s	Permeability, cm^2	Pore Pressure Increase ΔP, bars
50	10	2	5	0.1	10^{-14}	150
50	10	10	1	0.1	2.5×10^{-13}	30
50	100	2	50	1.0	10^{-15}	1500
50	100	10	10	1.0	2.5×10^{-14}	300
500	10	2	5	0.1	10^{-14}	1500
500	10	10	1	0.1	2.5×10^{-13}	300
500	100	2	50	1.0	10^{-15}	15000
500	100	10	10	1.0	2.5×10^{-14}	3000
500	100	100	1	1.0	2.5×10^{-12}	300

9

zone (6) such that no heat escapes from the center of the zone during fault slip may be applied to estimate the hydraulic diffusivity below which the fluid pressure increase will not be dissipated by fluid flow in the same time.

In Table 1, we assume various average shear stresses, displacements, and fault widths and compute the maximum allowable permeability such that no fluid leaks away from the center of the shear zone during the time over which slip occurs. The maximum fluid pressure increase generated is also given. To achieve these fluid pressures for the parameters chosen, hydraulic diffusivities need not be less than 1-2.5×10^{-3} cm^2 s^{-1}, which corresponds to permeabilities in the neighborhood of 10^{-12}-10^{-15} cm^2(10^{-4}-10^{-7} darcies). We note that the Eleana argillite, a fine-grained, clay-rich rock of 5-10% porosity and having similar properties to fault gouge from the San Andreas fault, has permeability between 10^{-18} and 10^{-19} cm^2 for intact samples at around 200 bars effective confining pressure [Lin, 1978]. Taking the viscosity of water to be 10^{-3} P with a compressibility of 10^{-10} cm^2 dyne^{-1} at 1000 bars and 300°C, the hydraulic diffusivity h is between 2×10^{-4} and 10^{-5} cm^2 s^{-1} given the porosities measured. As the thermal diffusivity of rock is around 10^{-2} cm^2 s^{-1}, pore fluid pressure generated thermally would outlive the thermal pulse by a comfortable margin, provided the fluid pressure does not exceed the least principal compressive stress. Admittedly, these measured permeabilities are quite low, but the techniques have only recently been applied to argillaceous rocks [Lin, 1978]. In any case, permeability higher by a factor of 10^7 (as shown in Table 1) would still permit large pore pressures to be generated by seismogenic heating. Provided shear strains are appreciably greater than unity and water is present, the conclusion that near-total stress drops take place in moderate to large earthquakes seems inescapable. The average ambient shear stress must, therefore, be approximately equal to the stress drop, which in large earthquakes is 100 bars or so.

The maximum pore fluid pressure that can be sustained by the fault zone is the magnitude of the total least principal compressive stress S_3 plus whatever tensile strength

the rocks might have. Taking the reasonable assumption that tensile strength is negligible, tensile fractures will open when $p = S_3$, allowing the fluid to escape from a heated fault zone. There remains an effective normal stress σ_n acting on the fault due to the maximum principal effective stress σ_1, and the shear stress will, therefore, not relax completely. The residual shear stress is less than half the initial shear stress if the dynamic friction is about 0.8 of the static frictional coefficient.

Further Data Relative to Creeping Portion of San Andreas Fault

T. McEvilly (personal communication, 1979) has conducted Vibroseis seismic reflection profiles from several kilometers west of the San Andreas fault to several kilometers east of it. Though the final calculations are not yet completed, results of significance to our discussions have been obtained and have been made available for our use in this paper.

Figure 2c indicates the generalized route of the profile. Actually, two separate investigations were conducted. The first consisted of a single common-depth point section on either side of the fault, while the latest investigation was a continuous common-depth-point profile from 15 km or so west of the fault to 6-8 km east of the fault. To the west of the fault the terrain is composed of granitic rocks of mid-Cretaceous age that have been translated approximately 300 km northward relative to the rocks east of the fault. All outcrops display strongly sheared and shattered rock, and such characteristics persist to the bottom of the 600-m deep Stone Canyon well. To the east of the fault the basement rock is composed of somewhat metamorphosed Franciscan sandstone of Jurassic age, overlain by some Tertiary and Cretaceous sediments. The prevalence throughout northern California of abnormally high fluid pressures in Franciscan sandstones beneath a blanket of the Great Valley sequence of sedimentary rocks has been proposed by Irwin and Barnes [1975].

The first item of interest from McEvilly's data is the marked change in two-way time to the Mohorovicic discon-

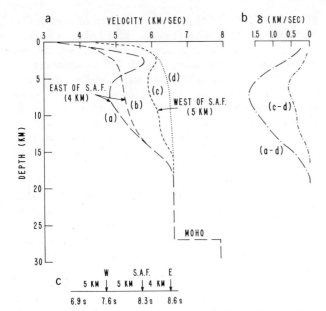

Fig. 2. Vibroseis profiles of creeping portion of San Andreas
fault.

tinuity as a function of distance west of the fault. Coherent reflections are observable just beyond a kilometer-wide fault zone; there the two-way time is 8.3 s. This time decreases westward, so the time is 7.6 s 5 km further west and as low as 6.9 s 10 km west of the fault. Beyond this distance, the two-way travel time remains at 6.9 s. Thus there is clear evidence of a broad region displaying increasing travel times as the fault is approached. On the east side of the fault, the only two-way data presently available indicate a two-way time of 8.6–8.9 s 4 km east of the fault, i.e., a travel time that is about a second greater than for a comparable offset west of the fault.

The data obtained during the first investigation have been analyzed to yield estimates of velocity as a function of depth at 5–6-km distance from the fault on both sides of the fault. Both of these locations are certainly within the zone of abnormally high travel times. Estimates of these velocity versus depth curves are given in Figure 2. Curve a is the velocity structure east of the fault by direct analysis of the data. Since it displays such a drastic structure, McEvilly has chosen to use in recent calculations a velocity structure similar to curve b. Curve c is McEvilly's best guess through various possible models for the west side of the fault. Curve d indicates very roughly the approximate expected velocity given no fluid pressure and a response of velocity to load stress. Curves a–d and c–d in Figure 2b are plotted to illustrate that the low velocity effect is much more exaggerated than simple inspection of curves a and c might suggest.

A final item of importance is that when travel time is converted to depth via the models a and c, the depths to the Mohorovicic discontinuity on the two sides of the fault are calculated to be identical, even though two-way times to the Moho differ by a second or more. Thus all time dif-

ferences appear to be due to velocity structure above the Moho, i.e., all two-way times of greater than 6.9 s on the west side of the fault imply a zone of reduced velocity in the crust with reduction in velocity becoming more pronounced as the fault is approached. *Healy and Peake* [1975] found a similar deep wedge of low-velocity rock along the San Andreas Fault from arrivals of local earthquakes.

The most likely explanation of the low velocity zones concentrated near the fault is high near-lithostatic fluid pressure in rocks of extremely low permeabilities, permeabilities being required to be exceedingly low to prevent depressuring of the zone. Though somewhat surprising, such low permeabilities are suggested by *Lin's* [1978] data noted elsewhere within this paper, the prevalence of very high pore pressure throughout Franciscan terrains far from the San Andreas fault, and the observed values of V_p at 700-m depth in fault gouge of the San Andreas fault (3 km/s).

As a complement to the data of McEvilly, we note the data of *Savage and Burford* [1973] and *Thatcher* [1979] on rates of deformation along the same portion of the San Andreas fault as a function of distance away from the fault. They found that the measured creep displacements along the San Andreas fault accounted for all the deformation observed. Both triangulation data (obtained over a 40-yr period) and laser ranging data were analyzed. The data show that averaged over times of a few years, fault creep takes place without elastic distortion of the adjacent blocks of crust out to distances of 30 km from the fault, i.e., to the maximum distance of triangulation data.

Although these results offer no information as to the absolute level of stress in the crust, the lack of strain accumulation accompanied by nearly steady-state fault creep suggests a weak fault zone. As shown experimentally by *Byerlee and Brace* [1968] and *Dieterich* [1978], stable sliding is favored by low normal stress on the experimental fault. This implies high pore fluid pressure and low driving stress.

To the north and south, where no creep is presently observed but where great earthquakes occurred in 1857 and 1906, the so-called 'locked' or stick slip sections of the San Andreas fault have been generally considered to have markedly different material properties and/or driving stresses from the creeping section. However, as noted by *Lachenbruch and Sass* [1980], there is no anomaly in heat flow over these sections of the fault, implying that the driving stress is low here also. This apparent paradox is readily resolved in light of the fact that creep and small earthquakes are closely associated, thus suggesting that the central section of the fault is under conditions which are close to those at the boundary between stick slip and stable sliding behavior [*Dieterich*, 1978]. Thus only a small difference in effective normal stress, rigidity of the adjacent crust, or the distribution of asperities on the fault in the locked sections would be required for these areas to be within the stick slip field.

Subduction Zones

Though the following arguments may seem quite subjective, we feel their content affords further compelling evidence of low values of operative shear stress and associated low effective stress. The arguments are concerned with the following facts: (1) very low tectonic stress in the hanging wall block of the Aleutian subduction zone, (2) low to vanishing deformation of sedimentary/volcanic rocks forming hanging wall blocks in Aleutians, South America, (3) slow earthquakes, (4) occurrence of deep-focus earthquakes and distribution of earthquakes versus depth, (5) failure of cumulative moment of earthquakes to explain more than 20–30% of movement in subduction zones.

The first fact was extensively documented at the time of large nuclear explosions on Amchitka Island by observations of strain at points on the island and of spectral amplitudes of long-period surface waves at points throughout the world. Both of these data sets confirmed the exceedingly low level of tectonic stress release associated with these explosions.. Radiation patterns of explosion-induced Rayleigh waves were little perturbed by waves resulting from associated tectonic energy released via relaxation subsequent to generation of a zero-strength volume by the explosion. This behavior is to be placed in contrast with the effects of tectonic release for explosions in granite at Nevada test site. For those explosions the phase of the 20-s Rayleigh waves in some azimuths was reversed from that of the explosion-induced waves themselves. In other words, the waves resulting from tectonic release were higher in amplitude than those of the waves induced directly by the explosion.

A truly remarkable aspect of the hanging wall blocks in many areas of strong convergence (i.e., subduction zones) is the near-total lack of compressive deformation. Tension (as evidenced by pervasive volcanism) is a more typical regime, not strong compression. Another aspect of this is the exceedingly minor uplift and deformation of Pleistocene terraces in the Aleutians. It seems clear that displacements measured in many tens of kilometers can occur without concomitant compressive stress and associated deformation within the subduction zone. We would suggest that stress conditions across this interface have little or nothing to do with rates of subduction. Rate of delivery of crust to the zone dictates the rate of subduction in an environment of low effective stress.

The occurrence of slow earthquakes may well be an expression of low effective stress. Of even greater interest, however, is the fact that such quakes are only a component in a broad spectrum of slip scenarios, varying from conventional earthquakes through the slow earthquakes of *Geller and Shimazaki* [1978] and the very slow earthquakes of *Pfluke* [1978] through creep events to near-steady state slippage. Until recently, it was tacitly assumed that all slippage on fault surfaces was related to conventional earthquakes. However, this assumption has been shown to be inconsistent with numerous observations. In addition to the specific detection of earthquakes implying low stress environments, other bodies of data are suggestive of the same thing. The failure of seismic moments to account for more than 20–30% of apparent displacement in some subduction zones implies that most movement is aseismic, probably continuous or episodic creep at low effective stress. A fact not widely known and appreciated is the pattern of numbers of earthquakes as a function of depth in a well-developed subduction zone. Such data have been published for the Kurile-Kamchatka zone [*Evernden*, 1970], a zone in which the horizontal component of the principal compressive stress is nearly perpendicular to the entire zone from one end to the other. The recurrence relationship (m_b versus log $N_{cumulative}$) has an identical slope for shallow-focus quakes and for deep-focus quakes. However, the number of deep quakes is only one-third as great as the number of shallow quakes, even though the area of 'fault surface' is several times larger for the zone of deep quakes. Another important fact for this area is that there are virtually no earthquakes with depths of less than 25 km ($m_b \geq 4.0$). Therefore failure via stick slip processes is not usually occurring in the upper 25 km of that subduction zone, nor is it the dominant mode of slippage at great depth. The distribution of earthquakes as a function of depth for worldwide seismic activity (most activity being in subduction zones) has the following distribution pattern [*Evernden*, 1970]: 68% of the total number of earthquakes occur at depths ≤60 km, 27% at 60–300 km, and 5% at depths >300 km. All these data can be explained via a model presuming low driving stress operating within an environment of low effective stress induced by high fluid pressure. We know of no other proposed hypothesis that achieves the integrated explanation of observed data.

A final point of possibly critical significance is the fact that nearly invariably, first-motion solutions for thrust earthquakes in subduction zones have the implied slip surface far steeper than the dip of the subducting slab, implying an inclined principal compressive stress nearly parallel to the dip of the subduction zone. In other words, the maximum principal compressive stress is generally inclined toward the direction of subduction. We consider that earthquakes are a result of failure in a zone near (and above?) that of actual subduction induced by low levels of shear coupling between the hanging wall block and the subducting plate, modified by compression along the slab due to body-force-induced stresses, not by push from a distant rear.

Therefore we conclude that a low-stress environment is consistent with observations in subduction zones. As elsewhere, we accept within such a model the occurrence of scattered several-hundred-bar asperities that localize points of abnormal stress and serve as foci for conventional earthquakes.

Intraplate Earthquakes

Even if it is given that earthquakes along the plate boundaries occur in an environment of low ambient shear

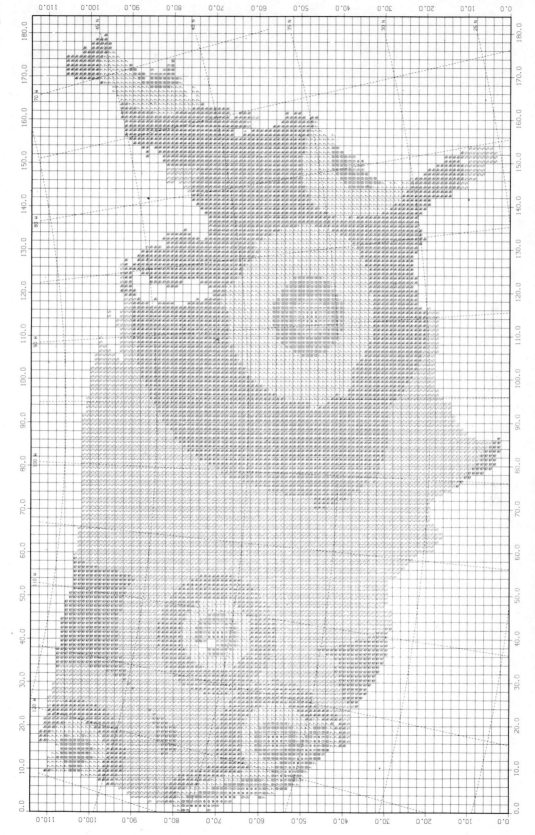

Fig. 3. Variation of attenuation parameter k throughout the United States. For ease of reading, the value actually plotted on the figure is $4k$.

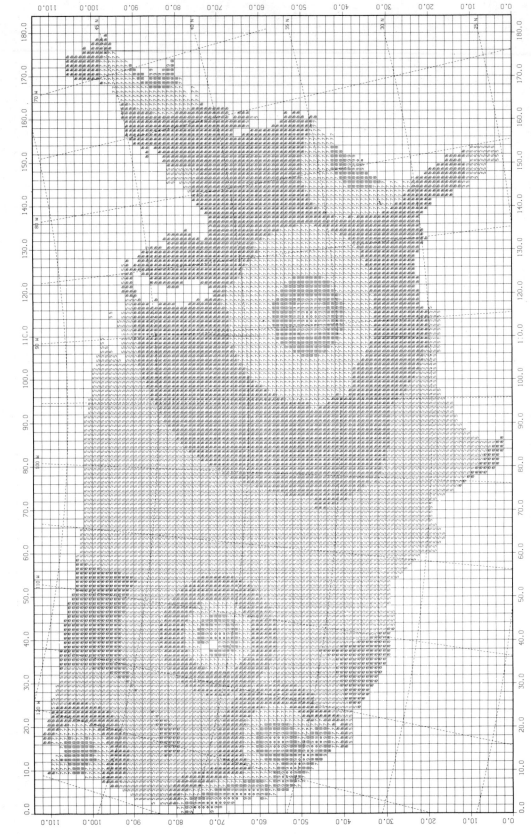

Fig. 4. Predicted maximum potential Rossi-Forel intensities throughout the United States as a result of the following set of past or potential U.S. earthquakes (lengths of break [2L] are indicated in square brackets): Cape Ann, Massachusetts (1755) [10]; Charleston, South Carolina (1886) [15]; New Madrid, Missouri (1811) [20]; Wasatch Fault near Salt Lake City, Utah [60]; Seattle, Washington (1949) [40]; Owens Valley, California (1872) [60]; Fort Tejon, California (1857) [320]; San Francisco, California (1906) [400]; Long Beach, California (1933) [22]. Ground condition is assumed to be saturated alluvium.

TABLE 2.   Observed and Calculated Parameters of $k = 1\frac{3}{4}$ (Region 7) and $k = \frac{1}{2}$ (Region 6) Earthquakes

Earthquakes	Intensity Scale of Published Data[a]	Attenuation Value (k), 4k	Date Year	Month	Day	Magnitude (M_L)	Seismic Moment (M_0), 10^{25} dyne cm	Area Included in Intensity VI Contour (A_{VI}), 10^{24} cm²	Length of Break (2L), km	Maximum Shaking Intensity (I (MX))	Calculated Lengths of Break (2L), km Using A_{VI} Values	Using All Intensities
Hemet	MM	1 3/4	1963	9	23	5.3	0.02	0.24		VI	1.2	
Lytle Creek	MM	1 1/2	1970	12	9	5.4	0.10	0.22		VII	1.1	1.1
Coyote Mountain	MM	1 3/4	1969	4	28	5.9	0.50	0.61		VII	3.6	
Parkfield	MM	1 3/4	1966	6	28	5.5	1.30	0.54	3	VII	3.1	
Desert Hot Springs	MM	1 3/4	1948	4	12	6.5	1.00	2.60		VII	27	10
Long Beach	MM	1 3/4	1933	3	11	6.3	2.00	1.20		VIII	9	22
Santa Rosa Mt.	MM	1 3/4	1934	3	19	6.2	4.00	2.30		VI	23	
San Fernando	MM	1 3/4	1971	2	9	6.4	4.70	2.30	16	IX	23	19
Borrego Mt.	MM	1 3/4	1964	4	8	6.5	6.0	3.4		VII	38	
Imperial Valley	MM	1 3/4	1940	5	18	7.1	20	3.3		IX	9	60
San Francisco	RF	1 3/4	1906	4	18	8.2	850	16	400	IX	130	400
Santa Barbara	RF	1 3/4	1925	6	29	6.3	20	4.4		IX		40
Lompoc	RF	1 3/4	1927	11	4	7.3	6.5			IX		70
Fort Tejon	RF	1 3/4	1857	1	9	8	900		320	IX		320
Wheeler Ridge	MM	1 1/2	1954	1	12	6.6	0.33	1.75		VII+	1.8	
Truckee	MM	1 1/2	1966	9	12	6.5	0.5	1.61		VII	1.6	
Bakersfield	MM	1 1/2	1952	8	22	5.8	0.55	0.32		VIII	0.3	2.1
Fairview Peak	MM	1 1/2	1954	12	16	7.0	90	14	40	VIII	31	
Kern County	MM	1 1/2	1952	7	21	7.7	170	17	60	IX	40	60
Oroville	MM	1 1/2	1975	8	1	5.9	0.2	0.48	1.5	VII	0.4	1.5
Pocatello Valley	MM	1 1/2	1975	3	28	6.0	0.65	1.4	3	VIII	1.4	3.0

[a] MM, modified Mercali; RF, Rossi Forel.

stress, it might still be proposed that intraplate earthquakes are high-stress events and, in addition, may occur in an environment of significantly higher ambient shear stress than do earthquakes along plate boundaries. The higher values of ambient shear stress in surficial hard rocks in the eastern US (few hundred bars) might well be interpreted to imply kilobar-level stresses at depth near the fault. The 1 order of magnitude greater moment for a given length of break in the Basin and Range area of the United States than in western California (see below) also might be interpreted as symptomatic of higher stress drop and higher ambient stress in the fault zones of this area.

However, we will suggest that such an interpretation is incorrect and that a quite different and possibly unexpected interpretation is consistent with available data, while one based on high shear stress in fault zones is not.

We begin this explanation by establishing three presently little appreciated facts. The first fact is that the attenuation rates of horizontally propagating seismic waves with frequencies around 1–4 Hz (i.e., frequencies of relevance in intensity observations) are grossly different throughout the United States. Figure 3 indicates the approximate pattern, the 4k values plotted being based on analysis of numerous isoseismal maps of U.S. earthquakes. The value 4k is related to ground motion via the relation:

ground motion proportional to Δ^{-k}. Attenuation rates vary in the conterminous U.S. from Δ^{-1} to $\Delta^{-1.75}$ [Evernden, 1975]. The impact of such variation in attenuation rates is shown on Figure 4, where the isoseismal maps for maximum size earthquakes in each region are plotted. The patterns for San Francisco (k = 1.75), Charleston (k = 1.25), and New Madrid (k = 1.00) are essentially as reported, while the pattern for a Salt Lake City quake is as predicted when using a length of break appropriate to the maximum earthquake in regions of k = 1.50.

The increasing spacing of isoseismals as k decreases is clearly shown. The implication of these intensity patterns and the resultant k value as regards the lengths of break of earthquakes in different regions has not been understood until recently. The tabulation at the side of Figure 4 indicates the length of break required by our model to explain absolute values of observed intensity maps when the k value has been established by the spacing of consecutive isoseismals. It is vital to understand how we calculate length of break in the model. Given the k value, we determine the energy required to explain the I versus Δ data of each quake. We then ascertain the length of break required in western California (k = 1.75) to provide that amount of energy. We then compare lengths so calculated with other data on actual lengths of break. Tables 2 and 3

TABLE 3.   Observed and Calculated Values of Length of Break 2L for Additional Earthquakes of $k = 1\frac{3}{4}$ and $k = 1\frac{1}{2}$ Regions

Earthquake	Year	Month	Day	Latitude, °N	Longitude, °W	Magni-tude$_{ob}$	2L, km Observed	2L, km Predicted
Cedar Mountain, Nevada	1932	12	20	38.8	118.0	7.2	61	66
Excelsior Mountain, Nevada	1941	1	30	38.0	118.5	6.3	1.4	3.5
Hansel Valley, Utah	1934	3	12	41.5	112.5	6.6	8	10
Manix, California	1947	4	10	35.0	116.6	6.4	4	10
Fort Sage Mountains, California	1950	12	14	40.1	120.1	5.6	8.8	1
Kern County, California	1952	7	21	35.0	119.0	7.7	60*	60
Rainbow Mountain, Nevada	1954	7	6	39.4	118.5	6.6	18	18
Rainbow Mountain, Nevada	1954	8	23	39.6	118.4	6.8	31	36
Fairview Peak, Nevada	1954	12	16	39.3	118.2	7.1	48	40
Hebgen Lake, Montana	1959	8	17	44.8	111.1	7.1	24	28†
Galway Lake, California	1975	5	31	34.5	116.5	5.2	7	1
Pocatello Valley, Idaho	1975	3	28	42.1	112.6	6.0	3‡	3
Oroville, California	1975	8	1	39.4	121.5	5.7	4 (1.5‡)	1.5

*33 km of fracture in bedrock. However, epicenter was about 30 km away under Quaternary deposits. See text.
†By use of data to south of epicenter. See text.
‡2L values determined from short period seismograms. See text.

give such data for U.S. quakes with known lengths of break. The somewhat surprising result is that our mode of calculating length of break, though seemingly highly arbitrary, accurately predicts observations in regions of $k = 1.75, 1.50,$ and 1.25. For the New Madrid quake, the largest known quake in $k = 1.0$, we calculate a length of break of 2.5–5 km (20 km if $k = 1.25$ to the very tip of the Mississippi Embayment).

All of the above is sketchily described, but details are given elsewhere [*Evernden*, 1975; *Evernden et al.*, 1981].

We have demonstrated our second important fact, i.e., the energy in intensity-relevant frequencies for earthquakes throughout the conterminous U.S. is dependent solely on fault length and has no relation to k value. This is really quite surprising but is supported by analysis of numerous earthquakes and is not denied by any such analyses. Thus the data give no support for the idea of increased stress drop for quakes in regions of low k values. Subsidiary facts requiring emphasis are the apparently strong dependence of maximum permissible length of break on k value and the total inability to casually determine size of an earthquake (i.e., source parameters) from an uncritical inspection of size of the region of sensible shaking.

The conclusion that we draw from these results is that the fault zones in all regions of the U.S. are very similar. Asperities must be of essentially equal strength and equal mean distribution on fault surfaces in all regions. Whatever the effective stress conditions are in one region, they are duplicated in all others. The inhomogeneity in stress on fault surfaces in $k = 1\frac{3}{4}$ regions (San Andreas fault), expressed by asperities with several-hundred-bar stress drops, while average stress drop is a few tens of bars, is now common knowledge. What is new here is our suggestion that this pattern is similar in all regions of the U.S. The mean stress can rise somewhat without affecting high frequency energy but cannot rise an order of magnitude.

Tables 2–4 provide the data relevant to our third fact,

while Figures 5 and 6 present the relevant data in graphical form. All lengths of break are the observed values when such data are available (adequate seismological data other than intensities can classify a length of break as 'observed') or are the calculated values via our model. All moments are those reported by other authors. The fact we wish to point out is apparent in Figure 5, i.e., for fixed length of break, there is approximately a tenfold increase in seismic moment with each decrease of $\frac{1}{4}$ in k value (see later discussion of Parkfield data). We calculate that an earthquake with length of break equal to 1 km in a region of $k = 1.00$ has a seismic moment of 10^{26} dyne cm. There is no way that such a moment can be explained with the conventional physical model of seismic moment, i.e., $M = \mu LDH$.

These gross differences in lengths of break versus k value versus short- and long-period energy is quite unexpected. It appears that even moment values have to be explained within heterogeneous earth models.

Simple attenuation models are clearly impossible for an explanation of the data of Figures 5 and 6. Extensive observations of surface waves over all kinds of crust/mantle types establish there to be only minimal differences in attenuation of long-period waves. This fact is exploited in order to achieve intercalibration of worldwide sites of underground nuclear testing.

What then can the explanation of facts 2 and 3 be? We suggest a model based on soft-inclusion concepts, as originally treated by *Eshelby* [1957]. To be more specific, we suggest that earthquakes of the eastern U.S. are along fault zones that constitute soft inclusions in an otherwise highly rigid and strong crust/mantle system.

We elaborate. Following *Eshelby* [1957], we consider the following situations: (1) uniform elastic medium of shear modulus μ_0, and (2) soft inclusion (sphere) of shear modulus μ_1 imbedded in an otherwise uniform elastic medium of shear modulus μ_0.

Presuming shear stress to be applied at distances that

TABLE 4. Data for Eastern United States Earthquakes

Earthquake	Year	Month	Day	Latitude, °N	Longitude, °W	Magnitude	Maximum Intensity	2L, km	Area Within Intensity IV Contours 10^{14} cm^2	Seismic Moment,[a] 10^{24} dyne cm
Grand Banks[b]	1929	11	18	44.5	55	···	···	1.3 (1)	20	63
East Missouri[b]	1965	10	21	37.9	91.1	5.2	VI	0.30–0.04 (1)	56	0.1
Cornwall/Massena	1944	9	5	45.0	74.8	···	VIII	1.0 (1¼)[c]	6.4	2.5
Illinois[b]	1968	11	9	38.0	88.5	5.3	VII	0.08 (1)	4.8	1.0
New Hampshire[d]	1940	12	20	43.8	71.3	···	VII	0.016 (1)	0.78	1.0
								0.16 (1¼)	1.2	1.0
Timiskaming[b]	1935	11	1	46.8	79.2	···	VII	0.11 (1)	6.8	3.2
Missouri[b]	1963	3	3	36.7	90.1	4.5	VI	0.022 (1)	1.1	0.1

[a]*Herrmann et al.* [1978].
[b]Earthquakes used in Figures 5 and 6.
[c]$k = 1$ not permitted by data. Result in agreement with *Evernden* [1975] and Figure 3.
[d]$k = 1¼$ solution in agreement with local magnitude, but earthquake in k 1 region of Figure 3. Uncertain interpretation.

are large compared with dimensions of the soft inclusion, let us investigate shear stress τ_1 at the center of the space and at the center of the sphere in terms of the remotely applied shear stress τ_A:

In case (1),

$$\tau_1 = \tau_A$$

In case (2),

$$\tau_1 = \frac{\tau_A \mu_1}{[\mu_0 - \beta(\mu_0 - \mu_1)]}$$

where $\beta_0 = 0.1333 \ (4 - 5\nu_0)/(1 - \nu_0)$, and ν_0 equals Poisson's ratio outside the inclusion. For $\nu = 0.25$,

$$\beta = 22/45 \simeq 1/2$$

Then, we find the following relationships:

μ_1/μ_0	1	0.50	0.10	0.05	0
τ_1/τ_A	1	0.67	0.18	0.10	0

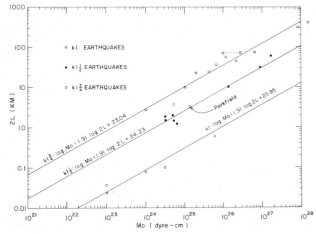

Fig. 5. 2L versus seismic moment for all k regions of conterminous United States.

Thus the stress within the soft inclusion is less than the distantly applied stress. Or in order to achieve a given level of shear stress on a fault within the inclusion, it will be necessary to apply a greater distant shear stress. If μ_1 is one-tenth of μ_0, then the externally applied stress must be 5.5 times the stress required on the fault surface. If μ_1 is one-twentieth of μ_0, then the externally applied stress must be 10 times that required on the fault surface. Thus if mean shear stress required for failure is 100 bars, regional stresses outside the inclusion must be 550–1000 bars, i.e., a low stress-drop quake in a highly stressed regional environment.

Now, consider the comparative changes in strain energy associated with fault zone failure at shear stress τ_1. Since strain energy change is a measure of moment, this analysis will be relevant to Figure 5.

Consider the following situations, assuming total stress drop on the fault: As before, i.e., uniform space:

$$\Delta E_a = 8r^3\tau_1^2/7\mu_0$$

where r is the radius of circular fault patch. There exists a soft region with radius R and shear modulus μ_1:

$$\Delta E_b = 8r^3\tau_1^2/7 \mu_1 \quad r \ll R:$$

As in the second case (above) except that total stress drop is presumed to take place within the entire volume of the soft inclusion:

$$\Delta E_b = \pi R^3 (\mu_0 + \mu_1) \tau_1^2/3\mu_0 \mu_1 \quad \text{for } \beta = \tfrac{1}{2}$$

We then obtain the following values:

μ_1/μ_0		1	0.5	0.10	0.05	0
E_b/E_a		1	2.0	10.0	20.0	∞
$E_c/E_a R/r =$	2.5	29	43.0	157.0	300.0	∞
$E_c/E_a R/r =$	5.0	229	343.0	1259.0	2405.0	∞
$E_c/E_a R/r =$	10.0	1832	2749.0	10079.0	19252.0	∞

It is apparent that nearly any desired ratio of 2L and M_0 is possible in concept, the asperities on the fault surface providing nearly all the high frequency energy, while inhomogeneous relaxation of much lower average stress

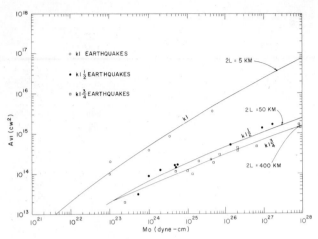

Fig. 6. A_{V1} versus seismic moment for all k regions of conterminous United States.

level provides the long-period energy. The questions that arise are: Why should one hypothesize total volume-relaxation? What are reasonable values of R (given a satisfactory answer to the first question)?

The basis for a total volume-relaxation hypothesis is founded on: (1) The conclusion, based on intensity and $2L$ data, that all fault zones are similar, i.e., are weak and are at low effective stress. (2) The fact of comparatively high values of measured ambient stress in many eastern United States rocks along with the fact of pervasive fracturing of rocks in the epicentral region of the New Madrid earthquake (F. McKeown, personal communication, 1979). This highly fractured mass may then relax partially or entirely with release of the fault surface (and may keep on relaxing for decades, a la Charleston?).

We suggest that the difference between the environments of earthquakes of western United States (WUS) and eastern United States (EUS) is dominantly the differences away from the fault zone. We must hypothesize that the relevant μ_1 is not that associated with propagation of shear waves through the inclusion but a μ_1 related to stress storage, i.e., a pseudo μ related to nonlinear deformation of a highly fractured mass.

Probably no more than a factor of 100 increase in M_0 is required to overcome the limitation of standard models for estimation of M_0. Thus given a μ_1/μ_0 ratio of 0.1, only very limited volumes of total relaxation are required, or only partial relaxation in a larger volume is needed. The required dimensions do not seem denied by any available data.

The Parkfield data are of the greatest interest in this regard. The Parkfield quake is clearly anomalous on Figures 5 and 6, it having a moment and observed length of break far greater than predicted by the intensity pattern and the usually observed correlation of intensity pattern and moment in regions of $k = 1\frac{3}{4}$. It is vital to note that the implied discrepancy between long and short periods is clearly observable on the seismograms, it having been shown [*Lindh and Boore*, 1981] that the high frequencies for this quake derived nearly entirely from the 3–4 km of the south end of the break. They show that the fault broke slowly from the north end southward and then broke rapidly for the final 3–4 km. Such a pattern of breakage would lead to long periods consistent with the long break and short periods with the short break. Interestingly, when the isoseismals are used to predict the location and length of break yielding the high frequencies, they predict 3.5 km located essentially exactly where Lindh and Boore show it to have been. Therefore we have an example in WUS of a quake having abnormal relative amounts of long- and short-period energy, the mechanism for achieving this abnormality being related to that proposed above for EUS quakes.

It may be useful to point out that the high moment for a short break is achieved only in association with a large volume relaxation. Thus the conventional μLDH formula for moment must be modified to express the effective dimensions of the relaxed volume. By this device, the more highly stressed rocks beyond the inclusion are allowed to relax far more than if there were no weak inclusion. There is thus no physical anomaly, just a more heterogeneous earth than usual models hypothesize.

Some interesting relationships are suggested. A major implication is that one seeks sites of potentially damaging eastern United States earthquakes by seeking zones of low ambient stress. High stress implies absence of high fluid pressure and little or no chance of fault failure. Low ambient stress implies high fluid pressure and extensive fracturing. These latter conditions should typify seismic zones in the eastern United States.

Therefore we conclude that fault zones in all regions of the U.S. are sites of low ambient shear stress and low stress drop, i.e., all fault zones are weak and are, in all probability, sites of much free water and low effective stress. Recent observations in the New Madrid fault zone are interesting in this regard. M. Zoback, when drilling the zone for making in situ stress measurements via hydrofrac, found the zone so shattered that no stress measurements were possible. Also, F. McKeown has pointed out the omnipresent mineralized waters in the zone as well as alkalic intrusions.

New Fault Breaks

The final refuge of high-stress earthquakes is in generation of new breaks through virgin kilobar-strength rocks. We now proceed to saw off even that small branch of the high stress hypothesis. Three lines of evidence seem to be pertinent:

1. Inspection of multistation m_b/M_s values of thousands of earthquakes indicates total or near-total absence of paired values, implying kilobar-level stress drops. Such events, if they exist, should occur once in a while on even a short temporal sample (few years). An argument that could be proposed here is that such new-break earthquakes

may invariably be very small earthquakes and thus may not appear in present-day data lists of quakes with paired M_s/m_b values. However, it was pointed out by *Evernden* [1970] that the slope of the mean M_s/m_b is uniformly 1.0 from m_b 5.5 to and below 4.0, implying no deflection of the Ms/m_b curve down to m_b 4.0 due to changes in stress drop versus magnitude. Therefore any effect must be at very small magnitudes.

2.   When the United States Weapon Laboratories made elaborate efforts, both experimental and theoretical, to predict observed ground motion of underground nuclear explosions, they found it impossible to use elastic coefficients observed on small specimens of intact rock in their computer codes. Ultimately, it was realized that the explosion itself was the experiment providing information of mean strength of granite, etc. in Nevada and the Aleutians. It was found that effective strengths of about 100 bars had to be assumed for granite if observations were to be predicted. Therefore the concept of kilobar-strength rock terrains in zones of deformation may be totally incorrect.

3.   If low effective stresses are typical of the crust, the formation of new faults in previously unfractured rock must then occur at low shear stress. Two observations suggest this to be so. Firstly, great overthrust sheets must have vanishly small cohesive strength at the time of their formation, or they could never be driven by the small stresses implied by their small thickness-to-length ratio [*Rubey and Hubbert*, 1959]. Secondly, large bedrock landslides take place in intact rock under gravitational stresses of 10-30 bars. Especially where the rupture surface does not follow an obvious pre-existing zone of weakness, one is compelled to the view that some mechanism, perhaps stress corrosion, operates to degrade the cohesive strength of rock.

Conclusions

In our view, there is a preponderance of evidence to indicate that faulting occurs under conditions of low (100 bars) average shear stress. The principal supposed difficulty in calling for fault motion at such low shear stress is that ambient high pore fluid pressures are required to reduce the frictional resistance of faults. The fluid pressure transients developed by thermal expansion of occluded water during seismogenic faulting will dissipate long before a recurrence of another great earthquake on the same fault. Therefore mechanisms for generating the high pore pressure and for maintaining the pressure once established, require explanation. The difficulty lies principally with transcurrent plate boundaries rather than thrust boundaries. The ocean floor sediments conveyed continuously down the subduction zone provide a renewable source of water which, through metamorphism and dehydration, can provide high fluid pressure to very considerable depths. A mechanism which provides for high pressure fluid in transcurrent fault zones is less apparent. Water-bearing rocks entrained in the fault zone may experience a rise in pore pressure through shear compaction of fluid-filled pore spaces. However, maintaining the fluid pressure through a few million years in a zone the order of a kilometer or so in width requires either exceedingly low permeability (10^{-11} darcy) or a renewable source for the fluid. Such low permeabilities have been measured in argillaceous rocks [*Lin*, 1978] at an effective confining pressure of 200 bars, and if the fault zone is composed of such fine-grained argillites, then the pore pressure may remain elevated for very long times.

Regardless of the mechanism for maintaining high ambient fluid pressure on the San Andreas fault, the implications for earthquake prediction are very important. Firstly, for shear failure to occur at 100 bars or so, the effective normal stress must be about 150 to 200 bars. In this regime, rock properties are sensitive to small changes in stress, because cracks remain open. Thus in the fault zone, variations in properties associated with dilatancy, such as seismic velocity, might be anticipated, but only in the narrow zone of failure where the fluid pressures are highest.

Second, geodetic strain measurements in southern California indicate that extensional horizontal strain of 2×10^{-6} perpendicular to the San Andreas fault may accumulate in less than 6 months [*Savage et al.*, 1981]. A corresponding reduction in effective normal stress of about 1 bar/yr inferred from these measurements is an insignificant fraction of 1000 bars of effective normal stress but a significant reduction if the effective normal stress is 100 bars, as we have concluded. A direct test of our conclusions through in situ measurements is desirable.

References

Brace, W. F., and J. B. Walsh, Some direct measurements of the surface energy of quartz and orthoclase, *Am. Mineral.*, *47*, 1111-1112, 1962.

Brune, J. N., T. L. Henyey, and R. F. Roy, Heat flow, stress, and rate of slip along the San Andreas fault, California, *J. Geophys. Res.*, *74*, 3821-3827, 1969.

Byerlee, J. D., and W. F. Brace, Stick slip, stable sliding, and earthquakes—Effect of rock type, pressure, strain rate, and stiffness, *J. Geophys. Res.*, *73*(18), 6031-6037, 1968.

Dieterich, J. H., Time-dependent friction and the mechanics of stick slip, *Pure Appl. Geophys.*, *116*, 791-806, 1978.

Eshelby, J. D., The determination of the elastic field of an ellipsoidal inclusion and related problems, *Proc. R. Soc. London, Ser. A*, *241*, 376-396, 1957.

Evernden, J. F., Study of regional seismicity and associated problems, *Bull. Seismol. Soc. Am.*, *60*, 393-446, 1970.

Evernden, J. F., Seismic intensities size of earthquakes and related parameters, *Bull. Seismol. Soc. Am.*, *65*, 1287-1313, 1975.

Evernden, J. F., W. M. Kohler, and G. D. Clow, Seismic intensities of earthquakes of conterminous United States: Their prediction and interpretation, *U. S. Geol. Surv. Prof. Pap.*, in press, 1981.

Geller, R. J., and K. Shimazaki, The June 10, 1975 Kurile Islands tsunami earthquake, in Fault Mechanics and its Relation to Earthquake, Prediction-Conference III, *U. S. Geol. Surv. Open File Rep., 78-380,* 213-226, 1978.

Healy, J. H., and L. Peake, Seismic velocity structure along a section of San Andreas fault near Bear Valley, California, *Bull. Seismol. Soc. Am., 65* (5), 1177-1197, 1975.

Irwin, W. P., and I. Barnes, Effect of geologic structure and metamorphic fluids on seismic behavior of the San Andreas fault system in central northern California, *Geology, 3,* 713-716, 1975.

Kennedy, G. C., and W. T. Holser, Pressure-volume-temperature and phase relations of water and carbon dioxide, in Handbook of Physical Constants, *Geol. Soc. Am. Mem., 97,* 371-383, 1966.

Lachenbruch, A. H., Frictional heating, fluid pressure, and the resistance to fault motion, *J. Geophys. Res., 85* (11), 6097-6112, 1980.

Lachenbruch, A. H., and J. H. Sass, Thermo-mechanical aspects of the San Andreas fault system, in *Proceedings of the Conference on Tectonic Problems of the San Andreas Fault System, Stanford Univ. Publ. Geol. Sci., 13,* 192-205, 1973.

Lachenbruch, A. H., and J. H. Sass, Heat flow and energetics of the San Andreas fault zone, *J. Geophys. Res., 85* (11), 6185-6222, 1980.

Lin, W., Measuring the permeability of Eleana argillite from area 17, Nevada test site, using the transient method, *Rep. UCRL, UC-11, 70,* Lawrence Livermore Lab., Livermore, Calif., 1978.

Lindh, A., and D. Boore, Control of rupture of fault geometry: The 1966 Parkfield Earthquake, *Bull. Seismol. Soc. Am.,* in press, 1981.

McGarr, A., and N. C. Gay, State of stress in the earth's crust *Ann. Rev. Earth Planet. Sci., 6,* 405-436, 1978.

McKenzie, D. P., and J. N. Brune, Melting of fault planes during large earthquakes, *Geophys. J. R. Astron. Soc., 29,* 65-78, 1972.

Pfluke, J., Slow earthquakes and very slow earthquakes, in Fault Mechanics and Its Relation to Earthquake Prediction-Conference III, *U.S. Geol. Surv. Open File Rep., 78-380,* 447 469, 1978.

Raleigh, C. B., Frictional heating, dehydration and earthquake stress drops, in *Proceedings of Conference II: Experimental Studies of Rock Friction with Application to Earthquake Prediction,* pp. 291-304, U.S. Geological Survey, Menlo Park, Calif., 1977.

Raleigh, C. B., and M. S. Paterson, Experimental deformation of serpentinite and its tectonic implications, *J. Geophys. Res., 70,* 3965-3985, 1965.

Rubey, W. W., and M. K. Hubbert, Role of fluid pressure in mechanics of overthrust faulting, *Geol. Soc. Am. Bull., 70,* 167-205, 1959.

Savage, J. C., and R. O. Burford, Geodetic determination of relative plate motion in central California, *J. Geophys. Res., 78,* 832-845, 1973.

Savage, J. C., W. H. Prescott, M. Lisowski, and N. E. King, Strain on the San Andreas fault near Palmdale, California: Rapid seismic change, *Science, 211* (4477), 56-58, 1981.

Sibson, R. H., Interactions between temperature and pore-fluid pressure during an earthquake faulting and a mechanism for partial or total stress relief, *Nature, 243,* 66-68, 1973.

Sibson, R. H., Power dissipation and stress levels on faults in the upper crust, *J. Geophys., Res., 85* (11), 6239-6247, 1980.

Thatcher, W., Systematic inversion of geodetic data in central California, *J. Geophys. Res., 84,* 2283-2295, 1979.

Waring, G. A., Springs of California, *U.S. Geol. Surv. Water Supply Pap., 338,* 1915.

Zoback, M. D., H. Tsukahara, and S. Hickman, Stress measurements at depth in the vicinity of the San Andreas fault: Implications for the magnitude of shear stress depth, *J. Geophys. Res., 85* (11), 6157-6173, 1980.

The Origin of the Measured Residual Strains in Crystalline Rocks

EARL R. HOSKINS AND JAMES E. RUSSELL

Department of Geophysics, Texas A & M University, College Station, Texas 77843

The effects of residual stresses and strains on the behavior and response of rock to imposed additional forces and on attempts to measure the in situ tectonic and gravity stresses have been a matter of controversy for over 10 years. In this paper we attempt to support the hypothesis that the directions and relative magnitudes of the measured microresidual strain in these rocks are often controlled and dominated by the microfractures that are present in the rock. The model we propose for the development of microresidual stresses in crystalline rock is as follows. First, during the emplacement of the rock mass, a random, microresidual stress field is generated by a thermal or other mechanism. Next, sets of microfractures are developed in the rock in response to external forces. These external forces may be due to temperature gradients, to stress relief, or the action of tectonic or other geologic processes. The orientation, spacing, and extent of the microfracture array will reflect the mechanism which caused it. The formation of the array of microfractures relieves a portion of the component of stored microresidual stress normal to the microfractures. The stress relief in this direction then defines a direction of minimum contained residual stress. This can be demonstrated by numerical models. When an investigator makes a strain relief test on a 'stress-free' block of this rock, he will find the direction of minimum strain relief to be perpendicular to the microfractures (because a portion of the initially random strain in this direction has already been relieved by the microfractures themselves), and the direction of maximum indicated strain relief must then be perpendicular to this direction. In the presence of in situ stresses acting normal to the microfractures, however, the maximum relieved strain may be measured perpendicular to the microfractures because of a greater compliance of the rock in this direction. Several field and laboratory test results are given along with numerical modeling results to support the hypothesis.

Introduction

The problem that we address in this paper is the origin of macroresidual stresses in crystalline rocks. Residual stress in rock has been recognized for nearly 200 years [*Varnes*, 1970]. Since the mid 1960's the subject has received an increasing amount of attention in field, laboratory, and theoretical studies. A partial list of the relevant references known to the authors is given in the reference list. In spite of this increased level of interest in the subject, controversy and uncertainty still remain as to the origin and significance of macroresidual stresses in igneous and metamorphic rock masses. In this paper we propose and attempt to support the hypothesis that the directions and relative magnitudes of the measured macroresidual stresses in these rocks are often controlled and dominated by the presence of microfractures. This is not a completely new idea, since it has been considered previously in one form or another by *Preston* [1966], *Norman* [1970], *Nich-*

ols [1975], and *Tullis* [1977] and investigated in some detail for specific rocks by *Engelder et al.* [1977], *Engelder and Sbar* [1977], and *Ciampa* [1980]. Our objective is to pull as much of the existing field and laboratory data together as we can and to show that the hypothesis is consistent with numerical model results.

The terminology and background information on residual stresses in rocks has largely been borrowed from the mechanical and metallurgical engineering literature. Most recent rock-mechanics-related residual stress discussions have referred to some standard reference such as *McClintock and Argon* [1966; cf. *Friedman*, 1972; *Russell and Hoskins*, 1973; *Nichols*, 1975; *Holzhausen and Johnson*, 1979]. We continue in this general line and define residual rock stress as a stress system acting in a block of rock that satisfies internal equilibrium when there are no external forces acting on the block and no temperature gradients present. If the residual stresses are consistent in magnitude and direction over 'large' volumes of rock (much larger than the individual grains), they are called 'macroresidual stresses.' If these domains of consistent stress orientation are very 'small' (the order of size of the individual grains), then we say that we have a 'microresidual stress.' This imprecise and somewhat unsatisfactory pair of definitions causes no great difficulty in mechanical metallurgy, since the residual stress magnitudes and directions are routinely controlled by standard forming and processing techniques. If an undesirable state of residual stress exists (micro or macro), it is simply either decreased (e.g., annealed or normalized) or modified to a more advantageous macrostate (by forging, cold working, heat treating, surface hardening, machining, etc.).

In comparison to most natural rock masses, most metals and ceramics are isotropic, homogeneous, fine grained, and unfractured. The problems related to isotropy, homogeneity, and grain size are distracting and complicating factors, but they do not pose serious conceptual difficulties. These terms describe ranges of material characteristics, and the ranges (for rocks versus metals and ceramics) obviously overlap to some extent. The problems that result from the prefractured nature of rock masses are another matter. It is the recognition of this characteristic of rock masses that has required the development of engineering rock mechanics as a separate branch of engineering mechanics. Many workers have contributed to the study of the effect of fractures on the mechanical properties of rocks, only a few of whom are listed here: *Brace* [1960], *Walsh* [1965a, b], *Hoek and Bieniawski* [1965], *Rosengren and Jaeger* [1968], and *Nur* [1971]. There is an additional long list of authors that could be given relating to work on the transport properties of fractured rocks. The point of this discussion is that it has become clear that the nature of the fracturing that is present in virtually all rock masses influences, and in many cases dominates, both the physical property measurements made in and of the rock mass as well as its response to imposed additional forces. We will suggest that it can also dominate the magnitudes and directions of the measured residual stresses.

The Origins of Residual Stresses in Rock

Gallagher [1971] and *Friedman* [1972] have proposed a mechanism for the development of macroresidual stresses in granular aggregates which seems to apply well to sandstones. Their model involves loading a mass of loose sand, cementing the grains together while they are still loaded, and then removing the applied stresses. The resulting sandstone will contain compressive residual stresses in the grains and tensile residual stresses in the cement and will continue to reflect the stresses originally imposed, even though the boundaries of the sample are now free of the applied stresses. Conceptually, however, their model does not describe the origin of macroresidual stresses in igneous and metamorphic rocks. *Savage* [1978] and A. Gangi (private communication, 1978) have proposed models for the development of residual stresses in granitic rocks

based on volume changes in quartz and feldspar during the cooling process. *Holzhausen and Johnson* [1979] discuss the origin of residual stresses in rocks from the point of view of inclusion theory. These analyses lead easily and directly to the development of microresidual stresses in igneous rocks. The extension of them to describe the origins of macroresidual stresses is more difficult. *Savage* [1978] treats the entire cooling granite pluton as a spherical inclusion in an infinite country rock matrix. This leads to radial and tangential residual stresses that appear to compare reasonably well with results reported by *Engelder et al.* [1977]. *Russell and Hoskins* [1973] reviewed the mechanics literature regarding the origin of macroresidual stresses and described several forming processes such as heat treating, rolling, and other nonelastic deformations which may have geologic analogs. *Holzhausen and Johnson* [1979] and others discuss several possible mechanisms for locking a macroresidual stress into a body of rock, including the plastic yielding of a bent beam and the effects of initial external stresses, temperature changes, and plastic deformation on a multilayered rock mass. All of these concepts are based on sound mechanical principles and lead to plausible and, in some cases, observable results. None of them, however, deal in any way with the fracturing that is observed in most rock masses from less than grain size to continental in scale. Furthermore, none of them are able to describe the origin of macroresidual stresses in relatively undeformed igneous and metamorphic rocks.

Preston [1966], *Norman* [1970], *Nichols* [1975], *Engelder et al.* [1977], *Engelder and Sbar* [1977], and *Ciampa* [1980] have discussed the effects of fracturing on residual stress measurements. Engelder et al., in particular, have correlated the opening of microfractures in Barre granite with the measured strain relaxation in residual stress relief tests. They (Engelder et al.) found that in general, in their field measurements the orientation of the direction of maximum strain relief was perpendicular to the strike of the predominant orientation of microfractures. Ciampa was able to demonstrate in a more complexly fractured granitic rock from a geothermal site in New Mexico that there was a set of microfractures aligned parallel to his laboratory-measured direction of maximum relieved strain. *Preston* [1966] performed an extensive series of field residual stress measurements on several rock types and in many locations in the United States and concluded that 'In every instance but one, the principal (relieved) strains appear to be consistently related to the orientation of the dominant fracture set.' *Norman* [1970] has shown that microfractures in crystalline rocks from Georgia are preferentially oriented normal to the direction of maximum strain relief observed in in situ stress measurements. Nichols, on the other hand, found that a parallel alignment of most microfractures with the maximum principal stress did not hold in his block of Barre granite, and he therefore concluded that stress relief microfracturing did not appreciably influence the measurement of residual stresses.

There is then an apparent contradiction in the observational data. Many of the residual strain measurements made in situ yield a maximum principal relieved strain approximately perpendicular to the most dominant set of microfractures, whilst those measured in the laboratory indicate a maximum strain relief approximately parallel to the microfractures. This apparent contradiction is discussed and at least partially explained later in this paper.

It has been generally accepted that brittle fractures in rocks proceed from preexisting flaws or microcracks or their intersections and that the initial direction of crack growth is parallel to the maximum (most compressive) principal stress; see *Kranz* [1979] for a recent review. Continued deformation of the rock leads to the growth of these microcracks and to the coalescing of some of them into zones of shear deformation inclined at 30° to 40° to σ_1. Since laboratory or field testing for rock strength is usually continued until 'failure' (to or past the maximum stress that the sample can withstand), the most obvious evidence of this failure is the throughgoing macroshear zone. It is our contention that the most important event in this process relative to residual stresses, however, is the development of the oriented array of microfractures parallel to the direction of the maximum principal applied stress.

The initial microcracking begins at about half to two thirds of the ultimate strength of the rock in laboratory compression tests. It is an irreversible process and leads to a relaxation of the microresidual stress field perpendicular to the microfractures by allowing strain relief in this direction adjacent to the stress-free crack boundaries. This process then converts an essentially random microresidual strain field (as can be developed by any of the various models given by Russell and Hoskins, Savage, Gangi, and Holzhausen and Johnson) into an apparent macroresidual strain field with its major axis parallel to the direction of the maximum principal stress (parallel to the microfractures). Clearly then, if the fracturing is not as well developed or as well ordered, a less consistent macroresidual strain field will be observed. Furthermore, if the microfracturing forms not in response to tectonic stresses but rather in response to thermal changes or to stress relief due to weathering or to excavation processes, then the orientation of the microfractures and thus the direction of easy strain relief will reflect these processes rather than the regional tectonics. It follows that in a rock with a complex thermal and tectonic history, we should expect complex and apparently inconsistent strain relief response. In a rock mass with minimal microfracturing we would expect other mechanisms and processes (such as Gallagher's and Friedman's) to dominate the residual strain behavior. A rock mass subjected to imposed additional forces such as gravity or tectonics will reflect a combination of the applied and residual stresses, remembering, however, that the residual stresses are controlled by the array of microfractures which can be closed for mechanical purposes by a relatively small compressive stress component acting normal to their long axis. Since this is an extremely nonlinear

process, separation and interpretation of their effects will, in general, be difficult.

Numerical Model

The purpose of this section is to demonstrate that the hypothesis we propose for the origin of residual stresses in rocks is consistent with a simple numerical model. Precise modeling of (1) the mechanism (s) that induced the initial 'locked in' strain distribution, (2) the process of microfracturing, and (3) the details of the strain relief measurement are beyond the state-of-the-art of modeling. However, with the judicious use of simple models we can improve our understanding of the phenomena and of the results of our residual strain measurements.

We assume a thin rock plate model in which the scale of the microfracturing is very small in comparison with the scale of the strain gage. In addition, we assume a simple checkerboard type pattern of elements where alternate elements are assumed to contract, thus inducing a pattern of initial strain. Again we assume that the element size is small in comparison with the gage length. The assumptions of the relatively small scale microfracturing and initial strain pattern allow us to assume continuum behavior. We further assume linear elastic behavior and that strain relief cuts in the rock plate are made without damaging the adjacent material.

The finite element model is composed of 81 quadratic elements that represent one fourth of the total rock plate because of the inherent symmetry. The one-quarter model is shown in Figure 1 and is assumed to be thin so that plane stress analysis is applicable. The dashed line in the lower left quadrant of the model represents the portion of the rock plate cut from the whole. We assume that our strain gage rosette completely covers the area to be removed, and we concern ourselves with only average strains and strain changes in this region.

Several simple numerical experiments can now be conducted on the model. In the first case, assume that the plate has been cut from a rock mass before any fracturing has occurred but after some mechanism (change in temperature, for example) has caused alternate elements to contract isotropically. The average strain components in the lower left quadrant of the quarter model are then calculated. Since the strain relief cuts are assumed to occur without damage to the material, we can now consider a model consisting of only the material removed from the plate and again calculate the average strain components in this partially strain-relieved region. The changes in strain that would be registered by the gages on the inner (partially strain-relieved) region during the strain relief process may then be calculated as simply the average strain in the inner model minus the average strain in the outer model (for the elements coinciding with the inner model). Note that this simple means of calculating the change in strain is possible because of the linear-elastic assumption that assures us that there is no path dependence in going from one state to another.

The changes in strain components for the model de-

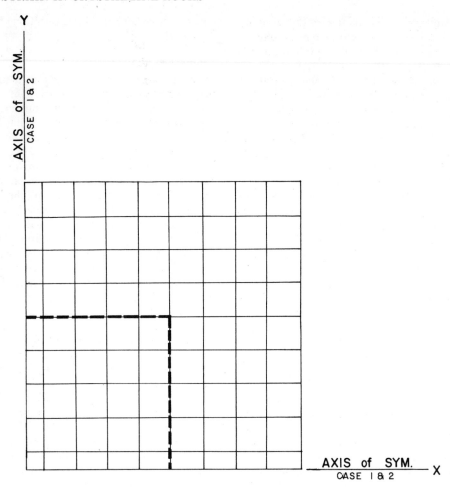

Fig. 1. Plane stress, elastic, finite element model of a rock plate having 81 elements and 100 nodes. Dashed lines indicate region removed for strain relief measurement.

scribed above allow us to calculate principal components of strain change and the principal directions. In the first numerical experiment described above, we found that the changes in average strain components were contractional and equal in the x and y directions and the shear strain components were essentially zero. In this case, there is no preferred direction, and every direction is a principal direction for the change in average strain.

Two points should be noted about the results of this first numerical experiment. First, since the change in average strain is contractional, we conclude that the outer portion of the plate was partially restraining the inner portion from contracting. Thus when the inner portion was cut free, it contracted further, leading to contractional strain changes. Note that if there is no influence of the outer region on the inner region, there would be no change in average strain upon cutting the inner region free.

The second point to be noted about the first numerical experiment is that although the average strain changes

are contractional, it is erroneous to imply a uniform tensile residual stress state as is normally done in in situ stress measurements using strain relief methods. In fact, both tensile and compressive stresses exist in both the outer and inner regions, both initially and after the strain relief cuts have been made. These tensile and compressive stress components are in balance in the sense that the integral of the stress components acting in any particular direction over the area of any hypothetical plane passing through the model must vanish in order that overall equilibrium be maintained when the surface tractions are zero and no unbalanced body forces exist. We note further that the details of the stress and strain states within the models we use may not be exactly correct because of numerical noise and a relatively coarse mesh, but the average values should be representative, and they appear to be consistent.

The second numerical experiment assumes that the rock plate is taken from a rock mass that has been subjected to some tectonic or thermal process that induced an array of microfractures that are preferentially oriented and closely spaced. The second numerical experiment was conducted by assuming that the microfracturing had relieved part (80%) of the initial strain in the direction normal to the strike of the microfractures. We further assume that the ubiquitous fracturing has induced anisotropy in the elastic properties and that the new material may be adequately modeled as a transversely isotropic elastic material. Again the boundaries of the model are traction free.

In the first case we assume that the orientation of the strike of the microfractures parallels the x axis. Then if our hypothesis is correct, the change in average strain in the x direction should be larger in magnitude than the change in average strain in the y direction, and the x and y directions should be the principal directions in this two-dimensional case. The material properties and initial conditions assumed are given in Table 1 and shown in Figure 2 along with the results. The resulting changes in average strain are consistent in both magnitude and direction with what we would expect from the hypothesis and the assumptions. Note that the ratio of the relieved principal strains is about 5, which is the ratio of the initial strains.

There should be no loss in generality by assuming that the material axes are coincident with the geometric axes. However, because of the slight deviation (2°) in the orientation of the principal axes, we felt it was advisable to try some other orientations of microfracturing. The next simplest case is to rotate the material axes and initial strains by 90° so that the strike of the microfracturing parallels the y axis. The results are again summarized in Table 1 and Figure 2. Arbitrary orientations between 0° and 90° are not possible with the mesh used in cases 1 and 2 because of the assumed symmetry. Therefore we modified the mesh by releasing the boundaries on the x and y axes. The modified model is therefore one quarter of the original model, and strain is relieved by cutting out the lower left quarter of the model. The material angle chosen was 15°, and again the results are consistent with our hypothesis in that the principal directions are rotated approximately 14.8° in the same direction and that the largest principal contractional average strain change parallels the strike of the microfractures. The ratio of the principal strains is again approximately equal to the ratio of the initial strains. The results are shown in Table 1 and Figure 2. The deviations of the numerical model principal angles from the hypothetical angles are probably due to noise in the calculations.

In summary, our numerical models have demonstrated that for a certain class of residual strain problems the principal directions of the relieved strain are controlled by the preferential direction of microfracturing. Furthermore, the model results have demonstrated that for this class of residual strain problems the sense (contractional or extensional) of the relieved average strain can tell us very little about the distribution and sense (compression or tension) of the residual stress distribution in general; see also *Friedman* [1972]. However, A. Gangi (private communication, 1980) has correctly pointed out that the

TABLE 1.   Properties, Initial Conditions, and Results of Second Numerical Experiment

Case	Initial Strains			Principal Average Strain Changes		
	$\varepsilon_{x^1}{}^0$, %	$\varepsilon_{y^1}{}^0$, %	β, deg	$\Delta\varepsilon_1$, 10^{-6}	$\Delta\varepsilon_2$, 10^{-6}	α,* deg
1	1.0	0.2	0	61.7	12.0	+ 2.1
2	0.2	1.0	90	61.7	12.0	+87.9
3	1.0	0.2	15	61.6	12.1	+14.8

Transversely isotropic material properties (all cases): $E_{x^1} = 3 \times 10^6$ psi, $E_{y^1} = 1 \times 10^6$ psi, $G_{x^1y^1} = 1 \times 10^6$ psi, $\nu_{x^1} = 0.3$, and $\nu_{y^1} = 0.1$. The x^1-y^1 axes are rotated counterclockwise by an angle of β from x-y axes (see Figure 2). E_{x^1} and E_{y^1} are moduli of elasticity associated with the x^1 and y^1 directions. Parameter ν_{x^1}/E_{x^1} is the strain in the y^1 direction due to a unit stress in the x^1 direction. Parameter ν_{y^1}/E_{y^1} is the strain in the x^1 direction due to a unit stress in the y^1 direction. $G_{x^1y^1}$ is the shearing modulus in the x^1-y^1 plane [*Lekhnitskii*, 1963].

*The angle α is measured counterclockwise from the x axis to the direction of maximum principal strain; contraction is positive (see Figure 2).

sense of the relieved strains (contractional or extensional) does give us information about the sources of the residual strain; i.e., contractional sources cause the overall dimensions of the body to decrease and the stresses within the source to be tensile. If a portion of the rock is cut free, further contraction takes place. In the case of extensional or dilating sources the overall dimensions increase and the stresses within the source are compressive. If a portion of the rock is cut free, further expansion takes place.

In the process of making these calculations we became acutely aware of the importance of properly averaging the strain components. Our first attempt at calculating strain components was simply to take the change in length divided by the original length of a line along the x axis, the y axis, and diagonally across the model. This is analogous to measuring strain changes in the laboratory by measuring the distance from the center of the plate to three points (0°, 45°, and 90°) both before and after cutting out the center region of the plate. This attempt failed and led to inconsistent strain change ratios and directions. Examinations of the displacement pattern showed that this procedure was not appropriate for determining average strain components. The finite element program was then revised to use the calculated average strain components in each element to average the strain components over the entire inner region. The average strain components were calculated as weighted averages with element areas as the weighting factors consistent with

$$\bar{\epsilon}_{ij} = \frac{1}{A} \int_A \epsilon_{ij}(x, y) \; dx \, dy$$

where $\bar{\epsilon}_{ij}$ is the average strain component over area A.

The sensitivity of our numerical results to the manner in which the strain components are averaged is due, at least in part, to the relatively few elements in the models

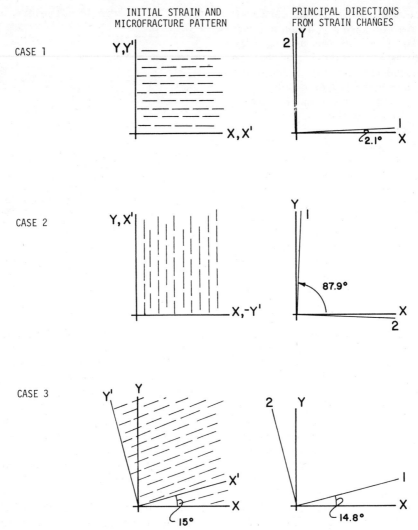

Fig. 2. Graphical presentation of results from numerical model demonstrating the correlation between the direction of microfracturing and the principal directions (1 and 2) calculated from relieved strain.

and the inhomogeneous nature of the initial strain pattern. Obviously, the same problem can exist in laboratory or field measurements, and care should be taken to insure that the gage dimensions and properties are adequate to provide a true average of the strain changes upon cutting out a portion of a rock sample.

One final note with respect to modeling. Numerical modeling was chosen rather than analytical modeling for two reasons: (1) For comparison with laboratory and field results, only average values for strain component changes are required. (2) The inhomogeneous nature of the initial strain pattern assumed and the transversely isotropic nature of the material would make the analysis very difficult. The numerical models proved to be adequate for our purposes.

Field Observations

One of the most puzzling aspects of the results of field measurements of residual strains has been their apparent consistency (at least with regard to the directions of maximum and minimum strain relief). It is relatively easy to conclude that residual strains should be inconsistent, random, sample size and shape dependent, material dependent, temperature dependent, etc. [cf. *Emery*, 1964; *Varnes*, 1970; *Russell and Hoskins*, 1973; *Holzhausen and Johnson*, 1979]. It is much more difficult to explain the regional consistency that has been found by *Hooker and Johnson* [1969] in the Appalachians, *Hoskins and Bahadur* [1971] and *Bahadur* [1972] in the Black Hills, *Hoskins et al.* [1972] in northern California, *Engelder and*

TABLE 2.   In Situ and Residual Elastic Strain Field in the Black Hills

Site	In Situ Strain Field		Residual Strain Field	
	Principal Strains	Direction	Principal Strains	Direction
3	+147.35 −198.35	N60°W	+129.84 −119.84	N81°W
4	+87.39 −17.39	N27°E	+66.40 −9.40	N12°E
19	+128.7 −41.7	N27°E	+81.50 −36.50	N30°E
26	+99.09 +7.91	N55°W	+84.78 +1.22	N56°W
34	+301.10 −77.09	N70°W	+115.48 −27.48	N55°W
37	+90.04 −130.04	N89°E	+65.48 −63.48	S78°E

From *Bahadur* [1972].

Sbar [1977] in upstate New York, and *Sbar et al.* [1979] in southern California. These data have been collected in terrains with various rock types, ages, and tectonic histories. It is possible to argue that the currently active tectonic stresses in the various regions are overprinting any residual stresses that may be present, and that, of course, was the point of making many of the field measurements in the first place. In the studies in the Black Hills and in northern California, however, this is clearly not the case, since some of the measurements were first made 'in situ' on outcrops and then subsequently repeated in the laboratory on oriented blocks of rock taken from the same or nearby outcrops. Table 2 shows the comparison of some of the in situ and laboratory results from the Black Hills. While the agreement is not perfect, the results are certainly similar, and the residual strain field comprises most of the total strain relieved in situ. Thirty-seven field measurements were made in the investigation [*Bahadur*, 1972], all by strain relief trepanning using electrical resistance strain gage rosettes cemented to smoothed horizontal surfaces, and the results show a general conformity with the regional geologic fabric as shown in Figure 3. All of the measurements were in Precambrian igneous and metamorphic rocks that are complexly folded and faulted. There is a general northwesterly trending schistocity that dips steeply and wraps around several domal structures that have granitic cores. The Black Hills is an area of moderate to rough topography with about 4000 feet (1220 m) of vertical relief. An in situ stress measurement has also been made in the northern Black Hills at a depth of 6200 feet (1890 m) in the Homestake Mine near Lead, South Dakota [*Bond*, 1970]. This site is well below the zone of in-

fluence of any topographic irregularities and was carefully selected to avoid the influence of adjacent mine openings. 'Doorstopper' strain relief techniques were used, and they gave results indicating that the principal stresses were vertical and horizontal, with the maximum stress vertical and equal to the overburden pressure, the intermediate principal stress acting horizontally in a northeasterly direction, and the minimum principal stress acting horizontally in a northwesterly direction. The maximum horizontal stress is approximately perpendicular to the direction of schistocity at the measurement site.

The investigations made in California by *Hoskins et al.* [1972] consisted of 6 in situ and 15 laboratory measurements of residual strain in coast range rocks north of San Francisco. The in situ tests were all performed at sites within 10 miles (16 km) of the Warm Springs Dam project near Cloverdale in Sonoma County. The laboratory tests were made on oriented samples collected in the general area between Cloverdale and Pt. Reyes. All of the measurements were made by a trepanning strain relief technique using strain gage rosettes as the strain sensing elements. The measured (relieved) maximum principal strains were aligned generally in a north-south direction (Figures 4 and 5) and were extensional. The east-west component was generally much smaller, and in some cases the relieved strain was contractional. The rock types ranged from the granites at Pt. Reyes and Bodega Head to the metavolcanics, metasediments, and ultrabasics of the Franciscan formation inland. Although the number of tests is small, there does not appear to be any systematic difference in the orientations of the direction of maximum strain relief in situ and the laboratory tests.

Ciampa [1980] investigated residual stresses in granitic cores from the Los Alamos Scientific Laboratories hot dry rock geothermal well GT-2. He determined the velocity anisotropy and the 'static' elastic properties and studied oriented thin sections taken from the rock, looking for direct and indirect evidence of microfracturing. While this rock has a very complex thermal and mechanical history (including its recent removal by coring from a 2700-m depth and 200°C environment to the laboratory), he found general agreement between the strike of the most dominant set of microfractures (as determined by velocity anisotropy and direct observation) and the direction of the maximum principal strain relieved by overcoring.

Discussion

The model we propose for the development of macroresidual stresses in crystalline rocks then is as follows. First, during the emplacement of the rock mass a microresidual stress state is generated by any or all of the mechanisms proposed by Savage, Gangi, or Holzhausen and Johnson. Next, sets of microfractures are developed in response to external forces. These external forces may be due to temperature gradients, to stress relief (from weathering, rapid erosion, and unloading or man-made excavations), or to the action of tectonic or other geologic processes. The

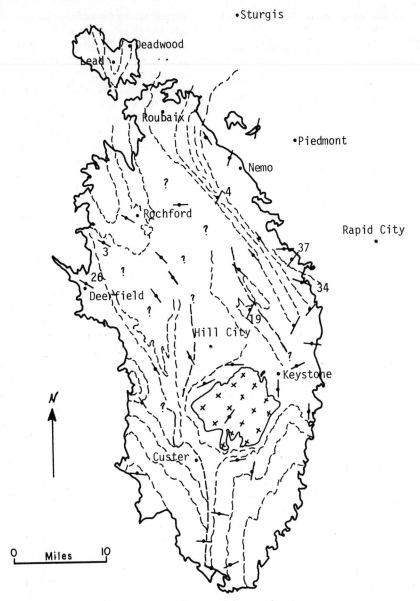

Fig. 3. Directions of the maximum principal elastic strain relieved in the Black Hills [from *Bahadur*, 1972]. The solid outline indicates the area of outcrop of Precambrian rocks. The dashed lines show the generalized direction of schistocity in metamorphic rocks. The dots with short lines through them indicate the locations of residual strain measurement sites and the direction of the maximum relieved residual strain. The numbered sites are those given in Table 2.

orientation, spacing, and extent of the microfracture array will reflect the mechanism which caused it (as per the usual rock mechanics dogma). The formation of the array of microfractures relieves the component of stored microresidual stress normal to the microfractures. The stress relief in this direction then defines a direction of minimum contained residual stress as shown in the numerical models. When an investigator makes a strain relief test on a 'stress-free' block of this prefractured rock in a plane perpendicular to the plane of the microfractures, he will find the direction of minimum strain relief to be perpendicular to the microfractures (because a large part of the strain in this direction has already been relieved by the microfractures themselves), and the direction of the maximum indicated strain relief must of course then be perpendicular to this direction, i.e., parallel to the strike of the array of microfractures. Figure 6 summarizes this discussion. If the test is made in situ in the presence of a

Fig. 4.   Map of a portion of the California coast north of San Francisco showing the recognized major faults in the area and the location of the residual strain measurement sites. The line through the measurement site indicates the direction of the measured maximum relieved strain. FR is the location of Fort Ross, SR is Santa Rosa, and CD is Cloverdale.

compressive stress component normal to the microcracks, then the strain gages during the strain relief test will sense and faithfully indicate the opening of the previously closed array of microfractures plus any elastic strain relief in the grains due to the compressive prestress. Since the opening of the microfractures is likely to be a relatively large part of the total indicated strain, this direction would be interpreted as the direction of maximum strain relief and therefore the direction of the maximum principal stress.

This argument offers an explanation for some of the otherwise conflicting results from field and laboratory measurements. In the Black Hills, for example, the surface in situ and laboratory residual strain measurements indicated an apparent maximum principal residual stress in a northwesterly direction parallel to the schistocity (which is the plane containing the preponderance of microfractures), while the deep underground in situ stress measurements indicated that the maximum horizontal principal stress was acting northeast and perpendicular to the schistocity. The surface measurements were made on apparently stress-free outcrops in a deeply eroded area, and as a result, they were not appreciably different from the laboratory tests made on small oriented samples taken

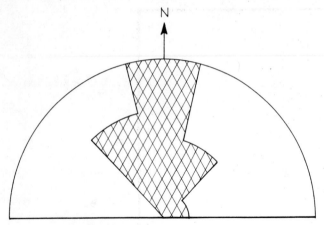

Fig. 5. Diagram showing the directions of the maximum measured relieved strains in the California study.

from the same outcrops. The underground measurements, on the other hand, were influenced by the in situ regional stresses and so indicated a maximum principal relieved strain perpendicular to the schistocity (the direction of greatest compliance).

The California surface residual strain results, in which the field and laboratory measurements were in agreement, can also be interpreted by this model. From Figure 4 it is apparent that there is an extensive array of right lateral strike slip faults in the area that strike in a northwesterly direction; yet the residual strain analyses indicated the maximum relieved strain direction to be generally north-south. The sites for the in situ tests, as well as the collection locations for the laboratory samples, were as 'good' as could be found in the area. By this we mean that the rock at these outcrops was the most sound, unweathered, and unfractured rock that we could locate. In general, it meant

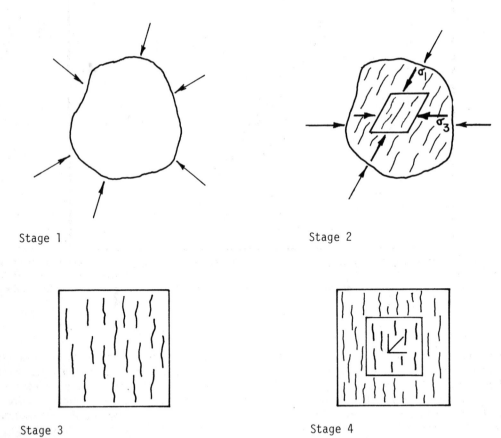

Fig. 6. Stages in the development of residual strain and its measurement. Stage 1: Self-equilibrating residual stress and 'tectonic' (or other) stress. Random microfracture orientation. Stage 2: Total stress state has increased, and microfractures are now oriented parallel to σ_1. Residual stresses in σ_3 direction have been partially relieved. Stage 3: A sample of the rock is removed from the rock mass, thus freeing it from the in situ stress field. Only residual stresses remain. Stresses due to gravity are ignored. Stage 4: Strain gages are placed on the sample, and a region around the gaged area is 'cut out,' thus relieving part of the remaining residual stresses in the block.

that we picked outcrops between the major faults, since the rocks in and around the fault zones are extensively fractured and altered and also, of course, are usually located in the valleys, where no 'good outcrops' exit anyway. If, as seems geologically reasonable, the faults are the result of an approximately north-south maximum compressive stress, we would expect from rock fracture mechanics to have a family of north-south trending microfractures present in these otherwise sound samples. The response of an unstressed surface in situ or laboratory sample to a strain relief experiment would be to show a miminum strain relief east-west (perpendicular to the microfractures) and a maximum strain relief north-south, and this is in fact what was observed. The data taken in situ in this case were also made on stress-free rocks because of the extensive weathering and 'stress relief' of the near-surface rocks in this area. We never did directly measure the tectonic stress in this area.

From the above discussion it seems clear that both 'residual strain' and 'in situ stress' tests in rocks made by surface strain measurements cannot be fully interpreted without some detailed knowledge of the state of microfracturing in the rock. It is not, in general, a simple elastic stress relief process, and a full interpretation of the field data, if possible at all, will require a site-by-site rock fabric analysis.

Acknowledgments. We owe John Handin much more than this simple acknowledgment for his dedicated efforts in rock mechanics throughout his career, his encouragement of our work for many years, and his support since we have come to Texas A&M University. For all of this we are extremely grateful to him. Our understanding of residual strain, such as it is, has evolved in the past 10 to 15 years through discussions with many co-workers and students, and we are indebted to them as well. Finally, Tony Gangi and Mel Friedman read earlier versions of this paper, and their comments and suggestions have immeasurably improved it.

References

Bahadur, S., In situ stress measurements in the Black Hills, South Dakota, PhD. dissertation, 169 pp., S. Dak. Sch. of Mines and Technol., Rapid City, 1972.

Bond, P. H., The direction and magnitudes of the principal stresses at the 6200 foot level of the Homestake mine, Lead, South Dakota, M.S. thesis, 32 pp., S. Dak. Sch. of Mines and Technol., Rapid City, 1970.

Brace, W. F., An extension of Griffith theory of fracture to rocks, *J. Geophys. Res.*, *65*, 3477-3480, 1960.

Ciampa, J. D., Microcracks, residual strain, velocity and elastic properties of igneous rocks from a geothermal test-hole at Fenton Hill, New Mexico, M.S. thesis, 125 pp., Texas A&M Univ., College Station, 1980.

Emery, C. L., Strain energy in rocks, in *State of Stress in the Earth's Crust*, pp. 235-260, Elsevier, New York, 1964.

Engelder, T., and M. L. Sbar, The relationship between in situ strain relaxation and outcrop fractures in the Potsdam Sandstone, Alexandria Bay, New York, *Pure Appl. Geophys.*, *115*, 41-55, 1977.

Engelder, T., M. L. Sbar, and R. Kranz, A mechanism for strain relaxation of Barre Granite: Opening of microfractures, *Pure Appl. Geophys.*, *115*, 27-40, 1977.

Friedman, M., Residual elastic strain in rocks, *Tectonophysics*, *15*, 297-330, 1972.

Gallagher, J. J., Photomechanical model studies relating to fracture and residual elastic strain in granular aggregates, PhD. dissertation, 127 pp., Texas A&M Univ., College Station, 1971.

Hoek, E., and Z. T. Bieniawski, Brittle fracture propagation in rock under compression, *Int. J. Fracture Mech.*, *1*, 137-155, 1965.

Holzhausen, G. R., and A. M. Johnson, The concept of residual stress in rock, *Tectonophysics*, *58*, 237-267, 1979.

Hooker, V. E., and C. F. Johnson, Near surface horizontal stresses including the effects of rock anisotropy, *Rep. Invest. U.S. Bur. Mines*, *7224*, 29 pp., 1969.

Hoskins, E. R., and S. Bahadur, Measurements of residual strain in the Black Hills (abstract), *Eos Trans. AGU*, *52*, 122, 1971.

Hoskins, E. R., J. E. Russell, K. Beck, and D. Mohrman, In situ and laboratory measurements of residual strain in the Coast Ranges north of San Francisco Bay (abstract), *Eos Trans. AGU*, *53*(11), 1117, 1972.

Kranz, R. L., Crack growth and development during creep of Barre granite, *Int. J. Rock Mech. Mining Sci.*, *16*(1), 23-25, 1979.

Lekhnitskii, S. G., *Theory of Elasticity for an Anisotropic Elastic Body*, translation from Russian by P. Fern, Holden-Day, San Francisco, 1963.

McClintock, F. A., and A. W. Argon, *Mechanical Behavior of Materials*, 770 pp., Addison-Wesley, Reading, Mass., 1966.

Nichols, T. C., Deformation associated with relaxation of residual stresses in a sample of Barre granite from Vermont, *Geol. Surv. Prof. Pap. U.S.*, *875*, 32 pp., 1975.

Norman, C. E., Geometric relationships between geologic structure and ground stresses near Atlanta, Georgia, *Rep. Invest. U.S. Bur. Mines*, *7365*, 24 pp., 1970.

Nur, A., Effects of stress on velocity anisotropy in rocks with cracks, *J. Geophys. Res.*, *76*, 2022-2034, 1971.

Preston, D. A., Stored elastic strain measurement in rocks by photoelastic technique: Its geologic value, *Tech. Progr. Rep. EPR 45-66-F*, Shell Oil Co., Houston, Tex., 1966.

Rosengren, K. J., and J. C. Jaeger, The mechanical properties of an interlocked low-porosity aggregate, *Geotechnique*, *18*, 317-326, 1968.

Russell, J. E., and E. R. Hoskins, Residual stresses in rock, in *New Horizons in Rock Mechanics, Proceedings of the 14th Symposium on Rock Mechanics*, pp. 1-24, American Society of Civil Engineers, New York, 1973.

Savage, W. Z., The development of residual stress in cooling rock bodies, *Geophys. Res. Lett.*, *5*(8), 633-636, 1978.

Sbar, M. L., T. Engelder, R. Plumb, and S. Marshak, Stress pattern near the San Andreas Fault near Palmdale, California from near-surface in situ measurements, *J. Geophys. Res.*, *84*(B1), 156-164, 1979.

Tullis, T. E., Reflections on mesaurement of residual stress in rock, *Pure Appl. Geophys.*, *115*(1/2), 57-68, 1977.

Varnes, D. J., Model for simulation of residual stress in rock, in *Rock Mechanics Theory and Practice, The 11th Symposium on Rock Mechanics*, pp. 416-426, Society of Mining Engineers of AIME, New York, 1970.

Walsh, J. B., The effect of cracks on the uniaxial elastic compression of rocks, *J. Geophys. Res.*, *70*, 399-411, 1965a.

Walsh, J. B., The effects of cracks in rocks on Poisson's ratio, *J. Geophys. Res.*, *70*, 5249-5257, 1965b.

Stress Measurements Via Shallow Overcoring
Near the San Andreas Fault

Terry E. Tullis

Department of Geological Sciences, Brown University, Providence, Rhode Island 02912

Twenty four successful measurements of absolute stress have been made at 10 sites in the area of the southern California uplift by using the U. S. Bureau of Mines stress relief method of overcoring a three-component borehole deformation gage. The measurements were made within 3 m of the ground surface. At most sites, repeated measurements were made and show excellent agreement, so that the measurements represent a good determination of the actual absolute stress in the rocks sampled. The total areal coverage and density of measurements are not adequate to define the stress trajectories in the region, but general agreement with the overcoring measurements in the same area by Sbar et al. (1979) further supports the correctness of the stress directions as representative of regions several kilometers in dimension. Comparison with absolute stress measurements made by hydraulic fracturing to depths of several hundred meters by Zoback et al. (1980b) shows no indication that the near-surface overcoring measurements give incorrect stress directions, but in most cases the stress magnitudes measured near the surface are too low. The direction of greatest principal compressive stress near to and southwest of the San Andreas fault is NNW, with good reproducibility, whereas in the Mojave block northeast of the fault the compression directions range from NW to NE and show more scatter. Careful comparison of alternative methods of reducing the data on the assumptions of plane stress or plane strain and isotropic rock behavior or orthotropic behavior show that although the plane stress or strain approximation makes little difference, consideration of rock anisotropy is important. An approximate treatment of this anisotropy gives nearly as accurate results as the much more involved orthotropic treatment; the former is suggested as the preferred method of initial data reduction in the field.

Introduction

The southern California uplift reported by *Castle et al.* [1976, 1977] is widely regarded as a potential area for a large earthquake. The region is a tectonically active one, in which the San Andreas Fault ruptured in a great earthquake in 1857. The 9.5-m maximum slip at the surface for that earthquake [*Sieh*, 1978a] and the rates of relative motion for this plate boundary of 6 cm/yr make it seem possible that a similar earthquake could occur at any time. This is also consistent with the 100- to 230-yr recurrence interval inferred by *Sieh* [1978b] for large earthquakes in this region. An unusual earthquake swarm in 1976-1977 along the San Andreas near Palmdale, reported by

McNally et al. [1978], may be some type of premonitory activity to a major earthquake.

Direct measures of stress or strain may be of use in helping to assess the likelihood and time of occurrence of a major earthquake. Studies by *Prescott and Savage* [1976] and *Savage and Prescott* [1979] of the strain of geodetic nets spanning the San Andreas fault in this area indicate a strain rate for the time span 1973.8 to 1977.4 that is essentially a north-south compression of 0.33×10^{-6}/yr and a less well determined shear strain accumulation for the period 1932-1963 that would be consistent with a NNW compression of 0.50×10^{-6}/yr. It is not clear whether this general rate of strain accumulation has been occurring uniformly since the 1857 earthquake or whether it may be

Fig. 1. A borehole deformation gage being read. The strain indicator is attached to the gage in the hole. John Handin for scale. This measurement was made only before and after overcoring, since the configuration of the drill did not allow the cable to pass through the drill chuck and water swivel. To obtain reliable measurements it is important to take readings during the drilling. This requires three strain indicators, one for each component of the gage.

niques must be employed. Ideally such measurements should be made at a variety of locations along the length of the fault and over a range of depths that includes focal depths. Although some valuable work of this type has begun with the use of hydraulic fracturing [*Zoback and Roller*, 1979; *Zoback et al.*, 1980a, b], it is a very difficult and expensive undertaking to get the type of coverage that would be desirable. Measurements by overcoring at the surface have the potential of being less expensive and more rapid than hydraulic fracturing. Although they cannot now be made at depths of over a few tens of meters, the relative ease of extensive areal coverage makes the method potentially attractive as one component of a program to understand the present-day tectonics of the region.

This paper reports a preliminary attempt to measure the stresses in the rocks near the ground surface in the vicinity of the southern California uplift, using the U.S. Bureau of Mines overcoring method [*Hooker and Bickel*, 1974]. The region is a large one, and the rocks are generally not well suited for making such stress measurements. Consequently, attention was concentrated on checking the reproducibility of measurements in a restricted geographic area, although at the same time including enough geographic coverage to test for homogeneity of stress. The number of sites and their distribution are inadequate to allow any tectonic inferences, except of a very general nature. However, the study does provide worthwhile information on the suitability of the overcoring technique for tectonic purposes. In addition, comparison with similar overcoring data of *Sbar et al.* [1979], and with results of shallow hydraulic fracturing stress measurements of *Zoback and Roller* [1979] and *Zoback et al.* [1980a, b], all in the same region, allows an appraisal of the usefulness of various methods of measuring stress in a complex region. Finally, some attention is given to the consequences of a variety of different assumptions and methods used in reducing the data from overcoring a borehole deformation gage.

an increased rate of straining that may precede another large earthquake. If this rate has been uniform since 1857, then the total strain accumulated in that period would be $4-6 \times 10^{-5}$, which would correspond to an increase in stress of 2-3 MPa. Although this is much less than the stress necessary to cause frictional sliding on faults at mid-crustal depths, it is of the order of the average stress drop that accompanies large earthquakes [*Hanks*, 1977, pp. 443-445]. However, earthquake stress drops span roughly 2 orders of magnitude, so that this type of argument is not very useful in predicting the likelihood of occurrence of the next major earthquake in this region.

Knowledge of the absolute magnitude of stress on the San Andreas fault would also be very desirable in trying to determine how likely it is that sudden slip may occur. Such information cannot be deduced from changes in strain, and so hydraulic fracturing and stress relief tech-

Method

General Measurement Procedures

The U. S. Bureau of Mines overcoring method for measuring stress consists of drilling a 38-mm-diameter hole in the rock with an EX-size diamond drill bit, placing a borehole deformation gage in the hole, and then drilling a concentric 152-mm-diameter slot around the inner borehole and gage, while monitoring the resultant change in diameter of the inner borehole [*Hooker and Bickel*, 1974]. The borehole deformation gage is capable of measuring the changes in diameter simultaneously along three diameters oriented at 60° to each other in the plane perpendicular to the axis of the hole. These diameter changes result from the fact that the annular volume of rock between the gage and the outer slot is freed from the stresses that existed on it prior to the overcore. If the rock deforms elastically, then a knowledge of its elastic constants allows one to convert the diameter changes into stresses that existed prior to drilling the inner hole. The Young's modulus of the rock is measured after overcoring by using a biaxial pressure chamber to pressurize the outside diameter of the sample, while the change in its inner diameter is measured with the borehole deformation gage [*Fitzpatrick*, 1962]. If the rock does not behave in a reversible linearly elastic manner, then the stresses cannot be computed exactly, although they may be estimated by using the elastic equations and elastic constants that represent a linear approximation to the rock behavior averaged over the strain that occurred on overcoring and the time duration of overcoring. Discussion of the method can be found in a number of U. S. Bureau of Mines publications [*Merrill and Peterson*, 1961; *Panek*, 1966; *Merrill*, 1967; *Hooker and Bickel*, 1974; *Hooker et al.*, 1974; *Obert*, 1964; *Becker and Hooker*, 1967; *Becker*, 1968; *Hooker and Johnson*, 1969; *Hooker et al.*, 1972; *Hooker and Duvall*, 1971]. In particular, the paper by *Hooker and Bickel* [1974] contains a very good description of the equipment and procedures used in making the field measurements. An important aspect of the technique is that the changes in the gage output be recorded simultaneously with the advance of the overcoring bit. This allows the detection of erratic behavior of the gage that might result from coring past a fracture or shifting of the gage in the inner hole. If readings are taken only before and after overcoring, such sources of spurious data can go undetected and result in the reporting of incorrect stresses.

Figure 1 shows a well-known and distinguished scientist contemplating an overcoring measurement on the Teton anticline in Montana, July 1964.

Care in Modulus Measurements

In order to get good measurements of the moduli of the overcores it is important that they be measured in a state as close as possible to the one in which the overcoring was done. This includes both the physical state of the core and the magnitude of stress exerted on it during the measurements.

Modulus measurements with the biaxial chamber should be made in the field immediately after overcoring if possible. This procedure was followed in the present study. In one instance the modulus was also measured several days later, after the core dried out (and perhaps 'aged' in other unknown ways). The moduli for the dry rock were higher by a factor of approximately 1.4 in each of the three directions measured by the gage. If it is not possible to make the modulus measurements soon after overcoring, the cores should at least be wrapped in some fashion to prevent their drying.

Ideally the modulus measurements should be made by using a pressure in the biaxial chamber that approximates the stress level in the rock prior to overcoring. The best way to select the appropriate pressure for the modulus

test is to pressurize until the deformation of the borehole is about the same as in the just completed overcoring operation. Because of the nonlinear response of the rocks (see discussion below and Figure 3), it is important that the secant modulus be measured from the peak pressure to zero pressure upon the depressurization stroke. Given the downwardly convex shape of the hysteresis loops (see Figure 3), an abnormally low modulus will be calculated if a secant modulus is taken between some lower than peak pressure and zero pressure.

Data Reduction by Assuming Isotropic Rock

If the rock behaves isotropically, then the equations that are used to determine the stress magnitudes and directions from the diameter changes are (from *Merrill and Peterson*, 1961; *Obert and Duvall*, 1967, pp. 413–416):

$$P = \frac{E}{6d\,(1-\nu^2)} \quad (R+S) \qquad (1)$$

$$Q = \frac{E}{6d\,(1-\nu^2)} \quad (R-S) \qquad (2)$$

and

$$\theta_p = \tfrac{1}{2}\tan^{-1}\frac{\sqrt{3}\,(U_2-U_3)}{2U_1-U_2-U_3} \qquad (3)$$

where P and Q are the magnitudes of the secondary principal stresses in the plane that is perpendicular to the borehole. (They are true principal stresses if the borehole is parallel to a principal direction of stress.) U_1, U_2, and U_3 are the increases in diameter of the borehole upon overcoring in directions 60° to one another; θ_p is the angle to the direction of P from the direction of U_1, the positive sense of θ_p being toward the direction of U_2, E is Young's modulus, ν is Poisson's ratio, and d is the diameter of the borehole. R and S are given by

$$R = U_1 + U_2 + U_3 \qquad (4)$$

and

$$S = \frac{\sqrt{2}}{2}\,[(U_1-U_2)^2+(U_2-U_3)^2+(U_3-U_1)^2]^{1/2} \qquad (5)$$

The ambiguities in the arctangent for θ_p are resolved by

if $U_2 > U_3$ and $U_2 + U_3 \ < \ 2U_1$, then $0 < \theta_p < 45°$

if $U_2 > U_3$ and $U_2 + U_3 \ > \ 2U_1$, then $45° < \theta_p < 90°$

$$(6)$$

if $U_2 < U_3$ and $U_2 + U_3 \ > \ 2U_1$, then $90° < \theta_p < 135°$

and

if $U_2 < U_3$ and $U_2 + U_3 \ < \ 2U_1$, then $135° < \theta_p < 180°$

Equations (1) through (6) assume that the rock is in a state of plane strain in the plane perpendicular to the borehole (i.e., these planes remain plane) and that the strain parallel to the borehole axis, ϵ_3, is zero. If ϵ_3 is non-zero, the plane strain solution can be obtained by using the same equations as for $\epsilon_3 = 0$, except instead of using the measured U_1, U_2, and U_3, modified values of these are used; the modified values being created by adding $\nu\epsilon_3 d$ to each measured value. This may be done directly if ϵ_3 is measured. If, instead, σ_3 is known or estimated (1) and (2) are used to obtain preliminary estimates for P and Q, which are then used to obtain an estimate for ϵ_3 from

$$\epsilon_3 = \frac{1}{E}\left[\sigma_3 - \nu(P + Q)\right] \qquad (7)$$

With a preliminary value for ϵ_3, the measured U_1, U_2, and U_3 are then modified by adding $\nu\epsilon_3 d$, and these modified values are then used in (1), (2), (4), and (5) as a second iteration for P and Q. Additional iterations can be made until P and Q stabilize.

The plane stress solution with the axial stress σ_3 equal to zero is given by (1) and (2) with Poisson's ratio set equal to zero. Neither the plane strain nor the plane stress solutions can be exactly correct. Since the measurements were made within 2.5 m of the stress-free ground surface, the plane strain assumption that planes perpendicular to the borehole remain plane is probably incorrect, but so is the plane stress approximation that these planes have no shear stress at the depth of the gage. It is best to make the calculations both ways so that the results may be compared. The resulting differences are in fact small, as is shown below. Furthermore, the two approaches give different stress magnitudes, but for an isotropic material they both give the same direction of the secondary principal stresses (3).

Young's modulus is determined from the thick-walled cylinder formula for the change in inner diameter that results from external pressurization in the biaxial chamber tests:

$$E = 2dP_0D^2/U(D^2 - d^2) \qquad (8)$$

where D is the outer diameter of the thick-walled cylinder produced by overcoring, P_0 is the applied pressure, and U is the increase in diameter of the inner hole upon release of the pressure. For the situation in which the stress parallel to the axis of the cylinder is zero (as is the case in simple biaxial chamber tests), (8) holds for both plane stress and plane strain (with nonzero ϵ_3, since σ_3 is zero). Thus the value of E is obtained with no assumptions required.

Data Reduction by Assuming Orthotropic Rock

For the rocks in this study, a certain amount of anisotropy in Young's modulus was found by using the biaxial chamber and measuring the change in diameter along the three diameters. Thus inaccuracies in both direc-

tions and magnitudes would result from using the isotropic equations (1-8). *Hooker and Johnson* [1969] have shown that the use of the isotropic formulas resulted in errors no larger than 25° in orientation and 25% in magnitude. However, such errors depend upon the amount of anisotropy. Consequently, an analysis of the data has also been made by using the appropriate anisotropic equations presented by *Hooker and Johnson* [1969], *Becker* [1968], and *Becker and Hooker* [1967]. I summarize their approach in this section since otherwise all three of these papers must be consulted in order to understand their approach. A much simpler approach that gives an approximate treatment of the anisotropy was also attempted in the initial field reduction of the data, and as will be seen, it eliminates most of the errors in directions that result from the isotropic treatment.

In order to deal with the response of anisotropic rock upon overcoring, the analyses are considerably simplified if the rock is assumed to be orthotropic, that is, if it is assumed to have three mutually perpendicular planes of elastic symmetry, with the borehole perpendicular to one of these planes. This is the approach taken by *Hooker and Johnson* [1969], *Becker* [1968], and *Becker and Hooker* [1967]. In an orthotropic material, there are nine independent elastic constants. In order to simplify the analysis and reduce the number of elastic constants that must be measured, one can assume a relationship between some of the elastic constants [*Becker*, 1968, p. 7]. This relationship has been approximately verified in at least one study [*Becker and Hooker*, 1967, pp. 5–6].

Using these assumed relations between the elastic constants, the relevant equations for the treatment of overcoring data in orthotropic rock, assuming plane stress with the axial stress $\sigma_3 = 0$, are:

$$P = \frac{L - K\cos^2\delta}{\sin^2\delta - \cos^2\delta} \qquad (9)$$

$$Q = K - P \qquad (10)$$

and

$$\delta = \tan^{-1}\left[T + (T^2 + 1)^{1/2}\right] \qquad (11)$$

where δ is the angle measured from the direction of the smallest Young's modulus E_1 in the plane perpendicular to the borehole to the direction of the secondary principal stress P. The positive sense for δ is that which goes from the direction of U_1 toward the direction of U_2. The arc tangent for δ in (11) is always taken to be in the first quadrant, and (9) and (10) determine whether this direction for P is that of the larger or smaller of the secondary principal stresses. In these equations, L, K, and T are given by

$$L = \frac{A\alpha_{22} + C(2(\alpha_{11}\alpha_{22})^{1/2} + \alpha_{22})}{2d(\alpha_{22} + (\alpha_{11}\alpha_{22})^{1/2})^2} \qquad (12)$$

$$K = \frac{A(\alpha_{11}\alpha_{22})^{1/2} + C\alpha_{11}}{\alpha_{11}d(\alpha_{22} + (\alpha_{11}\alpha_{22})^{1/2})} \qquad (13)$$

and

$$T = \frac{C\alpha_{11} - A\alpha_{22}}{B(\alpha_{11}\alpha_{22})^{1/2}} \qquad (14)$$

where A, B, and C are the components of a 3×1 matrix N

$$N = \begin{bmatrix} A \\ B \\ C \end{bmatrix} \qquad (15)$$

which is obtained from

$$N = M^{-1}U \qquad (16)$$

Here M is a 3×3 matrix that is defined by the directions to U_1, U_2, and U_3 from the directions of elastic symmetry:

$$M = \begin{bmatrix} \cos^2 \theta & \cos \theta \sin \theta & \sin^2 \theta \\ \cos^2 (\theta + 60) & \cos (\theta + 60) \sin (\theta + 60) & \sin^2 (\theta + 60) \\ \cos^2 (\theta + 120) & \cos (\theta + 120) \sin (\theta + 120) & \sin^2 (\theta + 120) \end{bmatrix} \qquad (17)$$

where θ is the angle from the direction of E_1 to the direction of U_1, and the positive sense for θ goes from the direction of U_1 toward the direction of U_2. U is a 3×1 matrix whose components are the expansions that occur upon overcoring in the three directions oriented 60° from one another:

$$U = \begin{bmatrix} U_1 \\ U_2 \\ U_3 \end{bmatrix} \qquad (18)$$

For reduction of the data via the use of the plane stress assumption, α_{11} and α_{22} are defined by

$$\alpha_{11} = 1/E_1 \qquad \alpha_{22} = 1/E_2 \qquad (19)$$

where E_2 is the largest Young's modulus in the plane perpendicular to the borehole. For the plane strain treatment of overcoring data in orthotropic material, the same equations (9-18) are used, but α_{11} and α_{22} are defined by

$$\alpha_{11} = \frac{1 - \nu_{13}\nu_{31}}{E_1} \qquad \alpha_{22} = \frac{1 - \nu_{23}\nu_{32}}{E_2} \qquad (20)$$

where ν_{ij} is the Poisson's coefficient that relates a change in length in the j direction to a uniaxial stress of opposite sign in the i direction.

The above approach is adequate for plane strain if the axial strain in zero. If the axial strain is nonzero (for example, if the axial stress is zero), then the approach must be somewhat modified, as in the similar isotropic situation discussed earlier. This is done by replacing A and C in (12), (13), and (14) by A_c and C_c, where

$$A_c = A + \nu_{31}\epsilon_3 d \qquad (21)$$

and

$$C_c = C + \nu_{32}\epsilon_3 d \qquad (22)$$

As in the isotropic case, (21) and (22) may be used directly with no iteration if ϵ_3 is measured. If ϵ_3 must be estimated from knowledge of σ_3, P, Q, and δ, then iteration is required, and P, Q, and δ are first approximated from (9), (10), and (11) by using A and C in (12), (13), and (14). Then ϵ_3 is estimated from

$$\epsilon_3 = \frac{1}{E_3} \left[\sigma_3 - \nu_{31} (P \cos^2 \delta + Q \sin^2 \delta) \right.$$
$$\left. - \nu_{32} (P \sin^2 \delta + Q \cos^2 \delta) \right] \qquad (23)$$

where E_3 is the Young's modulus parallel to the axis of the core. This estimate of ϵ_3 from (23) is used in (21) and (22), and then A_c and C_c replace A and C in (12), (13), and (14) during the second iteration for P, Q, and δ. Additional iterations can be made until P, Q, and δ stabilize.

In the anisotropic case the plane stress and plane strain solutions will give slightly differing values for δ, the orientation of the secondary principal stresses, as well as for their magnitudes P and Q. However, as shown below, those differences seem to be less than 1° for the amount of anisotropy in the rocks studied.

In order to determine all of the orthotropic elastic constants that must be known in order to convert the overcoring strains into stresses (even making the simplifying assumptions of *Becker* [1968, p. 7]) it is necessary to perform more complex tests on the recovered core than the biaxial chamber tests that are used for isotropic material. *Becker* [1968] shows that it is desirable to do a test in which a stiff plate applies the proper axial load on the hollow core in combination with a pressure on the outer surface in just such a way that no axial strain occurs. This is the simple plane strain situation. In this case the diametral deformation of the inner hole is sufficient to determine the plane strain values (20) of α_{11} and α_{22}. Application of the axial load alone, together with measurements of the axial strain and diametral deformations, is enough to determine E_3, ν_{31}, and ν_{32}. If these tests were performed in plane strain with the proper equipment, all of the needed elastic constants, α_{11}, α_{22}, E_3, ν_{31}, and ν_{32}, would be obtained for reducing the overcoring data via the use of the plane strain approximation. In order to use the plane stress approximation in reducing the overcoring data, E_1 and E_2 are needed for the plane stress definitions of α_{11} and α_{22} (19). These may be found from the plane strain values of α_{11} and α_{22} that were obtained in the plane strain modulus test just described by using (20) together with the known relations between elastic constants [*Becker*, 1978, p. 3]:

$$\frac{\nu_{31}}{E_3} = \frac{\nu_{13}}{E_1} \qquad \frac{\nu_{32}}{E_3} = \frac{\nu_{23}}{E_2} \qquad (24)$$

Additional Assumptions Made for Orthotropic Treatment

Unfortunately the only equipment available in this study for the elastic constant measurements was a biaxial chamber, and so the measurements could not be performed in plane strain with $\epsilon_3 = 0$. However, the anisotropic diametral deformations that occurred upon external pressurization of the overcore in the biaxial chamber give the most significant information for obtaining an anisotropic solution. Nevertheless, information about the Poisson's coefficients and the Young's modulus in the axial direction does not exist, and assumptions must be made concerning them. I made the assumption that $\nu_{31} = \nu_{32}$. In this case the biaxial chamber tests with free ends correspond to both plane stress and plane strain with $\sigma_3 = 0$, just as in the isotropic case. This allows the determination of E_1 and E_2 if measurements of the diametral deformation are made along three or more diameters. The relevant equation is:

$$U_\phi = \frac{2dD^2P_0}{D^2 - d^2}\left(\frac{\cos^2\phi}{E_1} + \frac{\sin^2\phi}{E_2}\right) \qquad (25)$$

where ϕ is the angle between the direction of E_1 and the diameter along which U_ϕ is measured. This equation may be obtained for $\sigma_3 = 0$, either for the plane stress situation, using Becker's equation (32) and plane stress definitions of α_{11} and α_{22}, or for the plane strain situation, using Becker's equation (43) together with the plane strain definitions of α_{11} and α_{22} and the assumption that $\nu_{31} = \nu_{32}$. For the case where $E_1 = E_2$, (25) reduces to the isotropic case (8).

In order to invert (25) to obtain E_1 and E_2 from measurements of U_ϕ at three values of ϕ, it is convenient to use the equation for an equiangular (or delta) strain rosette. Using diametral deformations rather than strains, the greatest expansion U_p upon depressurization of the biaxial chamber will be parallel to the lowest Young's modulus E_1, and the least expansion U_q will parallel E_2. From the strain rosette equations [e.g., *Obert and Duvall*, 1967. p. 411]

$$U_p = F + G \qquad (26)$$

$$U_q = F - G \qquad (27)$$

and

$$\phi_p = \tfrac{1}{2}\tan^{-1}\left|\frac{\sqrt{3}\,(U_2 - U_3)}{2U_1 - U_2 - U_3}\right| \qquad (28)$$

where

$$F = \tfrac{1}{3}\,(U_1 + U_2 + U_3) \qquad (29)$$

and

$$G = \frac{\sqrt{2}}{3}\,[(U_1 - U_2)^2 + (U_2 - U_3)^2 + (U_3 - U_1)^2]^{1/2} \qquad (30)$$

The angle ϕ_p is measured from the direction of U_1 to the direction of U_p with the positive sense being toward the direction of U_2. The ambiguities in the arc tangent in (28) are resolved as they were for θ_p, by using inequalities (6). Since U_p corresponds to E_1 or $\phi = 0$ and U_q to E_2 or $\phi = 90°$, (25) yields

$$E_1 = \frac{2dD^2P_0}{(D^2 - d^2)\,U_p} \qquad (31)$$

and

$$E_2 = \frac{2dD^2P_0}{(D^2 - d^2)\,U_q} \qquad (32)$$

Notice that if the directions of U_1, U_2, and U_3 relative to the rock are the same for the overcoring as for the biaxial chamber measurements, then θ (see (17)) $= -\phi_p$.

Data Reduction Via the Approximate Anisotropic Approach

The rather involved analysis represented by (9)–(32) for the orthotropic case suggests that a more approximate treatment of the anisotropic situation might be worth investigating, especially for preliminary field reduction of the data. I attempted a simple approximate treatment of this type, and it yields a good approximation, as is shown below.

The method consists of modifying the observed expansions upon overcoring by a multiplicative factor which takes into account the relative stiffness of each direction as compared to the average. The borehole deformation gage is placed in the same orientation and position in the rock for the modulus determination as for the overcoring measurements. For each of the three diameters, (8) is used to obtain a value for E_i in that direction. The average $\bar{E}$ of those three E_i is formed. Then the overcoring expansions U_i are each modified to U_i^m by (no summation on repeated subscripts)

$$U_i^m = U_i(E_i/\bar{E}) \qquad i = 1, 2, 3 \qquad (33)$$

This modification of the actual expansions increases those that were 'too small' because of a large modulus in that direction and decreases those that were 'too large' because of a small modulus. These values of U_i^m together with the average modulus $\bar{E}$ are then used as U_i and E in (1)–(7), as if the rock were isotropic.

Averaging Several Measurements

In order to check reproducibility of the measurements it is desirable to perform at least three measurements in each borehole. If these are relatively consistent, then they must be averaged in some way in order to obtain the best estimate for the stress at the site. One possible solution is to express the components of the stress tensor in a common coordinate system, average the components, then find the principal values and directions from these averaged components. However, this procedure combines the orientation and the magnitude information from each measurement in an undesirable way. It is often observed that the stress directions from repeated measurements are more consistent than their magnitudes. Consequently, the Bureau of Mines [e.g., *Hooker and Johnson*, 1969] has adopted the procedure of averaging the directions of the individual measurements and separately averaging the magnitudes of each of the principal stresses. If some of the measurements to be averaged have ratios of the greatest to the least principal stresses much closer to unity than others, then they may be weighted less strongly in forming the average of the stress directions [*Hooker and Johnson*, 1969, pp. 13-14]. In the present study, this ratio of the principal stress is similar enough between measurements that such a weighting makes a difference of less than 2° in the worst case, and so it is not done.

Site Descriptions

Unfortunately, for the purposes of making overcoring stress measurements, the region of the southern California uplift is not a semi-infinite homogeneous linearly elastic half space. The region to the south of the San Andreas fault is dominated by the topographically rugged San Gabriel mountains, whereas much of the Mojave block to the northeast of the fault consists of basin fill. Thus, to the south, most places are too steep for stress measurements made near the surface because the horizontal stresses pendicular to the local or regional slope will be reduced below their values at depth. In the Mojave desert, bedrock exposures are not abundant, and they are only accessible at the surface in the low mountains that rise above the alluvium. The bedrock topography in the desert is similar in ruggedness to that in the San Gabriel mountains, and consequently, it may be less meaningful to make a measurement on the shallow flanks of these desert mountains than the surface topography would suggest. Without detailed local shallow seismic reflection study, it is not clear in many such locations whether a steep bedrock slope may exist only a few meters from the site and be obscured by colluvium or alluvium.

An additional serious problem in the region is that the rocks tend to be highly fractured on all scales. Microfractures that reduce the elastic moduli are pervasive, as are larger open macrofractures. These macrofractures are found in all outcrops, and in many they are spaced so closely that it is quite difficult to obtain the 0.5 m long unfractured overcores that are necessary for successful stress measurements.

Many of the rocks tend to be quite weathered, frequently to the extent that intact cores cannot be obtained. For the granitic rocks, very recent road cuts show just as pronounced weathering as old outcrop surfaces. In some road cuts, spheroidal weathering has left a few boulders unweathered, while the bulk of the outcrop is quite weathered in patterns that suggest motion of water along major and minor fractures. In planning future stress measurements at greater depths in boreholes, an important consideration will be the depth to which such weathering extends.

The combination of the high density of fractures in the rocks and their degree of weathering raises some doubts as to how well the rocks at the surface are coupled to the underlying rocks at a depth of several kilometers where the focus of a large earthquake is likely to be. However, it seems quite possible that although the stress magnitudes are likely to be lower near the surface, the stress directions are likely to be similar to those at depth if a sufficient number of measurements is taken to allow an average over the local variations induced by individual nearby fractures.

Insofar as possible, sites for this study have been selected where the local topographic effects are not thought to affect the measured stress. Effects upon measured stresses from topography can come both from geometric amplification from a notch, such as a valley bottom, or geometric reduction from a protrusion, such as a hill [*Jaeger and Cook*, 1969, pp. 361-362; *Harrison*, 1976], as well as from the gravitational loading or unloading from a nearby hill or valley, respectively [*Jaeger and Cook*, 1969, pp. 356-358]. Making a correction to measured stresses to eliminate the effects of such contributions is a difficult and uncertain procedure. A quantitative analysis of the effects at the sites in this study has not been made, but it is felt that corrections from such effects would be small.

The locations of the sites are shown in Figure 4 and are given more exactly in Table 3. The rock types in which overcoring was done in this study include sandstone (sites 1, 2, 7, and 10), quartz monzonite (sites 3, 4, 8, 9 and 11), and a tuff (site 5). Sites 1, 7, and 10 are in the arkose of the Punchbowl Formation of Miocene and Pliocene age quite close to the San Andreas fault. The rock is massive, with relatively few fractures and virtually no bedding plane partings. Site 2 is in a 2 to 3-m-thick massive sandstone bed that was separated from adjacent thick sandstone beds by interbedded shales and thin sandstones. The site is located about 15 km SSW of the San Andreas fault, in the broad regional drainage of the Mint and Soledad canyon area. Sites 3, 4, 8, 9, and 11 are located in quartz monzonite of Mesozoic age in the Mojave block northeast of the San Andreas fault. The outcrops at sites 3 and 9 are somewhat weathered and are on the flanks of some low hills that rise from the desert. The rocks at sites 4, 8, and 11 are less weathered. Site 4 is on the gentle flank of a hill that rises out of the desert, site 8 is in a broad stream valley, and site 11 is on the Victorville pediment. The rock is quite fractured at site 4, while the results of the drilling at site 8 suggest that the measurements may have been made in an unweathered large boulder that was surrounded by more weathered material. At site 11 the entire area seems unusually unweathered and free of fractures. Site 5 is in a gently dipping tuff on the side of a low rise located about 30 km NNE of the San Andreas fault.

Results

Attempts were made to make stress measurements at 11 sites, with the aim of obtaining at least three measurements at different depths in the same hole at each site as a

TABLE 1. Measurement Data

Core	Site	Depth, m	Azimuth U_1, deg	Overcoring U_i, μm			Biaxial Chamber P_0, MPa; U_i, μm			
				U_1	U_2	U_3	P	U_1	U_2	U_3
1	1C	1.07	−64	61.6	32.3	55.4	2.79	79.4	111.1	93.8
2	1D	1.42	−64	75.0	43.2	66.0	3.47	90.9	135.6	102.8
3	1E	1.78	−64	77.3	45.1	59.9	3.11	85.0	122.8	98.2
4	2A	0.69	−65	5.9	0.2	21.2	1.74	51.6	33.5	28.0
5	2B	1.10	−65	16.7	−5.3	24.6	3.47	59.0	73.1	35.6
6	3A	0.66	65	13.1	9.3	10.5	0.58	38.3	34.6	39.6
7	3B	1.07	65	22.7	6.1	6.1	0.57	30.6	30.0	36.1
8	3D	1.71	65	20.6	7.0	1.9	0.56	22.3	34.7	40.5
9	4A	0.25	54	−2.7	−2.2	−13.4	0.55	7.6	6.0	7.6
10	4B	0.56	54	−1.4	−0.9	0.2	0.16	1.8	1.4	0.9
11	5A	0.18	4	2.0	9.0	6.2	0.43	9.4	9.0	9.2
12	7A	0.65	−2	134.6	111.6	104.6	0.29	107.0	105.3	110.5
13	7B	1.05	−3	64.9	55.3	52.2	0.43	87.3	75.8	92.2
14	7C	1.46	−1	50.6	69.9	55.2	0.43	81.1	88.0	70.5
15	8C	0.90	−45	4.5	5.9	6.4	4.16	6.8	7.1	6.1
16	9A	0.56	80	9.9	10.3	5.7	0.30	10.4	10.4	11.2
17	9C	1.23	80	10.9	7.8	11.8	0.58	14.7	10.4	9.0
18	9D	1.56	80	6.0	12.2	10.4	0.58	14.1	8.9	14.1
19	9E	1.87	80	6.2	9.9	9.8	0.59	12.6	10.6	11.2
20	9F	2.22	80	6.4	9.5	9.3	0.58	10.2	9.5	8.7
21	10A	0.61	−32	71.5	36.4	51.7	2.10	72.0	122.4	123.8
22	10B	0.95	−32	66.5	14.8	36.9	2.07	72.8	127.5	143.1
23	10C	1.28	−32	75.0	13.5	31.7	2.11	74.6	114.1	137.4
24	11A	0.18	55	4.0	1.8	2.5	1.77	5.8	6.1	4.5

check on reproducibility. However, because of a variety of problems at some of the sites, it was not always possible to make all three measurements. At one site, no measurements were possible; at three sites, only one was made; at two sites, only two were made, while at one site, five measurements were made because of poor reproducibility in the first three. This resulted in a total of 24 separate measurements at 10 sites.

The measurement and computational results of this study are presented in Tables 1, 2, and 3 and in Figures 2, 3, and 4. Table 1 gives the data for each overcore and corresponding biaxial chamber measurement in terms of the change in diameter of the inner 38.4-mm-diameter borehole. The depths given are distances between the ground surface at the measurement site and the location of the buttons on the gage. The directions of U_2 and U_3 are successively counterclockwise from that of U_1 when looking down the hole. In the case of the overcoring measurements the three U_i given are the changes between the relatively constant values of gage output before and after overcoring. All of these values have been taken from plots like the one for sites 9E and 10B, shown in Figure 2. Positive values indicate expansions upon overcoring caused by release of compressive stress. All of the values are expansions except for site 4, which behaved anomalously in giving tensions and in giving directions that are contradictory for the two measurements 4A and 4B.

In the subsequent biaxial chamber measurements the gage was positioned in the same orientation and depth in the inner hole as for the overcoring. The U_i correspond to secant moduli drawn for the depressurization cycle, from

the maximum pressure indicated (P_0) to zero pressure on plots like that in Figure 3. Inspection of this figure shows that the rocks are nonlinear. The calculated moduli shown in Table 2 are mostly on the order of a few GPa, which is about an order of magnitude lower than is typical for fresh, unfractured granites and sandstones [Clark, 1966]. In two cases, sites 8 and 11, the moduli are a few tens of GPa, as expected for fresh rock. In the field these higher modulus rocks can easily be detected qualitatively because they emit a crisp ringing sound when hit with a hammer. The low moduli and nonlinear nature of most of the rocks is thought to be due to the presence of micro-

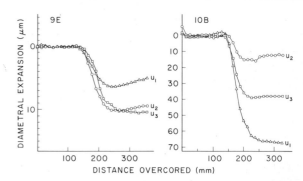

Fig. 2. Examples of the nature of the raw data from which the deformation of the borehole upon overcoring is determined. The smooth variation from a relatively constant level before the drill bit has passed the position of the gage to the level after it has passed indicates that the measurement is successful. Information from such plots is adjusted very slightly for nonlinearity in the response of the gage and is reported in Table 1.

cracks and alteration products. At site 6 it was not possible to make an overcoring measurement because the rock was not sufficiently coherent. Its modulus was on the order of 1 MPa. Outcrops with this degree of weathering and microcracking give a dull thud when struck with a hammer.

All of the rocks studied show a certain degree of elastic anisotropy as can be seen from Figure 3 and by inspecting the three biaxial chamber U_i columns in Table 1. Much of this anisotropy may be due to a preferred orientation of microcracks, although no thin section studies were made to investigate this.

The calculated stresses are shown in Table 2. Results are shown for each of the three approaches discussed in a preceding section (assumption that rock is isotropic; assumption that rock is orthotropic; assumption that rock anisotropy can be approximately treated by using the isotropic equations if the observed U_i upon overcoring are 'adjusted' to take into account the differing response in the three directions to the biaxial chamber tests). Table 2 also shows columns for the isotropic and orthotropic assumptions that correspond to the plane stress and plane strain assumptions, respectively.

It is instructive to compare the plane stress and plane strain results for stress directions. Even in the orthotropic case the directions of the principal stresses are virtually identical for the two approaches. For 7 of the 24 successful cores the plane stress and plane strain directions differed by 0.1°. The other orthotropic results gave the same answer for both assumptions with 0.1° accuracy. The isotropic plane stress and plane strain results, of course, give identical answers for stress directions. Thus for all practical purposes it makes no difference whether plane stress or plane strain assumptions are used in deriving stress directions, even when using the orthotropic equations. The plane stress procedure is somewhat simpler as discussed above.

The plane stress and plane strain results for stress magnitudes differ by only a few percent for the present data in which the axial stress is very low because the measurements are within 2 m of a free surface. In the next to the last row of Table 2 are shown the averages of the amount by which each of the plane strain results differs from the corresponding plane stress result. On the average, for both the isotropic and orthotropic assumptions the plane strain results give a 1% higher value for the maximum compressive stress σ_1. The least compressive stress σ_2 is lower for plane strain in both cases by 4% and 5%, as shown. Again the differences between the plane stress and plane strain results are not significant, considering the other sources of uncertainty in the stress magnitudes.

The comparison between the plane strain and plane stress results in Table 2 is valid for an assumed value of 0.25 for Poisson's ratio. As (21), (22), and (23) show, the difference between plane stress and plane strain is related to Poisson's ratio. A Poisson's ratio of zero would give the same results for both directions and magnitudes. Al-

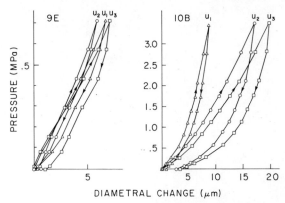

Fig. 3. Representative plots of the data from the biaxial chamber. These help determine the elastic modulus of the overcore. In each case the three components of the gage are positioned and numbered in the inner borehole in the same way as the overcoring measurements of Figure 2. The arrows show the pressurization and depressurization paths. The moduli used are secant moduli on the depressurization path from peak pressure to zero pressure. The diametral deformations are chosen to be approximately the same for the biaxial chamber tests as were found for the overcoring. For both samples the anisotropy of the rock behavior is clearly shown.

though the actual value of Poisson's ratio for these rocks has not been measured, the effect of using a Poisson's ratio larger than 0.25 is small. Thus for the orthotropic case and a Poisson's ratio of 0.5 the largest difference in angle between the plane stress and the plane strain directions for any of the cores is 0.7°, and for most cores it is only 0.1° or 0.2°. The differences in the magnitudes also increase as Poisson's ratio increases, but the differences are still small, considering the uncertainties in the magnitudes from other sources. If Poisson's ratio is 0.5, the values for σ_1 for plane strain average 7% higher than for plane stress, and for σ_3 the values average 26% lower.

The differences in results for the three assumptions made about anisotropy can be seen in Table 2. The orthotropic treatment is the most accurate available for these measurements in the absence of measurements of the anistropic Poisson's ratios. Consider these as a standard for comparing the plane stress results. The proposed anisotropic approximation does a much better job of estimating the proper stress directions and magnitudes than does the isotropic treatment. The largest discrepancy in directions, using the approximate anisotropic treatment, is 5.7° while it is 27.7° for the isotropic treatment. The averages of the absolute values of the discrepancies are shown in the last row of Table 2 and are 1.9° and 6.8° for the approximate anisotropic and isotropic treatments, respectively. The averages in the discrepancies in stress magnitudes are also shown in the last row of Table 2, and it is seen that although neither approach is too far off in magnitude, the approximate anisotropic approach does better by a factor of 2 than does the isotropic approach. Thus it is clear from this comparison that there is no point in ever

TABLE 2. Calculated Stress, MPa, and Moduli, GPa

Core	Site	Rock Type/ Dip & Strike	Isotropic Assumption Plane σ			Plane ε		Orthotropic Assumption Plane ε			Plane σ		Anisotropic Approximation Plane σ			$\bar{E}$	E_1	E_2	$A_z\sigma_1$, deg
			σ_1	σ_2	$A_z\sigma_1$, deg	σ_1	σ_2	σ_1	σ_2	$A_z\sigma_1^*$, deg	σ_1	σ_2	σ_1	σ_2	$A_z\sigma_1$, deg				
1	1C	Arkose/ N 65 W, 40 SW	1.89	1.32	−39.8	1.91	1.29	2.13	1.19	−42.6	2.10	1.21	2.08	1.23	−45.2	2.47	2.03	3.01	42.5
2	1D	Arkose/ N 65 W, 40 SW	2.48	1.82	−41.9	2.49	1.79	2.88	1.60	−41.7	2.83	1.63	2.77	1.67	−44.2	2.68	2.10	3.44	48.5
3	1E	Arkose/ N 65 W, 40 SW	2.35	1.73	−50.3	2.37	1.70	2.68	1.54	−47.6	2.65	1.57	2.64	1.58	−50.7	2.57	2.07	3.21	45.9
4	2A	Sandstone/ N 45 W, 29 SW	0.82	0.15	−12.5	0.84	0.13	0.97	0.15	− 5.3	0.94	0.16	0.99	0.11	−8.8	4.08	2.77	6.14	−71.4
5	2B	Sandstone/ N 45 W, 29 SW	1.53	0.22	−27.7	1.57	0.18	2.02	0.24	−24.55	1.95	0.28	1.95	0.30	−18.9	5.62	3.69	8.43	74.1
6	3A	Quartz Monzonite	0.20	0.16	74.4	0.20	0.16	0.20	0.16	63.45	0.20	0.17	0.20	0.17	64.8	1.28	1.18	1.38	−77.8
7	3B	Quartz Monzonite	0.33	0.12	65.0	0.33	0.10	0.34	0.10	61.8	0.34	0.12	0.34	0.11	63.4	1.46	1.29	1.65	−57.5
8	3D	Quartz Monzonite	0.30	0.08	57.4	0.30	0.07	0.38	0.07	57.0	0.37	0.08	0.40	0.07	60.3	1.50	1.06	2.10	−34.0
9	4A	Quartz Monzonite	−0.20	−0.83	22.9	−0.18	−0.85	−0.18	−0.80	25.9	−0.22	−0.78	−0.21	−0.77	24.1	6.52	5.62	7.54	83.5
10	4B	Quartz Monzonite	−0.03	−0.16	−57.2	−0.02	−0.16	−0.01	−0.12	−57.8	−0.02	−0.12	−0.01	−0.12	−61.7	10.35	6.91	15.32	34.9
11	5A	Tuff	0.39	0.19	−74.4	0.40	0.18	0.40	0.18	−74.2	0.40	0.18	0.40	0.19	−73.9	3.84	3.74	3.94	18.8
12	7A	Arkose/ N 37 E, 30 SE	0.36	0.31	−8.5	0.36	0.31	0.37	0.31	−12.9	0.37	0.31	0.37	0.31	−12.8	0.22	0.22	0.23	48.6
13	7B	Arkose/ N 37 E, 30 SE	0.34	0.30	−9.8	0.34	0.29	0.34	0.28	−30.05	0.35	0.29	0.34	0.29	−31.0	0.42	0.38	0.48	35.4
14	7C	Arkose/ N 37 E, 30 SE	0.38	0.31	−67.5	0.38	0.30	0.37	0.31	−87.35	0.37	0.32	0.37	0.32	−89.5	0.45	0.40	0.51	−42.4
15	8C	Quartz Monzonite	4.17	3.40	37.3	4.19	3.37	4.33	3.27	28.0	4.29	3.31	4.32	3.29	27.3	51.79	47.12	56.90	−83.4
16	9A	Quartz	0.30	0.22	47.8	0.30	0.21	0.31	0.21	48.1	0.31	0.21	0.31	0.21	48.0	2.30	2.19	2.42	−40.0

17	9C	Quartz Monzonite	0.65	0.51	−63.7	0.65	0.50	0.65	0.46	−36.05	0.69	0.47	0.70	0.46	−39.6	4.38	3.23	6.00	73.4
18	9D	Quartz Monzonite	0.60	0.41	−2.0	0.60	0.40	0.68	0.35	7.9	0.68	0.36	0.70	0.36	10.7	4.06	3.02	5.39	−69.6
19	9E	Quartz Monzonite	0.54	0.41	−9.0	0.54	0.40	0.58	0.38	−7.1	0.57	0.39	0.57	0.40	−6.5	4.24	3.84	4.68	88.3
20	9F	Quartz Monzonite	0.62	0.49	−8.4	0.62	0.48	0.65	0.46	−13.45	0.65	0.47	0.65	0.48	−14.5	5.08	4.64	5.57	62.9
21	10A	Arkose/N 40 W, 40 SW	1.44	0.98	−19.1	1.46	0.96	1.72	0.83	−25.7	1.69	0.85	1.78	0.80	−27.4	1.74	1.24	2.41	57.3
22	10B	Arkose/N 40 W, 40 SW	1.15	0.52	−19.4	1.17	0.49	1.41	0.44	−26.25	1.38	0.46	1.50	0.41	−27.2	1.62	1.09	2.37	51.9
23	10C	Arkose/N 40 W, 40 SW	1.30	0.49	−23.6	1.32	0.46	1.58	0.43	−29.4	1.54	0.46	1.66	0.39	−28.7	1.71	1.20	2.42	47.2
24	11A	Quartz Monzonite	1.20	0.76	64.3	1.22	0.75	1.23	0.73	75.5	1.21	0.75	1.19	0.77	76.3	27.17	22.64	32.52	20.8
Average difference from plane stress			—	+3%	0°	+1%	−5%	+1%	−4%	0°	—	—	+1%	−4%					
Average difference from orthotropic			−8%	±6.8°	−9%	+2%	±6.8°	−4%	±1.9°										Note: ν assumed = 0.25

* A_z shown to ± 0.1 if plane σ and ε agree to that significance and shown to ± 0.01 as average if they do not.

reducing overcoring data by using the isotropic approach if a three-component gage is used in making the biaxial chamber measurements. If one does not wish to do the rather complex calculations involved in the orthotropic approach, a very good approximation can be made by using the approximate anisotropic calculation described in the previous section. This involves virtually no more computation than the isotropic approach.

Discussion of Results
Evaluation of the Measurements

In general, the reproducibility at each site is excellent. This can be seen by comparing the separate rows for a given site in Table 2, by studying the standard deviations in Table 3, and, perhaps most easily, by studying Figure 4. For the several measurements in a given hole the stress directions tend to be quite similar, and the ratios between the principal stress magnitudes are also quite similar. The magnitudes show somewhat more variation, but not a large amount.

At site 4 the results are anomalous; the reproducibility is very poor, and the stresses were tensions. Possibly, some type of residual stress was being measured for which the equilibrium volume was larger than the separation between the two measurements [Tullis, 1977]. Alternatively, some type of measurement error may have resulted from inadvertent shifting of the gage, motion on macrofractures, or unusual thermal effects. In the field, nothing unusual was noted about the behavior of the instruments during overcoring, except that the holes contracted rather than expanded. The internally inconsistent tension data from site 4 cannot represent the regional stresses, and hence they have been omitted from Figure 4.

Apart from site 4, the greatest lack of reproducibility came from site 9, in which the upper three measurements gave divergent results for the stress directions and magnitudes. Consequently two more measurements were made further down in the hole, and they gave directions closer to each other and to the one just above them (Table 2, Figure 4). During drilling, a fracture was encountered at a depth of 0.79 m when trying to make the second overcore, which would have been 9B. Another fracture was found at 1.35 m depth at the bottom of core 9C, whereas no further fractures separate the mutually consistent cores 9D, 9E, and 9F. It thus appears that the stress states are different on opposite sides of each of the two fractures encountered in this hole. This situation appears to be unusual, since in many of the other holes, reproducible measurements were separated by fractures that seemed as prominent as those in this hole. However, it does suggest that the coupling of the stresses across some fractures may not be very good and that a statistically significant number of measurements is necessary in order to have confidence in the stress directions.

The only other site where reproducibility is not excellent is site 7, in which the stress direction from the bottom core diverges from those of the top two. However, the difference between the principal stresses is small at this site for all three measurements, and so the principal stress directions are not well determined.

TABLE 3. Average Stress Magnitudes and Directions*

Core	Site	Location	Maximum Principal Compressive σ_1, MPa	Minimum Principal Compressive σ_2, MPa	Stress Ratio, σ_1/σ_2	Azimuth, deg CW from N
1	Punchbowl Acres	SE¼, SW¼, NE¼, Sec 13, T.4N. R. 10W., L. A. Co.	$2.53 \pm .38$	$1.47 \pm .23$	$1.72 \pm .03$	-44.0 ± 3.2
2	Vasquez Rocks	NW¼, SE¼, SE¼, Sec 26, T.5N., R. 14W., L. A. Co.	$1.45 \pm .72$	$0.22 \pm .08$	$6.36 \pm .95$	-15.0 ± 13.6
3	Lovejoy Buttes	NE¼, NW¼, SE¼, Sec 19, T.6N., R. 9W., L. A. Co.	$0.30 \pm .09$	$0.12 \pm .04$	2.86 ± 1.63	60.7 ± 3.3
4	Adobe Mt.	SE¼, NE¼, NW¼, Sec 14, T.7N., R. 8W., L. A. Co.	-0.11 ± 0.13	-0.45 ± 0.47	0.20 ± 0.08	-16.0 ± 59.2
5	Rosamond Airport	NE¼, SW¼, SW¼, Sec 7, T.9N., R. 12W., Kern Co.	0.40	0.18	2.14	-74.2
7	Cajon	SW¼, NE¼, SW¼, Sec 36, T.3N., R. 6W., San Bernadino Co.	0.36 ± 0.01	0.31 ± 0.01	1.18 ± 0.01	-43.5 ± 39.1
8	Silverwood Lake	SW¼, SE¼, SE¼, Sec 25, T.3N., R. 5W., San Bernadino Co.	4.29	3.31	1.30	28.0
9	Piute Butte	SW¼, NW¼, SE¼, Sec 32, T.7N., R. 9W., L. A. Co.	0.58 ± 0.16	0.38 ± 0.11	1.53 ± 0.19	-0.1 ± 31.2
9	Piute Butte	(same, but only bottom three measurements in hole)	0.63 ± 0.05	0.41 ± 0.06	1.56 ± 0.26	-4.2 ± 11.0
10	Deep Springs Ranch	NE¼, SW¼, SW¼, Sec 18, T.4N., R. 9W., L. A. Co.	1.53 ± 0.15	0.59 ± 0.22	2.77 ± 0.70	-27.1 ± 2.0
11	Victorville Pediment	NW¼, SE¼, SW¼, Sec 28, T.6N., R. 4W., San Bernadino Co.	1.21	0.75	1.62	75.5

*Using orthotropic assumption.

Only one measurement was made at site 8 because of difficulties with fracturing in the deeper parts of the hole. The surface rock was much more coherent and unaltered than the rock below about 1.1-m depth. Thus there is reason to suspect that the measurement at site 8 may represent the stress in a higher-modulus surface rock that is not well coupled to its substrate and is possibly not representative of the regional stresses.

The arkoses of the Punchbowl Formation at sites 1 and 10 are the most satisfactory of all the rocks in which measurements were made because of their massive character, their relative freedom from fractures, their apparent good coupling to the underlying rock, and their relatively high stresses. Site 11 on the Victorville pediment also appears good, but there is no check on reproducibility. At sites 3 and 9 the fractured and weathered character of the granite, together with the difficulty in knowing the details of the local bedrock topography under the adjacent colluvial cover, make it difficult to know how representative the measured stresses are of the regional stresses. At site 5, montmorillonite in the tuff allowed only one measurement because it caused the drill to bind and the water circulation to become blocked; thus, there is no check on reproducibility at this site. The measurements at site 2 are reproducible and are felt to accurately represent the stress in the sandstone layer in which they were made, but because that layer is separated from adjacent sandstone layers by nearly equal thicknesses of shale, the stress measurement may not represent the average stress orientation for that formation. It is possible that the higher-modulus sandstones may act as stress guides and that the av-

erage stress on the formation could be somewhat more north-south than measured since the bedding strikes N 45 W and the direction of the greatest principal compressive stress in the sandstone is N 15 W. Quantitative evaluation of the importance of this effect is difficult without knowing more about the mechanical properties of the two layers, but it probably does not represent a serious error in the stress directions.

Thus for the sites (1, 2, 7, and 10) in which the features of the local setting seem to suggest that the results should give reasonably representative directions for the regional stresses, the directions all tend to be somewhat west of north. The other sites (3, 5, 9, and 11) in which there is no reason to definitely discard the results, but for which there is either no check on reproducibility or some uncertainty about the subsurface topography or rock coupling at the site, show a larger scatter of stress directions. The remaining sites (4 and 8) have features about them which make their data seem bad or doubtful.

The foregoing discussion has principally concentrated on the stress directions because surface stress directions are more likely to be representative of the directions at depth than are surface stress magnitudes of the magnitudes at depth. This is because the magnitudes are likely to be more affected by imperfect coupling to the underlying rock than are the directions. Because of the characteristics outlined above, sites 2 and 10 in the Punchbowl formation are likely to be the ones for which the magnitudes are most representative of the true shallow-level stresses. Most of the other measurements probably give magnitudes that are lower than would be obtained at depths of a few tens of meters in the same location. However, it is quite possible that the stress magnitudes may be relatively low down to depths of several tens of meters in much of the area because of the fractured and altered character of the bedrock.

Comparison With Other Surface Overcoring Data

At the same time as these measurements were made (May and June 1977), another set of overcoring measurements was made in the same area by *Sbar et al.* [1979], using the doorstopper method. In general their measurements give similar directions, magnitudes, scatter, and uncertainty as those reported here. In most cases their measurement sites are far enough removed from those reported here that direct comparison of the results is not warranted, since the stresses can be expected to have some spacial variation. However, some of the sites are sufficiently close that direct comparisons are valid. The comparisons are quite favorable and lend support to the notion that each set of measurements is an imperfect but meaningful sampling of a regionally varying stress field.

The best combined data exist for the measurements made in the Punchbowl Formation, about 2 km southwest of the San Andreas fault. My sites 1 and 10 (T1 and T10) and their sites 1 and 2 (S1 and S2) are all within the same unit and are within 3 km of each other. With the exception of site S1 they are all within 750 m of each other. From WNW to ESE the sites are S1, T1, S2, and T10, and the azimuths of σ_1 are $-57°$, $-44°$, $0°$, and $-27°$, respectively (negative azimuths are W of N). The average of these azimuths is $-32°$. The general consistency of these measurements at the surface over an area of these dimensions suggests that the stress directions at a depth of a few kilometers are likely to be quite similar.

The other area where the two sets of data were taken in close proximity is at Lovejoy Buttes, about 16 km to the northeast of the San Andreas fault. My site 3 (T3) and the *Sbar et al.* [1979] sites 6 and 7 (S6 and S7) are within 3.5 km of each other; sites T3 and S6 are separated by only about 50 m. The azimuths of σ_1 at these sites T3, S6, and S7 are $60°$, $78°$, and $32°$, respectively. The general consistency of these results suggests that more significance may be attached to the measurement at my site T3 than the uncertainties referred to in the discussion of site T3 above might indicate.

The general pattern that emerges from consideration of my Figure 4 and Figure 4 of *Sbar et al.* [1979] is that near the San Andreas fault, and on its southwest side, the direction of σ_1 is north northwest, whereas in the Mojave block northeast of the fault, the direction of σ_1 is northeast. As a first approximation this result seems mechanically unlikely since it results in a right-lateral sense of shear stress on the fault from the Pacific plate side and zero shear stress or perhaps a left-lateral sense of shear stress on the fault from the North America plate side. However, if the stress trajectories are sufficiently curved, perhaps because of earlier inhomogeneous distribution of

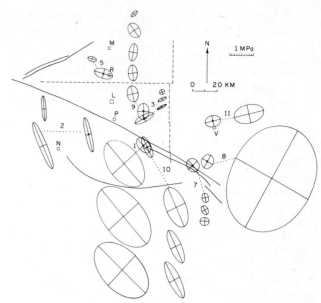

Fig. 4. Map of a portion of the southern California uplift, showing all of the data and the averages of the stress directions and aspect ratios of the stress ellipses. The location of each site is shown by a dot in the center of the average stress ellipse axes. The site numbers are shown adjacent to the dotted line that connects the average measurement to the individual ones that contribute to it. For each site the individual measurements are arranged with the one closest to the top of the hole above the ones that were successively below it. The scale for the magnitudes of the stress applies to the individual measurements, but the average measurements have been scaled so all the average ellipses would have the same area. The towns of Newhall, Palmdale, Lancaster, Rosemond, Mojave, and Victorville are shown, as are the county lines (dashed) that separate Los Angeles, Kern, and San Bernadino counties. The San Andreas and associated faults trend WNW across the center of the figure, and a portion of the Garlock fault is in the northwest corner.

slip on the San Andreas fault, then the measured distribution of stress directions could be correct. A larger number of measurements, involving both a higher density and a larger area of coverage, is necessary to determine how the existing measurements fit into the complete stress pattern. Unfortunately, the number and distribution of suitable outcrops for surface stress measurements of this type is not very conducive to obtaining the additional measurements that are needed.

Comparison With Hydraulic Fracturing Measurements

A series of hydraulic fracturing measurements has also been made in the area by USGS personnel [*Zoback and Roller*, 1979; *Zoback et al.*, 1980a, b]. Two of their measurement sites are very close to two of my sites, so it is instructive to compare the results. Unfortunately, the data on stress directions from the hydraulic fracturing measurements are not complete, owing to troubles with the borehole televiewer that was used to image the fractures and trouble with debris in one of the holes.

Again the best site for comparison is in the Punchbowl Formation just south of the fault. My site T10 is only a few meters from site Mojave 1 (Z1) of *Zoback et al.* [1980*a*, *b*]. The stress directions reported by *Zoback et al.* [1980*a*] are incorrect for the top portion of hole Z1 caused by an equipment malfunction. The correct azimuths reported by *Zoback et al.* [1980*b*] are −20°, −23°, and −4° at depths of 80, 185, and 218 m, respectively. The average of these azimuths, −16°, compares favorably with the average of my near-surface measurements at this site of −27°.

The other site at which nearby stress measurements were made by both hydraulic fracturing and surface overcoring is at Puite Butte in the Mojave block 22 km northeast of the San Andreas fault, sites Z4 and T9. The only stress directions that exist for this site from hydraulic fracturing are 'poor' in quality. They are at depths of 73 and 104 m and give azimuths for σ_1 of 104° and 93°, respectively. The surface overcoring site T9 is about 500 m from the Z4 well and gives a σ_1 direction of −4°. Thus, at this location, the direction of compression from surface overcoring is a tectonically more reasonable one than that given by the hydraulic fracturing measurements. It seems reasonable to dismiss the hydraulic fracturing measurements at this site because it is quite possible that fractures designated as being of poor quality may not even be hydraulic fractures [*Zoback et al.*, 1980*b*].

The stress magnitudes from the hydraulic fracturing technique seem more useful and reliable than the magnitudes from surface overcoring. This is in part because the measurements in wells allow the determination of a vertical stress gradient which may be extrapolated to greater depth. Also the stress magnitudes near the surface are likely to be reduced by fracturing to a lower value than the upward extrapolation of such a gradient would give. In general, in this study the magnitude of the stresses measured in the igneous rocks decreases as their modulus decreases, and this modulus decrease seems to correlate with the extent of fracturing and of alteration. However, the coherency of the arkosic rocks of the Punchbowl Formation at sites 1 and 10 is better and seems similar to what might be expected at depth. The values of σ_1 for these sites are 2.5 and 1.5 MPa, respectively, whereas the values for σ_2 are 1.5 and 0.6 MPa. Upward extrapolation of the hydraulic fracturing data for this site Z1 [*Zoback and Roller*, 1979, Figure 2; *Zoback et al.*, 1980*b*, Figure 7] would be quite compatible with these ranges of σ_1 and σ_2 magnitudes at the surface. However, at site T9 in the fractured granodiorite the surface overcoring magnitudes of 0.6 and 0.4 MPa for σ_1 and σ_2, respectively, are considerably lower than the upward extrapolated values from hydraulic fracturing in hole Z4 of perhaps 5 and 2.5 MPa.

Conclusions

If adequate precautions are taken in site selection and field procedures, the results of this study and comparisons with other stress measurements made in the same area suggest that it is possible to get tectonically useful measurements of stress directions from overcoring near the ground surface. If the rock types are suitable, significant values for the stress magnitudes at the surface can also be obtained. The present set of measurements, even when combined with other recently available data on absolute stress measurements in the same area, are too few in both density and total areal coverage to be very useful in helping to understand the present day tectonics of the region or to aid in earthquake prediction. Near the San Andreas fault, on the southwest side, the greatest compression direction is north northwest. The measurements in the Mojave block northeast of the fault are less reliable and show more scatter; the directions range from northwest to northeast.

A greatly increased program of surface overcoring measurements would have the potential to add to our understanding of this important region. However, because of the poor quality of many of the rock types, especially their high density of fractures, and because of the large areas of mountainous topography on the south side of the San Andreas fault and the extensive areas of basin fill on the northeast side, such a program of surface measurements can never give a very complete picture of the stress pattern. A combination of surface measurements, where possible, and a program of measuring stresses in deep wells is the best way to approach the determination of the tectonic stress distribution in this area.

Acknowledgments. A large number of people contributed significantly to this project, and I am grateful for all their help. Barry Raleigh arranged for me to be a temporary employee of the U. S. Geological Survey during most of the work and offered encouragement and assistance. Verne Hooker and Dave Bickel of the U. S. Bureau of Mines loaned me nearly all of the overcoring equipment used in the project and offered much valuable advice. Richard Goodman kindly loaned the gasoline-powered drill used in all of the work. Pat Oswald-Sealy and Marty Gothberg provided invaluable cheerful assistance in the field work. Marc Sbar, Terry Engelder, Richard Plumb, and Steve Marshak aided in some field logistics and shared their data. Ron Shreve, Tom Dibblee, Al Barrows, Fred Miller, and Dennis Burke provided much valuable advice as to appropriate areas to search for sites. Al Vaughn aided in obtaining permission from property owners to drill on their land. Gene Burbank, Luis Carle, the San Bernadino County Museum, and the Antelope Valley Indian Museum kindly allowed drilling sites on their property. Marc Parmentier offered helpful suggestions in reducing the data, and Jan Tullis, Hans Swolfs, and Terry Engelder offered helpful suggestions for improving the manuscript. Special thanks are also due to John Handin and Dave Stearns, who first introduced me to overcoring (see Figure 1).

References

Becker, R. M., An anisotropic elastic solution for testing stress relief cores, *Rep. Invest. U. S. Bur. Mines*, *7143*, 15 pp., 1968.

Becker, R. M., and V. E. Hooker, Some anisotropic considerations in rock stress determinations, *Rep. Invest. U. S. Bur. Mines, 6965*, 23 pp., 1967.

Castle, R. O., J. P. Church, and M. R. Elliott, Aseismic uplift in southern California, *Science, 192*, 251-253, 1976.

Castle, R. O., M. R. Elliott, and S. H. Wood, The southern California uplift (abstr.), *Eos Trans. AGU, 58,* 495, 1977.

Clark, S. P., Handbook of Physical Constants, *Mem. Geol. Soc. Am., 97*, 587 pp., 1966.

Fitzpatrick, J., Biaxial device for determining the modulus of elasticity of stress-relief cores, *Rep. Invest. U. S. Bur. Mines, 6128*, 13 pp., 1962.

Hanks, T. C., Earthquake stress drops, ambient tectonic stresses, and stresses that drive plate motions, *Pure Appl. Geophys., 115*, 441-458, 1977.

Harrison, J. C., Cavity and topographic effects in tilt and strain measurement, *J. Geophys. Res., 81*, 319-328, 1976.

Hooker, V. E., and C. F. Johnson, Near-surface horizontal stresses, including the effects of rock anisotropy, *Rep. Invest. U. S. Bur. Mines, 7224*, 29 pp., 1969.

Hooker, V. E., and W. I. Duvall, In situ rock temperature: Stress investigations in rock quarries, *Rep. Invest. U. S. Bur. Mines, 7589*, 12 pp., 1971.

Hooker, V. E., and D. L. Bickel, Overcoring equipment and techniques used in rock stress determinations, *Inf. Circ. U. S. Bur. Mines, 8618*, 32 pp., 1974.

Hooker, V. E., D. L. Bickel, and J. R. Aggson, In situ determination of stresses in mountainous topography, *Rep. Invest. U. S. Bur. Mines, 7654*, 19 pp., 1972.

Hooker, V. E., J. R. Aggson, and D. L. Bickel, Improvements in the three-component borehole deformation gage and overcoring techniques, *Rep. Invest. U. S. Bur. Mines., 7894*, 20 pp., 1974.

Jaeger, J. C., and N. G. W. Cook, *Fundamentals of Rock Mechanics*, 515 pp., Chapman and Hall, London, 1969.

McNally, K., H. Kanamori, J. C. Pechman, and G. Fuis, Earthquake swarm along the San Andreas fault near Palmdale, southern California, 1976-1977, *Science, 201*, 814-817, 1978.

Merrill, R. H., Three-component borehole deformation gage for determining the stress in rock, *Rep. Invest. U. S. Bur. Mines, 7015*, 38 pp., 1967.

Merrill, R. H., and J. R. Peterson, Deformation of a borehole in rock, *Rep. Invest. U. S. Bur. Mines, 5881*, 32 pp., 1961.

Obert, L., Triaxial method for determining the elastic constants of stress relief cores, *Rep. Invest. U. S. Bur. Mines, 6490*, 22 pp., 1964.

Obert, L., and W. I. Duvall, *Rock Mechanics and the Design of Structures in Rocks*, 650 pp., John Wiley, New York, 1967.

Panek, L. A., Calculation of the average ground stress components from measurements of the diametral deformation of a drill hole, *Rep. Invest. U. S. Bur. Mines, 6732*, 41 pp., 1966.

Prescott, W. H., and J. C. Savage, Strain accumulation on the San Andreas fault near Palmdale, California, *J. Geophys. Res., 81*, 4901-4908, 1976.

Savage, J. C., and W. H. Prescott, Geodometer measurements of strain during the southern California uplift, *J. Geophys. Res., 84*, 171-177, 1979.

Sbar, M. L., T. Engelder, R. Plumb, and S. Marshak, Stress pattern near the San Andreas fault, Palmdale, California, from near-surface in situ measurements, *J. Geophys. Res., 84*, 156-164, 1979.

Sieh, K. E., Slip along the San Andreas fault associated with the great 1857 earthquake, *Bull. Seismol. Soc. Am., 68*, 1421-1448, 1978a.

Sieh, K. E., Prehistoric large earthquakes produced by slip on the San Andreas fault at Pallett Creek, California, *J. Geophys. Res., 83*, 3907-3939, 1978b.

Tullis, T. E., Reflections on measurement of residual stress in rock, *Pure Appl. Geophys, 115*, 57-68, 1977.

Zoback, M. D., and J. C. Roller, Magnitude of shear stress on the San Andreas fault: Implications from a stress measurement profile at shallow depth, *Science, 206*, 445-447, 1979.

Zoback, M. D., J. C. Roller, J. Svitek, and D. Seeburger, Hydraulic fracturing stress measurements and the magnitude of shear stress on the San Andreas fault in Southern California, *U. S. Geol. Surv. Open-File Rep. 80-625*, 490-518, 1980a.

Zoback, M. D., H. Tsukuhara, and S. Hickman, Stress measurements of depth in the vicinity of the San Andreas fault: Implications for the magnitude of shear stress at depth, *J. Geophys. Res., 85*, 6157-6173, 1980b.

Understanding Faulting in the Shallow Crust: Contributions of Selected Experimental and Theoretical Studies

D. W. STEARNS,[1] G. D. COUPLES,[2] W. R. JAMISON,[2] AND J. D. MORSE

Department of Geology and Center for Tectonophysics, Texas A & M University, College Station, Texas 77843

Our understanding of faulting in the shallow crust has been greatly enhanced by the efforts of both experimentalists and theoreticians. The laboratory analysis of fault mechanics has been advanced by research directed toward empirical testing of the validity of the generalized Coulomb criterion, for both fracture and sliding on existing surfaces. This general criterion appears to hold for cases of plane strain, but it must be modified for more complex deformational settings. Rock models of faults have suggested a number of testable hypotheses concerning faulting, and the work directed toward hypothesis verification has further advanced our understanding of the ramp regions of thrust faults and upthrust faulting. Theoretical concepts dealing with the results of boundary value problems have aided in explaining how large regions of the earth's crust can be deformed by several distinct fault types during a single deformation episode. Consideration of both vertical and horizontal components of fault displacement, which vary with depth when a fault is curved in cross section, allows prediction of the location and type of secondary, compensatory structures. Field experience to date suggests that these idealized experimental and theoretical studies constitute a basis upon which we can construct useful frameworks for thinking about shallow faulting.

Introduction

Faulting is a subject that both fascinates and puzzles many geoscientists. Field geologists, including subsurface geologists, are interested in faulting primarily from a kinematic viewpoint, which stems from their desire to make historical inferences and palinspastic restorations. The experimentalist, on the other hand, has a different approach to faulting and is driven to study the mechanisms involved in the faulting process. The theoretician's curiosity is best satisfied when general expressions relating the dependent and independent variables important in the faulting process can be written. The purpose of this paper is to summarize selected efforts of the experimentalists and theoreticians that have significantly aided field geologists in their search for answers to faulting problems. The background data for this paper have been published in several journals and books. Many of these publications are not commonly used except by the structural specialist, but many other geoscientists must also deal with faulting. Therefore we will attempt to draw together the results of experimental and theoretical studies that can be helpful to anyone who works with faults in the shallow crust. The paper is not intended for geoscientists who are already experts in the subject.

A working definition of a geological fault is 'a discontinuity in the displacement field of deformed rock masses, where the component of differential displacement parallel to the discontinuity has a nonzero value.' In stating the aims of this paper, it is important to note that for theoreticians and experimentalists, this definition may be unsuitable, but field geologists could work with this definition, even though each individual may prefer a different wording. To the field geologist the word 'fault' represents a feature, not a process. However, experimentalists and theoreticians, in their quest to understand the process of faulting, have aided the field geologist who works with those features in the shallow crust called 'faults.' Thus our purpose is to discuss how certain experimental and theoretical studies have contributed to the understanding of features in the shallow crust that geologists have classified as faults for over two hundred years.

We will summarize ideas prevalent before 1950, but we will concentrate on post-1950 studies. The selection of this date is by design: about this time, organized, systematic laboratory investigations of the physical properties of shallow crustal rocks were initiated. Previous experimental studies of rock behavior were done under environmental conditions that represented either surface rocks or

---

[1] Now at University of Oklahoma, Norman, Oklahoma.
[2] Now at Cities Service Oil Company, Tulsa, Oklahoma.

very deeply buried rocks. In the 1950's and 1960's, John Handin influenced and guided the investigation of physical properties of rocks in the shallow crust, an investigation that changed our way of thinking about faulting.

Owing to the authors' experience, our discussion is directed primarily toward faults that occur in or contribute to the deformation of layered sedimentary rocks in the shallow crust.

The experimental and theoretical work considered here falls into three general categories: (1) attempts to establish criteria for faulting (experimental and theoretical studies), (2) rock models of certain types of faulting, and (3) analytical solutions to specific problems that have aided the field geologist. Each category is discussed separately. Their connecting fiber is that they all have contributed to our ability to understand natural faults.

General Attitudes Toward Faulting Prior to 1950

In 1942, E. M. Anderson published a book that significantly influenced our approach to faulting. In this work, Anderson related the common fault types to the orientation of the three principal stresses in the shallow crust at the time of faulting. The premise upon which this work was based stemmed from Anderson's observation that many faults intersect the earth's surface. Because this surface is an air-rock interface and thus cannot support a shear stress, it must be a principal plane, with one of the three principal stresses vertical. Where overburden is the maximum principal stress σ_1, normal faults occur. Thrust faults result when the least principal stress σ_3 is overburden. If the intermediate principal stress σ_2 is vertical, then wrench faults result.

Anderson then assumed that the orientation of the three principal stresses near the earth's surface would not change with depth (as a universal condition, this assumption has not withstood the test of time). Several implications of this supposition were defended vehemently by many geoscientists over the next few decades. One implication is that all fault planes should be linear in cross section. A second implication is that the stress fields causing deformation in the shallow crust should be uniform, thus producing a single fault type pervading large regions. Acceptance of this assumption further implies that the presence of more than one fault type in a region requires more than one period of deformation.

Owing to the unquestioned acceptance of these ideas, inappropriate questions have been posed. For example, a common question is: what is the fault type for this or that area? There is no doubt that in some regions of the earth's crust, one fault type seems to be ominipresent, thus implying a uniform stress field throughout a large volume of rock. However, there are also regions in the earth's crust where nonuniform stress fields have existed and multiple fault types are now present in one area as the result of a single period of deformation.

Criteria for Faulting

Coulomb Criterion

According to *Hempel* [1966], a theory is introduced in an attempt to explain a certain class of phenomena as the necessary consequences of other phenomena that are regarded as more primitive. Hempel states, 'Theories seek to explain regularities and, generally, to afford a deeper and more accurate understanding of the phenomena in question.' According to this definition, we do not have an adequate, universal theory to explain faulting, because we do not clearly understand the relationships between such entities as friction, loss of cohesion, failure, and slip in rock material. To circumvent the problem of incompletely understanding the primitive phenomena necessary for a theory, *Coulomb* [1776] coined the term 'internal friction' in his early studies of shear fracture and developed not a fracture theory but, as *Handin* [1969] correctly points out, a fracture criterion:

$$\tau = \tau_0 + \sigma \tan \phi \qquad (1)$$

This expression states that shear fracture (fault initiation) will occur whenever the shear stress τ on any plane exceeds the cohesive shear strength of the material τ_0 plus the product of the normal stress across the plane σ and the coefficient of 'internal friction,' $\tan \phi$. *Coulomb* [1776] showed that if this criterion holds, then when a material for which ϕ is nonzero is subjected to a stress difference large enough to cause shear fracture, the fracture planes that develop will be inclined less than 45° to σ_1 (maximum principal stress). The departure from 45° varies directly with the value of ϕ. If θ is the angle between σ_1 and the shear fracture plane, then

$$\theta = 45° - \phi/2 \qquad (2)$$

The Coulomb criterion was intended to apply only to shear fracture initiation; but faulting, in the geological sense, requires that considerable slip occur after initiation. Therefore, even if it could be shown that the Coulomb criterion holds true for rock materials, some other functional relationship is required to describe the post-initiation slip necessary to produce geological faults. The first step in this direction was taken by *Rankine* [1857], who modified the Coulomb criterion for materials with no tensile strength (sands) to

$$\tau = \sigma \tan \phi \qquad (3)$$

This relationship could apply equally to slip occurring after fracture initiation, but in such a case, $\tan \phi$ (internal friction) becomes μ (coefficient of sliding friction), and the slip condition is

$$\tau = \sigma \mu \qquad (4)$$

If between the time of creation and the occurrence of the slip necessary to make it a significant fault, some cohesion is reestablished across the fractured zone, then the criterion returns to an altered form of (1):

$$\tau = \tau_0{}^* + \sigma \tan \phi^* \qquad (5)$$

where $\tau_0{}^*$ is the new cohesive shear strength and $\tan \phi^*$ is the new coefficient of internal friction.

Although an adequate theory of faulting that can be tested by field data does not exist, the Coulomb criterion and its modifications still serve as relationships by which faulting data can be organized. Therefore we completely agree with *Handin* [1969] when he states: '... the Coulomb criterion can be useful even though its basis, the concept of internal friction, is unsatisfying. This criterion has certainly been widely and successfully applied to engineering problems involving both soils and rock mechanics of the shallow crust, and doubtless its use will continue until an adequate mechanistic theory of fracture is developed.'

Empirical Tests of the Coulomb Criterion

In using the Coulomb criterion as an aid in understanding faulting, it is necessary first to consider the range of processes that constitute our concept of faulting. A distinction is made here between the creation of a material discontinuity in rock and the motion of rocks past each other along an existing discontinuity. The first process, creation of a surface, is most appropriately called 'fracture,' and in a strict sense, the Coulomb criterion (1) applies only to this process. The second process, movement along a discontinuity, leads to 'faulting' (4). At the outset of any consideration of faulting, it is extremely important that the field geologist completely separate the data and theory that pertain only to creation from those that relate to the large slip necessary for a shear fracture to become a significant fault.

An early benefit to the field geologist was the empirical testing of the Coulomb criterion in the laboratory. *Handin and Hager* [1957] systematically and convincingly showed that the criterion adequately describes shear fracture in sedimentary rocks that have been deformed under confining pressures within the range expected in the shallow crust. In 1963, Handin and others showed that it is also valid to use the Coulomb criterion when considering pore-fluid pressures in the rocks. Furthermore, their papers, along with those of *Jaeger* [1959] and *Paterson* [1958], demonstrate that even though internal friction has not been physically identified, its effects can be measured in the laboratory.

On the basis of this experimental work, *Stearns* [1967] studied natural fractures associated with faulting. Stearns found that the laboratory experiments were good replicas, on a small scale, of natural faulting of sedimentary rocks. Therefore the type of faulting and the strike of the faults can be delineated in a faulted area by studying outcrop-scaled fractures. Two attitudes, one for each of the conjugate shear fractures, can be established. However, it takes other geologic information (for example, west side down-dropped) to determine which of these two planes parallels the actual fault or, in some cases, whether there are faults parallel to both planes. On the basis of the same experimental work, *Friedman* [1969] established predictive tools for use in determining fault attitudes by studying the microscopic deformation characteristics in close proximity to actual faults.

Effect of Shear and Normal Stresses on Faulting

Faulting occurs more readily when cohesive strength, normal stress, or friction are reduced. Because normal stress σ and the coefficient of friction μ occur as a product term in (4), it follows that if either σ or μ approaches zero, then the value of the other becomes relatively unimportant.

The first efforts to make use of this obvious statement involved applying the law of effective stress to thrust faulting. This law states that the effective normal stress σ_e on any plane is equal to the normal stress due to burial and tectonism minus the pore fluid pressure. *Handin et al.* [1963] demonstrated experimentally that this law is valid for many sedimentary rocks. *Hubbert and Rubey* [1959] applied the concept of effective stress in an analysis of overthrusting. By assuming negligible cohesion ($\tau_0 = 0$) they determined that in the presence of high pore pressure, movement of very large thrust plates is greatly facilitated.

Other geologists posed valid questions about some of Hubbert and Rubey's assumptions. *Hsü* [1969] objected to their neglect of cohesion, and *Forrestal* [1972] pointed out that nonuniform stress conditions might be expected at the base of a sliding sheet. Nonetheless, the basic premise still holds: high pore pressure makes faulting easier.

Hubbert and Rubey [1959] applied the concept of effective stress specifically to thrust faults caused by gravity sliding. The concept is equally valid, however, for any type of fault. Furthermore, elevated pore pressures may permit reactivation of stable faults, as was convincingly demonstrated by the Denver [*Evans*, 1966] and Rangely [*Raleigh et al.*, 1971] earthquake studies. In these examples, artificially elevated pore pressures caused slip on old faults that had been stable before the local fluid pressures were increased.

When considering pore pressure effects related to faulting, the time duration of the elevated pressures becomes important. There is sometimes a tendency to include or exclude pore pressure considerations solely on the basis of long-term geologic conditions that might or might not produce regional zones of high pore pressure. Certainly, however, there are many instances when in low-permeability rocks the time duration of the slip (faulting) is shorter than the time necessary to regain fluid pressure equilibrium. In these cases, even though pore pressure may return to normal long after the event, it could have been high during the actual faulting. The following circumstances are most likely: (1) when the fault conditions cause rapid loading of low-permeability rocks, such as many shales, (2) when temperature elevation causes greater fluid expansion than rock expansion [*Magara,* 1978], or (3) when there is a rapid dewatering of minerals containing bound water.

Handin and Hager [1957] and *Handin et al.* [1963] also showed that the shear fracture strength of sedimentary rocks is substantially reduced under conditions of low normal stress. It is no wonder, then, that faulting is so prolific, even under small tectonic forces, in areas of rapid clastic deposition (hence high pore pressure), as in the Texas and Louisiana Gulf Coast regions.

It is convenient for extrapolation to nature that the law of effective stress is expressed as a simple equation. However, in the field it is difficult, and often impossible, to determine whether high pore pressure existed during deformation. Creating high pore pressure, either in a mathematical analysis or in an experimental situation, may be fairly simple, but it is difficult to make historical inferences from field observations as to whether and where high pore pressures were associated with deformations that occurred millions of years ago.

Frictional and Strength Effects on Faulting

In considering the relative motion of two rock masses past one another, it is possible to distinguish two types of slip [*Stearns,* 1978]. One type of slip (type A) involves a discrete discontinuity across which slip occurs, as in the sliding of one layer of material directly past another identical layer. Another type of slip (type B) does not involve a discrete slip surface, but the yielding of some intervening material, such as the displacement of one layer of rock relative to another when there is an intervening weak shale. Type A slip is most frequently visualized when the term 'fault' is mentioned. However, type B slip is, perhaps, just as important in contributing to the total deformation observed in faulting, especially in thrust fault terrains.

In type A slip, the frictional characteristics of the sliding surface determine the ease of faulting. In type B slip, the bulk behavior of the weaker material is important, and ease of faulting is most dependent, therefore, on the yield strength of this material.

It is reasonable to suppose that the maximum displacement along a fault at the time of its formation does not greatly exceed the offsets noted during individual seismic events on active faults. Yet many faults of interest to field geologists have displacements that are many times larger than the offsets noted during seismic events. It is concluded, therefore, that a significant portion of the displacement along such faults occurred at a time, or times, subsequent to the initiation of faulting.

If only cases of type A slip are considered, then faulting must proceed (after exceeding cohesion) by a process related to the frictional sliding of two rock masses past one another. Laboratory investigations reviewed by *Logan*

[1975] demonstrate that friction, in the sense illustrated by (4), is an inadequate tool for explaining all the microscopic processes by which rocks slide past one another. This is true for 'clean' interfaces and also when 'gouge' is present. Thus, from the standpoint of micromechanics, the concept of friction may not be particularly useful. However, 'friction' is helpful in other senses. While the concept of friction per se does not describe the microprocesses of slip, it implicitly incorporates these effects in its description of the macroscopic stress conditions for sliding. From a practical standpoint, 'friction' synthesizes the effects of processes that are too complicated and numerous to be incorporated directly into a macroscopic analysis.

The field geologist is seldom able to determine which microprocesses have operated and at which point or points in the history of slip on a fault they were active. Moreover, these determinations, even if possible, would then need to be summarized to be useful in determining the physical conditions under which slip had occurred. The concept of friction meets this need.

Experimental studies by *Logan et al.* [1972] have demonstrated that even when type A slip is involved, variations in the coefficient of friction μ that are dependent on lithology are very important to the field geologist. Experiments show that dolomite sliding past dolomite has a lower value of μ than does limestone sliding past limestone. Furthermore, when testing dilithologic samples of limestone and dolomite, the resulting value of μ is closer to that for two limestone surfaces than it is for two dolomite surfaces. The lesson learned by the field geologist is that not all interfaces between carbonate rocks have the same potential to become slip surfaces. In fact, from the standpoint of faulting, we should not use the term 'carbonate,' but we should carefully determine and separate dolomite-dolomite interfaces from limestone-limestone or limestone-dolomite interfaces.

Laboratory studies have shown that in general, brittle materials sliding on brittle materials have a lower value of μ than do ductile materials sliding on ductile materials. Furthermore, if the two materials, one on each side of the interface, are of contrasting ductilities, then the more ductile material seems to control the ease of slip. Is it any wonder then that the Heart Mountain detachment surface [*Pierce,* 1957] occurs on the first bedding plane in the layered dolomites of the Bighorn Formation (Ordovician) rather than along the shale-dolomite interface (3 m below) that marks the Cambrian-Ordovician contact?

Our intuitive concept of rock friction sometimes makes 'friction' a cover-up term for processes that do not even involve friction. For example, everybody 'knows' that wet shales are 'slippery,' and therefore any shale in the stratigraphic section is frequently regarded as a potential detachment, or slip, horizon. The image seems to be one of the shale acting as a lubricant, thus reducing 'friction.' However, laboratory studies now provide information that allows field geologists to be far more discerning when selecting probable shale detachments. *Wilson* [1970] showed

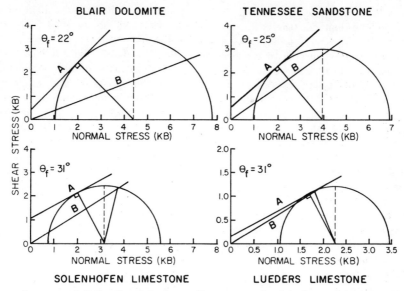

Fig. 1. Mohr envelopes A and sliding line B for four sedimentary rocks; θ_f is the angle between the shear fracture (fault) and σ_1. Stress circles are all for $\sigma_3 = 1$ kbar [after *Handin*, 1969].

that shales containing water-bound clays are much more likely to be the locations of detachment horizons (type B slip) than are kaolinitic-type shales. Even then, detachment has more to do with the release of water during slip, which causes the material to become a mud slurry with a low yield stress, than it does with friction. Other weak layers, such as salt, can also serve as easy detachments. However, it is not low friction that makes detachment relatively easy across a salt layer, but low-yield stress that allows type B slip to occur. Likewise, laboratory studies by *Logan and Shimamoto* [1976] show that a gouge zone with even small quantities of silica or calcite is less likely to become a zone of type B slip than is a gouge zone without either of these constituents.

Zone of Weakness Concept

A casual reading of geological literature indicates a widespread misuse of the concept of zone of weakness (that is, the reactivation of old faults in a new deformation period). Many geologists believe that once a fault exists in a rock mass, any new deformation phase, no matter what the orientation of the stress field, will surely cause reactivation of the old fault.

In rock, a zone of weakness might be a fracture or other discontinuity, or it might be an actual zone of weak material, perhaps a zone of preexisting gouge. However, such a zone of weak material can be treated as a frictional zone, or as an existing discontinuity. We will restrict our discussion as though the zone were, in fact, an existing discontinuity.

Handin [1969] illustrated the basic method used to determine whether under a particular loading condition, a preexisting discontinuity would experience slip or whether a new fault would be created. In this paper,

Handin presented new experimental data but pointed out that it agreed with the work of others [*Jaeger*, 1959; *Handin and Stearns*, 1964; *Byerlee*, 1967], so there was substantial testing to support the principles Handin described.

Is it mechanically easier to create a new fault, or is it easier to reactivate an old one? To reactivate an old fault, the shear stress along the fault must satisfy the conditions of (4), but to create a new fault, the conditions of (1) must be satisfied. By examining these equations, it is readily apparent that the two cases involve different material properties. In (1) the shear stress must exceed the cohesive shear strength τ_0 of the material plus the product of normal stress σ across the plane times the coefficient of internal friction for the material, tan ϕ. In the case of reactivation, τ_0 has already been overcome (or at least changed to τ_0^*), and the shear stress must exceed the product of the normal stress across the old fault σ and the coefficient of friction μ. Not only are there different material properties involved in the two cases, but when the new principal stresses do not have the same orientation as those that caused the original fault, the σ terms in (1) and (4) will have different values.

Handin [1969] addressed this subject by presenting experimental data for four different rocks deformed under 1-kbar confining pressure (Figure 1). In each case, line A represents (1), and line B represents (4). The slopes of lines A and B represent tan ϕ and μ, respectively. It can be seen that in some cases the two slopes are nearly equal (Solenhofen Limestone), while in other cases the slopes are quite different (Blair Dolomite).

To facilitate discussion of these data in relation to fault reactivation, assume that σ_1 (maximum principal stress) is horizontal. For Lueders Limestone (Figure 1), it would be easier to form a new fault than to reactivate an old one dipping steeper than 34° or less than 27° (the two intercepts of the stress circle by line B). A second deformation phase would have to be very nearly parallel to the first phase in order to reactivate an old fault. However, in the case of Blair Dolomite, if an old fault plane existed with a dip anywhere between 4° and 65°, then the fault would be reactivated (slip would occur on it) instead of forming a new fault (dipping 22°). For an old fault with a dip less than 4° or greater than 65°, a new fault would form rather than reactivating the old one. For each of these rocks, should the old fault reheal so that there is a τ_0^* for reactivation (line B would then intercept the τ axis at some positive value), slip on old faults would occur over a narrower range of dips than when zero cohesion was assumed.

In the earth's crust, reactivation of an old fault would probably involve characteristics of the gouge material and would not be as simple as the mono-lithologic cases treated by *Handin* [1969]. Modern experimental work addresses the deformation of gouge zones [*Logan et al.*, 1979] and emphasizes that field geologists should be cautious when considering the possible reactivation of old faults.

The Coulomb Criterion and Fault Orientation in Bulk Deformation

The Coulomb criterion predicts that fault surfaces will form with a specific orientation θ relative to the principal compressive stress axis. The angle θ is determined directly from ϕ through (2). Early experimental work [*Handin and Hager*, 1957; *Handin et al.*, 1963] shows that tan ϕ decreases with an increase in the confining pressure. Consequently, θ should increase as σ_3 increases, with a limiting value of 45°. However, in experimentally induced faults produced under conditions expected in the shallow crust, the values of θ cluster around 30°.

The Coulomb criterion allows for two shear fracture orientations when the stresses required for failure are obtained. These faults, referred to as conjugate faults, have opposing senses of offset and are parallel to σ_2, with σ_1 at an angle of θ to these fracture planes. A point not addressed by the Coulomb criterion is the ability, or lack of ability, of the rocks to achieve any necessary deformation (strain) by displacement along the Coulomb-predicted fractures.

There are other expressions, termed collectively 'constitutive relationships', that relate the bulk deformation (strain) of rock to the state of stress. In rock that deforms primarily by motion along faults, the orientation of and sense of motion along the faults must allow the required bulk deformation to occur, as well as satisfy the Coulomb criterion. In some instances, it would appear that these two conditions cannot be mutually satisfied. By specifying the deformation of the rock, kinematic constraints are imposed on the deforming material. These kinematic constraints are most likely to be incompatible with Coulomb-predicted fault orientations when relatively large deformations occur and/or when the deformation is distinctly three-dimensional (that is, non-plane strain).

Some basic theoretical considerations indicate when and why the Coulomb-specified fault systems are inadequate for the achievement of certain bulk deformations. *Bishop* [1953] has shown that any volume-constant, three-dimensional strain may be accommodated by five independent slip systems. Because any single plane can contain only two of these independent slip systems, at least three slip planes are required to allow complete freedom to accommodate three-dimensional strain. *Reches* [1978] extended this analysis to a body wherein both strain and rotation are specified and has shown that at least four slip planes (faults) are required to satisfy these conditions. For any specific, volume-constant, three-dimensional state of strain, four preferred fault sets are derived [*Reches*, 1978]. Only in the case of plane strain (two-dimensional deformation) does the number of fault sets reduce to two.

A simple but well-controlled set of experiments on faulting in clay shows, in fact, that four, rather than two, fault systems are activated as the material deforms in a three-dimensional manner [*Oertel*, 1965]. Oertel finds that the observed fault development is incompatible with the Coulomb criterion because (1) the attitude of fault planes depends upon intermediate stress magnitude, (2) the number of fault planes is four rather than two, and (3) the intersect angle between fault sets of opposing sense of shear is much less than that predicted by the Coulomb criterion (or by the analysis of *Reches* [1978]. The last observation appears to be dependent upon the material under consideration, but the first two, and especially the second, are compatible with some field observations (primarily deformed sandstones of the Colorado Plateau [*Reches*, 1978; *Jamison*, 1979]). It would appear that use of the Coulomb criterion (1) should be restricted to situations of plane strain.

Faulting in the Basin and Range Province of the western United States provides an example of the multiplicity of fault orientations required for large, three-dimensional deformation. Horizontal elongation in this region is accomplished by displacement along normal faults that have a general north-south trend. This system of east- and west-dipping normal faults forms, in an east-west cross-section, an apparent conjugate set, implying a vertical σ_1. In map view, however, the normal faults do not strike exactly north-south, rather they vary in strike from about N30°W to about N30°E [*Donath*, 1962]. For both the east- and west-dipping normal faults, two average strikes are seen, providing the four fault planes required for unrestricted three-dimensional strain. If σ_1 were vertical, then the angle θ for this region is close to the value predicted by the Coulomb criterion, but the fact that the fault orientations do not indicate a regionally consistent orientation for σ_2 is not in accordance with the Coulomb criterion.

Another important aspect of the deformation accomplished by motion along Coulomb-predicted faults, again derived primarily from theoretical considerations, is that the material is explicitly required to dilate as it deforms [*Drucker and Prager*, 1952]. The amount of dilatancy that occurs depends upon, among other things, the amount of

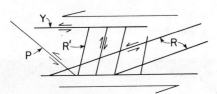

Fig. 2.   Geometric relationship between Riedel shears R conjugate Riedel shears R', and principal displacement shear Y. R_1 and R_2 of *Bartlett's* [1980] nomenclature are equivalent to R and R', respectively. Bartlett's P is also illustrated.

strain and the value of tan ϕ. Only when tan ϕ is zero ($\theta = 45°$) is there no dilatancy of the material. In certain well-studied systems that have been subjected to large shear strains, both natural [*Tchalenko and Ambraseys*, 1970] and experimental [*Tchalenko*, 1970; *Wilcox et al.*, 1973; *Mandl et al.*, 1977], the failure surfaces do deviate from the Coulomb-predicted orientation of 30° after a limited amount of strain. In systems that are subject to direct shear, such as wrench fault systems, the initial fault fabric that develops is an en échelon system of faults, which are inclined at a specific angle to the direction of wrenching. These en échelon faults are termed Riedel shears (R), and are often accompanied by a second system of faults that are termed conjugate Riedel shears (R'). The R and R' fault systems have an orientation and sense of slip that may be interpreted in terms of the Coulomb criterion, where the maximum compressive stress is at 45° to the fault zone (Figure 2).

Direct shear experiments in clay [*Tchalenko*, 1970; *Wilcox et al.*, 1973] and in granular materials [*Mandl et al.*, 1977; *Friedman and Shimamoto*, 1978] have shown that Riedel and conjugate Riedel shears develop in the early stages of deformation. However, as shearing of the body increases, a new shearing surface, referred to as the principal displacement shear (Y in Figure 2), develops parallel to the maximum-shear-stress surface. Continued shearing of the material is accomplished almost exclusively by motion along this principal-displacement shear. The maximum-shear-stress surface, which is the surface of the principal-displacement shear, is oriented 45° to σ_1 (that is, $\theta = 45°$), hence, as indicated above, no dilatancy of the material is associated with motion along this zone.

Strike slip faults are the appropriate field example of a direct shear system. In a study of the surface deformation associated with the 1968 Dasht-e Bayēz earthquake in Iran, *Tchalenko and Ambraseys* [1970, Figure 5] found that many small-offset R and R' faults developed. Additionally, numerous small faults parallel to the zone of wrenching coalesced to form a major Y shear zone. The largest amount of offset is accomplished along the Y faults. The R and R' faults serve to indicate that σ_1 is 45° to the major shear zone (and Y), but the kinematic constraints on this system prohibit large amounts of offset from occurring along the R and R' faults.

Bartlett [1980] studied the development of wrench faults in rock materials under confining pressure. In experiments, Bartlett found that the angular relationships

among the small-offset faults (R$_1$, R$_2$, and another set P) could not all be explained simply by development of Coulomb shear fracture sets. However, Bartlett did find that after relatively small displacements had occurred on all fault systems, subsequent shearing occurred along the Y zones.

Not all deformational systems, even though they may involve large, three-dimensional deformation, are actually subject to these kinematic constraints. For example, in many regions of normal and reverse faulting, much of the deformation is accomplished by motion perpendicular to the air-rock interface. Since this is a free surface, the dilatancy associated with motion along Coulomb-predicted fault systems may be absorbed at the interface. The wrench fault system does not have this convenience, and thus motion along the Coulomb-predicted faults (R and R') is suppressed.

The Coulomb criterion applies well in the case of plane strain (two-dimensional deformation). However, for constrained, three-dimensional deformation, the fault orientations derived from the Coulomb criterion are inadequate for accomplishing the required deformation. In these instances, the kinematic constraints of the system demand that other faults form in order to accomplish the required motions.

Rock Models of Faults

The laboratory and theoretical studies discussed to this point have been concerned in some form or another with criteria of faulting. There is another experimental approach that has contributed to the understanding of faults in the shallow crust. This approach involves the creation and analysis of geometrical look-alikes for certain large-scale natural fault structures. During the past several years, these look-alikes have been produced by deforming specimens machined out of rock materials. Such studies have been concentrated on curved reverse faults [*Friedman et al.*, 1976a, b; *Gangi et al.*, 1977] and ramp regions of overthrust faults [*Morse*, 1977, 1978]. The overall resemblance of these deformed rock specimens to their natural counterparts is obvious and striking (Figure 3). However, the following questions are prompted by studying these specimens: (1) Are correlations and comparisons to nature legitimate? (2) What are the limits to such activity? (3) What can we expect to learn about large-scale structures by studying these small specimens?

Contemporary philosophy of science [*Hempel*, 1966] helps answer these questions: All scientific knowledge is established by a process composed of two distinct but complementary activities: inventing hypotheses and testing hypotheses (justification). Invention and justification are distinct acts of mind, because completely different thought processes are required for each. Hypothesis invention does not arise by performing logical operations on data; it is not a result of deduction. Hypothesis invention does depend on those thought processes that belong to psychology, not logic. In contrast, once the hypothesis exists, its

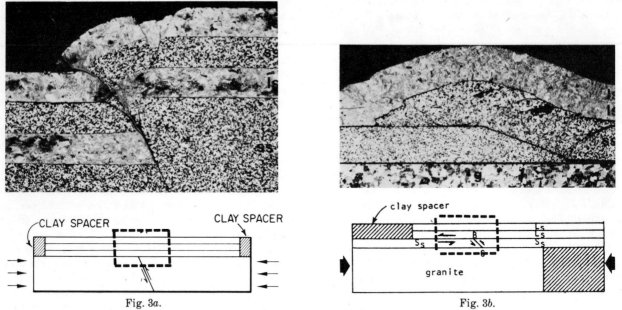

Fig. 3a.   Fig. 3b.

Fig. 3. Photomicrographs of rock models of an upthrust fault (a), and a thrust fault ramp region (b). The line drawing below each photomicrographs shows the cross section of the undeformed rock package. Sense of motion along pre-cuts is indicated in the cross section. Dashed outline indicates the approximate region of the photomicrograph above.

verification–the actual act of establishing a piece of scientific knowledge–depends greatly on deductive logic.

If this view is correct, and if the experimental studies in question are to play any role in the scientific process, then they must contribute either to the invention of hypotheses or to their verification. Here we subscribe to the view of *Morse* [1978] that these experimental studies have very little to do with verifying or testing hypotheses. Hypotheses concerning structural processes and geometries at the large scale in nature cannot be tested by examining experimentally deformed rock specimens, because there are important differences between the experimental and natural systems. We refer to the rock specimens in this type of experimentation as 'rock models', much the same as other workers refer to physical models composed of clay as 'clay models.' They certainly are not scaled models in *Hubbert*'s [1937] sense, so any hypotheses that arise from the rock models are testable only in the field. Therefore, these experiments are important to the field geologist in the invention of hypotheses.

The role of hypothesis invention played by these experiments and their analysis is very important, because it is not always clear, a priori, what features to look for in the field, where to look for them, or what role they play in the overall deformation process. This is particularly true if the process is unfamiliar. On the other hand, in the experiments the relationship of the minor structures to the histories of the stress and displacement fields (or to the overall process) is easily seen, because almost an entire mechanical system is available for inspection. Once the evolution of the rock specimens is described, the large-

scale rock body can then be examined in the appropriate locations for similar features. Thus the role that this experimentatal activity plays is one of providing an environment that is conducive to the invention of hypotheses about large-scale structural processes. Once formulated, these hypotheses still need to be tested because of the distinction that exists between the discovery of an hypothesis and its justification. Furthermore, the hypotheses can be tested only in the field, because the experimental and natural systems are only analogous and are not identical.

There is no doubt that these studies have already greatly benefited the field geologist. Many features of faulting in the Rocky Mountains foreland were observed but not understood for years, until the first experimental rock models were created. However, as *Stearns and Weinberg* [1975] point out, the role of these features could not be appreciated, primarily because of scale, until the experimentalist had shown similar features on a smaller scale. Likewise, by studying rock models, *Vaughn* [1976] and *Friedman et al.* [1976a] confidently interpreted structural geometries of the Rocky Mountains foreland that had previously defied interpretation. Subsequently, *Morse* [1977] and *Serra and Morse* [1979] employed rock models to better understand the deformation of the ramp regions in thrust faults.

Such experiments are valuable to the field geologist in another way, which involves an intermediate step between the experiment and the field application. From layer-thickness changes, calcite twin lamellae, microfractures, and faults, *Friedman et al.* [1976a, b] have measured the strains and deduced the stress fields in the specimens.

These deduced stress fields have been shown, in turn, to be similar to the stress field determined from linear, isotropic, elastic mathematical representations of the rock model [*Min*, 1975; *Friedman et al.*, 1976a; *Gangi et al.*, 1977]. Both the petrofabric and the mathematical studies of the experiments give rise to an understanding of details of the deformation that the experiments alone cannot provide. Like the experimental geometries themselves, the petrofabric and mathematical analyses serve as yet another basis for hypothesis invention that would be impossible from field studies alone.

Problem Solutions Pertinent to Natural Faulting

Boundary Value Problems and Physical Models

Hubbert [1951] performed a series of model experiments with constrained sand, which contained horizontal marker streaks of plaster of Paris. Hubbert loaded the sand to produce horizontal compression. This, in turn, produced offsets of the marker streaks, which very much resembled the assumed geometry of thrust faults. However, in cross section, many of the thrust faults had a curved, rather than straight, trajectory. Then by a mechanical analysis, in which for the most part the Coulomb criterion was assumed to be valid, Hubbert inferred the stress conditions at the boundaries of the sand and presented a solution to the qualitively stated boundary value problem. The solution contained curved stress trajectories that would explain the curved thrust faults observed in the models.

The results of this study helped establish an important challenge to *Anderson*'s [1942] assumption that all faults would have the same dip at all depths. Because the stress conditions in Hubbert's models were geologically reasonable, Hubbert provided a basis for believing that field observations that indicated curved thrust faults could, under certain circumstances, reflect the expected response and would not require special explanation. Certainly, seismic, subsurface, and surface studies over the following three decades showed that many faults are curved in cross section.

A companion paper to that by Hubbert was written by *Hafner* [1951]. This work stands as a hallmark in the literature, contributing to our understanding of natural fault systems, because it spurred the investigation of many different aspects of the faulting problem. In addition, it introduced into the geological literature the powerful technique of applying specific boundary value solutions to studies of faulting (Hubbert only inferred a solution).

In the companion paper, Hafner states stress conditions at the boundaries of a finite volume of material that he thought were geologically reasonable for the earth's crust. Starting with these boundary conditions and continuing through a series of simplifying assumptions, Hafner solved for the stress field that would exist within the bounded volume of material. Knowing the stress field, Hafner then was able to plot the potential shear fracture field by using the Coulomb criterion (shear fractures form

in rock materials at about 30° to the trajectory of the maximum principal stress).

Hafner determined the stress states resulting from several different sets of boundary conditions and made new statements about faulting, as can be seen by examining one typical solution (Figure 4). It is readily apparent that laterally variable boundary loads can produce a nonuniform stress field with curving trajectories (Figure 4a), which gives rise to a potential shear fracture field that is also nonuniform (Figure 4b). It is important in understanding faulting that these boundary conditions (which are geologically reasonable) produce a potential fault field that contains: (1) many faults that are curved in cross section, and (2) many different types of faults (at the surface they would be classified anywhere from standard normal faults to low-angle overthrusts).

Hafner's solutions are significant, because unlike *Anderson*'s [1942] conclusions they indicate that many different fault types can form in the same region during a single deformation episode. However, this solution does not indicate that various fault types can form in a haphazard manner. Instead the solution demands specific interrelationships of fault types as a function of the loading conditions.

Few other boundary value problems dealing with faulting appeared in the literature for the next 25 years. *Couples* [1977] employed mathematical techniques that were similar to those of Hafner but used boundary conditions that Couples felt were more compatible with specific regions in the Rocky Mountains foreland. The power of this analytic technique was demonstrated once again by showing that slight changes in the boundary conditions are capable of generating quite different potential fault fields. Couples demonstrated that the particular fault patterns could show considerable variation and that changes in the boundary conditions could produce areas of preferential faulting. Within Hafner's solution, most of the potential faults are as likely to form as the others. However, in Couples' solutions, certain regions are more likely to fault, because the stress difference in these regions increases more rapidly with loading than it does in other regions. *Withjack* [1979] considered the third dimension in solutions and demonstrated that from one set of boundary conditions that are geologically reasonable, several different fault types can be produced, but their spatial relationship is fixed.

Gangi et al. [1977] solved a boundary value problem that represented a single, experimentally created fault system. Not only were their analytical results in good agreement with the final geometry of the experiment, but they also gave insights into what sorts of deformation might be expected to accompany the faulting. In addition, their work demonstrated that boundary value problems could be applied to single fault zones as well as to fault fields.

Sanford [1959] solved a series of mixed boundary value problems (boundary conditions stated in terms of both

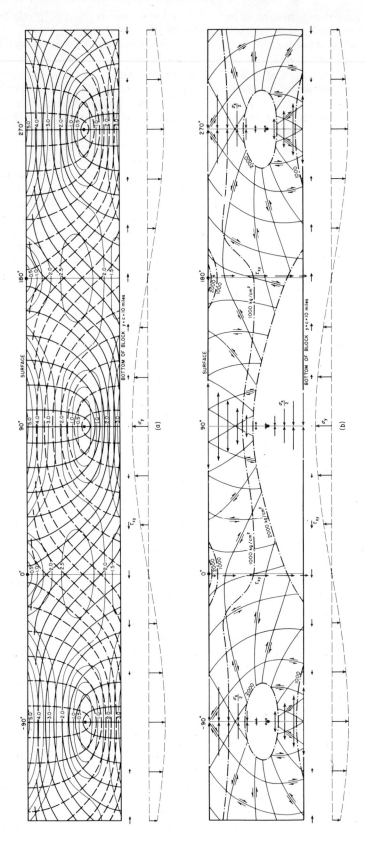

Fig. 4. Solution to a boundary value problem where $\tau_{xy} = 0$ on the upper surface, $\sigma_x = cy$ on the ends, and σ_y is a sine function at the base. (a) σ_1 trajectories (dark solid lines) and σ_3 trajectories (dashed lines). (b) Potential shear fracture field [after *Hafner*, 1951].

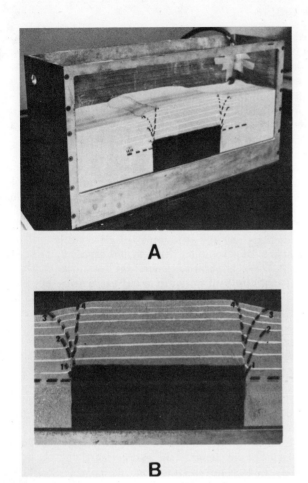

A

B

Fig. 5. Faults produced in a sandbox experiment. Overview (a) and close-up (b). Horizontal dashed lines indicate the initial position of the piston (solid). Horizontal, light lines are displacement markers, not layers, in the sand. The numbers on the faults (nonhorizontal dashed lines) in b refer to the sequential formation of the faults. Total piston displacement was 1.8 cm.

displacements and stresses) and attempted to verify them by physical models. In these now familiar models, Sanford subjected a mixture of sand and clay (in a glass-fronted box) to various loading conditions. These models somewhat resembled those of *Hubbert* [1951] but Sanford produced uplift by using a hydraulic piston at the base of the sand. The 'faults' that form in such a sandbox and their sequence of formation can be studied visually (Figure 5). As the sand mixture is deformed, the curved reverse faults experience slip and small normal faults form behind them (Figure 5).

It is useful to compare the models of *Sanford* [1959] to the results of *Hafner* [1951] and *Couples* [1977]. The uplift of the piston can be considered as an extreme case, in contrast to the smooth sinusoid used by Hafner. In each of these models, both reverse and normal faults form, but because of the high stress concentrations at the piston edges, regions of preferred faulting develop in the sandbox, much like those seen in Couples' solutions.

From the standpoint of final geometry, the analytical solutions are static and only apply up to the point of rupture: they do not apply during displacement along the faults. Yet the normal faults in the sandbox do not form until after there has been perceptible displacement on the curved reverse faults; that is, the normal faults are a consequence of the displacement on the reverse faults. As the reverse faults flatten, they locally shorten the material. For a constant-volume system, there must be an equal elongation elsewhere, and the normal faults in the sandbox accomplish just this. The normal faults 'die out' downward where the reverse faults are essentially vertical. At this point there is no more shortening that needs to be compensated. In the sandbox, a type of faulting is seen that is secondary and results from movement on other faults, not from original loading conditions. The sandbox approach is valuable to the field geologist, because it demonstrates faulting that results from changing boundary conditions.

Boundary conditions other than those of *Sanford* [1959] have also been studied by using sandbox models. For example, *Stearns et al.* [1978] investigated the effects of horizontal compression when it is added to vertical uplift. The reverse faults are much flatter (have greater curvature) near the surface of the sand under these conditions than when horizontal compression is not added. Also, much more uplift was accomplished by motion along reverse faults before the normal faults occurred than by pure uplift. Certainly, there are many other kinematic systems in the earth pertinent to faulting that can be modeled by the techniques developed by Sanford. The value of this approach is that it is an easy way to obtain a solution for the case of changing boundary conditions, a situation more difficult to address analytically.

Applicability of Boundary Value Problems

Each of the boundary value solutions discussed above is based upon several assumptions about material properties and rheological response. The geological appropriateness of these simplifying assumptions (necessary to make the solutions mathematically tractable) can be argued abstractly, but such arguments are futile. The appropriateness of the assumptions can only be determined by checking the predictions of the solution against an actual occurrence in the shallow crust. Like any test of a solution, the goodness of fit between prediction and actuality determines the confidence in the solution and its implications. The work of *Hafner* [1951], *Sanford* [1959], and *Couples* [1977] has been tested in the field by several workers [e.g., *Prucha et al.*, 1965; *Howard*, 1966; *Stearns*, 1975; *Palmquist*, 1978; *Matthews and Work*, 1978; *Couples and Stearns*, 1978; *Couples*, 1978]. From this field testing, certain domains of confidence and caution have evolved.

All the analytical solutions discussed assume that the body is a linearly elastic material and that only infinitesimal strains occur. Furthermore, they are static solutions and only strictly apply before the formation of the first rupture. Certainly, in many instances, rocks were considerably strained before faulting, and some faults do form after others were created, observations that clearly violate the assumptions of the solutions. Nonetheless, many workers have found a good fit between geometries predicted by the solutions and those seen in nature, which should give the field geologist confidence that these assumptions, which are not always correct in detail, do not rule out the applicability of such solutions.

The material assumptions that require the body to be homogenous, isotropic, and continuous definitely place restrictions on the regions of applicability of the solutions. As pointed out by *Stearns et al.* [1978], the solutions probably are most applicable to shallowly buried crystalline rocks, deeply buried, thick, massive carbonate sections, or rapidly deposited clastic sections, such as those in the Tertiary basins of California. The solutions are probably least applicable to shallowly buried, layered sedimentary rocks of varying composition. Any time that the depositional history of an area produces rock sequences with highly varied compositional layering and where normal stresses across this layering are low (shallow burial), bedding plane slip becomes such an active mechanism that the material assumptions behind the solutions do not give a good description of the actual behavior. Likewise, this type of solution should not be applied indiscriminately when considering the deformation of any thick, ductile sequence, such as salt or shale, because of the difference in material behavior, although in certain circumstances the correspondence principle could be used to make predictions about the deformation.

All the solutions discussed, except those of *Withjack* [1979], assume that a two-dimensional approach is appropriate for faulting problems. The solutions should not be applied where faults of diverse trend intersect or in regions of large, three-dimensional strain. However, they are suitable where parallel cross sections are unchanged over a considerable lateral distance.

Solutions of the type presented by Hafner, Couples, and Withjack are also valuable, because they address the faulting of a large volume of rock. Such solutions consider an entire fault field, often spanning several kilometers, rather than single faults within the earth's crust. Certainly, the presence of several faults of different types, in predicted positions relative to the boundaries of the area and each other, provides greater confidence in a solution than would be the case for a solution that predicts just a single feature. In fact, it is dangerous to apply these solutions to a single fault, because virtually all fault types are predicted to occur by these solutions; applying these solutions to a single fault virtually assures a fit even before the test is made.

It must also be remembered that the mathematical solutions discussed above apply only up to fault formation. That is, any of the potential shear fractures can eventually become a major fault with large throw, and this will determine the position of the structural mountains and basins. Therefore when the boundary conditions are stated in stress components, there is no necessary relationship between the apparent 'uplift' in the boundary statement and the subsequent position of mountains or basins.

The restrictions here placed upon the analytical solutions should not be considered as drawbacks. On the contrary, it was the precise statements that are necessary to the formulation of the mathematical solution that help emphasize the parameters important in understanding faulting. A well-stated problem and solution is valuable, not only in what it says positively, but also in establishing what restrictions are necessary in order to use it.

From the preceding discussion, we conclude that when the assumptions and conditions are geologically realistic, boundary value solutions are valuable, especially from the standpoint of hypothesis generation. It is evident, therefore, that the field geologist could use a more complete suite of such solutions than now exists. The engineer has available specific solutions to a multiplicity of structural and loading conditions. The structural geologist also needs to have a greater choice of solutions so that mathematical conditions can be more closely matched to specific geological conditions.

A Classification of Faulting

To the field and subsurface geologist, the most important contribution of these theoretical studies is that they have forced us to reconsider our classification of faults that shorten or elongate the earth's crust. If we restrict ourselves, as older classifications do, only to the terms 'normal' (crustal elongation; see Figure 6) and 'reverse' (crustal shortening; see Figure 6), we emphasize only the horizontal component of the displacement produced by the fault, totally ignoring the vertical component. Studies by *Hafner* [1951], *Sanford* [1959], *Couples* [1977], and *Withjack* [1979] emphasize the importance of tracking both the horizontal and the vertical components of the displacement field throughout the volume of faulted crust being studied.

Even for constant-dip faults, a finer grouping than just 'normal' and 'reverse' is required. We must also ask what effect the fault system has on the elevation (vertical component) of the crust and whether the vertical changes are greater or less than the horizontal changes. These questions require that the two general categories, normal and reverse, be further divied into four categories by considering, as category boundaries, faults that have equal components of horizontal and vertical movement (dashed lines, Figure 6). Even though all normal faults elongate the crust, some cause greater changes in its elevation then in its length (area II, Figure 6), while others cause only slight changes in elevation but large changes in horizontal lengths (area I, Figure 6).

These considerations are not absolutely new in statement, but they are emphasized, primarily because they are not usually considered in practice. For example, in the ex-

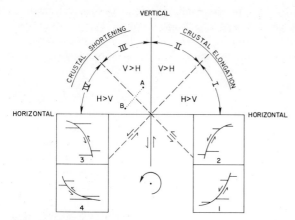

Fig. 6. Suggested cross-sectional classification of normal and reverse faults emphasizing the relationship between the vertical V and horizontal H component of the displacement. Only faults that produce a counterclockwise sense of shear are considered. Crustal elongation is produced by faults in zones I and II, and crustal shortening in zones III and IV. Insert drawings illustrate curved faults that flatten with depth (1 and 4) or steepen with depth (2 and 3). The path of the dashed line from A to B represents the continuous relationship of V and H along the fault depicted in insert 3 from depth to the surface, and along the fault in insert 4 from the surface to depth.

perience of the authors, regions of high-angle reverse faulting (area III, Figure 6) are frequently used indiscriminately to 'prove' points about overthrust faulting (area IV, Figure 6). On the other hand, reverse and normal faults are placed in separate categories, even though in some cases, normal and reverse faults have more in common with each other with respect to displacement (areas II and III, Figure 6) than they do with either other normal faults (area I, Figure 6) or with other reverse faults (area IV, Figure 6).

This fourfold breakdown of normal and reverse faults has another important advantage over one that is only concerned with horizontal movement. Both the theoretical and field studies mentioned here emphasize the importance of faults that are curved in cross section. When such a fault forms, the relationship between the vertical and horizontal components of the displacement field changes with depth. This demands compensation within the system in order to maintain constant volume. For example, consider the curved reverse fault that forms above the piston in a Sanford-type model (Figure 2; and Figure 6, part 3). At depth, most of the displacement is vertical (point A, Figure 6). Near the surface there is more horizontal displacement than vertical displacement on this fault (point B, Figure 6), and in between there is a changing relationship between the vertical and horizontal components of displacement (dotted line AB, Figure 6). Because there is more horizontal displacement at shallow levels than at depth, some compensating structure must form that produces an elongation that diminishes with depth. In the sandbox the normal faults that die out downward accom-

plish just this (Figure 5). This is why the normal faults do not form until after there has been some displacement on the curved reverse fault.

It is common knowledge that there are many other types of curved faults in the earth's crust, for example, growth faults (Figure 6, part 1) or even many overthrust faults (Figure 6, part 4). Each curved fault requires the formation of a secondary compensational system that will not be vertically continuous. The compensation does not have to be faulting, as is the case in the sandbox; it could be accomplished by changes in thickness, folding, faulting, or any combination of these, depending upon the make-up, geometry, and environment of the deforming body. What is important is that if a fault curves significantly in cross section, then compensation is required. Using the classification presented here (Figure 6), the needed compensation is obvious, and the type of compensation (shortening or elongation) and the approximate depth where it is needed can be estimated. The specific structural geometry that the compensation will assume can be somewhat anticipated by applying the concepts of rock behavior that have evolved from laboratory measurements. A cross section that contains a curved fault but no compensation automatically tells the viewer that something is wrong: either the fault is interpreted improperly, or other structures developed within the area of the cross section have been overlooked.

Conclusions

The understanding of faulting in the shallow crust has been advanced greatly through experimental and theoretical studies over the past 30 years. There may be some field geologists who refuse to use these results, because they do not exactly represent reality. In our opinion, they are rejecting very valuable tools. There are other geoscientists who are perhaps overly zealous in the direct application of simplified studies to a more complicated reality. However, it is the responsibility of field geologists to correct these inappropriate applications and, in some instances, to redesign a better experiment or better specify a theoretical problem, especially appropriate boundary conditions. Some laboratory studies, using rock models, generate many testable hypotheses concerning large-scale structures, hypotheses that would otherwise remain undiscovered. By considering both the horizontal and vertical components of movement along curved faults, field geologists may both construct appropriately balanced cross sections and recognize the presence of important, compensating structures.

One conclusion is inescapable: many geoscientists who neither wear field boots nor employ electric logs as everyday tools have, nonetheless, aided those geoscientists who do in making better fault interpretations.

References

Anderson, E. M., *The Dynamics of Faulting*, 183 pp., Oliver and Boyd, Edinburgh, 1942.

Bartlett, W. L., Experimental wrench faulting at confining pressure, M.S. thesis, Texas A&M Univ., College Station, Tex., 1980.

Bishop, J. F. W., A theoretical examination of the plastic deformation of crystals by glide, *Philos. Mag.*, *44*, 51-64, 1953.

Byerlee, J. D., Frictional characteristics of granite under high confining pressure, *J. Geophys. Res.*, *72*, 3639-3648, 1967.

Coulomb, C. S., Sur une application des regles maximis et minimis a quelques problemes de statique relatifs a l'architecture, *Acad. Sci. Paris Mem. Math. Phys.*, *7*, 343-382, 1776.

Couples, G., Stress and shear fracture (fault) patterns resulting from a suite of complicated boundary conditions with applications to the Wind River Mountains, *Pure Appl. Geophys.*, *115*, 113-134, 1977.

Couples, G., Comments on applications of boundary-value analyses of structures of the Rocky Mountains foreland, Laramide Folding Associated With Basement Block Faulting in the Western United States, *Geol. Soc. Am. Mem.*, *151*, 337-354, 1978.

Couples, G., and D. W. Stearns, Analytical solutions applied to structures of the Rocky Mountains foreland on local and regional scales, in Laramide Folding Associated With Basement Block Faulting in the Western United States, *Geol. Soc. Am. Mem.*, *151*, 313-336, 1978.

Donath, F. A., Analysis of Basin-Range structure, south-central Oregon, *Geol. Soc. Am. Bull.*, *72*, 985-990, 1962.

Drucker, D. C., and W. Prager, Soil mechanics and plastic analysis of limited design, *Q. Appl. Math.*, *10*, 157-165, 1952.

Evans, D., The Denver area earthquakes and the Rocky Mountain Arsenal well, *Mt. Geol.*, *3*, 23-36, 1966.

Forrestall, G. F., Stress distributions and overthrust faulting, *Geol. Soc. Am. Bull.*, *83*, 3073-3082, 1972.

Friedman, M., Structural analysis of fractures in cores from Saticoy field, Ventura, California, *Am. Assoc. Petrol. Geol. Bull.*, *53*, 367-398, 1969.

Friedman, M., and T. Shimamoto, Fracture patterns in simulated fault gouge (abstract), *EOS Trans. AGU, 69*, 1208, 1978.

Friedman, M., J. Handin, J. M. Logan, K. D. Min, and D. W. Stearns, Experimental folding of rocks under confining pressure: Pt. III, Faulted drape folds in multi-lithologic layered specimens, *Geol. Soc. Am. Bull.*, *87*, 1049-1066, 1976a.

Friedman, M., L. W. Teufel, and J. D. Morse, Strain and stress analysis from calcite lamellae in experimental buckles and faulted drape folds, *Phil. Trans. R. Soc. London Ser. A.*, *283*, 87-107, 1976b.

Gangi, A. F., K. D. Min, and J. M. Logan, Experimental folding of rocks under confirming pressure: Pt. IV., Theoretical analysis of faulted drape-folds, *Tectonophysics*, *42*, 227-260, 1977.

Hafner, W., Stress distributions and faulting, *Geol. Soc. Am. Bull.*, *62*, 373-398, 1951.

Handin, J., On the Coulomb-Mohr failure criterion, *J. Geophys. Res.*, *74*, 5343-5348, 1969.

Handin, J., and R. V. Hager, Jr., Experimental deformation of sedimentary rocks: Tests at room temperature on dry samples, *Am. Assoc. Petrol. Geol. Bull.*, *41*, 1-50, 1957.

Handin, J., and D. W. Stearns, Sliding friction of rocks (abstract), *EOS Trans. AGU*, 103, 1964.

Handin, J., R. V. Hager, Jr., M. Friedman, and J. N. Feather, Experimental deformation of sedimentary rocks under confirming pressure: Pore pressure tests, *Am. Assoc. Petrol. Geol. Bull.*, *47*, 717-755, 1963.

Hempel, C. G., *Philosophy of Natural Science*, 116 pp., Englewood Cliffs, N. J., 1966.

Howard, J. H., Structural development of the Williams Range Thrust, Colorado, *Geol. Soc. Am. Bull.*, *77*, 1247-1264, 1966.

Hsü, K. J., Role of cohesive strength in the mechanics of overthrust faulting and of landsliding, *Geol. Soc. Am. Bull.*, *80*, 927-952, 1969.

Hubbert, M. K., Theory of scale models as applied to the study of geologic structures, *Geol. Soc. Am. Bull.*, *48*, 1459-1520, 1937.

Hubbert, M. K., Mechanical basis of certain familiar geologic structures, *Geol. Soc. Am. Bull.*, *62*, 355-372, 1951.

Hubbert, M. K., and W. W. Rubey, Role of fluid pressure in mechanics of overthrust faulting: I. Mechanics of fluid-filled porous solids and its application to overthrust faulting, *Geol. Soc. Am. Bull.*, *70*, 115-166, 1959.

Jaeger, J. C., The frictional properties of joints in rocks, *Geofis. Pura Appl.*, *48*, 148-158, 1959.

Jaeger, J. C., and N. G. W. Cook, *Fundamentals of Rock Mechanics*, 515 pp., Chapman and Hall, London, 1969.

Jamison, W. R., Laramide deformation of the Wingate Sandstone, Colorado National Monument: A study of cataclastic flow, Ph. D. thesis, 171 pp., Texas A&M Univ., College Station, Tex., 1979.

Logan, J. M., Fracture in rock, *Rev. Geophys.*, *13*, 358-361, 1975.

Logan, J. M., and T. Shimamoto, The influence of calcite gouge on the frictional sliding of Tennessee sandstone (abstract), *EOS Trans. AGU*, *57*, 1011, 1976.

Logan, J. M., T. Iwasaki, M. Friedman, and S. A. Kling, Experimental investigation of sliding friction in multilithologic specimens, in Geological Factors in Rapid Excavation, *Geol. Soc. Am. Rev. Eng. Geol.*, *Case History 9*, 55-67, 1972.

Logan, J. M., M. Friedman, N. G. Higgs, C. Dengo, and T. Shimamoto, Experimental studies of simulated gouge and their application to studies of natural fault zones, in *Proceedings of the 8th Conference on Analysis of Actual Fault Zones in Bedrock*, pp. 305-343, National Earthquake Hazards Reduction Program, Menlo Park, California, 1979.

Magara, K., *Compaction and Fluid Migration: Practical Petroleum Geology*, 319 pp., Elsevier, New York, 1978.

Mandl, G., L. M. H. de Jong, and A. Maltha, Shear zones in

granular material: An experimental study of their structure and mechanical genesis, *Rock Mech.*, *9*, 95-144, 1977.

Matthews, V., III, and D. R. Work, Laramide folding associated with basement block faulting along the northeastern flank of the Front Range, Colorado, in Laramide Folding Associated With Basement Block Faulting in the Western United States, *Geol. Soc. Am. Mem.*, *151*, 101-124, 1978.

Min, K. D., Analytical and petrofabric studies of experimental faulted drape folds in layered rock specimens, Ph. D. thesis, 89 pp., Texas A&M Univ., College Station, Tex., 1975.

Morse, J. D., Deformation in ramp regions of overthrust faults: Experiments with small-scale rock models, *Wyo. Geol. Assoc. Guideb. Ann. Conf.*, *29*, 457-470, 1977.

Morse, J. D., Deformation in ramp regions of thrust faults: Experiments with rock models, M.S. thesis, 138 pp., Texas A&M Univ., College Station, Tex., 1978.

Oertel, G., The mechanism of faulting in clay experiments, *Tectonophysics*, *2*, 343-393, 1965.

Palmquist, J. C., Laramide structures and basement block faulting: Two examples from the Big Horn Mountains, Wyoming, in Laramide folding associated with basement block faulting in the Western United States, *Geol. Soc. Am. Mem.*, *151*, 125-138, 1978.

Paterson, M. S., Experimental deformation and faulting in Wombeyan marble, *Geol. Soc. Am. Bull.*, *69*, 465-475, 1958.

Pierce, W. G., Heart Mountain and South Fork thrusts, Park County, Wyoming, *Am. Assoc. Petrol. Geol. Bull.*, *45*, 591-626, 1957.

Prucha, J., J. A. Graham, and R. P. Nickelson, Basement controlled deformation in Wyoming province of Rocky Mountains foreland, *Am. Assoc. Petrol. Geol. Bull.*, *49*, 966-992, 1965.

Raleigh, C. B., J. O. Bredehoeft, and J. P. Bohn, Earthquake control at Rangely, Colorado (abstract), *Eos Trans. AGU*, *52*, 344, 1971.

Rankine, W. J. M., On the stability of loose earth, *Phil. Trans. R. Soc. London*, *147*, 9-27, 1857.

Reches, Z., Analysis of faulting in three-dimensional strain field, *Tectonophysics*, *47*, 109-129, 1978.

Sanford, A. R., Analytical and experimental study of simple geologic structures, *Geol. Soc. Am. Bull.*, *70*, 19-52, 1959.

Serra, S., and J. D. Morse, Deformation in ramp regions of overthrust faults: Field and experimental studies (abstract), *Mem. Can. Soc. Petrol. Geol.*, *83-84*, 1979.

Stearns, D. W., Certain aspects of fracture in naturally deformed rocks, in NSF Advanced Science Seminar in Rock Mechanics, Spec. Rep., Air Force Cambridge Res. Lab., Bedford, Mass., 1967.

Stearns, D. W., Laramide basement deformation in the Bighorn Basin-The controlling factor for structure in the layered rocks, *Wyo. Geol. Assoc. Guideb. Ann. Conf.*, *27*, 82-106, 1975.

Stearns, D. W., Faulting and folding in the Rocky Mountains foreland, in Laramide Folding Associated With Basement Block Faulting in the Western United States, *Geol. Soc. Am. Mem.*, *151*, 1-38, 1978.

Stearns, D. W., and D. M. Weinberg, A comparison of experimentally and naturally formed drape folds, *Wyo. Geol. Soc. Guideb. Annu. Conf.*, *27*, 159-166, 1975.

Stearns, D. W., G. Couples, and M. T. Stearns, Deformation of non-layered materials that affect structures in layered rocks, *Wyo. Geol. Assoc. Guideb. Ann. Conf.*, *30*, 213-225, 1978.

Tchalenko, J. S., Similarities between shear zones of different magnitudes, *Geol. Soc. Am. Bull.*, *81*, 41-60, 1970.

Tchalenko, J. S., and N. W. Ambrasseys, Structure analysis of the Dasht-e-Bayez (Iran) earthquake fractures, *Geol. Soc. Am. Bull.*, *81*, 41-60, 1970.

Vaughn, P. H., Mesozoic sedimentary rock features resulting from volume movements required in drape folds at corners of basement blocks Casper Mountain area, Wyoming, M.S. thesis, 93 pp., Texas A&M Univ., College Station, Tex., 1976.

Wilcox, R. E., T. P. Harding, and D. R. Seely, Basic wrench tectonics, *Am. Assoc. Petrol. Geol. Bull.*, *57*, 74-96, 1973.

Wilson, R. C., The mechanical properties of the shear zone of the Lewis overthrust, Glacier National Park, Montana, Ph. D. thesis, 81 pp., Texas A&M Univ., College Station, Tex., 1970.

Withjack, M., An analytical model of continental rift patterns, *Tectonophysics*, *59*, 59-81, 1979.

Probabilistic Treatment of Faulting in Geologic Media

FRED A. DONATH[1]

University of Illinois, Urbana, Illinois 61801

ROBERT M. CRANWELL

Sandia Laboratories, Albuquerque, New Mexico 87185

A method is presented for assessing the probability of faulting within geologic media. For specified fracture or frictional sliding criteria and rock properties, the probability of a new fault developing or of movement recurring on an existing fault can be calculated as conditional on the state of stress. For an assumed stress variation with time, the probability of the stresses reaching critical values by any time t can be determined. Coupling of the conditional probability of faulting with the probability of achieving a given state of stress by some specified time t allows calculation of the probability of faulting by time t.

Introduction

A major concern in the siting of a geologic repository for high-level nuclear wastes is the possibility that either an undetected fault might be present or that a fault might develop within the repository system during the period of time that the buried wastes are hazardous. Depending on the location and characteristics of the fault, such a feature could divert the movement of groundwater along less desirable pathways in the subsurface, thus increasing the potential for rapid return to the biosphere of radionuclides that might be dissolved in the groundwater. Three important questions that need to be addressed, therefore, are (1) what is the probability that an undetected fault might be present; (2) given an existing state of stress and the rock properties, what is the probability that a fault might develop or that movement along an existing fault might occur; and (3) given that a fault is present, what are the consequences of its presence in the repository system?

In conventional risk analysis, risk is defined as: risk = (probability) × (consequences). By this definition, comparative risks associated with different potentially disruptive events or features can be assessed once the probability of occurrence of each feature or event has been ascertained and the consequences of its occurrence have been evaluated. A large number of variables must be considered in evaluating the performance of the undisrupted or normal repository system; in addition, the effects of a large number of possible disruptive events might have to be evaluated for a given system. Hence, the risk assessment of a geologic repository system is a complex undertaking. The repository system is defined here as (1) the depository with its engineered features and radioactive waste, (2) the geometric arrangement of geologic units and features surrounding the depository, (3) the spatial distributions of hydraulic, thermal, mechanical, and geochemical properties of these units and features, and (4) the boundary conditions and all processes and events that can influence transport of radionuclides within the defined region. More succinctly, the repository system is a specified region in the earth's crust that contains the depository with its buried waste and the hydrogeologic and geochemical environment that exists in that region.

A repository system in bedded salt differs in various ways from one in granite, shale, dome salt, or basalt. As a result, the probabilities and consequences (hence, risks) associated with different events or disruptive features (e.g., a fault) will also differ among the repository systems in different host media. A general discussion of geologic and other factors that can affect the isolation of high-level nuclear waste in geologic repositories is presented in a report prepared for the American Physical Society [*APS*, 1978]. Further elaboration can be found in a subgroup report for the Interagency Review Group [*IRG*, 1978; *Donath et al.*, 1979] and *Donath* [1979, 1980]. An introduction to the risk methodology for geologic disposal of radioactive waste, being developed for the U.S. Nuclear Regulatory Commission, is presented in *Campbell et al.* [1978].

If the consequences of an event or feature are not significant, determination of the probability of occurrence of the feature or event becomes an academic rather than a relevant exercise. Thus, consequences are normally evaluated before the probabilities are considered. For the case of a fault zone in a geologic repository system, the con-

---

[1] Now at CGS, Inc., Urbana, Illinois 61801.

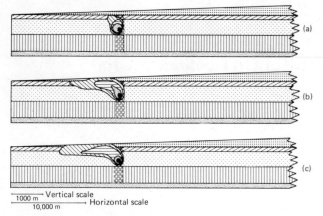

1000 m ⌐ Vertical scale
10,000 m ⌐ Horizontal scale

Fig. 1. Cross section of hypothetical sequence of layered rocks with typical hydraulic characteristics in which a vertical disruptive zone (dashed lines) of high hydraulic conductivity is present in the discharge portion of the basin. Distribution pattern of contaminant particles transported from their source (small dot) by a steady state groundwater system is shown after (a) 5,000 years, (b) 10,000 years, and (c) 15,000 years. The intermediate (horizontally lined) area includes between 2 and 10% of the particles within the system at the times represented. For the representation shown, 9% of the particles in the system are released to the surface after 10,000 years, 25% reach the surface after 15,000 years. The reader is referred to the APS [1978] report, from which this figure is taken, for further discussion.

sequences could be very significant. For example, Figure 1 illustrates the influence of a vertical zone of high hydraulic conductivity on groundwater transport within a layered rock sequence having typical hydraulic properties and boundary conditions. This example could represent a fault that passes through a high-level waste depository. The contours shown in each illustration indicate the distributions at three different times of contaminants supplied continuously to the groundwater system at the position of the dot. After 10,000 years of steady state transport in this system, 9% of the particles supplied to the system

have reached the surface; the number reaching the surface after 15,000 years exceeds 25%.

Any release from a given repository system that includes a fault is, however, highly dependent upon the location and characteristics of the fault zone. Figures 2 and 3, taken from a CGS [1979] interim report to the U.S. Nuclear Regulatory Commission, show that the effects of a fault zone on the repository system could be either very localized or far reaching, depending upon the location and characteristics of the zone. The contours in these figures give the logarithm of the ratio of (1) time for a contaminant particle to leave the host medium in the repository system with a disruptive feature to (2) time for a particle to leave the host medium in the normal or undisrupted system. The influence in this repository system of a vertical disruptive zone of high hydraulic conductivity is restricted to the immediate area of the zone (Figure 2), whereas the influence of a zone of low hydraulic conductivity is distributed over a large part of the system (Figure 3).

Preliminary analysis indicates that fault zones within groundwater systems can be of great consequence. Thus, the probability that an undetected fault might be present or that faulting might occur is a matter of concern in the evaluation of sites for nuclear waste isolation. The probability that an undetected fault might be present in a repository system has been addressed by Cranwell and Donath [1980]. This paper discusses the probability that a fault might develop or that movement along an existing fault might occur.

Failure Criteria for Faulting

The conditions for formation of a new fault and for renewed movement on an existing fault can both be represented by a linear failure criterion of the form

$$\tau = \tau_0 + \sigma\psi \qquad (1)$$

where τ and σ are the shear and normal stresses on the potential fault plane or on an existing fault plane and ψ is the tangent of the angle of internal friction for intact rock

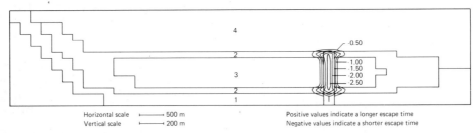

Horizontal scale ├───┤ 500 m   Positive values indicate a longer escape time
Vertical scale ├───┤ 200 m   Negative values indicate a shorter escape time

Fig. 2. Cross section of hypothetical sequence of layered rocks in which a vertical disruptive zone of high hydraulic conductivity is present in the discharge portion of the basin. Contours give logarithms of ratio of exit time for particles in the disrupted system to move out of the host medium relative to exit time in the undisrupted system. Most of the repository system remains unaffected by the presence of the high permeability zone, but particles located within the zone could exit the host medium (unit 3) more than 300 times faster than similarly located particles in the undisrupted system.

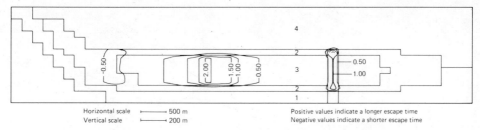

Horizontal scale ├──────┤ 500 m
Vertical scale ├──────┤ 200 m

Positive values indicate a longer escape time
Negative values indicate a shorter escape time

Fig. 3.   Hypothetical layered sequence identical to that shown in Figure 2 except that a vertical disruptive zone of low hydraulic conductivity is present in the discharge portion of the basin. The presence of the low permeability zone affects transport throughout much of the repository system. Particles located within the zone would take approximately 10 times longer to exit unit 3 than similarly located particles in the undisrupted system. Particles located elsewhere in the system could require 100 times as long to exit or could exit 3 times faster, depending upon their location.

or the coefficient of sliding friction for displacement on an existing fault. The term τ_0 represents the cohesive strength for intact rock or the shearing resistance along an existing fault. Different symbols will be used later to distinguish between the parameters for intact and faulted material, respectively.

The formation of a new fault or the displacement along an existing fault can be predicted to occur when the existing state of stress in the rocks satisfies the stated criterion, that is, when

$$\tau - \tau_0 \geq \sigma\psi \qquad (2)$$

where

$$\tau = (S_1 - S_3)\sin\theta\cos\theta \qquad (3)$$

and

$$\sigma = S_1\sin^2\theta + S_3\cos^2\theta \qquad (4)$$

S_1 and S_3 are the maximum and minimum principal stresses, respectively, and θ is the angle between the fault plane and the direction of maximum compression, S_1 (Figure 4).

For a given state of stress in the earth's crust, the criterion for development of a new fault might not be satisfied, but the criterion for sliding could be met for certain orientations of the sliding surface. This is illustrated diagrammatically in Figure 5. The (fracture) criterion for the formation of a new fault is represented on the $\tau - \sigma$ diagram by the line $\tau = C + \sigma\tan\phi$; the criterion for sliding on an existing fault is represented by the line $\tau = R + \sigma\mu$. The existing state of stress defines the size and position of the Mohr circle. In this instance the circle lies below the fracture criterion and no potential plane of fracture exists for these stress levels. In contrast, any fault already present that is inclined more than θ_a or less than θ_b to the direction of maximum compression would satisfy the sliding criterion and would therefore be subject to recurrent movement. The specified range is strictly valid only for planes parallel to the existing intermediate principal stress. Planes that are not parallel to S_2 can also sat-

isfy the sliding criterion; which of these might be activated depends upon the magnitude of S_2.

Given a state of stress and the fracture criterion (or criteria) for the rocks in a repository system, a prediction regarding the formation of a new fault and its orientation is straightforward and rather trivial. However, of considerable interest for this particular situation are the probabilities associated with achieving a given state of stress that would lead to fracture. This aspect will be discussed below. For the situation in which faults already exist in the system, the problem is to identify the range of orientations that satisfy the sliding criterion (or criteria) for a given state of stress and to determine the probability that an existing fault surface falls within this range. The probabilities associated with achieving the given state of stress are naturally of considerable interest in this instance as well.

Fault Orientations Favorable for Sliding

The analysis we present can be adapted to any absolute orientation of the principal stresses and to existing faults of any orientation; however, because our primary objective is demonstration of the methodology, we shall assume two special conditions that simplify the analysis. First, we

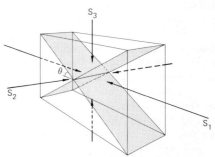

Fig. 4.   Relationship between existing low-angle conjugate faults and existing principal stresses for the situation analyzed in this paper.

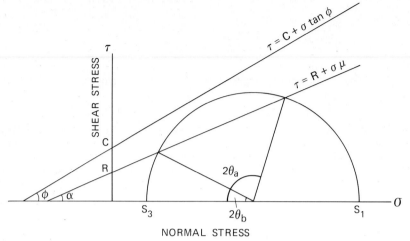

Fig. 5. Criteria for formation of a new fault, $\tau = C + \sigma \tan \phi$, and for sliding along an existing fault, $\tau = R + \sigma\mu$. For intact rocks, C represents the cohesive strength and ϕ the angle of internal friction. For an existing fault surface, R represents the shearing resistance and μ the coefficient of sliding friction. S_1 and S_3 represent the existing maximum and minimum principal stresses, respectively, and points on the circle represent the values of shear and normal stress on planes parallel to S_2 that are inclined to S_1 at an angle θ (see Figure 4).

will evaluate only the probabilities for movement on existing faults that strike parallel to the intermediate principal stress, S_2. Second, we will restrict our discussion to that state of stress in which the minimum principal stress is vertical.

The first condition would actually obtain only if the faults had formed at an earlier time in the existing stress system when the magnitudes were much higher. That a later, causally unrelated stress system might develop such a relationship with older existing faults would, in fact, be highly fortuitous. Nevertheless, faults with strikes not parallel to S_2 would be even less likely to undergo sliding for a given state of stress, and this assumed condition therefore provides a conservative estimate of the potential for sliding in the existing environment.

Our reasons for choosing the second condition for illustration are that (1) this state of stress favors the enhancement of thrust faults, which are typically characterized by crush zones of low permeability which, in turn, appear to have the greatest influence on groundwater systems, and (2) in situ stress measurements indicate this state of stress exists for about 70% of the measurements made at depths of 600 m or less [McGarr and Gay, 1978, Table 1]. As supported by the data summarized by McGarr and Gay [1978], we will assume that the vertical (minimum) principal stress is equal to the weight of the overburden. For a repository depth of 600 m and for an assumed average rock density of 2.7 g/cm³, the vertical stress at this depth would therefore be 16 MPa (2300 psi). Thus the relationship to be analyzed for purposes of illustration is that depicted in Figure 4, in which each of two possible conjugate low-angle faults is inclined at an angle θ to the direction of maximum (horizontal) principal stress, S_{1h}.

As stated above, and illustrated diagrammatically by Figure 5, for movement to occur along an existing fault, the shear stress along it must satisfy the criterion for sliding along that particular fault. This is dependent both on the orientation of the fault (θ) and on its frictional properties (R, μ). The most favored orientation for movement (θ_m) on an existing fault is presumed to be a value of θ identical to that which held when the fault initially formed (θ_f).

For all common rocks θ_f is typically about 30°, but the value will vary depending upon rock type and conditions existing at the time of fault formation. Both experimental and theoretical evidence indicate that the angle θ_f cannot exceed 45° in isotropic media, and it will therefore be assumed in our discussion that the maximum variation from the mean value is ±15°. Unknown, but of considerable importance, is the nature of the distribution about the mean which faults might have as a consequence of material properties and environmental conditions. If the distribution is uniform, then the probability of $\theta_m = \theta_f = 45°$ is equal to the probability of $\theta_m = \theta_f = 30°$. If the distribution is normal or log normal, the probabilities are more heavily weighted toward the value of the mean (i.e., 30°).

Byerlee [1978], in summarizing the frictional properties of rocks, concluded that at normal stresses up to 200 MPa the shear stress required to cause sliding is given approximately by the equation $\tau = 0.85\sigma$, and that at normal stresses above 200 MPa the friction is given approximately by $\tau = 0.5 + 0.6\sigma$. The large amount of variation in reported values and the stated lack of dependence on rock type appears to be, in part, a consequence of the definitions of friction parameters that were used. Jaeger [1959] showed that contact surfaces in sliding can display an inherent strength that is constant over a given pressure range. This was systematically demonstrated for several rock types at different temperatures and displacement rates by *Donath et al.* [1972], and the reader is referred to that paper for further discussion of this point.

The experimental results of *Donath et al.* [1973] suggest that the coefficient of sliding friction μ is a constant at higher confining pressures (20 MPa and above) but has characteristic values for different rock types that range from 0.47 for slate to 0.73 for limestone, with the mean of 0.60 being characteristic of sandstone. Similarly, these investigators found that the shearing resistance R was characteristic of rock type, at the pressures, temperatures, and displacement rates evaluated, and ranged from 15 MPa for the 30° orientation of slate to 24 MPa for a lithographic limestone, with the mean of 19.5 being nearly that (19 MPa) determined for sandstone. It might be noted, in passing, that these values for the frictional parameters are not significantly different from the values of the internal friction and cohesive strength for the intact rocks. This suggests that at these higher confining pressures, sliding on the faults can only be effected by shearing through asperities of the intact rock along the fault zone.

The magnitude of the principal stress just sufficient to induce sliding is

$$S_1{}^* = S_3(1 + \sin \alpha)/(1 - \sin \alpha) + 2R \cos \alpha/(1 - \sin \alpha) \quad (5)$$

where S_3 is the minimum principal stress, R is the frictional resistance in the absence of normal stress, and the angle α is derived from the relationship $\tan \alpha = \mu$. The relationship given does not take into account the effect of pore pressure. Because any fluid pressure in existing pore space reduces all normal stresses equally, (5) can represent the relationship where pore pressure exists if $S_1{}^*$ is taken to be $(S_1 - p)^*$ and S_3 is taken to be $(S_3 - p)$, that is, the principal stresses are replaced by the effective stresses.

When $S_1 = S_1{}^*$, the Mohr circle representing the state of stress is just tangent to the sliding criterion (cf. Figure 5) and the stress is favorable for slip only on a single plane that makes an angle $\theta^* = 45° - \alpha/2$ with S_1. Thus in a hypothetically ideal situation, given the values of the frictional parameters (μ and R) and the depth (S_3), one can calculate the lowest value of the maximum principal stress ($S_1{}^*$) that would induce sliding on a fault of this preferred orientation. Inasmuch as μ appears to be a constant for a given rock type [*Donath et al.*, 1973] whereas R might be expected to vary depending upon the nature of the fault zone and environmental conditions, the value of $S_{1h}{}^*$ is shown in Figure 6 as a function of the frictional resistance R for different coefficients of sliding friction. For reference, maximum measured in situ values of $S_{1h}{}^*$ in Africa and Canada for depths up to 600 m, taken from *McGarr and Gay* [1978], are also shown.

As stated, for $S_1 = S_1{}^*$ only a single orientation of a fault would be favorable for sliding. As S_1 exceeds $S_1{}^*$, other orientations become favorable, namely, all orienta-

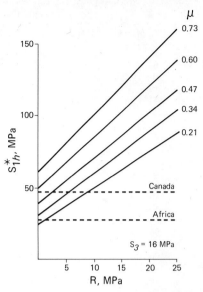

Fig. 6.   Magnitude of horizontal stress required to induce sliding ($S_{1h}{}^*$) at a depth of 600 m ($S_3 = 16$ MPa) is shown as a function of shearing resistance (R) on the fault plane for different coefficients of sliding friction (μ). Maximum measured in situ values of S_{1h} at depths of 600 m or less in Africa and Canada are indicated.

tions that lie between the angles θ_a and θ_b to the direction of S_1 (cf. Figure 5). Table 1 lists calculated values of the limiting angles θ_a and θ_b with increasing S_{1h} for the parameter values selected for this analysis: $S_{3v} = 16$ MPa; $\mu = 0.47$, 0.60, and 0.73; and $R = 10.0$, 15.0, 19.5, and 24.0 MPa. The range of orientations susceptible to sliding clearly enlarges rapidly with increasing S_1 (Table 1). For mean values of $\mu = 0.60$ and $R = 19.5$ MPa, the critical value of S_{1h} to induce sliding would be 119 MPa, and sliding would occur only on a fault inclined to S_1 at an angle of about 30°. With an increase of S_{1h} to 125 MPa, any fault inclined 30° ± 7 to S_1 would undergo slip.

Probability of Sliding on Existing Faults

Recall that for purposes of illustration, we assume in this analysis that some previously formed fault strikes parallel to the existing intermediate principal stress and that the maximum principal stress is horizontal. As stated previously, the first assumption yields a conservative result. Therefore, for a given magnitude of S_{1h}, sliding cannot occur unless θ lies within the range $\theta_b \leq \theta \leq \theta_a$. Thus determination of the probability that sliding will occur is the same as the probability of an existing fault lying within the range (i.e., θ_a to θ_b) that would permit sliding at a given magnitude of S_1.

In actual field relationships the orientation of the fault plane itself might vary about some mean value of θ_f because of rock or environmental factors, as stated above, or the orientation of S_{1h} in some later stress field might vary about the horizontal. For this analysis we shall assume

TABLE 1. Limiting Angles for Sliding (θ_a and θ_b) and Probabilities of Sliding for Uniform (P_U), Normal (P_N), and Log Normal ($P_{\ln}$), Distributions of Θ for Different Coefficients of Friction (μ) and Shearing Resistance (R), as a Function of S_1

S_1, MPa	S_3, MPa	μ	R, MPa	θ_a	θ_b	$\Delta\theta$	P_U	P_N	$P_{\ln}$
71.2*	16	0.47	10.0	32.4	32.4	0.0	0.03	0.08	0.06
75				40.3	24.6	15.7	0.52	0.89	0.82
80				43.9	20.9	23.0	0.76	0.98	0.97
85				46.3	18.5	27.8	0.92	0.996	0.995
90				48.1	16.7	31.4	1.00		
95				49.5	15.3	34.2			
100				50.7	14.1	36.6			
105				51.7	13.1	38.6			
110				52.6	12.3	40.3			
115				53.3	11.5	41.8			
120				54.0	10.9	43.1			
125				54.5	10.3	44.3			
130				55.1	9.8	45.3			
86.9*	16	0.47	15.0	32.4	32.4	0.0	0.03	0.08	0.06
90				38.7	26.2	12.5	0.42	0.80	0.69
95				42.3	22.6	19.7	0.66	0.96	0.93
100				44.6	20.2	24.4	0.81	0.99	0.98
105				46.4	18.4	28.0	0.93	0.996	0.996
110				47.8	17.0	30.8	1.00		
115				49.0	15.8	33.2			
120				50.0	14.8	35.2			
125				50.9	14.0	36.9			
130				51.7	13.2	38.5			
101.1*	16	0.47	19.5	32.4	32.4	0.0	0.03	0.08	0.06
105				38.9	26.0	12.9	0.43	0.82	0.71
110				41.9	22.9	19.0	0.63	0.95	0.91
115				44.0	20.8	23.2	0.77	0.98	0.97
120				45.6	19.2	26.4	0.88	0.99	0.99
125				47.0	17.9	29.1	0.97	0.997	0.998
130				48.1	16.8	31.3	1.00		
115.3*	16	0.47	24.0	32.4	32.4	0.0	0.03	0.08	0.06
120				39.0	25.9	13.1	0.44	0.83	0.72
125				41.6	23.2	18.4	0.61	0.94	0.90
130				43.5	21.3	22.2	0.74	0.98	0.97
85.2*	16	0.06	10.0	29.5	29.5	0.0	0.03	0.08	0.05
90				36.7	22.3	14.4	0.48	0.86	0.74
95				39.5	19.5	20.0	0.66	0.96	0.92
100				41.4	17.6	23.8	0.79	0.99	0.98
105				42.9	16.1	26.9	0.90	0.994	0.99
110				44.2	14.9	29.3	0.98	0.998	0.998
115				45.2	13.8	31.4	1.00		
120				46.1	13.0	33.1			
125				46.8	12.2	34.6			
130				47.5	11.5	36.0			
102.9*	16	0.60	15.0	29.5	29.5	0.0	0.03	0.08	0.05
105				33.9	25.2	8.7	0.29	0.63	0.47
110				37.3	21.7	15.6	0.52	0.89	0.79
115				39.4	19.6	19.8	0.66	0.96	0.92
120				41.0	18.0	23.1	0.77	0.98	0.97
125				42.3	16.7	25.6	0.85	0.99	0.99
130				43.4	15.6	27.8	0.93	0.996	0.995
118.8*	16	0.60	19.5	29.5	29.5	0.0	0.03	0.08	0.05
120				32.6	26.5	6.1	0.20	0.47	0.33
125				36.3	22.8	13.5	0.45	0.84	0.70
130				38.4	20.6	17.8	0.59	0.93	0.87
101.3*	16	0.73	10.0	26.9	26.9	0.0	0.03	0.08	0.05
105				32.2	21.7	10.6	0.35	0.72	0.53
110				34.8	19.0	15.8	0.53	0.90	0.77
115				36.6	17.3	19.4	0.65	0.95	0.89
120				38.0	15.9	22.1	0.74	0.98	0.95
125				39.1	14.8	24.4	0.81	0.99	0.98
130				40.1	13.8	26.3	0.88	0.99	0.99
121.0*	16	0.73	15.0	26.9	26.9	0.0	0.03	0.08	0.05
125				31.9	22.0	9.9	0.33	0.69	0.50
130				34.2	19.6	14.6	0.49	0.87	0.72

*Critical value of maximum principal stress, S_1^*.

that S_{1h} is horizontal and that all variation is represented in $\theta_f = \theta_m$. We will consider the probabilities for uniform, normal, and log normal distributions on the range of θ. If the endpoints of this range are arbitrarily taken as the lower 0.1% and upper 99.9%, then the corresponding parameters needed to define the normal and log normal distributions on this range can be determined.

Uniform Distribution

For $\mu = 0.60$ and $\Theta = 29.5°$, suppose that Θ is uniformly distributed on the interval (14.5, 44.5). Then for any θ in (14.5, 44.5), the distribution function of Θ is defined by

$$F_\Theta(\theta) = P(\Theta \le \theta) = \frac{\theta - 14.5}{44.5 - 14.5} = \frac{\theta - 14.5}{30}$$

The probability that movement will occur on an existing fault surface is given by

$$P(\text{movement} \mid S_1) = P(\theta_b \le \Theta \le \theta_a)$$

$$= P(\Theta \le \theta_a) - P(\Theta \le \theta_b)$$

$$= \frac{\theta_a - 14.5}{30} - \frac{\theta_b - 14.5}{30}$$

$$= \frac{\theta_a - \theta_b}{30} = \frac{\Delta\theta}{30} \tag{6}$$

Where θ_a and θ_b are determined by the value of S_1 for given values of μ and R. Thus if we want the probability of movement to be less than or equal to 0.5, then by (6),

$$\Delta\theta \le 15° \tag{7}$$

Identical arguments apply for other values of μ.

Normal Distribution

Assume $\mu = 0.60$, $\Theta = 29.5°$, and 14.5 and 44.5 are the lower 0.1% and upper 99.9%, respectively, for a normal distribution of Θ about $\bar{\Theta}$. Then,

$$0.001 = \frac{1}{s\sqrt{2\pi}} \int_{-\infty}^{14.5} \exp - [(x - 29.5)^2/2s^2]\, dx$$

$$0.999 = \frac{1}{s\sqrt{2\pi}} \int_{-\infty}^{44.5} \exp - [(x - 29.5)^2/2s^2]\, dx$$

Using the change of variable, $z = (x - 29.5)/s$, to put the above in standard normal form, we have

$$\Phi(z_1) = \frac{1}{\sqrt{2\pi}} \int_{-\infty}^{z_1} \exp - (z^2/2)\, dz = 0.001$$

$$\Phi(z_2) = \frac{1}{\sqrt{2\pi}} \int_{-\infty}^{z_2} \exp - (z^2/2)\, dz = 0.999$$

where $z_1 = (14.5 - 29.5)/s$ and $z_2 = (44.5 - 29.5)/s$. From

any standard normal distribution table, we see that $z_1 = -3.1$ and $z_2 = 3.1$. By using either of these, the standard deviation s can be determined to be 4.84. Thus Θ is normally distributed with mean 29.5 and standard deviation 4.84.

The probability that movement will occur on an existing fault is given by

$$P(\text{movement} \mid S_1) = P(\theta_b \le \Theta \le \theta_a)$$

$$= \frac{1}{\sqrt{2\pi}} \int_{z_b}^{z_a} \exp - (z^2/2)\, dz \tag{8}$$

where $z_a = (\theta_a - 29.5)/4.84$ and $z_b = (\theta_b - 29.5)/4.84$. Using a standard normal distribution table, we find that if we want this probability to be, say, 0.5, then,

$$z_a = 0.674$$
$$z_b = -0.674 \tag{9}$$

Thus

$$\theta_a - \theta_b = 6.52 \tag{10}$$

If we want the probability of movement to be less than or equal to 0.5, then from (10) we see that

$$\Delta\theta \le 6.52° \tag{11}$$

Identical arguments apply for other values of μ.

Log Normal Distribution

Assume $\mu = 0.60$, $\bar{\Theta} = 29.5$, and 14.5 and 44.5 are the lower 0.1% and upper 99.9%, respectively, but that Θ has a log normal (base e) distribution about $\bar{\Theta}$. By definition, ln Θ has a normal distribution with ln $14.5 = 2.67$ and ln $44.5 = 3.80$ as the lower 0.1% and upper 99.9%, respectively. Using arguments like those for the normal distribution, we find that ln Θ is normally distributed with mean $(2.67 + 3.80)/2 = 3.23$ and standard deviation $(3.80 - 3.23)/3.1 = 0.18$.

Again, if we want the probability of movement to be, say, 0.5, then

$$0.5 = P(\text{movement} \mid S_1)$$

$$= P(\theta_b \le \Theta \le \theta_a)$$

$$= \frac{1}{0.18\sqrt{2\pi}} \int_{\theta_b}^{\theta_a} 1/x \exp - [(\ln x - 3.23)^2/2(0.18)^2]\, dx$$

$$= \frac{1}{\sqrt{2\pi}} \int_{z_d}^{z_c} \exp - (z^2/2)\, dz$$

where $z_c = (\ln \theta_a - 3.23)/0.18$ and $z_d = (\ln \theta_b - 3.23)/0.18$. Using equations similar to (9), we have

$$z_c = 0.674$$
$$z_d = -0.674$$

Thus

$$\ln \theta_a - \ln \theta_b = 0.243$$

$$\theta_a/\theta_b = 1.28$$

Also,

$$(\theta_a + \theta_b)/2 = 29.5$$

Therefore,

$$\theta_b = 25.88$$

$$\theta_a = 33.13$$

Thus

$$\theta_a - \theta_b = 7.25 \tag{12}$$

If we want the probability of movement to be less than or equal to 0.5, then from (12), we see that

$$\Delta\theta \leq 7.25° \tag{13}$$

Variation of μ

Because logarithms to any base can be related to natural logarithms by means of a constant, the above results are not affected by the choice of the base. However, if other values of μ are assumed, the $\Delta\theta$ for 0.5 probability changes for log normal distributions. For example, for $\mu = 0.47$, $\Delta\theta \leq 7.03°$; $\mu = 0.60$, $\Delta\theta \leq 7.25°$; $\mu = 0.73$, $\Delta\theta \leq 7.32°$. Thus, as μ varies from 0.47 to 0.73, the 0.5 probability cutoff in the log normal case varies from 7.03 to 7.32. In the uniform and normal cases, the 0.5 probability cutoff remains the same. The variation for the log normal case reflects the fact that the density function for a log normal random variable is equal to zero for $x \leq 0$.

The distribution functions for uniform and normal distributions are represented in Figure 7, which shows graphically why the probability is much higher for a given $\Delta\theta$ with a normal distribution than with a uniform distribution. For a 30° range of Θ, the probability of a normally distributed random Θ falling within a $\Delta\theta = 16°$ (i.e., $\Theta \pm 8°$) is 90%, whereas it is only 53% for the uniformly distributed random Θ. If the distribution of possible orienta-

tions of a fault about the mean is normal, the probability of movement on a fault for $\Delta\theta = 7.5°$ (i.e., $\Theta \pm 3.75°$) is greater than 50%, whereas it is only 25% for a uniform distribution.

Returning to a consideration of the magnitudes of S_{1h} required to induce sliding on faults with different values of the frictional parameters, we can now correlate these magnitudes with the probabilities derived for different assumed distributions. Figure 8 indicates the $\Delta\theta$ that would permit sliding, as a function of S_{1h}, for different values of μ and R. For a uniform distribution, there is a 95% probability of recurrent movement on any low-angle thrust fault in the repository system when the $\Delta\theta$ dictated by S_{1h} reaches or exceeds 28.5°. This condition cannot exist for S_1 below 130 MPa if $\mu = 0.73$ unless R is less than 10 MPa. If $\mu = 0.60$, R must be less than 15 MPa; for $R = 10$ MPa, S_1 would need to be 108 MPa. For $\mu = 0.47$, the probability of recurrent movement would be 95% at $S_1 \geq 124$ MPa for $R = 19.5$ MPa or 95% at $S_1 \geq 106$ MPa for $R = 15$ MPa. For a uniform distribution there is a 50% probability of sliding when the $\Delta\theta$ reaches or exceeds 15° and a 25% probability when the $\Delta\theta$ reaches or exceeds 7.5°. For comparison, the 0.50 and 0.95 probability levels are indicated on Figure 8 for a normally distributed random Θ.

In the example cited earlier of $\mu = 0.60$ and $R = 19.5$, we noted that with an increase of S_{1h} to 125 MPa, a fault inclined 30° ± 7 to S_1 would undergo slip. From (6) and (8) (see also Table 1) we can determine that the probability of a fault lying within ±7° of the mean value of Θ is 45% for a uniform distribution but 84% for a normal distribution. Thus the probability of sliding on the fault for $S_{1h} = 125$ MPa would be 45 or 84%, respectively, depending upon the assumed distribution.

Probability of S_1 Reaching or Exceeding S_1*

It is reasonable to assume that for a given repository system the state of stress in that system will change with time. Thus for any assumed distribution (with time) of stresses for the system, the statistical parameters associated with that distribution would also vary with time. Moreover, at any given time, variance might be associated with the inability to measure accurately or with natural variations in the media. For purposes of illustration we

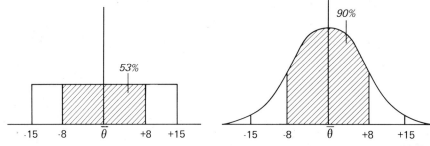

Fig. 7.   Distribution functions for uniform (left) and normal (right) distributions of Θ about $\bar{\Theta}$ for a 30° range. The probability of Θ falling within ±8° of Θ is 53% for a uniform distribution and 90% for a normal distribution.

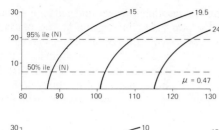

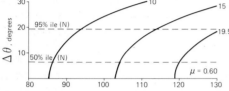

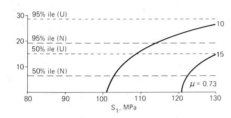

Fig. 8. $\Delta\theta$ that permits sliding, as a function of S_{1h}, for different values of μ and R, and $S_3 = 16$ MPa. Probability of sliding indicated for uniform (U) and normal (N) distributions.

will assume that at any time t the maximum principal stress S_1 is normally distributed with mean m and standard deviation s. We will assume, further, that the parameters m and s have varied linearly with time, such that

$$m(t) = at + b$$

$$s(t) = ct + d$$

Thus the probability of S_1 reaching or exceeding, at some time t', the magnitude S_1^* that would lead to shear fracture or sliding is given by

$$P(S_1 \geq S_1^*) = \frac{1}{s(t')\sqrt{2\pi}} \int_{S_1^*}^{\infty} \exp - [(x - m(t'))^2/2s(t')^2]\, dx$$

As an example, assume that at time zero (e.g., closure of the depository) in situ measurements indicate that the maximum principal stress, S_1, is normally distributed with mean $m = 20$ MPa and standard deviation $s = 5$ MPa. Assume further that the parameters m and s vary linearly with time (in years), $0 \leq t \leq 5 \times 10^5$, according to

$$m(t) = 20 + (22 \times 10^{-5})t$$

$$s(t) = 5 + (13 \times 10^{-5})t$$

If S_1^* is taken to be 85 MPa, then the probability of achieving a state of stress just sufficient to induce faulting on the most favored fault orientation by, say, time $t = 2 \times 10^5$ years is

$$P(S_1 \geq 85 | t = 2 \times 10^5)$$

$$= \frac{1}{31\sqrt{2\pi}} \int_{85}^{\infty} \exp - [(x - 64)^2/1922]\, dx$$

$$= \frac{1}{\sqrt{2\pi}} \int_{0.68}^{\infty} \exp - (x^2/2)\, dx$$

$$= 1 - \Phi(0.68) = 0.25$$

Similarly, the probability of achieving a state of stress just sufficient to induce faulting on the most favored fault orientation by time $t = 4 \times 10^5$ years is

$$P(S_1 \geq 85 | t = 4 \times 10^5)$$

$$= \frac{1}{57\sqrt{2\pi}} \int_{85}^{\infty} \exp - [(x - 108)^2/6498]\, dx$$

$$= \frac{1}{\sqrt{2\pi}} \int_{-0.40}^{\infty} \exp - (x^2/2)\, dx$$

$$= 1 - \Phi(-0.40) = 0.66$$

Thus, for example, the probability of $S_1 \geq S_1^*$ increases from 25% at 200,000 years to 66% at 400,000 years after closure of the depository. The distributions of S_1 at $t = 2 \times 10^5$ and $t = 4 \times 10^5$ years are depicted in Figure 9, along with the assumed S_1^* of 85 MPa.

Although we have assumed a normal distribution of S_1, and a linearly varying m and s, the above arguments hold for other assumed distributions and for other assumed degrees of variance of the statistical parameters with time.

Summary and Conclusions

Determination of the probability that a fault might develop or that movement along an existing fault might occur requires that the appropriate failure criteria and rock parameters be specified for the system being analyzed and that the state of stress be known or can be predicted for that system. Although representative failure criteria and rock parameters (e.g., frictional characteristics) can be determined with reasonable confidence from laboratory or

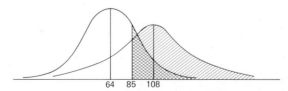

Fig. 9. The probability distribution of S_1 after 2×10^5 years ($m = 64$ MPa) and after 4×10^5 years ($m = 108$ MPa). An assumed S_1^* of 85 MPa is indicated for reference. The probability of achieving a state of stress just sufficient to induce faulting increases from 25% at 200,000 years to 66% at 400,000 years.

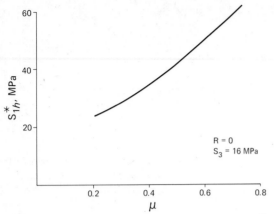

Fig. 10. S_{1h}^* required to induce sliding, as function of the friction coefficient μ. Shearing resistance $R = 0$ and $S_3 = 16$ MPa.

field investigations and the existing state of stress can be accurately measured, predictions about changes in stress levels cannot be made at the same level of confidence. Therefore, conditional probabilities will have to be calculated (i.e., given that a specified state of stress will be reached, the probability of fault development or of fault movement can be determined for that state of stress). For fault (shear fracture) development the probability that a fault will form is equal to 1.0 once the critical stress for the fracture criterion is reached. For displacement along an existing discontinuity, the probability that movement will occur is dependent upon the orientation of the discontinuity, relative to the direction of maximum principal stress, as well as on the magnitude of the stress. The conditional probability calculated for a given value of S_1 can be coupled with the probability that S_1 will reach that value, for an assumed distribution of stress variation with time, to determine the probability of movement along the sliding surface by time t.

Although the specific criteria and parameter values selected in this paper for illustration of the method would need to be reassessed before application to a real site, the numerical results presented do provide insight into conditions that could lead to displacement on existing faults in real systems. The maximum in situ values of S_1 reported for measurements in Africa and Canada for depths up to 600 m are 27.5 and 47.2 MPa, respectively [*McGarr and Gay*, 1978]. From Figure 6 it can be seen that for stress levels comparable to those reported in Africa, sliding would not occur on existing faults unless both the coefficient of sliding friction were very low (e.g., $\mu = 0.2$) and the shearing resistance were essentially zero. For higher stress levels, such as those reported for Canada, displacement on existing faults would be likely only on those characterized by relatively low shearing resistance and friction coefficients less than about 0.5. Figure 10 shows the dependence of S_1^* on the frictional coefficient for non-existent shearing resistance. These are, therefore, minimum values of S_1 that could induce sliding. It should be noted, however, that the results we present are in terms of the principal stresses S_1 and S_3 and do not take into account the possible effects of pore pressure. One could consider the values of S_1 presented here to be effective pressures if $S_3 = 16$ MPa is taken to be an effective pressure.

If the experimentally determined mean values for the friction parameters of $\mu = 0.60$ and $R = 19.5$ MPa were used and if one were to assume $S_{1h} = 47.2$ MPa, as for the Canadian measurements, an increase in pore pressure of 33.8 MPa would be required to induce sliding on the most favored fault orientation. If the shearing resistance (R) is zero for otherwise the same situation, however, only 1.3 MPa of pore pressure would induce sliding. Figure 11 shows the increase in pore pressure required to induce sliding as a function of increasing shearing resistance for this situation.

We have presented a method for assessing the probabil-

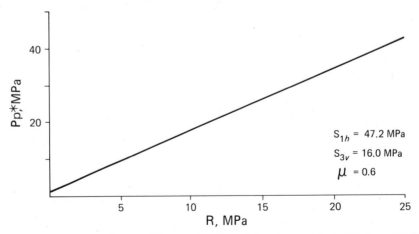

Fig. 11. Pore pressure required to induce sliding, as function of shearing resistance R. $S_{1h} = 47.2$ MPa, $S_{3v} = 16$ MPa, and $\mu = 0.60$.

ity of faulting in geologic media. The illustration of this method with typical parameter values and available in situ data suggests that the probability of faulting is zero until relatively high stress levels or moderately high pore pressures develop but that the probability rapidly approaches 1.0 once these conditions are achieved.

References

APS, Report to the American Physical Society by the study group on nuclear fuel cycles and waste management, *Rev. Modern Phys.*, *50*, S1-186, 1978.

Byerlee, J., Friction of rocks, *Pure Appl. Geophys.*, *116*, 615-626, 1978.

Campbell, J. E., R. T. Dillon, M. S. Tierney, H. T. Davis, P. E. McGrath, F. J. Pearson, Jr., H. R. Shaw, J. C. Helton, F. A. Donath, Risk methodology for geologic disposal of radioactive waste, Interim Rep. *NUREG/CR-0458*, *U.S. Nucl. Regul. Comm.*, Washington, D. C., 1978.

CGS, Scenario development and evaluation related to the risk assessment of high level radioactive waste repositories, *Interim Rep. NRC-04-79-185*, Urbana, Ill., 1979.

Cranwell, R. M., and F. A. Donath, An application of geometric probability to the existence of faults in anisotropic media, *Sci. Basis Nucl. Waste Manage.*, *2*, 787-794, 1980.

Donath, F. A., The isolation of high-level radioactive waste, in *Electrical Utilities in Illinois: Sixth Annual Illinois Energy Conference Proceedings*, pp. 234-240, University of Illinois, Chicago, 1979.

Donath, F. A., Post-closure environmental conditions and waste form stability in a geologic repository for high-level nuclear waste, *Nucl. Chem. Waste Manage.*, *1*, 103-110, 1980.

Donath, F. A., L. S. Fruth, Jr., and W. A. Olsson, Experimental study of frictional properties of faults, in *New Horizons in Rock Mechanics: Proceedings of the 14th Symposium of Rock Mechanics*, edited by H. R. Hardy, Jr. and R. Stefanko, pp. 189-222, American Society of Civil Engineers, New York, 1973.

Donath, F. A., T. Greenwood, C. Klingsberg, P. McGrath, J. Nichols, D. Stewart, and I. Winograd, Technical issues in nuclear waste isolation—Excerpts from Appendix A of the report of the IRG Subgroup on Alternative Technology Strategies, in *Waste Management '79*, edited by R. G. Post, pp. 145-160, University of Arizona, Tucson, 1979.

IRG, Isolation of radioactive wastes in geologic repositories: Status of scientific and technological knowledge, Appendix A, Subgroup Report on Alternative Technology Strategies for the Isolation of Nuclear Waste, *TID-28818*, Interagency Rev. Group, Washington, D. C., 1978.

Jaeger, J. C., The frictional properties of joints in rock, *Geofis. Pura Appl.*, *43*, 148-158, 1959.

McGarr, A., and N. C. Gay, State of stress in the earth's crust, *Annu. Rev. Earth Planet. Sci.*, *6*, 405-436, 1978.

Strain in the Ramp Regions of Two Minor Thrusts, Southern Canadian Rocky Mountains

J. H. SPANG[1] AND T. L. WOLCOTT[2]

Department of Geology and Geophysics, University of Calgary
Calgary, Alberta, Canada T2N 1N4

S. SERRA

Amoco Production Company, Research Center, Tulsa, Oklahoma 74102

Calcite twinning strain measurements have been made in the ramp regions of two well-exposed, small-scale overthrust faults in limestones in order to understand better two of the deformational styles observed in ramp regions. In the first ramp region the thickness of the bedding planes is small relative to the height of the ramp, and the intragranular twinning strains show plane strain in the transport plane. The maximum principal strain is in the transport direction and is either layer parallel or at a small angle to bedding in the counter dip direction. Thus the thrust fault ramps, which make an average angle of 25.5° with bedding, are planes of high resolved shear strain, while the bedding planes are planes of low resolved shear. The low magnitude of the layer parallel shear strains in this style of ramp region may mean that some mechanism, such as bedding parallel slip, is important in order for the rocks to bend in the hinge regions. In the second ramp region the thickness of the bedding is approximately equal to the height of the ramp. For two samples from the upper hinge and one in the ramp region the least principal strain is layer parallel and approximately perpendicular to the transport direction. The maximum principal strain is essentially perpendicular to the fault plane in the upper hinge, and in the ramp region both the ramp and bedding are in an orientation of high resolved shear strain. The high magnitude of the layer parallel shear strain may reflect the inability of the layers to slip along bedding planes in this style of ramp region.

Introduction

This study describes the orientation and magnitude of intragranular, calcite twinning strains in the ramp regions of two minor thrusts in limestone. The purpose of the study is to gain insight into the mechanism whereby rocks in the hanging wall of a thrust move over a thrust fault ramp. Based primarily on the field work of *Butts* [1927] in the Pine Mountain overthrust in the southern Appalachian Plateau, *Rich* [1934] developed a model to explain the development of hanging wall structures resulting from movement over a step or ramp. Rich's hypothesis suggests that the fault is parallel to bedding in in-

competent rocks above the ramp. Folds are created in the hanging wall as a result of movement over the ramp, and these folds are 'rootless' in that they do not extend below the fault (Figure 1a). Rich's hypothesis enabled a group of seemingly unrelated structures to be viewed as necessary consequences of a single, relatively simple system. *Serra* [1978] cites a large number of examples where the Rich model has been applied to similar structures in the southern Appalachian Valley and Ridge and Plateau provinces and in the North American Cordillera.

Recently, *Serra* [1977] has extended the Rich model to include the effects of variations in rock type, sequence of rock types, and thicknesses of individual rock types in the hanging wall. Figures 1b-1d summarize the three structural styles or modes described for ramp regions by *Serra* [1977]. Mode 1 involves imbricate thrusting in the ramp region where the dip of successive imbricates becomes

[1] Now at Center for Tectonophysics, Texas A&M University, College Station, Texas 77843.

[2] Now at Norcen Energy Resources Ltd., Calgary, Alberta, Canada T2P 2X7.

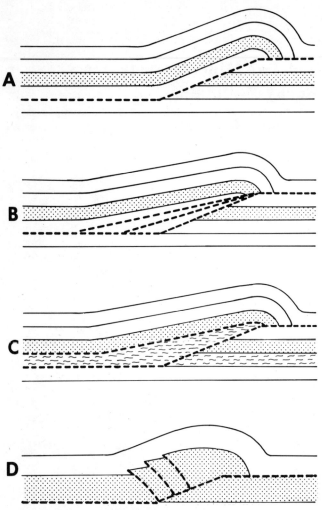

Fig. 1. Schematic representation of the deformation in ramp regions of thrust faults showing the final stage in the development of structures as hypothesized by *Rich* [1934] (Figure 1a) and modes 1 to 3 of *Serra* [1977] (Figures 1b-1d, respectively). Layers represent mechanical units which can consist of either a single bed or a group of beds which show little or no evidence of internal or interbed slip during thrusting [*Serra*, 1977]. Heavy dashed lines are faults. Thickened unit in Figure 1c represents mechanically weak rocks.

shallower going upsection in the hanging wall. Generally, the faults splay off the master sole fault, and this type of behavior has only been observed where the thickness of individual mechanical units is considerably smaller than the height of the ramp. Serra defines a mechanical unit as either a single bed or a group of beds which exhibit little or no evidence of interbed or internal slip. Mode 2 is characterized by thickening of mechanically weak rocks which are in contact with the sole fault. In addition to movement along the sole fault, movement also appears to occur along the upper contact of the thickened unit. Mode 3 is characterized by curved reverse faults in the hanging wall and is

observed only when the thickness of individual mechanical units is approximately equal to the height of the ramp.

Geologic Setting

The two minor thrusts examined in this study are located in the Front Ranges structural subprovince of the southern Canadian Rocky Mountains (Figure 2). The study area is within the McConnell thrust plate which is the easternmost of the four major overthrusts which make up the Front Ranges in this area. On a regional scale the overthrusts are listric and low angle. Locally, steeper dips are associated with steps, imbrications, and folded thrusts [*Dahlstrom*, 1970]. *Bally et al.* [1966] and *Price and Mountjoy* [1970] use seismic data and field mapping to infer that the thrusting is 'thin-skinned' and that the gently southwesterly dipping Hudsonian basement is not involved probably as far west as the Rocky Mountain Trench. Folding and thrusting migrated from SW to NE, and there is a corresponding decrease in the intensity of the deformation in the same direction. Unfortunately, entire ramp segments in the major overthrusts are not completely exposed, and it is for this reason that we have chosen to study two completely exposed outcrop scale ramps.

The two minor thrusts are exposed along the Trans Canada Highway just west of Lac des Arcs and at Pigeon Mountain, respectively. The Lac des Arcs outcrop (Figure 3) is in the Devonian Palliser Formation, which is a dark grey wackestone to grainstone with occasional dolomitic mottling. Based on repetitive measurements, the average angle between bedding and the minor thrust ramps is 25.5° adjacent to the samples examined in detail. One of the ramps (A in Figure 3) is composed of several faults. The thickness of bedding is substantially less than the

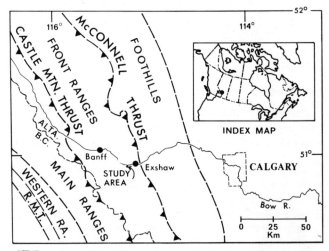

Fig. 2. Location of the study area in the Front Ranges of the southern Canadian Rocky Mountains. The Foothills, Front Ranges, Main Ranges, and Western Ranges are structure subprovinces of the Rocky Mountains, and they are bounded on the west by the Rocky Mountain Trench (RMT) and on the east by the orogenically undisturbed plains.

ramp height. The Pigeon Mountain outcrop (Figure 4) is in the upper Banff Formation (Mississippian) approximately 35 m below the contact with the overlying Livingstone Formation (Mississippian). The Banff Formation consists of dark grey, interbedded argillaceous lime mudstone and wackestone. The height of the ramp (h in Figure 4) is approximately equal to the thickness of bedding. Based on repetitive measurements in the ramp region, the average angle between the ramp and bedding is 22.5°.

The transport direction of the McConnell thrust is toward 063°, based on the local structural trends. This value has been confirmed by detailed field mapping in this area by J. H. Spang and his co-workers at the University of Calgary [e.g., *Brown and Spang*, 1978]. The displacement on the McConnell thrust is of the order of 30 km in a time period of post early Campanian to pre-Paleocene (10 m.y.), which means that the mean displacement rate is about 3 mm/yr [*Wheeler et al.*, 1974]. *Ghent and Miller* [1974] have examined authigenic minerals in Cretaceous clastics from structurally below the McConnell thrust, and they estimated the temperature to have been between a high of 250°-280°C and a low of 150°-180°C. Also, based on the lack of lawsonite, $P_{\text{lithostatic}}$ was less than 300 MPa, and using reasonable stratigraphic and structural reconstructions, $P_{\text{lithostatic}}$ is probably between 100 and 200 MPa. *Jamison and Spang* [1976] have developed a method of estimating differential stress using twin lamellae in naturally deformed carbonates. Using two samples from two localities within the McConnell thrust plate, they calculate that the differential stress may have been as high as 125 MPa. One of their samples (296) is from the same outcrop at Pigeon Mountain, where one of the thrust ramps examined in this study is exposed.

Method of Study

The calcite strain gage method [*Groshong*, 1972, 1974] has been applied to four samples of naturally deformed limestone from each outcrop. The calcite strain gage method is a least squares calculation which determines intragranular twinning strains. The method calculates the deviatoric strain by using measured shear strains obtained from all measurable twin sets in a population of

Fig. 3. View, looking south, of thrust fault ramps in the Palliser Formation (Devonian) along the south side of the Trans Canada Highway 0.3 km west of Lac des Arcs. The trend of the vertical roadcut is N70°E, which is slightly oblique to the local transport direction of the McConnell thrust. Bar scale is 2 m.

Fig. 4. View, looking south, of a thrust fault ramp (A) with bedding plane parallel segments (B) before and after the ramp. The vertical roadcut is in the upper Banff Formation (Mississippian) on the south side of the Trans Canada Highway at Pigeon Mountain. The thickness of bedding (h) is approximately equal to the height of the ramp. The trend of the outcrop is N80°E. Bar scale is 2 m.

grains. *Groshong* [1972, 1974] discusses the assumptions and limitations of the technique. All thin sections were cut from vertical sections parallel to the local transport direction of the McConnell thrust (063°).

Groshong [1974] has proposed several different methods for filtering the data used in the calculation. Figure 5 shows the orientation and magnitude of the principal strain axes calculated using all of the data as well as two of the filtering techniques. The first filtering technique involves removing all twin sets which should not have twinned, given an irrotational homogeneous stress field (minus all negative deviations), and the second technique involves removing a percentage of the twin sets which represent either too much or too little strain (minus 20% of the largest deviations). As can be seen in Figure 5, all of the methods calculate virtually the same orientation for the

principal axes with some variation in their magnitudes. In this study we have chosen to present all of our data, using data sets where all of the negative deviations have been removed.

Intragranular Strain

Figure 6 is a sketch of the Lac des Arcs outcrop (Figure 3) which shows bedding planes, fractures, and thrusts. Four oriented samples from this outcrop have been analyzed using the calcite strain gage technique. Using repetitive measurements from sample locations 1, 2, and 4, the average angle between the thrust fault ramps and bedding is 25.5°, while the fault is bedding plane parallel at sample 3. Given the upward and downward limitations of the outcrop and lack of correlatable layers between the hanging wall and footwall, it is not possible to estimate the net slip. Also, no grooves or slickensides have been recognized on the fault surfaces. Unfortunately, the best estimate that we can make for the direction of the net slip is to use the line of intersection of the ramps with bedding and assume that the slip direction was at right angles to this line.

The value of the square root of the second strain invariant ($\sqrt{J_2}$, Table 1) is a measure of the total deviatoric strain in a sample [*Jaeger*, 1962, p. 92]. Using the values for $\sqrt{J_2}$ for samples 1-4 (Table 1), all of the samples show relatively uniform values with a mean of 6.0% for the four samples. In general, the maximum and least principal

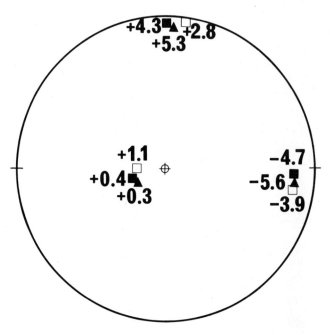

Fig. 5. Principal axes of intragranular twinning strain for sample 4 calculated using all data (solid squares), minus all negative deviations (solid triangles), and minus 20% of the largest deviations (open squares). Values are in percent, and compressive strain is negative.

strains (ϵ_1 and ϵ_3, respectively) are in the transport plane. The maximum principal compression is either layer parallel or slightly up dip, and the least principal strain is at a high angle to bedding. With the exception of sample 3 the magnitude of the intermediate principal strain is 16% of $\sqrt{J_2}$ or less, which means that the samples approximate plane strain. In sample 3 the intermediate principal strain is larger, and the sample tends toward uniaxial extension at a high angle to bedding and to the fault. In all of the samples, including the bedding plane parallel fault above sample 3, the fault plane is a plane of relatively high resolved shear strain. In contrast to this, the bedding planes are planes of very low resolved shear strain with the exception of sample 3. Fractures plotted on the stereonets and shown on the sketch in Figure 6 are present in several different orientations. There are fractures parallel to the thrust ramps (e.g., sample 1) as well as in the conjugate orientation (e.g., sample 4).

Figure 7 is a sketch of the fault, bedding, and fracture planes in the Pigeon Mountain outcrop. Four oriented samples from this outcrop have been analyzed using the calcite strain gage technique. Based on repetitive mea-

surements in the ramp region the angle between the ramp and bedding is 22.5°. In the lower ramp region there is an apparent thickening of the hanging wall which may be due to the presence of a fault beneath sample 7. This fault is shown by a solid line which is characterized by a slickensided fault surface. This fault is shown as turning into several parallel fractures (fault) surfaces indicated by parallel dashed lines. This fault has the effect of lowering the dip of the ramp by 9° in the lower hinge and may correspond to mode 1 behavior (Figure 1b). Similarly, there may be some thinning in the upper hinge possibly due to movement along the fractures labeled f.

The results of the twinning strain measurements are seemingly quite different from the earlier discussion of samples 1–4. In samples 5 and 6 the maximum principal strain is oriented at approximately right angles to the fault plane, which might account for some of the apparent thinning in the upper hinge region. In these same samples the least principal strain is approximately layer parallel at a very high angle to the transport direction. Sample 7 yields a least principal strain which is layer parallel and at a high angle to the transport direction. The maximum

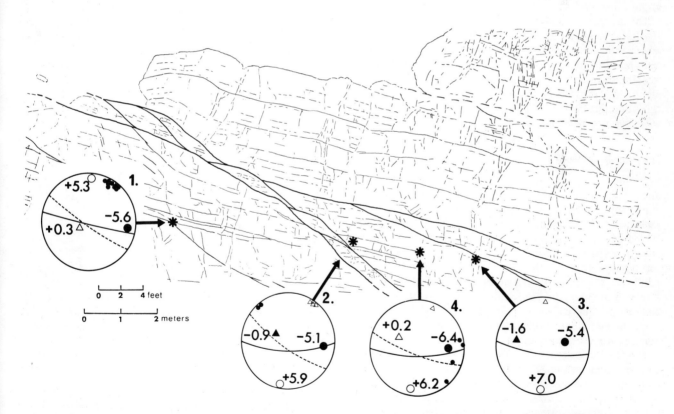

Fig. 6. Palliser Formation outcrop just west of Lac des Arcs (Figure 3) showing thrust fault ramps (heavy lines), bedding fractures, and sample locations (asterisks). Stereonets are lower-hemisphere, equal-area projections, and the primitive circles are vertical with a strike of 063°. Great circles represent bedding (solid line) and fault planes (dashed). In sample 3 the fault is bedding plane parallel. Poles to fractures (small solid circles) and poles to faults (small open triangles) are also shown. Principal axes for intragranular twinning strains are in percent and are calculated from samples where all negative deviations have been removed.

TABLE 1.   Additional Information From Strain Gage
Calculations

Sample	Number of Grains	Number of Twin Sets	Percent Negative Deviation	$\sqrt{J_2}$*
1	50	62	14.5	5.5
2	50	60	13.3	5.6
3	50	53	15.1	6.4
4	50	56	17.8	6.3
5	50	79	17.7	3.1
6	50	62	19.4	6.3
7	50	82	26.0	3.4
8	50	67	37.3	4.3

* $\sqrt{J_2} = (\frac{1}{2}(\epsilon_1^2 + \epsilon_2^2 + \epsilon_3^2))^{1/2}$

principal strain is in the transport direction and both the bedding plane and both fault orientations lie in orientations of high resolved shear strain. However, the fault which represents a lower angle cutoff is in a more favorable orientation for slip relative to the maximum principal strain. Sample 8 corresponds to plane strain in the transport direction with a near layer parallel maximum principal strain and a maximum principal extension at a high angle to bedding.

The average value of the square root of the second strain invariant ($\sqrt{J_2}$ in Table 1) for samples 5-8 is 4.3%, which indicates that the total strain in these samples is less than in samples 1-4. Also, there is considerably more 'scatter' in these data, as can be seen in the larger percentage of the data which represent negative deviations. Thus these samples may show some overprinting of strains, but we have not been able to separate out different strain fields using any meaningful criteria.

Discussion of Results

Various authors have discussed the complicated stress and strain history rocks undergo as they pass through a ramp region [e.g., *Morse*, 1977; *Wiltschko*, 1979]. The rocks are folded as they pass through the lower hinge, unfolded on the ramp, folded in the opposite sense as they pass over the upper hinge, and, finally, unfolded again after passing through the upper hinge. *Wiltschko* [1979] points out that in terms of bending stresses there is no net strain. Thus for a perfectly lubricated fault plane in the absence of gravity an element of rock would have the same shape after passing through a ramp region if only bending stresses were present.

The thrust fault ramps in the Lac des Arcs outcrop correspond to mode 1 of *Serra* [1977], where the thickness of bedding is much less than the height of the ramp. *Chapple and Spang* [1974] discuss a model for buckling of a multilayered sequence. Provided that bedding plane slip is sufficiently easy, the layer parallel shear strain is taken up by bedding plane slip, and the strain within layers will be dominated by layer parallel bending strains. The rootless folds above thrust fault ramps are clearly not buckle folds, but some analogies can be made to the model proposed by

Chapple and Spang. As the hanging wall rocks move through the ramp regions, we would expect that most of the layer parallel shear strain would be accommodated through bedding plane (interlayer) slip, provided that bedding plane slip is sufficiently easy. If this is the case, then deformation within the individual layers would be dominated by layer parallel bending strains. We might expect that measured strains would represent a combination of layer parallel shortening, (neutral surface) bending, and layer parallel shear. Examination of the calculated principal axis orientations in samples 1-4 shows that the layer parallel shear strain is low within the layers. The maximum principal strain is nearly layer parallel and

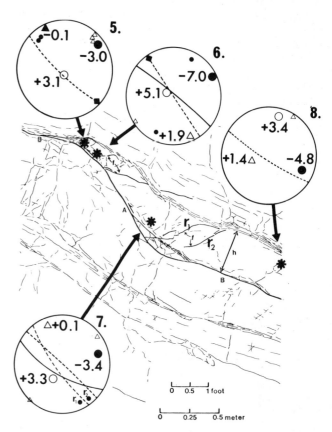

Fig. 7.   Pigeon Mountain outcrop (Figure 4) showing the thrust fault (heavy line) with bedding plane parallel (B) and ramp (A) segments. The height of the ramp (h) is approximately equal to the bedding thickness. Fractures r_1 and r_2 correspond to reverse faults. Sample locations are indicated by asterisks, and the stereonets are lower-hemisphere equal-area projections, and the primitive circles are vertical with a strike of 063°. Great circles represent bedding (solid line) and fault planes (dashed). In sample 8 the fault is bedding plane parallel. Poles to fracture planes (small solid circles) and poles to faults (small open triangles) are also shown. Principal axes for intragranular twinning strains are in percent and are calculated from samples where all negative deviations have been removed. Small solid squares on the fault planes in samples 5 and 6 indicate the orientation of slickensides on the fault planes.

therefore probably represents some layer parallel shortening, perhaps related to the formation of the ramps, and some layer parallel bending strains, perhaps related to bending in order for the hanging wall to move up the ramp.

Wiltschko [1979] used a linearly viscous, plane strain model to examine thrust sheet deformation at a ramp. Wiltschko found that the bending stresses generated at a ramp are a significant portion of the total stress field and that the resistance to movement due to internal deformation is of the same order of magnitude as fault zone drag at a ramp. Wiltschko's model shows three zones of potential normal faulting: (1) in the lower portion of the hanging wall at the upper hinge, (2) in the lower portion of the hanging wall at the lower hinge, and (3) in the upper portion of the layer in the central portion of the ramp. Outside of the three normal fault regimes, reverse faults are predicted. Going from the bottom of the layer to the top in the central ramp region, reverse fractures first increase in dip (i.e., concave upward) and then decrease in dip after passing through an inflection point (i.e., concave downward). Fractures labeled r_1 and r_2 to the right of sample 7 in Figure 7 show a similar geometry. Thus an element of rock along the top or bottom of a layer passing over a ramp will have been within both normal and reverse fault regimes, while an element in the center of the layer will remain within the reverse faulting regime [*Wiltschko,* 1979, p. 1100]. Wiltschko's theoretical model corresponds to mode 3 of *Serra* [1977] and the Pigeon Mountain outcrop of this study. *Rodgers and Rizer* [1980], using both a photoelastic model and a dislocation model of a thrust fault in a homogeneous elastic half-space show secondary fault trajectories which represent both synthetic and antithetic movement. The synthetic faults in the hanging wall are approximately parallel to the major thrust (dislocation) and may represent the master thrust extending itself with more or less the same dip. The antithetic faults are concave downward and similar to those described by other authors and to the Pigeon Mountain outcrop in this study.

Morse [1977, 1978] used nonscaled rock models to examine mode 3 behavior in carefully controlled experiments. The specimens consist of a layer of Coconino sandstone with a 20° precut in it. The sandstone is underlain by a shorter, rigid block of granite and overlain by one or more shorter layers of limestone, and all of the layers are lubricated. During deformation the hanging wall block is folded as it is forced over the rigid ramp. The overlying limestone is force-folded into conforming to the shape of the overthrust plate. Along the ramp the style of deformation in the sandstone is that of a series of curved, antithetic reverse faults. By running a series of experiments with different amounts of displacement, *Morse* [1977, p. 461] was able to show that the reverse faults form either at the lower hinge or only slightly up the ramp. Not surprisingly, the reverse faults thicken the sandstone layer along the ramp. However, in the present study the apparent thickening in the lower hinge beneath sample 7 is clearly not due to this type of antithetic reverse faulting but rather synthetic thrusting.

Samples 5 and 6 from the Pigeon Mountain outcrop show a maximum principal strain at a very high angle to both bedding and the fault plane. These samples may reflect neutral surface bending at the upper hinge, although the maximum principal extension is perpendicular to the transport direction instead of parallel to it as might be anticipated. *Spang et al.* [1975] and *Spang and Brown* [1980] have reported on the dynamic analysis of calcite, dolomite, and quartz from 57 samples within the McConnell thrust in this area. They found that in the internal portions of the thrust sheet the maximum principal compression was oriented consistently in the transport direction either parallel to bedding or at a small angle to bedding in the counter dip direction. The deformation usually corresponds to a uniaxial shortening or plane strain in the transport plane. Local reorientation of the stress field can be demonstrated close to minor structures developed within the thrust sheet. Such an orientation for the maximum principal strain, as in samples 5 and 6, is consistent with the work of *Spang and Brown* [1980].

Spang et al. [1979] report on the development of cleavage in the upper Banff Formation from the same outcrop at Pigeon Mountain. The cleavage makes an angle of 25° with bedding and is thus in essentially the same orientation as other fractures and thrust fault ramps in this study. *Spang et al.* [1979] have also made twinning strain measurements on one sample of limestone. The maximum principal strain (−5.0%) is oriented slightly up dip in the transport direction and makes an angle of 35° with the cleavage. The least principal strain (+3.4%) is parallel to both cleavage and layering and is at a right angle to the transport direction. This implies that in these outcrops not only are other fractures and thrust fault ramps in planes of high resolved shear strain but that the cleavage is as well.

Conclusions

Lac des Arcs Outcrop (Mode 1, Serra [1977])

1. The intragranular twinning strains show plane strain in the transport plane. The maximum principal strain is in the transport direction and is either layer parallel or at a small angle to bedding in the counter dip direction.

2. The thrust fault ramps represent planes of high resolved shear strain, while the bedding planes are planes of low resolved shear strain.

3. Using the square root of the second strain invariant as a measure of the total strain in each sample, the average strain in these samples is 6.0%.

Pigeon Mountain Outcrop (Mode 3, Serra [1977])

1. The intragranular twinning strain in two samples from the upper hinge shows a maximum principal strain

essentially perpendicular to the fault plane and a layer parallel extension perpendicular to the transport direction. This could account for some of the apparent thinning in the upper hinge.

One sample from the ramp region has a layer parallel extension perpendicular to the transport direction and a maximum principal strain in the transport direction such that both the ramp and bedding are in an orientation of high resolved shear strain. A mode 1 synthetic fault, which lowers the dip of the ramp by 9°, is more favorably oriented for slip than the main ramp. This synthetic fault also results in thickening of the lower hinge.

A sample in the hanging wall in the layer parallel segment before the lower hinge shows plane strain in the transport plane.

2. On the average the total strains in these samples, 4.3%, are lower than strains in the mode 1 samples from the Lac des Arcs outcrops.

References

Bally, A. W., P. L. Gordy, and G. A. Stewart, Structure, seismic data, and orogenic evolution of southern Canadian Rocky Mountains, *Bull. Can. Pet. Geol.*, *14*, 337–381, 1966.

Brown, S. P., and J. H. Spang, Geometry and mechanical relationship of folds to thrust fault propagation using a minor thrust in the Front Ranges of the Canadian Rocky Mountains, *Bull. Can. Pet. Geol.*, *26*, 551–571, 1978.

Butts, C., Fensters in the Cumberland overthrust block in southwestern Virginia, *Va. Geol. Surv. Bull.*, *28*, 12 pp., 1927.

Chapple, W. M., and J. H. Spang, Significance of layer-parallel slip during folding of layered sedimentary rocks, *Geol. Soc. Am. Bull.*, *85*, 1523–1534, 1974.

Dahlstrom, C. D. A., Structural geology in the eastern margin of the Canadian Rocky Mountains, *Bull. Can. Petrol. Geol.*, *18*, 332–406, 1970.

Ghent, E. D., and B. E. Miller, Zeolite and clay-carbonate assemblages in the Blairmore Group (Cretaceous) Southern Alberta Foothills, Canada, *Contrib. Mineral. Petrol.*, *44*, 313–329, 1974.

Groshong, R. H., Jr., Strain calculated from twinning in calcite, *Geol. Soc. Am. Bull.*, *83*, 2025–2038, 1972.

Groshong, R. H., Jr., Experimental test of least squares strain gage calculation using twinned calcite, *Geol. Soc. Am. Bull.*, *85*, 1855–1864, 1974.

Jaeger, J. C., *Elasticity, Fracture, and Flow*, p. 212, Methuen, London, 1962.

Jamison, W. R., and J. H. Spang, Use of calcite twin lamellae to infer differential stress, *Geol. Soc. Am. Bull.*, *87*, 868–872, 1976.

Morse, J., Deformation in ramp regions of overthrust faults: Experiments with small scale rock models, *Guideb. Wyo. Geol. Assoc. Annu. Field Conf.*, *29th*, 457–470, 1977.

Morse, J., Deformation in ramp regions of thrust faults: Experiments with rock models, M.Sc. thesis, 138 pp., Tex. A&M Univ., College Station, 1978.

Price, R. A., and E. W. Mountjoy, Geologic structure of the Canadian Rocky Mountains between Bow and Athabasca rivers: A progress report, in Structure of the Southern Canadian Cordillera, edited by J. O. Wheeler, *Spec. Pap. Geol. Assoc. Can.*, *6*, 7–39, 1970.

Rich, J. L., Mechanics of low-angle overthrust faulting as illustrated by Cumberland thrust block, Virginia, Kentucky, and Tennessee, *Am. Assoc. Petrol. Geol. Bull.*, *18*, 1584–1596, 1934.

Rodgers, D. A., and W. D. Rizer, Deformation and secondary faulting near the leading edge of a thrust fault, *Q. J. Geol. Soc. London*, in press, 1980.

Serra, S., Styles of deformation in the ramp regions of overthrust faults, *Guideb. Wyo. Geol. Assoc. Annu. Field Conf.*, *29th*, 487–498, 1977.

Serra, S., Styles of deformation in the ramp regions of overthrust faults, Ph.D. thesis, 83 pp., Tex. A&M Univ., College Station, 1978.

Spang, J. H., and S. P. Brown, Dynamic analysis of a small imbricate thrust and related structures, Front Ranges, southern Canadian Rocky Mountains, *Q. J. Geol. Soc. London*, in press, 1980.

Spang, J. H., W. R. Jamison, and J. A. Smith, Dynamic analysis of the McConnell thrust plate and associated structures, *Eos Trans. AGU*, *56*, 1062, 1975.

Spang, J. H., A. E. Oldershaw, and M. Z. Stout, Development of cleavage in the Banff Formation at Pigeon Mountain, Front Ranges, Canadian Rocky Mountains, *Can. J. Earth Sci.*, *16*, 1108–1115, 1979.

Wheeler, J. O., H. A. K. Charlesworth, J. W. H. Monger, J. E. Muller, R. A. Price, J. E. Reesor, J. A. Roddick, and P. S. Simony, Western Canada, in *Mesozoic-Cenozoic Orogenic Belts—Data for Orogenic Studies*, 809 pp., Geological Society, London, 1974.

Wiltschko, D. V., A mechanical model for thrust sheet deformation at a ramp, *J. Geophys. Res.*, *84*, 1091–1104, 1979.

Analysis of a Horizontal Catastrophic Landslide

E. G. BOMBOLAKIS

Department of Geology and Geophysics, Boston College, Chestnut Hill, Massachusetts 02167

An analysis of the Hartford Dike slide is made in terms of limit equilibrium principles and the impulse-momentum theorem of physics. This slide was triggered accidentally by artificially induced excess pore pressure within the upper layers of a horizontal varved clay when fill was being deposited above the varved clay. There are two particularly important features of this type of landslide. One is that the Hartford Dike slide is a special case of a translational type of landslide that appears to be not too uncommon if attention is focused strictly on the best documented landslide studies in the literature. For this reason the equations of sliding distance, travel time, and basal shear strength are derived in a form capable of being generalized. A second important feature of the Hartford Dike slide is that it involved an instability mechanism in which excess pore pressure was not dissipated rapidly along the glide zone. Data analysis indicates that the 'internal angle of friction' had been reduced to a value smaller than 3° or 4° along the basal shear zone. Therefore, in terms of basic principles, the mechanics of the Hartford Dike slide is potentially useful for future analysis of certain earthquake-induced landslides and of fundamental problems associated with the occurrence of certain overthrusts, decollements, and 'gravity tectonic' slides.

Introduction

Whether modern landslides or paleotectonic slides are to be analyzed, it is essential to focus special attention on the best documented studies of modern landslides. For example, if overthrusting is initiated by abnormally high pore pressures [*Hubbert and Rubey*, 1959], then the question of rapid dissipation of excess pore pressure during displacement poses a fundamental problem in slide mechanics. No mechanism yet has been documented as to how the shear resistance along glide surfaces is reduced sufficiently to account for the displacements in overthrusts, decollements, and low-angle 'gravity tectonic' slides [*DeSitter*, 1956; *Billings*, 1972; *Hobbs et al.*, 1976; *Voight*, 1976]. Neither is there any reliable method for predicting earthquake-induced landslides. During the past 30 years, however, several dozen landslide studies have been documented sufficiently to establish cause and effect within narrow limits of probability. Most of these studies are in the soil mechanics literature [e.g., *Skempton*, 1964; *Bjerrum*, 1967, 1974; *Wilson*, 1970; *Wilson and Mikkelsen*, 1978; *Seed*, 1968, 1979]. Some of the results are surprising, contrary to what might seem obvious from surface mapping or air photo interpretation. The Hartford Dike slide [*Terzaghi and Peck*, 1948; *Terzaghi*, 1950] illustrates this situation, and because of this slide's theoretical importance, a quantitative analysis of it is presented here.

A substantial number of the best documented landslides involve displacements that are principally horizontal. The Turnagain Heights landslide [*Seed and Wilson*, 1967; *Voight*, 1973] and the Hartford Dike slide are prime examples. Both slides were catastrophic. Both involved an instability mechanism within clay systems containing silt or sand layers. The probable instability mechanism at Turnagain Heights was either sand liquifaction or a sand-clay instability triggered by the 1964 Alaskan earthquake. Sand liquifaction characteristically is followed by intrusion of sand dikes and rapid dissipation of excess pore pressure. A few instances of sand intrusion were noted at Turnagain Heights [*Seed*, 1968], but uncertainty still exists as to which instability mechanism was the fundamental failure mode [*Scott*, 1977]. In the Hartford Dike case, however, there is no question that an instability was triggered by an increase in pore pressure. This increased pore pressure was transmitted through silt layers along the potential glide surface within upper beds of a horizontal varved clay. But the instability mechanism was not liquifaction in the ordinary sense. Abnormally high pore pressures were measured in the slide zone months after the slide occurred. The Hartford Dike slide therefore is of particular interest in the analysis of certain modern and paleotectonic slides because it involved an important type of instability mechanism, one in which excess pore pressure is not dissipated rapidly along the glide surface during displacement.

The quantitative portion of the following analysis is ap-

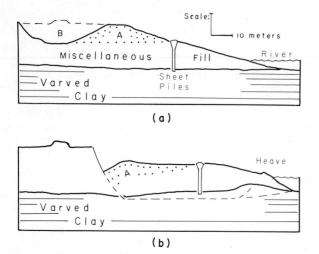

Fig. 1. The Hartford Dike landslide [after *Terzaghi*, 1950]. Semi-liquid sand fill in topographic depression B in 1*a* produced the slide in 1*b*.

proximate. It is made in terms of limit equilibrium principles and the impulse-momentum theorem of physics. Its primary purpose is to present and develop some important landslide concepts. The information published by Terzaghi is not sufficient to perform an 'exact' analysis of the Hartford Dike slide. This slide evidently was a consulting project, and so certain information and data were not published. For example, numerous borings and tests were performed under his direction but most of the data are not available in publications. However, Terzaghi's papers, project reports, and unpublished data were deposited in the Terzaghi Library of the Norwegian Geotechnical Institute after his death in 1963. Consequently, a more refined quantitative analysis might be possible in the future.

Analysis

The situation prior to landsliding of the Hartford Dike is illustrated in Figure 1*a*. The varved clay had been overlain for some time by miscellaneous fill deposited for flood control purposes. Sand dike A and the vertical sheet piles are the earlier key features. Later, a sluicing operation deposited sand in a semiliquid condition in topographic depression B. During the sluicing operation, sand dike A suddenly subsided, and the row of sheet piles, together with the foreland, moved horizontally over a distance up to almost 20 m, all within a few minutes. The row of sheet piles remained vertical and perfectly intact, as indicated in Figure 1*b*. Some heave occurred in the foreland. Therefore, if we did not know the details of this landslide, it could have been misinterpreted by air photo studies as a rotational slide in which the total weight of the landslide mass played the principal role as a gravitational driving force. A different explanation is required, however.

We begin the analysis with reference to Figure 2. Figure 2*a* represents a standard state of stress beneath a horizontal ground surface. The corresponding force represen-

tation is shown in Figure 2*b*. The horizontal and vertical normal stresses are σ_h and σ_v, respectively, where K_0 is the 'coefficient of earth pressure at rest' and γ is the unit weight of the material, usually expressed in pounds per cubic feet [*Lambe and Whitman*, 1969]. No shear stress exists along B-C. However, when slope G-E is created in Figure 2*b* by eroding the material within H-E-G-D-H', the total force on C-D is reduced, and so a shear stress is developed along B-C. B-C then becomes a potential glide surface. For convenience, therefore, the events leading to the Hartford Dike slide will be described qualitatively with free-body diagram A-B-C-E-G-A.

The Hartford Dike slide is a special case of the landslide system shown in Figure 3 [*Terzaghi and Peck*, 1948]. The potential landslide mass in that figure is delineated by N-B-C-H-E-G-A-N, where A-B-C-E-G-A is the free-body diagram described above. We consider first the likely manner of failure in Figure 3 from a general point of view before discussing the specific mechanics of the Hartford Dike slide. B-C is the potential glide surface that lies within a potentially weak layer of stratified earth material. Free-body diagram A-B-C-E-G-A undergoes creep or quasi-static displacement to the right in Figure 3. Simultaneously, the compressive stress on surface A-B diminishes, whereas the compressive stress on surface E-C increases. Consequently, the stress history in region A-B-M-N-A is analogous to triaxial extension, whereas the stress history in region E-C-I-H-E is analogous to triaxial compression.

These two types of stress history can lead to normal faulting and thrust faulting (or heave), respectively, whether the earth material is sediment [*Terzaghi and Peck*, 1948] or rock [*Griggs and Handin*, 1960; *Handin*,

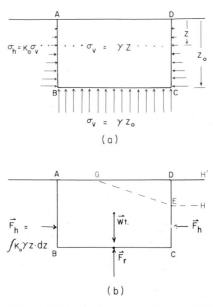

Fig. 2. Standard state of stress (*a*) beneath horizontal ground surface and (*b*) its force representation. Erosion of material in H-E-G-D-H' alters the standard state.

1966; *Bombolakis*, 1979]. Therefore, graben development at one end, and heave at the other end, can be regarded as 'end effects' of horizontal landsliding in Figure 3. This concept is important for two reasons: (1) This type of mechanical behavior in landsliding now has been documented in a few cases for rock as well as sediments [*Wilson*, 1970; *Burland et al.*, 1977; *Wilson and Mikkelsen*, 1978], and (2) this concept can be applied in appropriately modified form when glide surface B-C has a nonzero dip in the glide direction.

The traditional method of slope stability analysis for Figure 3 is the wedge method [*Lambe and Whitman*, 1969]. A principal objective, as in other methods of slope stability analysis [*Morgenstern and Sangrey*, 1978], is to predict whether failure is imminent. A fundamental problem, however, is how to predict quantitatively whether the type of landslide in Figure 3 will proceed gradually or catastrophically when normal faulting and/or heave are incipient. One way to attack this problem is to reconsider the safety factor employed in slope stability analysis. This safety factor can be defined to be the ratio of failure-resisting forces to failure-driving forces, as indicated in Figure 3. The failure-resisting forces include potential forces owing to the strength of the material. Consequently, landsliding is imminent when the safety factor approaches unity. If the safety factor is reduced only a small amount below unity, then Newton's laws require that the landslide proceed gradually. Therefore, a necessary condition to trigger a catastrophic slide suddenly is that there be a rapid reduction of sufficient magnitude below unity in the value of the safety factor.

In the case of the Hartford Dike system, precursory creep probably occurred along the varved-clay glide surface in response to preexisting shear stress and the subsequent increase of pore pressure. If the only effect of this gradually increasing pore pressure on the silt layers had been a strength reduction of silt in accordance with the effective stress principle, then the Hartford Dike slide would have proceeded gradually instead of catastrophically. Precursory creep evidently triggered the instability along the basal glide zone. The triggering mechanism is not known. Hypotheses on the mechanism are presented later because the following analysis by the impulse-momentum theorem does not require that the triggering mechanism be specified.

The impulse-momentum theorem can be represented by

$$I = \int F(t) \cdot dt = \int d(MV) \qquad (1)$$

where $F(t)$ is the resultant of all relevant external forces acting on a system of mass M, with MV the momentum of the center of mass. Internal forces may contribute to the deformation of the system in important ways during landsliding, but analysis by the impulse-momentum theorem is independent of the internal forces because of Newton's Third Law [e.g., *Den Hartog*, 1961]. Therefore, it is strictly for convenience that the slide mechanics in Figure

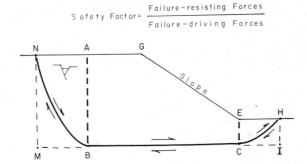

Fig. 3.   Translational landslide configuration [after *Terzaghi and Peck*, 1948]. See text.

3 is subdivided into the following three stages. The precursory stage is the stage where precursory phenomena lead to the initiation of catastrophic sliding at time $t = 0$. The first stage of sliding is defined for the time interval t_1, during which normal faulting occurs at the head of the slide. The second stage is defined for time interval t_2 for the remainder of the sliding episode. The total time for catastrophic sliding accordingly is $t_f = t_1 + t_2$.

The momentum of the landslide mass is taken to be nearly zero during the precursory stage for the following reasons. No precursory creep is mentioned in the published literature on the Hartford Dike slide. However, *Terzaghi* [1950] notes that precursory phenomena have to be almost universal in occurrence, the principal exception being for the case of earthquake-induced catastrophic slides. He found from study of various landslides that human observers frequently fail to detect precursory phenomena. Consequently, precursory creep displacement in the Hartford Dike case might have been relatively small and too slow to have been detected. Therefore, until unpublished information on this slide can be examined in the Terzaghi Library, the present assumption is that the momentum is nearly zero during the precursory stage.

During this stage the effective stresses in regions A-B-M-N-A and E-C-I-H-E of Figure 3 approach the active and passive Rankine states of limiting equilibrium, respectively [*Lambe and Whitman*, 1969]. The equations for the Rankine states are very similar in form to the equation for horizontal stress in Figure 2a. The active Rankine coefficient K_a and passive Rankine coefficient K_p are substituted for K_0 to calculate Rankine stresses at appropriate boundaries of the potential landslide mass. The relationship between these coefficients is $K_a < K_0 < K_p$. Integration of the Rankine stresses over the appropriate boundaries yields the active and passive Rankine thrusts. The active Rankine thrust is an 'external' driving force acting against the incipient normal fault surface in Figure 3, whereas the passive thrust is an external resisting force acting against region E-C-I-H-E. An external resisting shear also acts on B-C, as was stated previously.

Because of the special features of the Hartford Dike case, an additional external force acts against potential normal fault surface N-B in Figure 3 during the pre-

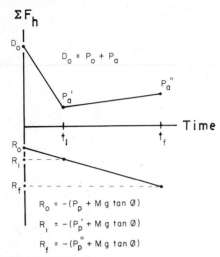

Fig. 4. Approximate time variation of external horizontal forces acting on landslide material during sliding. Sum of driving forces is graphed above time axis, with sum of resisting forces graphed beneath the time axis. Terms are defined in the text.

cursory stage. The sluicing operation at Hartford Dike produced an excess pore pressure in Figure 1a with respect to Figure 1b. The sluicing not only produced an increase in pore pressure along B-C in Figure 3, it also created an excess pore pressure that exerted an external driving force against the potential normal fault surface while the effective stresses in N-B-M-N-A were approaching the active Rankine state.

All external forces on the landslide mass satisfy the equations of mechanical equilibrium while the safety factor decreases toward unity during the precursory stage. However, when the instability is triggered at the end of this stage, a rapid drop in strength occurs along B-C in Figure 3. An important observation of the Hartford Dike slide is that it started suddenly. Therefore, an assumption in the analysis is that the decrease in basal shear strength occurs with sufficiently high speed that, in accordance with (1), a momentum MV_0 is imparted 'instantaneously' to the landslide mass at time $t = 0$. At this same instant of time, the landslide mass also is acted upon by driving forces owing to the excess pore pressure plus the effective

active Rankine stresses and by resisting forces owing to the passive Rankine thrust and the reduced basal shear strength. In the Hartford Dike case the basal glide zone material had to be in an undrained condition during dynamic sliding because excess pore pressures were measured in the glide zone months after the slide occurred. Therefore, it is assumed that the reduced basal shear resistance, beginning at $t = 0$, can be characterized by 'friction angle' ϕ whose tangent is the ratio of basal shear resistance to total normal load.

The sudden impulse at the end of the precursory stage is treated as horizontal, and angular impulse components during sliding are neglected for reasons that follow the next two statements. The locations of the incipient normal fault and glide surfaces in Figure 1a are indicated by reference to Figure 1b. Data from numerous test borings were not published, and so the centers of mass of the landslide material currently are indeterminate. However, the centroids of the cross sections of landslide material in Figures 1a and 1b are at nearly the same height above the glide surface, despite the change of shape of the landslide mass during landsliding. Furthermore, the sheet piles remained vertical and perfectly intact. Therefore, only horizontal driving and resisting force components are utilized in the quantitative analysis. In addition, the Hartford Dike slide had a width of 365 m perpendicular to the direction of sliding; namely, a width appreciably larger than the dimensions of the landslide cross sections in Figure 1. No additional information on the three-dimensional geometry of the slide has been published, and so quantitative analysis is made by considering the cross sections in Figure 1 and 3 to be of unit width, with the assumption that the normal stresses acting perpendicular to each cross section are principal stresses.

The force-time relations for first-stage and second-stage landsliding are approximated by the linear variations in Figure 4. The sum of all horizontal driving forces is graphed above the time axis, with driving forces reckoned positive, resisting forces negative. Two driving forces act against the incipient normal fault at the head of the slide at $t = 0$: the active Rankine thrust and the excess fluid pressure thrust. P_a and P_0 are their horizontal components, respectively. P_a decreases during normal faulting

TABLE 1. Values of Driving and Resisting Forces for the Hartford Dike Slide

Time	Driving Forces	Resisting Forces
$t = 0$	$P_a = 7.3 \times 10^5$ N/m to 1.0×10^6 N/m $P_0 = 7.1 \times 10^5$ N/m to 7.6×10^5 N/m	$P_p = 1.0 \times 10^5$ N/m to 1.6×10^5 N/m $Mg \tan \phi < 1.4 \times 10^6$ N/m
$t = t_1$ ($t_1 < 60$ s)	$P_a' < P_a''$	$P_p' = 9.5 \times 10^4$ N/m to 1.2×10^5 N/m
		$Mg \tan \phi < 1.4 \times 10^6$ N/m
$t = t_f$ ($10^2 < t_f < 10^3$ s)	$P_a'' = 1.8 \times 10^5$ N/m to 3.3×10^5 N/m	$P_p'' = 1.1 \times 10^5$ N/m to 1.5×10^5 N/m $Mg \tan \phi < 1.4 \times 10^6$ N/m

Total weight Mg of landslide material was between 2.0×10^7 N/m and 2.4×10^7 N/m, based on Figure 1 and data culled from sources specified in text. N/m is Newtons per meter width, where the unit width is measured perpendicular to the cross sections in Figure 1.

to P_a' at time t_1, and P_0 decreases to zero at t_1 for reasons to be discussed later. The resisting forces at $t = 0$ are the basal shear resistance and the horizontal passive Rankine thrust $(-P_p)$ at the toe. P_p changes in magnitude to P_p' at t_1, whereas $-Mg \tan \phi$ is the average basal shear from $t = 0$ until time t_f, g being the gravitational acceleration. The graphs between t_1 and t_f may have slopes different from those shown, depending on the force values at t_1 and t_f. When landsliding is terminating, the value of the active Rankine thrust at the head of the slide is P_a'', whereas the total resisting force is $-(P_p'' + Mg \tan \phi)$ at t_f. P_p'' at the toe is calculated on the basis of material strength and the increased thickness of landslide material caused by heave.

The combined sum of horizontal driving and resisting forces during first-stage sliding, and also during second-stage sliding, can be represented by force equations of the following form:

$$\Sigma F_h = mt + b = Ma \qquad (2)$$

where m and b are constants, M is the mass of landslide material, and a is the acceleration of the center of mass. The appropriate type of velocity-time and distance-time relations that follow from (2) are

$$V = \frac{m}{2M} t^2 + \frac{b}{M} t + V_0 \qquad (3)$$

and

$$S = \frac{m}{6M} t^3 + \frac{b}{2M} t^2 + V_0 t + S_0 \qquad (4)$$

respectively, where, for first-stage sliding, V_0 is the instantaneous velocity at $t = 0$ and $S_0 = 0$ when displacements during the precursory stage can be neglected.

Therefore, using Figure 4 and equations (1)-(4), we obtain the following results:

$$I_0 = MV_1 - \tfrac{1}{2}[(P_0 + P_a + P_a')$$
$$- (P_p + P_p' + 2Mg \tan \phi)]t_1 \qquad (5)$$

$$V_1 = \frac{S_1}{t_1} + [(P_0 + P_a + 2P_a' - P_p - 2P_p')/6M$$
$$- (\tfrac{1}{2}) g \tan \phi] t_1 \qquad (6)$$

and

$$t_f = t_1 + \frac{V_1}{[g \tan \phi + (P_p' + P_p'' - P_a' - P_a'')/2M]} \qquad (7)$$

and

$$S_f = S_1 + \frac{(\tfrac{1}{2}) MV_1^2}{Mg \tan \phi + P_p' - P_a'} \cdot A \qquad (8)$$

for which

$$A = 1 + [(P_a' - P_a'') - (P_p' - P_p'')]$$

$$\cdot \frac{(P_a' + 3P_a'') - (P_p' + 3P_p'') - 4Mg \tan \phi}{3[(P_a' + P_a'') - (P_p' + P_p'') - 2Mg \tan \phi]^2}$$

where I_0 is the magnitude of the sudden impulse that starts the slide at $t = 0$, V_1 is the velocity at time t_1, S_1 is the horizontal distance traversed by the center of mass during first-stage sliding, t_f is the total time duration of catastrophic sliding, and S_f is the total sliding distance. S_1 can be estimated from the net slip of the normal fault. Coefficient A reduces to unity for the special case in which the driving and resisting forces are constant during second-stage sliding. That is, when $A = 1$, constant deceleration is established after normal faulting is completed at the head of the slide.

Data Analysis and the Instability Mechanism

Numerical calculations with respect to the Hartford Dike slide are tabulated in Table 1. A distinctive feature of this slide is that heave at the toe raised landslide material to almost the same elevation as the down-faulted dike (see Figure 1). Dike A was constructed by compacting the sand layer upon layer after sheet-pile emplacement. A principal purpose of sheet piles is to prevent a spreading failure of soft fill when more fill (e.g., dike A) is to be placed [Trautwine, 1900]. The fill on the foreland side of the sheet piles was a soft silt, and so the unit weight of compacted sand at the head was greater than the unit weight of silt at the toe when sliding began. Several types of 'compacted' and 'soft' sediments are characterized by fairly representative values of unit weight, porosity, permeability, and strength parameters. These values were obtained from Terzaghi and Peck [1948, pp. 25, 29, 31, 86, 376], Lambe and Whitman [1969, pp. 31, 149, 440], Peck et al. [1974, pp. 20, 22, 28, 43, 87, 93], and from Attewell and Farmer [1976, p. 39] for the calculations tabulated in Table 1.

P_0 was calculated from the excess fluid pressure in Figure 1a with respect to Figure 1b, as stated previously. Therefore, P_a and P_p were calculated on the basis of the active and passive Rankine effective stresses, respectively, utilizing appropriate thicknesses of landslide material at the head and toe regions at time $t = 0$. The active Rankine thrust decreased during normal faulting to P_a', which is shown to be smaller than P_a'' in Figure 4. P_a' must have been smaller than P_a'' because cavitation undoubtedly occurred during the normal faulting episode (e.g., see discussion and data on sand cavitation by Lambe and Whitman [1969, pp. 339-341]). Sand dike A subsided suddenly, in less than 1 min [Terzaghi and Peck, 1948], whereas the total duration of landsliding amounted to several minutes [Terzaghi, 1950]. P_0 therefore had to decrease to zero by at least time t_1. The mechanical behavior of the soft silt in the foreland, however, had to be quite different from that of the sand when landsliding began. The coefficient of permeability of silt is of the order of 10^{-6} to 10^{-7} m/s, a value usually far below that of sand. Therefore, in order to evaluate P_p' and P_p'', the strength of landslide material in the

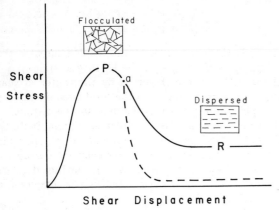

Fig. 5. Deformational behavior (solid-line graph) of an over-consolidated clay [after *Lambe and Whitman*, 1969]. Diagrams above P and R represent idealized internal clay structure, where P and R are the peak and residual strengths obtained during slow deformation under drained conditions. See text for discussion of how dashed-line graph might result from an instability.

foreland was estimated on the basis of the c/p ratio, $\phi = 0$ method described by *Peck et al.* [1974] for the deformation of soft fine-grained sediments under undrained conditions. The resisting thrust at the toe during t_1 and t_f was estimated with this method by utilizing the changes in thickness of foreland material indicated by the heave in Figure 1. The reason P_p' and P_p'' are close in value to P_p is that the average strength of wet silt is frequently smaller under undrained conditions than under drained conditions.

Terzaghi [1950] emphasized that '... the wedge-shaped body of silt, located on the river side of the sheet piles, could not possibly have advanced over a distance up to 60 feet without undergoing intense compression and shortening in the direction of movement ...' if the basal shearing resistance had not become extremely small. This basal shear can be evaluated by solving (6) and (7) for tan ϕ, namely,

$$\tan \phi = \left[\frac{2S_1}{gt_1^2} + \frac{P_0 + (P_a + 2P_a') - (P_p + 2P_p')}{3Mg} \right] \frac{t_1}{t_1 + 2t_2}$$
$$+ \left[\frac{(P_a' + P_a'') - (P_p' + P_p'')}{Mg} \right] \frac{t_2}{t_1 + 2t_2} \quad (9)$$

Although part of the landslide material slid almost 20 m, the actual total sliding distance, measured with respect to the centroid of the slide mass, is closer to 10 m, and so S_1 can be estimated conservatively to have been 5 m more or less. Therefore, by using the data in Table 1, (9) indicates that ϕ was smaller than 3° or 4°.

Precursory creep evidently was essential for the instability that created such a low value of ϕ. A prime possibility is that creep at a silt-clay layer interface can lead to a strain-weakening type of instability if the clay is overconsolidated or if the clay is a type for which the pore fluid at the interface chemically promotes dispersion of clay

platelets during creep. For example, the strain-weakening behavior of a saturated overconsolidated clay is illustrated in Figure 5. The solid-line graph represents deformational response to slow application of shear stress under drained conditions [*Lambe and Whitman*, 1969]. P is the peak shear strength and R is the residual shear strength. The idealized flocculated clay structure at P is transformed by shear to the dispersed type at R, where the clay platelets become parallel in the shear zone. The effective stress is supported by mineral grain contacts (not shown) within the clay and by the adsorbed water on the clay platelets [*Bombolakis et al.*, 1978]. The residual shear strength of saturated clay is accordingly a strong function of effective mean stress and total water content, which can vary over a wide range. For a specific water content, the quasi-static residual shear strength is

$$R = \sigma' \tan \phi' \quad (10)$$

where σ' is the effective normal stress and ϕ' is the friction angle for these conditions [*Lambe and Whitman*, 1969].

Slope stability analyses of the well-documented translational landslides in clays and compaction shales indicate that sliding frequently occurs when the peak shear strength is reduced toward the residual strength [e.g., *Skempton*, 1964; *Bjerrum*, 1967]. Where the descriptions are reasonably adequate, it is clear that these slides did not progress catastrophically when P was reduced to R. Therefore, one way in which a strain-weakening instability might produce catastrophic slides is indicated by the dashed-line portion of the graph in Figure 5. For example, suppose that premonitory creep has proceeded along the solid-line graph to point 'a,' where the bonds in the flocculated structure are being broken down. If the water con-

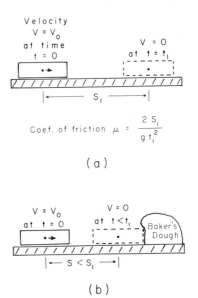

Fig. 6. Example problems of sliding friction to illustrate applicability of equations derived in text. The equations are valid for 6a as a special case, but not for 6b.

tent of the clay at point 'a' increases rapidly along an interface between the clay and a saturated permeable medium, then the shear strength of the clay at the interface will drop rapidly along the dashed-line graph to a value smaller than the quasi-static residual strength. This type of mechanism might explain why other similar slides, such as the 1915 Hudson River slide [*Terzaghi*, 1950], occurred catastrophically.

A problem for future consideration is how the relations between resisting and driving forces limit the validity of (6)-(9). This problem is illustrated briefly by the example of a block on a table in Figure 6. A sudden impulse in Figure 6a imparts an instantaneous velocity V_0 to the block at $t = 0$. The only horizontal external force acting on the block during sliding is a constant friction at the base, where the coefficient of sliding friction is $\mu = \tan \phi$. There is only one stage of sliding, with S_1 and t_1 the sliding distance and travel time, respectively. Since $t_2 = 0$ in this situation, (9) reduces to the equation in Figure 6a as a special case. In a duplicate experiment in Figure 6b, however, a batch of baker's dough happens to lie in the path of the sliding block. The sliding distance in Figure 6a accordingly is not attained in 6b, and the sense of basal shear is reversed as the block partially penetrates sticky plastic dough before coming to a full stop. Therefore, one way to test the validity of (5)-(8) is to analyze the parameters in (9) to determine, for example, when the sign of $-Mg \tan \phi$ is altered.

Concluding Remarks

There are several noteworthy features of the Hartford Dike slide that may prove helpful in the analysis of certain modern and paleotectonic slides. The Hartford Dike slide was horizontal, and a consideration of it may help eliminate confusion regarding the importance of certain types of gravitational driving forces in low-angle slides, overthrusts, and decollements. Furthermore, in contrast with some recent catastrophic landslides in sediments, the instability mechanism in the Hartford Dike slide was not sand liquefaction. The analysis of this slide indicates that in addition to the sand liquefaction research currently emphasized for prediction of earthquake-induced landslides, research also needs to be done on the stress-strain-time behavior of clay and compaction shale at interfaces with saturated permeable media.

The value of ϕ for the Hartford Dike slide is well below typical values of the friction angle reported for the residual strength in well-documented noncatastrophic translational slides in clay systems. Typical values for these slides are close to 10° to 15° [e.g., *Bjerrum*, 1967]. The instability mechanism that led to the low value of ϕ was a mechanism in which excess pore pressure was not dissipated rapidly during sliding. An important consequence is that relatively little deformation (apart from normal faulting at the head of the slide) occurred above or below the basal glide zone. This is the kind of observation that has been a puzzle for many years in the study of over-

thrusts in shale. For example, in a structural geology textbook used extensively almost half a century ago, we read that one of the remarkable features of a number of overthrusts in the Rockies is '... the frequent absence of slickensides, zones of brecciation, gouge, and other common phenomena associated with fault surfaces ...' [*Nevin*, 1936].

Acknowledgments. It is a pleasure to acknowledge B. Johnson of the Center for Tectonophysics, Texas A&M University, and H. J. Melosh of the Department of Earth and Space Sciences, SUNY at Stony Brook, for their helpful comments on a preliminary draft of this paper.

References

Attewell, P. B., and I. W. Farmer, *Principles of Engineering Geology*, 1045 pp., John Wiley, New York, 1976.

Billings, M. P., *Structural Geology*, 3rd ed., 594 pp., Prentice-Hall, New Jersey, 1972.

Bjerrum, L., Progressive failure in slopes of over-consolidated plastic clay and clay shales, *J. Soil Mech. Found. Div.*, *93*, 3-49, 1967.

Bjerrum, L., Problems of soil mechanics and construction on soft clays, State-of-the-Art Report to session 4, *Geotech. Eng. Environ. Control. Proc. Spec. Sess. Int. Conf. Soil Mech. Found. Eng.*, 8th, 1974.

Bombolakis, E. G., Some constraints and aids for interpretation of fracture and fault development, in *Proceedings of the Second International Conference on Basement Tectonics*, edited by M. H. Podwysocki and J. L. Earle, pp. 289-305, Basement Tectonics Committee, Inc., Denver, 1979.

Bombolakis, E. G., J. C. Hepburn, and D. C. Roy, Fault creep and stress drops in saturated silt-clay gouge, *J. Geophys. Res.*, *83*, 818-829, 1978.

Burland, J. B., T. I. Longworth, and J. F. A. Moore, A study of ground movement and progressive failure caused by a deep excavation in Oxford clay, *Geotechnique*, *27*, 557-591, 1977.

Den Hartog, J. P., *Mechanics*, 462 pp., Dover, New York, 1961.

DeSitter, L. U., *Structural Geology*, 552 pp., McGraw-Hill, New York, 1956.

Griggs, D. J., and J. Handin, Observations of fracture and a hypothesis of earthquakes, in Rock Deformation, edited by D. T. Griggs and J. Handin, *Mem. Geol. Soc. Am. 79*, pp. 347-364, 1960.

Handin, J., Strength and Ductility, in *Handbook of Physical Constants*, edited by S. P. Clark, Jr., *Mem. Soc. Am. 97*, pp. 223-290, 1966.

Hobbs, B. E., W. D. Means, and P. F. Williams, *An Outline of Structural Geology*, 571 pp., John Wiley, New York, 1976.

Hubbert, M. K., and W. W. Rubey, Role of fluid pressure in mechanics of overthrust faulting, *Geol. Soc. Am. Bull.*, *70*, 115-206, 1959.

Lambe, T. W., and R. V. Whitman, *Soil Mechanics*, 553 pp., John Wiley, New York, 1969.

Morgenstern, N. R., and D. A. Sangrey, Methods of stability analysis, in Landslides, Analysis, and Control, edited by R. L. Schuster and R. J. Krizek, *Spec. Rep.˙ 176*, Transp. Res. Bd., Nat. Acad. Sci., Washington, D. C., 1978.

Nevin, C. M., *Principles of Structural Geology*, 2nd ed., John Wiley, New York, 1936.

Peck, R. B., W. E. Hanson, and T. H. Thornburn, *Foundation Engineering*, 2nd ed., John Wiley, New York, 1974.

Scott, R. F., Hazards from landslides, in *Geological Hazards*, 2nd ed., edited by B. A. Bolt, W. L. Horn, G. A. Macdonald, and R. F. Scott, pp. 148–196, Springer-Verlag, New York, 1977.

Seed, H. B., Landslides during earthquakes due to soil liquefaction, *J. Soil Mech. Found. Div.*, SM5, 6110–6179, 1968.

Seed, H. B., Considerations in the earthquake-resistant design of earth and rockfill dams, *Geotechnique, 29*, 215–263, 1979.

Seed, H. B., and S. D. Wilson, The Turnagain Heights landslide, *J. Soil Mech. Found. Div., SM4*, 325–353, 1967.

Skempton, A. W., Long-term stability of clay slopes, *Geotechnique, 14*, 77–101, 1964.

Terzaghi, K., Mechanism of landslides, in *Engineering Geology*, edited by S. Paige, pp. 83–123, Geological Society of America, Boulder, 1950.

Terzaghi, K., and R. B. Peck, *Soil Mechanics in Engineering Practice*, 566 pp., John Wiley, New York, 1948.

Trautwine, J. C., *The Civil Engineer's Pocketbook*, 17th ed., 866 pp., John Wiley, New York, 1900.

Voight, B., The mechanics of retrogressive block-gliding, with emphasis on the evolution of the Turnagain Heights landslide, Anchorage, Alaska, in *Gravity and Tectonics*, edited by K. A. DeJong and R. Scholten, pp. 97–121, John Wiley, New York, 1973.

Voight, B. (Ed.), *Mechanics of Thrust Faults and Decollement*, 441 pp., Dowden, Hutchinson, and Ross, New York, 1976.

Wilson, S. D., Observational data on ground movements related to slope stability, *J. Soil Mech. Found. Div.*, SM5, 7508–7533, 1970.

Wilson, S. D., and P. E. Mikkelsen, Field instrumentation, in Landslides, Analysis, and Control, edited by R. L. Schuster and R. J. Krizek, pp. 112–137, *Spec. Rep. 176*, Transp. Res. Bd., Nat. Acad. Sci., Washington, D. C., 1978.

Tectonics of China: Continental Scale Cataclastic Flow

JOHN J. GALLAGHER, JR.

Basin Study Department, Cities Service Company, Tulsa, Oklahoma 74102

Stratigraphic, structural, and earthquake evidence indicates that cataclastic flow, that is, flow by brittle mechanisms (e.g., fracture and slip), was dominant in China from late Paleozoic. This process has operated over a range of scales including the continental scale. China is made up of large brittle basement elements immersed in ductile zones which are analogous to porphyroclasts (large, often brittle fragments) surrounded by fluxion (foliation or flow) structures in cataclastic rocks, respectively. This basement fabric for China is seen on Landsat imagery and on tectonic maps and is comparable to cataclastic rock fabrics seen in fault zones, on outcrops, and in thin sections. Brittle basement elements are broken into two or more large rigid blocks, and the dimensions of elements and blocks are within 1 order of magnitude of each other. Ductile zones are made up of fragments which are many orders of magnitude smaller than the ductile zones. Rigid blocks and fragments are identified, and their dimensions are measured through earthquake, fault, and fracture patterns. Rigid basement blocks are surrounded by earthquakes. The sedimentary rocks over the basement faults at the block boundaries seem to be affected by fault movements because they are characterized by facies changes, thickness changes, high-angle faults, and forced folds. Ductile basement zones are earthquake prone, and deformation of the ductile basement affects the overlying sedimentary rocks, as is demonstrated by unconformities and by a wide variety of structures. Thrust faults, buckle folds, and strike slip faults are common in and adjacent to western ductile zones. Structures are most intensely developed where ductile zones abut brittle elements. Both brittle elements and ductile zones are rifted and cut by strike slip faults in eastern China. The mechanical fabric of China and the boundary conditions acting on China are now and always have been determined by its plate tectonic history. This inference is made from recently published plate tectonic interpretations. Geologic maps show that there are six elements and that each element has a Precambrian, crystalline core which is surrounded by upper Paleozoic continental margin suites of rocks, including subduction complexes, among others. Geologic data on ophiolites demonstrate that the brittle elements and their margins were juxtaposed and then welded together along suture zones during Permian and Triassic time to make China. Cenozoic plate motions affecting China resulted in the collision with India where it converges with southwest China and the extension in eastern China where island arcs move away from the mainland and where grabens are actively forming. The juxtaposition to Siberia, which acts as a buttress against northern China, explains the compression of western China, and the absence of a buttress in the Pacific Ocean explains why eastern China can extend. Furthermore, laboratory data on the mechanical behavior of rock under conditions analogous to the shallow crustal conditions of interest in China show that all rocks are weaker in extension than they are in compression. Basement rock in western China is strong because it is compressed, but this same basement rock is weak in eastern China because it is in extension. The tectonics of China or, in mechanistic terms, the way in which the mechanical framework of China responds to Cenozoic boundary forces was a result of China's previous plate tectonic history. Crystalline cores are the rigid blocks that form brittle elements. Both the continental margin suites and the sutures are the ductile zones. The sutures and sediment patterns seen in the basins and ranges of China can be explained in terms of this tectonic scenario.

LATE CAMBRIAN

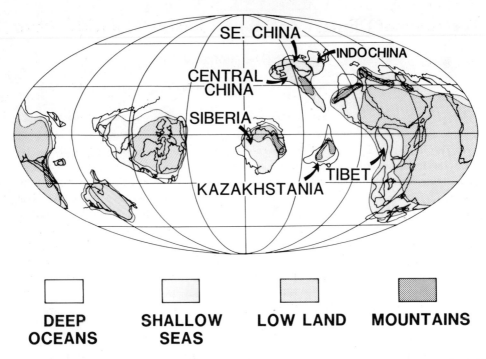

Fig. 1.   Continent distribution and gross physiography for Late Cambrian [after *Scotese et al.*, 1979].

Introduction

The tectonic provinces of China are well defined, their history is well established, and a scenario for their plate tectonic heritage has been proposed [*Chi-ching*, 1978; *Kenyon et al.*, 1979; *Scotese*, 1979]. Chinese scientists recognized that 'depth fractures,' which include a host of structures (among them are strike slip faults, geologic contacts between different basement types, and zones composed of various kinds of faults which cut through or nearly through the lithosphere), are important because they separate tectonic provinces [*Lee*, 1939; *Chi-ching et al.*, 1974]. These provinces have been categorized as 'stable blocks,' 'graben systems,' or 'foldbelts' [*Terman*, 1974]. Earthquakes in the foldbelts and graben systems and observed surface structures were used by *Tapponnier and Molnar* [1976] and *Molnar and Tapponnier* [1977] to show that at the continental scale, China behaves like a plastic body being deformed by a rigid indenter, which is the India plate. Their very creative application of two-dimensional, plane strain, plastic flow theory to China explained the types of structures observed and, in general, the regions where the structures were seen.

My work builds on theirs and takes into account in more detail the mechanical processes responsible for the deformation of China. The result is that the locations of structures are better explained and the effects of tectonics on the broad aspects of sediments and unconformities are explained. In particular, this paper shows that stable blocks

are brittle elements composed of rigid basement blocks of granitic rock, that foldbelts are ductile basement zones composed of fragments of continental margin and suture zone suites of metamorphic rock, and that the graben systems are rifts that cut across both basement types.

Stable blocks are surrounded by foldbelts and resemble fabrics of cataclastic rocks shown by *Higgins* [1971]. My work shows that these resemblances are both geometric and rheological. This leads to a general conclusion of the paper, which is that cataclastic flow, as defined by *Handin* [1966], exists over a range of scales from submicroscopic to global and that the process involves deformation of an aggregate which may be mechanically different from the deformation of the constituent elements.

Plate Tectonic and Orogenic History

Maps of continental drift based on paleomagnetic and paleoclimatic data show that six continental elements, which were separate during middle Paleozoic time, came together to make China (and adjacent Eurasia) by the end of Triassic time [*Scotese et al.*, 1979; *Scotese*, 1979]. These continental elements were (1) Siberia, (2) Kazakhstania, (3) North China-Korea, (4) Tibet, (5) South China, and (6) Indochina (Figure 1). India was a seventh continental element which was added during Eocene and younger time [*Scotese*, 1979]. Today these seven continental elements join together along suture zones (Figure 2). The evidence that these zones may be sutures or closed oceans

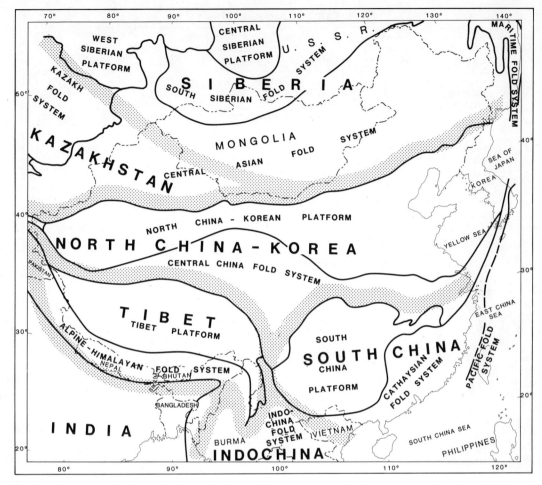

Fig. 2.   The seven plate tectonic elements of east Asia are connected along zones roughly shown by the stippling. (The base map is adapted from *Terman* [1974], and element boundaries are inferred from *Scotese et al.* [1979].)

includes the presence of ophiolites from Mesozoic and older oceans and typical geosynclinal suites of rocks [*Hutchison*, 1975; *Chi-ching*, 1978]. Further evidence is given by *Terman* [1974] and *Chi-ching* [1978], who take up the geology of the continental elements and suture zones and their history.

Each of the six continental elements composing China is cored by pre-Sinian (older than 1400 m.y.a.) crystalline basement rock which was intensely deformed during Caledonian and older orogenies [*Terman*, 1974]. Table 1 shows a comparison of terms commonly used to describe orogenies and includes terms used primarily for China like 'Sinian,' and selected geography is shown in Figure 3. The effect of younger pre-Cenozoic orogenies on these continental elements was limited to extension of the eastern China-Korean Platform and of the Central Siberian Platform (Figure 2).

There is a general youthening southward of suture zone ages from late Paleozoic in the north, to early Mesozoic in the south and northeast, and to Cenozoic along the Indus

suture. The central sutures and the adjacent foldbelts were most strongly compressed by the Hercynian orogeny, but the southern and northeastern foldbelts were most intensely deformed by the Variscan and Indosinian orogenies. These dates correspond well with those for collision of the continental elements obtained from plate tectonic data [*Scotese*, 1979]. Igneous rock related to cratonic breakup and sedimentary rock deposited during drifting were metamorphosed during the suturing process and form the basement rock in suture zones.

Present-day plate movements show that western China (and adjacent Eurasia) is being compressed by the continental impact of the India plate [*Molnar and Tapponnier*, 1977; *Hayes and Taylor*, 1979; *Kenyon et al.*, 1979]. From this it is inferred that Siberia acts as a buttress, and this explains why western China is compressed. Eastern and northern China (and adjacent USSR) are being extended in marginal seas and along rift systems like the Shansi grabens and the extension zone from Lake Baikal on to the west [*Hayes and Taylor*, 1979; *Chen et al.*, 1979]. Thus

TABLE 1. Orogenic Terms

Period	Isotopic Age, m.y.a.	China	North America	Europe	USSR
Cenozoic					
Quaternary	15	Himalayan	Pasadenian	Alpine	Kamchatkian
Tertiary	67	Yenshanian			
			Laramide		
Mesozoic					
Cretaceous	141			Kimmerian	Cimmerian
Jurassic	195	Indosinian			
Triassic	230	Variscan	Nevadan	Variscan	
Paleozoic					
Permian	280	Hercynian	Appalachian	Hercynian	Hercynian
Carboniferous	345				
Devonian	395	Caledonian	Acadian	Caledonian	Caledonian
Siberian	435				
Ordovician	500				
Cambrian	570	Hsingkaiian	Taconic		
Late Proterozoic					
Eocambrian	700	Yangtzeian		Assyntian	Baikalian
Chingpaikou	1000				
Chihsien	1400	Hsueh-feng	Grenville	Dalslandian	Hyperborean
(Sinian		Hutoian	Mazatzal	Gothian	Ovruchian
Subera)*				Laxfordian	Volynian
Changcheng	1950	Chungtiaoian	Hudsonian	Karelian	Krivorogian
Early Proterozoic					
Huto	2000	Wutaiian			
		Taishanian			Bug-Podolian
Wutai	2500	Fupingian			
Archaean					
Fuping			Kenoran	Saamian	Dneprovian

Timetable and orogenic terms are a combination of those presented by *Terman* [1974], *Van Eysinga* [1975], and *Chi-ching* [1978].
*Institute of Geochemistry, Academia Sinica [*Chi-ching*, 1978].

the Pacific plate does not appear to act as a buttress, despite convergence along the subduction zones of the northwest Pacific, because eastern China is moving eastward in relation to the rest of China. This observation is supported by *Lee et al.* [1980], who show the presence of extensional structures and the absence of compressional structures in trench walls for subduction zones of the western Pacific.

Consequently, China is an example of a continental scale body that owes its mechanical fabric to plate tectonic processes, and zones of weakness in this fabric have been reactivated by present-day crustal movements, as evidenced by earthquake activity. Reactivation of preexisting planes of weakness (or mechanical anisotropies) by plate movements is controlled by physical laws, such as Amontons law of friction [*Bowden and Tabor*, 1964], which, thus far, have been applied too infrequently to structural geology [*Bombolakis*, 1979]. Indeed, the observed association between earthquakes and faults, where the faults follow geologic contacts which juxtapose different rock types with different ages, suggests that this is happening in both compressing and extending regions of

China [*Molnar and Tapponnier*, 1977]. Reactivation is an important process which deserves more attention.

Earthquakes

Earthquakes in onshore China are less than 70 km deep [*York et al.*, 1976], so it seems reasonable to assume that the basement rock is brittle. Three types of basement behavior can be distinguished for China: (1) where earthquakes occur, behavior is brittle (i.e., fracturing, faulting, crushing, slipping, sliding, and rotating); (2) where earthquakes do not occur and there is evidence of active deformation, behavior is purely ductile (i.e., plastic or viscous or both); and (3) where there are no earthquakes and there is no evidence of active deformation but the region is surrounded by actively deforming zones, the undeformed region may be subjected to stress, but the region behaves rigidly (i.e., does not deform internally). This last criterion must be used cautiously for regions where historical earthquake data are sparse, such as parts of western China.

The absence of earthquakes suggests that rigid base-

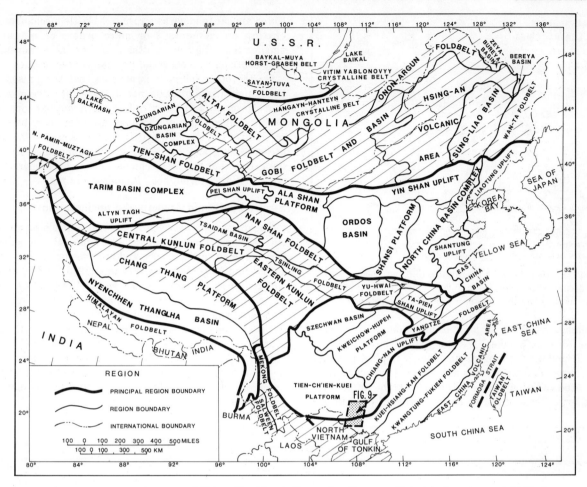

Fig. 3. Major tectonic elements of China. Hatching indicates ductile zones. Unhatched zones are rigid blocks or brittle elements. (The base map is adapted from *Terman* [1974].)

ment blocks underlie the basins and ranges of the six continental elements but not the foldbelts which surround them. For example, the Ordos Basin of the North China-Korea element is surrounded by earthquakes, but no large earthquakes are reported within the basin (Figure 4). Other rigid regions surrounded by earthquakes include the Dzungarian, Tarim, Szechwan, and parts of the North China Basin Complex in addition to the Ala Shan and Tien-Ch'ien-Kuei uplifts (Figure 3; see, for example, *York et al.* [1976]). Earthquake patterns show that the rigid blocks which underlie these regions probably have characteristic diameters of the order of 10^5–10^6 m. *Molnar and Tapponnier* [1977] have shown that many earthquakes in China can be related to individual faults. Some of these faults occur at the borders of rigid blocks like the blocks in the Ordos Basin.

Other earthquakes occur with individual faults, but there are so many faults that the regions are pervaded by earthquakes. The width of these earthquake zones [*York et al.*, 1976], number of structures, and geological com-

plexities [*Terman*, 1974] suggest individual rigid block sizes of the order of 10^3 m or smaller. Where they are subjected to horizontal compressional forces, the zones are marked at the surface by thrust faults, strike slip faults, anticlinoria, and synclinoria. Examples of these zones surround the Dzungarian, Tarim, and western Ordos basins and include the Tsaidam Basin and the Tsinling Foldbelt (Figure 3). Where they are subjected to extensional forces, the zones are characterized by rifts and strike slip faults. Examples of such zones are the graben systems adjacent to the eastern Ordos Basin and the basins of northeastern China.

In apparent contradiction to the brittle deformation for China, suggested by earthquakes, are flow patterns on maps [*Terman*, 1974]. This apparent dilemma between brittle and ductile behavior or flow can be resolved if the size of the brittle blocks in the zone of flowage is very much smaller than the size of the ductile zone itself. Irregularities of brittle block edges would not be obvious at the scale of the flowage zones. Zones of flow have character-

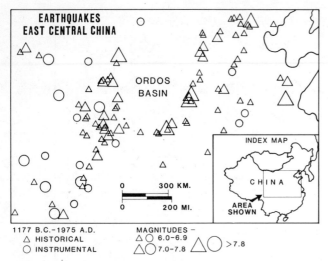

Fig. 4. Earthquakes with magnitudes of 6 and greater for the region around the Ordos Basin [after *York et al.*, 1976]. The depth to focus is less than 70 km for all earthquakes shown.

istic diameters of 10^5-10^6 m, and individual brittle blocks may be 10^3 m or smaller. *Terman* [1974] shows Cenozoic faults following foliation planes, which implies that the brittle blocks in ductile zones are bounded by faults or by reactivated foliation planes. In any event, the implication is that large pieces of the continental crust of present-day

China appear to move as a result of brittle mechanisms operating on elements 2 or more orders of magnitude smaller than the zone widths and that these movements may contribute to observed flowlike characteristics.

On the other hand, it is obvious that the present-day earthquakes in China do not demonstrate unequivocally that Mesozoic and older deformation mechanisms were brittle. Hence much of the flow character could be due to mechanisms other than brittle mechanisms and may have occurred long ago; however, stratigraphic and structural data show that brittle mechanisms have been operative in parts of China throughout Phanerozoic time.

Stratigraphic and Structural Considerations

The distinction between rigid basement blocks and ductile basement zones is supported by the data on sedimentary rocks, unconformities, and hiatuses presented by *Lee* [1939] and *Meyerhoff and Willums* [1976] and by data on surface structure shown by *Terman* [1974] and *Chi-ching* [1978] and seen on Landsat images. The total sediment isopachs (which are somewhat speculative owing to lack of data) and surface structures in the Tarim Basin Complex seem to be more or less typical of those in other China basins on rigid basement blocks. Basement structure and depth of subsidence are inferred from sediment isopachs and from surface structure by using techniques discussed by *Gallagher and McGuire* [1979].

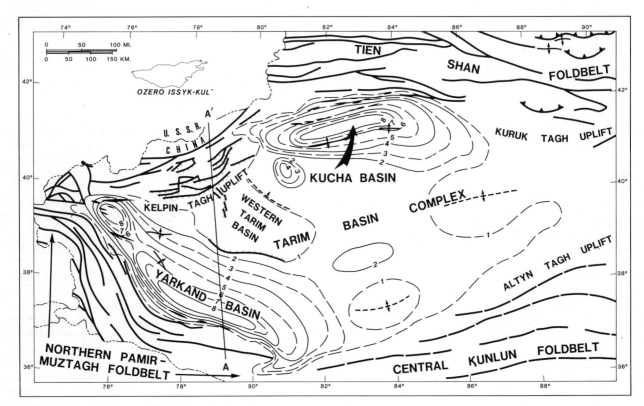

Fig. 5. Isopach map of Phanerozoic sediments (in kilometers) with standard symbols for surface structures in the Tarim Basin Complex [after *Terman*, 1974].

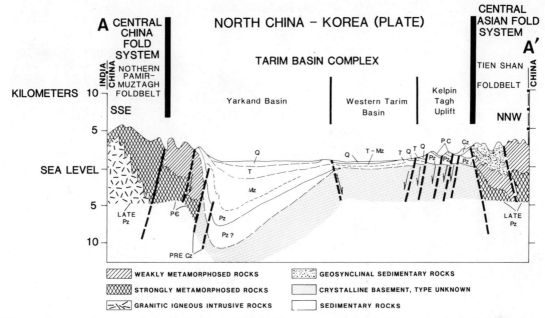

Fig. 6. Cross section through the Tarim Basin Complex [after *Terman*, 1974]. The section is about 500 km long. Faults were active during the Cenozoic.

Sediment thicknesses are uniform and total a few kilometers in the western Tarim Basin (Figure 5), and today the sedimentary rocks are near sea level (Figure 6). There are no major surface structures within the central part of the basin, but its limits are defined by faults, folds, and increased sediment thicknesses, all of which imply vertical movements of less than 1 km above or below sea level of a flat-topped, fault-bounded rigid basement block with map dimensions of a few hundred kilometers.

Just south of the western Tarim Basin is the Yarkand Basin, which is an asymmetrical basin with sediments that thicken southwestward to several kilometers. All but 1 km or so of these are below sea level (Figures 5 and 6). The southern limit of this basin is in thrust fault contact with the basement rocks of the Northern Pamir-Muztagh Foldbelt. These structural and stratigraphic data imply a fault-bounded, rotated basement block that subsided at one end to several kilometers below sea level. Map dimensions of the blocks are of the order of 200 × 600 km.

The Kelpin Tagh Uplift of the northwestern border of the Tarim Basin Complex is a major basement block which is broken by at least five major high-angle faults spaced at about 10-km intervals (Figure 5). Paleozoic carbonate and Mesozoic clastic rocks are exposed, and forced folds like those described in the U.S. Rocky Mountains by *Prucha et al.* [1965], *Hodgson* [1965], *Stearns* [1971, 1978], and *Gallagher and McGuire* [1979] should occur in this mechanical situation, but the structures have not been described in mechanical terms for this uplift in China.

The other basins of China, inferred from the same kinds of stratigraphic and structural data to be on rigid basement blocks in the Cenozoic, are the central and eastern

Tarim, the Dzungarian, the Ordos, the Szechwan and adjacent blocks to the south, and at least parts of the Sung-Liao and the North China basins (Figure 3). Uplifts of rigid blocks include the Ala Shan and the Kweichow-Hupeh and Tien-Ch'ien-Kuei platforms.

The sediment patterns and surface structures in the Tsaidam Basin typify those in other China basins on ductile crust. Faults, folds, unconformities, and facies changes are much more common in the Tsaidam Basin than they are in the Tarim Basin Complex [*Meyerhoff and Willums*, 1976]. Thus it is much more difficult to correlate from one stratigraphic section to another. Major folds are spaced at 10 km or less, and major faults are separated by 50 km or less. This suggests the possibility of a ductile basement with deforming brittle elements 10 km or smaller in size.

An alternative solution is that the ductile rock around the Tsaidam Basin is laterally squeezing the sediments and that the basin is actually on rigid blocks. I cannot deny this possibility but prefer to think that the Tsaidam Basin is underlain by ductile crust, mainly because other regions about the same size but on rigid crust, such as the easternmost part of the Tarim Basin Complex, do not display the numerous and intense structural and sedimentary features seen in the Tsaidam Basin.

Other basins interpreted to be on ductile crust on the basis of structures, stratigraphy, and unconformities similar to those in the Tsaidam Basin are small basins in the Tien-Shan and Nan Shan foldbelts (namely, the Turfan, Pre-Nan Shan, Min Ho, and Central Yunan basins). Nearly all the mountain ranges in the major China fold systems are on ductile crust, and these include the Central

CATACLASTIC ROCKS

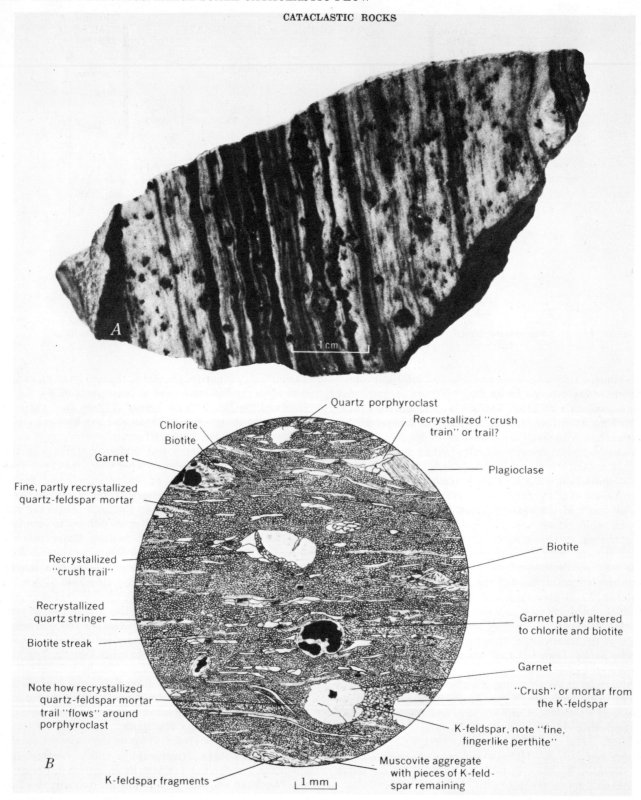

Fig. 7. (*a*) A typical cataclastic rock hand specimen. (*b*) Microdrawing of mylonite [after *Higgins*, 1971]. The flow character of the hand specimen is seen in thin section to be a result of grain crushing, sliding, and rotation.

Fig. 8a. Landsat mosaic of the region around the Tarim Basin (courtesy LEERCO, Salt Lake City, Utah).

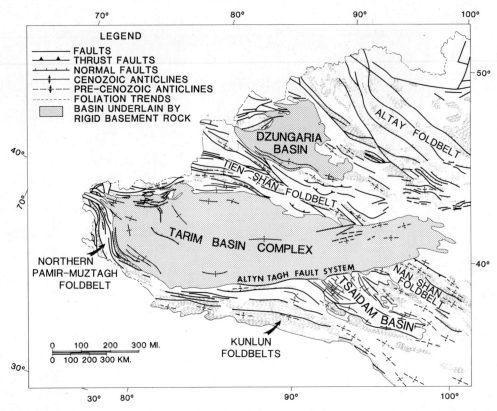

Fig. 8b.   Geologic sketch map of the Tarim Basin and surrounding region [after *Chi-ching*, 1978; *Terman*, 1974].

Asian, Central China, and Indochina fold systems (Figure 2).

Where ductile regions are adjacent to brittle elements, they are generally high and intensely deformed and occur in narrow zones tens of kilometers wide but hundreds of kilometers long. They are roughly coincident with suture zones and have been called depth fractures [*Chi-ching*, 1978]. In fact, they can be assigned to any one of three categories depending on whether they are dominated by thrust, strike slip, or normal faults. As *Tapponnier and Molnar* [1977] have pointed out, thrust faults are common in the western part, strike slip faults are common in the central part, and normal faults are common in the northeastern part of China, although other types of faults are found in each of these parts. They emphasize the relation of faults to the state of stress and boundary conditions, but I emphasize in addition the mechanical fabric of the basement rock of China.

Horizontally compressed regions of mixed basement, which is fragments of brittle basement thrust into ductile basement zones as seen on geologic maps such as the map by *Terman* [1974], are cut by thrust and strike slip faults and are characterized by episodic erosion since Triassic; thus Mesozoic and younger sediments are rare.

Regions of mixed basement subjected to extension are cut by high-angle faults, including strike slip faults, and generally have the characteristics of grabens. Examples of

thrust-faulted, mixed basement are in the Himalayas, the northern Tarim Basin, and the southern Dzungarian Basin [*Terman*, 1974]. Major strike slip faults are in the Altyn Tagh and central Kunlun regions. Graben systems extend from the Shansi region northward. Fault planes are seen on geologic maps to follow or parallel major geologic boundaries between different rock types with different ages [*Terman*, 1974] which are apparently not welded together (by igneous or metamorphic processes) and thus are probably planes of weakness. Other extensional regions are in the marginal basins offshore, and the faults in these regions cut across major geologic boundaries [*Chen et al.*, 1979; *Kenyon et al.*, 1979].

Cataclasis

Cataclasis is proposed as the major deformation mechanism for China because the plate tectonic, structural, stratigraphic, earthquake, and general geologic data for China and the resulting inferences discussed above support this proposition. The process of cataclasis was defined by *Handin* [1966] as 'crushing, granulation and fracture of individual grains or localized regions within the aggregate accompanied by intergranular adjustments.' He had the physical insight and foresight to generalize the definition. The term 'localized regions' leaves the way open for scale independence and suggests that the process of cataclasis may apply to very large as well as very small geo-

Fig. 8c. Thin section of a mylonite showing a fractured porphyroclast surrounded by fluxion structures [from *Higgins*, 1971].

logic bodies. This same latitude of scale is implicit in other widely accepted definitions of the term [e.g., *Higgins*, 1971]. Furthermore, the term 'aggregate' emphasizes an important distinction between the mechanical properties of the sum total of elements (which is the aggregate) and those for each individual element. Finally, this definition suggests not only that brittle fracture is important but also that rigid-body translation and rigid-body rotation are important mechanisms in the process of cataclasis. Rigid-body movements may be difficult to detect in thin section, but they can be observed in geologic bodies such as outcrops by using criteria like those for recognition of faults.

Stearns [1968, 1969] combined the *Handin* [1966] concepts of cataclasis with the *Griggs and Handin* [1960] concept of cohesion to amplify distinction between the properties of aggregates and those of individual elements. Most field and experimental studies of cataclasis emphasize the mechanical properties of the aggregate [e.g., *Borg and Maxwell*, 1956; *Borg et al.*, 1960], but *Gallagher et al.* [1974] studied the constituent elements of aggregates and related aggregate deformation to stress concentrations and rigid-body movements of elements as observed in photo-elastic models. These phenomena were compared with the behavior of aggregates like arrays of glass spheres, quartz sand, and sandstone, but the mechanical processes ob-

served should apply equally well to elements that are several orders of magnitude larger. These field and experimental studies provide further mechanical insight into the behavior of brittle elements in cataclastic fabrics like those observed by *Higgins* [1971] (Figure 7). Most important, there is a difference between the behavior of the cataclastic aggregate and the cataclastic elements; some groups of elements may be brittle, while others may be ductile.

Stearns [1969] referred to cataclasis observed in the field at the scale of beds or at the scale of force fold hinges as 'large scale cataclasis.' He documented field examples of cataclasis at this scale from structures in the Rocky Mountains in the western United States. He suggested that the process of brittle fracture spans at least 8 orders of magnitude; thus the process of cataclasis should occur over the same range of scales. Indeed, *Higgins* [1971] proposed that cataclastic structures in thin sections may reflect map patterns, which further supports the idea of cataclasis over a range of scales.

Moreover, the *Handin* [1966] definition implies that the size of the aggregate is so much larger (at least an order of magnitude) than the size of constituent elements that irregular or angular shapes of individual elements do not detract from the smooth outline and flow characteristics of the aggregate. However, neither the *Handin* [1966] definition nor other definitions [*Higgins*, 1971] mention absolute size, so the term 'cataclasis' and associated terms such as 'cataclastic flow' could apply to deformed continental aggregates with flow characteristics such as China where individual brittle elements are as large as major basins and major ranges.

Many of the deformation features of China, as seen on the *Terman* [1974] map or Landsat images, resemble those seen in the field, in thin sections of naturally and experimentally deformed rocks (Figure 8), and in photoelastic models [*Gallagher et al.*, 1974]. For example, ductile regions appear to flow around the Tarim Basin (Figures 8a and 8b) in much the same way that fluxion structures (cataclastic foliation or flow structures) seem to flow around porphyroclasts (relatively large fragments in a cataclastic rock) in mylonites (Figure 8c) or the way deformation bands [*Aydin*, 1977] separate undeformed regions in the Wingate Sandstone of Colorado [*Jamison*, 1979]. Specifically, the Cenozoic faults in the Northern Pamir-Muztagh, central Kunlun, and Tien-Shan foldbelts and the foldbelt north of the Kelpin Tagh Uplift follow the borders of the Tarim Basin Complex, and they appear to flow around it. *Terman* [1974] maps these Cenozoic faults parallel to foliation planes in the Paleozoic metamorphic rocks of these foldbelts (Figure 8b), and it is these faults and foliation structures that resemble fluxion structures (Figure 8c). The major faults (if sediments could be stripped off) which separate the rigid basement blocks within the Tarim Basin Complex might resemble fractures in porphyroclasts. Major basement faults separate the Yarkand Basin from the western Tarim Basin Complex, and similar basement faults probably underlie forced folds seen on the surface of the southern and eastern Tarim Basin Complex.

An example supporting (but not proving) pre-Cenozoic cataclasis in China is the rigid blocks (brittle elements) immersed in ductile material (ductile elements) in Triassic and older rocks of the South China Platform (Figure 9). Brittle elements are fractured with patterns resembling those predicted by *Gallagher et al.* [1974] from strain distributions observed in cemented photoelastic models and arrays of fractures observed pervading grains by *Friedman* [1963] in thin sections of calcite-cemented sandstone. Ductile elements appear to have flowed around the brittle elements. Both the brittle and the ductile elements were deformed during Mesozoic time [*Terman*, 1974].

A final example which suggests cataclasis for China is the zones of intense deformation seen in the Kunlun Foldbelt, which resemble intensely deformed zones reported in the field by *Stearns* [1978], the deformation bands seen in the field by *Jamison* [1979], and corresponding zones seen in experimentally deformed rocks by *Borg et al.* [1960] and *Handin* [1966] and in clay models of China by *Gallagher and Withjack* [1980]. Zones like these are observed in China on maps and on Landsat images, and they not only display cataclastic geometries but also show deforma-

tional characteristics such as fractures and flow phenomena.

These examples show that continental scale cataclastic features seen in various parts of China look like the smaller features seen in cataclastic rocks. It seems reasonable to conclude that they have similar mechanical properties and are related to the same kinds of mechanical processes, namely, cataclasis.

Summary

1. The present-day tectonics of China is a result of both ancient and modern global scale tectonic processes. The present-day boundary conditions are a result of plate movements. The present-day mechanical fabric of China is a Permo-Triassic and Cenozoic plate tectonic collage.

2. There are six rigid basement elements in China welded together by suture zones which follow along Paleozoic continental margins. Rigid basement blocks are continental, granitic material. Structures in sediments over rigid blocks are lacking except for high-angle faults and forced folds over block boundaries. Sediment patterns over rigid basement blocks are uniform, demonstrating that vertical movements or sediments thicken in such a way that they reflect rigid-body rotation of basement blocks.

3. Ductile zones are continental margin rock series and sutures which include oceanic basement rock and mantle-derived material. Structures in Cenozoic sediments over ductile basement are strike slip faults, buckle folds, and forced folds over ductile basement faults. Sediment patterns over ductile basement change rapidly, and unconformities are common, showing that the dimensions of the rigid blocks in these zones are small and that the resulting deformation features are closely spaced.

4. Intensely deformed zones are composed of ductile basement tectonically intruded by brittle basement. Structures in sediments over intensely deformed zones are high-angle thrust faults, anticlines cored by high-angle thrust faults, and buckle folds. Sediment patterns are like those over ductile basement.

5. China displays continental scale flow patterns on tectonic maps and on Landsat imagery, but in apparent contradiction, earthquakes show that China deforms by brittle mechanisms. The solution to this problem is a logical consequence of the plate tectonic, earthquake, structural, and stratigraphic data presented and is this: China deforms by cataclastic flow. Furthermore, comparison of field observations, rock mechanics data, and photoelastic and clay model studies with evidence for brittle deformation and for flow in China supports cataclasis.

Conclusion

China displays cataclastic processes from the continental scale down to probably the scale of thin sections. The evidence for cataclasis comes from earthquakes, mapped geology, stratigraphic data, structural data, and Landsat imagery. Interpretations are based on processes seen in

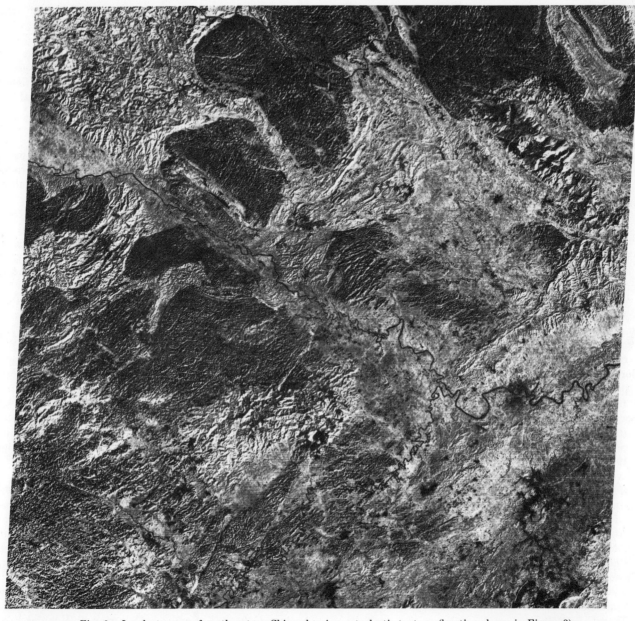

Fig. 9. Landsat scene of southeastern China showing cataclastic texture (location shown in Figure 3).

rock mechanics studies and on observed textures of cataclastic rocks.

At the continental scale there are regions of rigid basement blocks immersed in ductile zones which are analogous to porphyroclasts (large fragments) surrounded by fluxion structure (foliation or flow structure), respectively. Rigid continental crust appears to be broken into relatively large blocks with like compositions in much the same manner that cataclastic fault breccias are broken into large fragments with the same mineral or rock type composition. The aggregate and elements are of the same order of magnitude.

On the other hand, ductile zones have a wide spectrum of element sizes and compositions of elements. Aggregates are 1 or more orders of magnitude larger than the average element. These are analogous to protomylonites (crush breccia with conglomeratic or arkosic texture) and their finer-grained equivalents.

The regions of mixed basement are between rigid blocks and ductile zones and are mechanically similar to but coarser grained than the ductile zones.

At the continental scale the cataclastic process is intimately related to global scale tectonics. The initial fabric of China was rifted, drifted elements of continental crust which grew by continent margin tectonic, stratigraphic, igneous, and metamorphic processes and by subduction.

These allochthonous elements were sutured together. Later, horizontal compression and extension related to global tectonic processes reactivated some planes of weakness in the sutures, and a new period of cataclasis began. Perhaps these considerations could enlighten us about the mechanics of smaller cataclastic rocks.

Acknowledgments. John Handin introduced me to the very important process of cataclasis, and for this I am most grateful. Thanks to Martha Withjack, Bill Rizer, Dave Stearns, Bob Reed, Cindy Scheiner, Kevin Burke, and Jamie Jamison for important suggestions that materially improved the manuscript and for thought-provoking comments and discussions.

References

Aydin, A., Small faults formed as deformation bands in sandstone, in *Proceedings of Conference II: Experimental Studies of Rock Friction With Application to Earthquake Prediction*, pp. 591-616, U.S. Geological Survey, Menlo Park, Calif., 1977.

Bombolakis, E. G., Some constraints and aids for interpretation of fracture and fault development, *Proc. Int. Conf. New Basement Tectonics 2nd.*, 289-305, 1979.

Borg, I., and J. C. Maxwell, Interpretation of fabrics of experimentally deformed sands, *Am. J. Sci.*, *254*, 71-81, 1956.

Borg, I., M. Friedman, J. Handin, and D. V. Higgs, *Experimental Deformation of St. Peter Sand: A Study of Cataclastic Flow, Rock Deformation, a Symposium*, pp. 131-191, Geological Society of America, Boulder, Colo., 1960.

Bowden, F. P., and D. Tabor, *The Friction and Lubrication of Solids*, 2nd ed., Clarendon, Oxford, 1964.

Chen, P.-H., J. J. Gallagher, Jr., and C. S. Kenyon, Paired rift-spread systems of China (abstract), *Eos Trans. AGU*, *60*, 398, 1979.

Chi-ching, H., An outline of the tectonic characteristics of China, *Eclogae Geol. Helv.*, *71*(3), 611-635, 1978.

Chi-ching, H., J. Chi-shun, J. Chun-fa, C. Chih-meng, and C. Cheng-kun, Some new observations on the geotectonic characteristics of China, *Acta Geol. Sin.*, Engl. Transl., 1-52, 1974.

Friedman, M., Petrofabric analysis of experimentally deformed calcite-cemented sandstones, *J. Geol.*, *71*, 12-37, 1963.

Gallagher, J. J., Jr., and M. J. McGuire, Subsurface structure and stratigraphy revealed by surface lineaments, *Proc. Int. Conf. New Basement Tectonics 2nd*, 392-405, 1979.

Gallagher, J. J., Jr., and M. O. Withjack, Basement tectonics of China—Continental-scale cataclastic flow (abstract), *Am. Assoc. Pet. Geol. Bull.*, *64*, 710, 1980.

Gallagher, J. J., Jr., M. Friedman, J. Handin, and G. M. Sowers, Experimental studies relating to microfracture in sandstone, *Tectonophysics*, *21*, 203-247, 1974.

Griggs, D. V., and J. Handin, Observations on fracture and a hypothesis of earthquakes, *Mem. Geol. Soc. Am.*, *79*, 347-364, 1960.

Handin, J., Strength and ductility, *Mem. Geol. Soc. Am.*, *97*, 223-289, 1966.

Hayes, D., and B. Taylor, Tectonics, in *A Geophysical Atlas of the East and Southeast Asian Seas, Map Chart Ser.*, vol. MC-25, edited by D. E. Hayes, Geological Society of America, Boulder, Colo., 1979.

Higgins, M. W., Cataclastic rocks: *Geol. Surv. Prof. Pap. U.S.*, *687*, 1-97, 1971.

Hodgson, R. A., Genetic and geometric relations between structures in basement and overlying sedimentary rocks with examples from Colorado Plateau and Wyoming, *Am. Assoc. Pet. Geol. Bull.*, *49*, 935-949, 1965.

Hutchison, C. S., Ophiolite in southeast Asia, *Geol. Soc. Am. Mem.*, *86*, 797-806, 1975.

Jamison, W. R., Laramide deformation of the Wingate sandstone, Colorado National Monument: A study of cataclastic flow, Ph.D. dissertation, 170 pp., Texas A&M Univ., College Station, 1979.

Kenyon, C. S., P.-H. Chen, and J. J. Gallagher, Jr., Tectonics of China's basement rocks (abstract), *Eos Trans. AGU*, *60*, 398, 1979.

Lee, C.-S., G. C. Shor, Jr., L. D. Bibee, R. S. Lu, and T. W. C. Hilde, Okinawa Trough: Origin of a back-arc basin, *Mar. Geol.*, *35*, 219-241, 1980.

Lee, J. S., *The Geology of China*, Murby, London, 1939.

Meyerhoff, A. A., and J.-O. Willums, Petroleum geology and industry of the Peoples Republic of China, *Tech. Bull. U.N. Econ. Soc. Comm. Asia Pac. Comm. Coord. J. Prospect. Miner. Resour. South Pac. Offshore Areas*, *10*, 103-212, 1976.

Molnar, P., and P. Tapponnier, Relation of the tectonics of eastern China to the India-Eurasia collision: Application of slip-line field theory to large-scale continental tectonics, *Geology*, *5*, 212-216, 1977.

Prucha, J. J., J. A. Graham, and R. P. Nickelson, Basement controlled deformation in the Wyoming Province of Rocky Mountain Foreland, *Am. Assoc. Pet. Geol. Bull.*, *49*, 966-992, 1965.

Scotese, C. R., *Continental Drift*, University of Chicago Press, Chicago, Ill., 1979.

Scotese, C. R., R. K. Bamback, C. Barton, R. Van der Voo, and A. M. Ziegler, Paleozoic base maps, *J. Geol.*, *87*, 217-277, 1979.

Stearns, D. W., Certain aspects of fracture in naturally deformed rocks, in *Rock Mechanics Seminar*, vol. 1, pp. 97-116, available as *Rep. AD669375*, Natl. Tech. Inform. Serv., Springfield, Va., 1968.

Stearns, D. W., Fracture as a mechanism of flow in naturally deformed layered rocks, in *Proceedings of the Conference on Research in Tectonics*, pp 68-52, 79-96, Geological Survey of Canada, Ottawa, Ont., 1969.

Stearns, D. W., Mechanisms of drape folding in the Wyoming Province, *Guideb. Wyo. Geol. Assoc. Annu. Field Conf.*, *23*, 125-143, 1971.

Stearns, D. W., Faulting and forced folding in the Rocky

Mountains Foreland, *Mem. Geol. Soc. Am.*, *151*, 1–37, 1978.

Tapponnier, P., and P. Molnar, Slip-line field theory and large-scale continental tectonics, *Nature*, 264, 319–324, 1976.

Tapponnier, P., and P. Molnar, Active faulting and tectonics in China, *J. Geophys. Res.*, *82*, 2905–2930, 1977.

Terman, M., *Tectonic Map of China and Mongolia*, Geological Society of America, Boulder, Colo., 1974.

Van Esyinga, *Geologic Time Table*, Elsevier, New York, 1975.

York, J. E., R. Cardwell, and J. Ni, Seismicity and Quaternary faulting in China, *Bull. Seismol. Soc. Am.*, *6*, 1983–2001, 1976.

A Constitutive Equation for One-Dimensional Transient and Steady State Flow of Solids

ANTHONY F. GANGI

Department of Geophysics, Texas A&M University, College Station, Texas 77843

The constitutive equation for one-dimensional flow in solids is postulated to be given by a scalar function of the state variables of stress and strain (σ and e), their rates ($\dot\sigma$ and $\dot e$), temperature (T), and material structure (S) as $f(e, \dot e, \sigma, \dot\sigma, T, S) = 0$. This function represents a hypersurface in the state variable space and is the totality of equilibrium states of the material. It is assumed that the state variables remain on this surface as the body flows. This representation is assumed to be valid when only one flow mechanism is operative. When two or more mechanisms operate, two or more constitutive equations are necessary, one for each mechanism. When first-order kinetics are assumed to hold and the differential form of the constitutive equation is linear, the usual solutions for creep, constant-strain-rate, and relaxation experiments are obtained. These latter solutions have been obtained previously from a microstructural viewpoint (e.g., the generation and annihilation of dislocations and their contribution to the flow strain or stress), while the solutions obtained here are from a macroscopic viewpoint. While good fits of the data with these solutions are obtained, systematic deviations occur. It is found that a large part of these deviations (in many cases) can be explained or fitted if two or more independent flow mechanisms are assumed to operate. In this case, each mechanism has a constitutive equation which depends only upon the parameters associated with that mechanism. This approach gives solutions with multiple terms that are identical in form to the one-mechanism solution when first-order kinetics are assumed to hold. To explain oscillatory-flow characteristics and tertiary or accelerating flow, dependent multiple mechanisms are necessary if we continue to require that (1) first-order kinetics hold and (2) the differential forms of the constitutive equation be linear. If it is assumed that the constitutive equation for each mechanism depends only on the stress and strain experienced by the second mechanism, coupled linear differential constitutive equations result. The solution to this set of equations allows both damped oscillatory solutions and exponentially increasing solutions. The latter may be used to describe the accelerating or tertiary creep that occurs prior to failure in some cases. This solution shows that failure can be predicted by using linear systems if there is coupling between two flow mechanisms.

Introduction

Constitutive equations for the flow of solids under stress have been given since the first measurements were made. *Andrade* [1910] used an empirical one to describe the elongation l of lead, lead/tin, and copper wires under constant tensile stress:

$$l(t) = l_0(1 + \beta t^{1/3}) \exp (\dot e_s t) \qquad (1)$$

which in terms of true or natural strain gives

$$e(t) = \ln [l(t)/l_0] = \dot e_s t + \ln (1 + \beta t^{1/3}) \qquad (2)$$

The first term on the right describes, principally, the steady state flow (or creep), while the second term represents the transient creep (or, as Andrade called it, the 'beta flow'). Andrade found that this empirical equation would fit his experimentally determined elongations quite accurately when the wires were deformed at room temperature ($\sim$15°C) and at 162°C (for the lead wire). He found that accelerating flow, or tertiary creep, was suppressed up to failure if the load was varied (to compensate for the decrease in the cross-sectional area as the wires lengthened) so as to maintain constant stress. He showed that both $\dot{e}_s$ and β were strongly stress and temperature dependent.

McVetty [1934] proposed an empirical equation for the creep strain $e(t)$ (written in modern notation and form):

$$e(t) = e_0 + \dot{e}_s t + e_T[1 - \exp(-rt)] \qquad (3)$$

for high-temperature (400°C to 800°C) constant-tensile-stress (or creep) tests on iron alloys. In (3), e_0 is a constant strain, $\dot{e}_s$ is the steady state strain rate, e_T is the transient strain, and r is the rate constant. He found that both $\dot{e}_s$ and $(e_0 + e_T)$ were strongly temperature and stress dependent. *Lacombe* [1939] proposed an empirical equation that was similar to Andrade's for creep:

$$e(t) = e_0 + at^m + bt^n \qquad (4)$$

where $0 < m < 1$, $n \simeq 1$ (but, generally greater), and e_0, a, and b are constants. *Cottrell and Aytekin* [1947] modified Andrade's equation and applied it to creep of zinc crystals (11°C $< T <$ 60°C) in shear:

$$\gamma(t) = \gamma_0 + \dot{\gamma}_s t + \beta t^{1/3} \qquad (5)$$

Johnston and Gilman [1959], in a detailed experimental study of the densities and velocities of dislocations in lithium fluoride crystals that were undergoing plastic flow, proposed that the rate of change of dislocation density $\dot{\rho}(t)$, during plastic flow, could be expressed as

$$\dot{\rho}(t) = r\rho - k\rho^2 = r\rho(1 - \rho/\rho_s) \qquad (6)$$

where the first term $r\rho$ represented the 'multiplication' or creation rate of dislocations, while the second term $k\rho^2$ represented their 'attrition' or annihilation rate, and ρ_s is the dislocation density at steady state. The proportionality constants r and k were assumed to be functions of stress and temperature but not of strain or strain rate. They relate the strain rate to the dislocation density and dislocation velocity by using the Taylor-Orowan relationship:

$$\dot{e}(t) = gb\rho(t)v \qquad (7)$$

where g is a geometric factor (usually assumed to be 1 or of that order), b is the Burgers vector for the crystal, and v is the velocity of the dislocations (a function of stress and temperature). They used (6) and (7) along with an experimental relation ($\rho(t) \simeq 10^9 e(t)$) to construct stress curves for the lithium fluoride crystals and compare them with data obtained from a constant-strain-rate test. A reasonably good fit was obtained. This was one of the earliest attempts to describe, analyze, and understand the transient flow of solids on the basis of a detailed study of their microscopic properties.

Garofalo [1960] introduced the modern form of McVetty's empirical equation (see (3)) to describe creep deformation of austenitic stainless steel at high temperatures (600°C to 800°C). He found empirically that over this temperature range and for stresses between 90 to 210 MPa the steady state strain rate $\dot{e}_s$ and the rate constant r

were related by

$$\dot{e}_s/r = \text{a constant} = 1/K \qquad (8)$$

Li [1963] started with Johnston and Gilman's assumptions: (1) that the multiplication of dislocations satisfy first-order kinetics*, while their annihilation satisfy second-order kinetics* (as in (6)) and (2) that the Taylor-Orowan equation (see 7) holds. They obtained a time-varying strain equation for a creep test:

$$e(t) = e_0 + \dot{e}_s t + \frac{\dot{e}_s}{r} \ln\left[1 + \frac{\dot{e}_i - \dot{e}_s}{\dot{e}_s}(1 - e^{-rt})\right] \qquad (9)$$

where

$$\dot{e}_s/r = gbv/k \qquad (10)$$

and $\dot{e}_i$ is the initial strain rate $\dot{e}(0)$. From this expression it is seen that the transient strain e_T is

$$e_T = \lim_{t=\infty} [e(t) - e_0 - \dot{e}_s t] = (\dot{e}_s/r) \ln(\dot{e}_i/\dot{e}_s) \qquad (11)$$

Li shows that (9) gives an excellent fit to *Garofalo*'s [1960, 1965] data on austenitic stainless steel. See also *Akulov* [1964a, b] for a similar result that was obtained from an investigation of the statistics of dislocations and by analogy to the theory of ferromagnetic domains. Note that if second-order kinetics (6) holds and the empirical relationship $\dot{e}_s/r =$ constant (8) holds, then the proportionality constant for dislocation immobilization k must have the same temperature and stress dependence as the dislocations' velocity v, according to (10).

Conway and Mullikin [1966] tested the Andrade, Lcombe, Cottrell-Aytekin, and McVetty-Garofalo equations as well as one of their own, namely

$$e(t) = e_0 + at^{1/3} + bt^{2/3} + \dot{e}_s t \qquad (12)$$

by using creep data for arc-cast tungsten measured at 2400°C and 5.5 MPa. While they conclude that their equation, which is a four-parameter fit, gives a better fit than those of Andrade, McVetty-Garofalo, and Cottrell-Aytekin and almost as good a fit as Lacombe's, the errors given show there is little to choose between these equations in terms of fitting the data. That Lacombe's equation would give the best fit to the data is not surprising, considering that Lacombe's is a five-parameter equation while the others have four and fewer adjustable parameters to fit the data. They did not try to fit Li's equation, but Li showed his equation fitted Garofalo's creep data as well as the McVetty-Garofalo equation did.

Evans and Wilshire [1968] used the McVetty-Garofalo equation (3) to fit their high-temperature ($T \simeq 0.5\ T_M$; $T_M =$ the melting temperature) creep data for nickel, zinc, and iron for stresses between 75 to 220 MPa, 40 to 83 MPa,

---

* For first-order kinetics, the time rate of change of the dislocation density is proportional to the dislocation density itself, while for second-order kinetics, it depends on the dislocation density squared as well (see (6)).

and 31 to 69 MPa, respectively. They found, empirically, that (8) held (except at the lower stresses for the iron) and that the initial and final strain rates were proportional; that is

$$\dot{e}_i/\dot{e}_s = \text{a constant} \qquad (13)$$

But if (3) holds, we have for the initial strain rate:

$$\dot{e}_i = \dot{e}(0) = \dot{e}_s + r e_T \qquad (14)$$

and, if (8) and (13) are correct, then (14) indicates the total transient strain e_T must also be independent of temperature and stress. Evans and Wilshire found this to be approximately true for the temperature and the stress ranges used (except for the low stresses on iron).

Webster et al. [1969] started from an assumption of first-order kinetics (letting $k = 0$ in (6)) and the Taylor-Orowan equation (7) and derived (3). They argued that (8) should hold 'because the same process is assumed to be controlling the cooperative rearrangement of the dislocations as is controlling their individual motion.' They also argued that 'the same form of relationship exists between stress and dislocation density measured from tensile stress as exists between stress and dislocation density measured during steady-state creep' and, therefore, that it is reasonable for (13) to hold. While it is difficult to follow their arguments, it is clear from the experimental data that their conclusions are valid over the limited ranges of stress and temperature they used, that is, at high homologous temperatures ($T/T_M \simeq 0.5$) and at high stresses, where the contribution of grain-boundary sliding to the total creep is small. Following Dorn [1954], they introduced the 'stress/time/temperature' parameter

$$\theta = \dot{e}_s t \qquad (15)$$

and indicate that the total transient-strain and steady-state-strain curve is independent of stress and temperature when expressed as a function of θ:

$$e(t) - e_0 = \theta + e_T[1 - \exp(-K\theta)] \qquad (16)$$

where K is defined in (8). The same conclusion is reached and tested in greater detail by Amin et al. [1970]. They show that (16) is a master (universal) creep curve, independent of temperature and stress, for a number of different metals when $T/T_M > 0.5$ and the stress is high enough that grain-boundary sliding is suppressed. However, the ranges of temperature and stress used are fairly limited. Also, a large part of their data is steady state creep, where the universal curve should fit by definition.

Evans and Wilshire [1970] also fit their creep data for a copper-aluminum alloy to (3), but they find that the total transient-creep strain e_T is a linear function of the stress at high temperatures (483 and 581°C), while it is independent of stress at a lower temperature (413°C). They also found that, over this temperature range and for stresses between about 7 and 100 MPa, the ratio of the initial strain rate to the steady state strain rate was constant

(13); however, they find a large difference between the measured ratio (about 100) and the computed ratio (about 4, using (14)). They explain the difference as being due to the rapid increase of the strain hardening ($h = \partial\sigma/\partial e$) during the initial 10%-15% of the transient creep. They measured the strain hardening h and the rate of recovery ($R = \partial\sigma/\partial t$) by using small, incremental changes in the applied stress [see Mitra and McLean, 1966, and Ishida and McLean, 1967]. They also find that the measured strain deviates from that computed by using (3) (with parameters e_0, $\dot{e}_s$, e_T, and r determined so as to minimize this deviation) only in the same initial 10%-15% of the transient creep. They attribute this also to the rapid change in the strain hardening h during that part of the creep curve. They conclude, as Mitra and McLean [1966] did, that the strain rate is predicted by the Bailey-Orowan equation, namely

$$\dot{e}(t) = R/h \qquad (17)$$

The Bailey-Orowan equation is derived by assuming the stress is a function of strain and time only, $\sigma(e, t)$, and then setting its differential to zero (which holds for a creep test), that is [Bailey, 1926; Orowan, 1947]

$$d\sigma = h\,de - R\,dt = 0 \qquad (18)$$

This can be considered a constitutive equation for creep and has been used as such by Mitra and McLean [1966], Evans and Wilshire [1968, 1970], and Gittus [1971]. However, (18) is not in the proper form for a constitutive equation. It indicates there is an explicit time dependence to the constitutive equation

$$f(\sigma, e, t) = 0 \qquad (19)$$

and this is not a reasonable assumption, because the relationship between the state variables (stress, strain, temperature, material structure, etc.) does not change with time, even though the variables themselves may be time dependent. In any event, the Bailey-Orowan equation (17) may be obtained from a constitutive equation of the form $f(\sigma, e, \dot{e}) = 0$ or

$$\sigma = \sigma(e, \dot{e}) \text{ or } d\sigma = \frac{\partial\sigma}{\partial e}de + \frac{\partial\sigma}{\partial \dot{e}}d\dot{e} \qquad (20)$$

and assuming

$$\frac{\partial\sigma}{\partial \dot{e}} = \frac{\partial\sigma}{\partial t}\bigg/\frac{\partial \dot{e}}{\partial t} \simeq \frac{\partial\sigma}{\partial t}\bigg/\frac{d\dot{e}}{dt} \qquad (21)$$

Thus the Bailey-Orowan equation may not be rigorously valid; however, it is approximately correct, which may explain its successful use in describing creep experiments [Mitra and McLean, 1966; Evans and Wilshire, 1968, 1970; and Gittus, 1971].

Recently, Ajaja and Ardell [1979] proposed that the Taylor-Orowan equation (7) is incomplete and should include a term that represents the strain rate caused by dislocation annihilation. They assume their empirical relationship between dislocation velocity v and dislocation density holds:

$$v = A_0 \rho^{-5/2} \qquad (22)$$

and they use the same second-order kinetics assumed by *Johnston and Gilman* [1959], *Li* [1963], and *Akulov* [1964a, b] for the annihilation rate of dislocations $\dot{\rho}_a$ and first-order kinetics for the generation rate $\dot{\rho}_g$:

$$\dot{\rho} = \dot{\rho}_g + \dot{\rho}_a = r\rho\,(1 - \rho/\rho_s) \qquad (23)$$

where ρ_s is the steady state dislocation density, and ρ is the net dislocation density. They express the total strain $e(t)$ in terms of $e_n(t)$ and $e_a(t)$, the strains 'due to the glide motion of the net dislocations, and the annihilation process,' respectively. They find that $e_n(t)$ tends to give a concave-upward strain variation with time, while $e_a(t)$ gives a concave-downward contribution. As a result, both transient (decelerating) creep and pseudo tertiary (accelerating) creep are obtained from the same expression, while the dislocation density continuously decreases. They demonstrate an excellent fit of their measured creep data for type 304 stainless steel with their equation. They also show that the strain caused by dislocation annihilation, $e_a(t)$, is larger by a factor of the order of five than the glide-motion strain $e_n(t)$.

The constitutive equations used to determine the flow of ductile materials are primarily based on their microscopic behavior, that is, the generation, annihilation, and movement of dislocations, vacancies, impurity atoms, etc. In the above we have reviewed primarily the metals literature, but the same approach has been taken for the flow of ceramics and rocks (see *Carter and Kirby* [1978] for a recent detailed review with many additional references). In the following, we will take a different approach to the determination of the constitutive equation for ductile materials. We will ignore the microscopic details and look at the problem from a macroscopic viewpoint, taking a postulatory approach to the constitutive equation. While this approach does not give direct insight into the physical mechanisms involved in the flow of ductile materials, it does give general constraints and guides our investigation for the microscopic physical mechanisms. Such an approach has been taken earlier by *Hollomon* [1946] and *Zener and Hollomon* [1946].

Macroscopic Constitutive Equation

The constitutive equation for the flow of ductile materials is a relationship between the state variables which describe the material and its environment. For the case of one-dimensional flow (that is, where the stress and strain can be treated as scalar quantities), and when there is only one mechanism giving rise to the ductile flow (we will consider multiple mechanisms later; see also *Parrish and Gangi* [1977, 1981] for a treatment of multiple mechanisms in steady state or secondary creep), the constitutive

equation is expected to be a scalar function of the state variables:

$$f\,(e, \dot{e}, \sigma, \dot{\sigma}, T, S) = 0 \qquad (24)$$

where e and $\dot{e}$ are the strain and strain rate, σ and $\dot{\sigma}$ are the stress and stress rate, T is the absolute temperature, and S is the structure parameter. The structure parameter S contains such properties as the crystal structure, the dislocation densities, impurity and vacancy densities, and other structure parameters (porosity, fracture density, etc.). The constitutive equation is not an explicit function of position or time; it is not a function of position because the material is assumed to be homogeneous, and the physical location of the material is not expected to affect its properties directly. It is not a function of time because we assume the material does not undergo an irreversible change while in a constant environment. This is expected to be true for metals and most rocks; the minor time-varying properties of rocks (e.g., their radioactivity) are assumed to be negligible in this analysis.

The constitutive equation (24) represents a surface in the six-dimensional parameter space of the state variables e, $\dot{e}$, σ, $\dot{\sigma}$, T, and S. We assume all equilibrium states are on this surface and that during ductile flow the values of the state variables are constrained to remain on the surface. That this allows the material to exhibit the normal flow properties can be seen by considering a reduced constitutive equation $f\,(e, \dot{e}, \sigma) = 0$, which is shown in Figure 1. In this case we assume the temperature, the structure, and the stress rate do not affect the properties of the body (this is for convenience only, to allow a graphic representation of the constitutive surface; the illustration of a hypersurface in six-dimensional space is difficult to achieve).

We see from Figure 1 that even this simplified constitu-

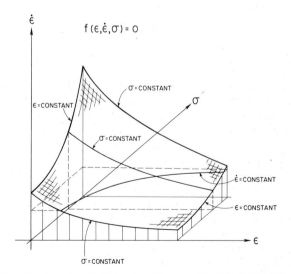

Fig. 1. A reduced dimensionality constitutive surface illustrating the paths of a creep experiment (σ = constant) and a constant-strain-rate experiment ($\dot{e}$ = constant).

tive surface can illustrate the behavior of a ductile material. For example, the variation of strain and strain rate in a creep experiment (σ = a constant) is given by the intersection of the plane, σ = constant, with the surface, and the resulting curve shows a high strain rate at small strains that decreases to a constant strain rate at large strains. This is the behavior noted in materials that show transient as well as steady state creep and no accelerating creep. The variation for a constant-strain-rate experiment ($\dot{e} = \dot{e}_0$, a constant) shows the expected behavior of the stress as a function of strain (or time, $t = e/\dot{e}_0$), that is, a decelerating, but continuously increasing, stress which approaches a steady state value. Figure 1 demonstrates that it may be possible to illustrate the behavior of a body under different test conditions by means of a single constitutive surface, at least if just one mechanism is operative.

If the body exists on the constitutive surface given by (24), and we make a small change in the state variables but remain on the surface, we obtain

$$df = 0 = f_e de + f_{\dot{e}} d\dot{e} + f_\sigma d\sigma + f_{\dot{\sigma}} d\dot{\sigma} + f_\mathrm{T} dT + f_\mathrm{S} dS \quad (25)$$

where

$$f_e = \partial f/\partial e \qquad f_\sigma = \partial f/\partial \sigma \qquad f_\mathrm{T} = \partial f/\partial T \qquad f_\mathrm{S} = \partial f/\partial S$$

Equation 25 is the differential form of the constitutive equation, and we will use it as a starting point. The partial derivatives of the constitutive equation f_e, f_σ, etc. are functions of all the state variables and can be determined by various experiments in which small changes are made in some state variables. This differential form of the constitutive equation should be the starting point for the characterization of material properties from a macroscopic point of view. Its usefulness will be illustrated by a few examples.

Single-Mechanism Cases

Creep Experiments

A creep experiment is one in which the stress on the body is raised rapidly to some value and held there, that is, $\sigma(t) = \sigma_0 H(t)$, where $H(t)$ is the Heaviside step or unit function: $H(t) = 0$ for $t < 0$, and $H(t) = 1$ for $t > 0$. If we assume the temperature is held constant during the experiment ($dT = 0$), then (25) gives (after dividing through by dt):

$$(d\dot{e}/dt) + R_e \dot{e} = a\delta'(t) + b\delta(t) + cH(t) \quad (26)$$

where

$$R_e(e, \dot{e}, \sigma, \dot{\sigma}, S, T) = f_e/f_{\dot{e}} \quad (27)$$

$$a = -\sigma_0 f_{\dot{\sigma}0}/f_{\dot{e}0} \qquad b = -\sigma_0 f_{\sigma0}/f_{\dot{e}0} \quad (28)$$

and we have assumed

$$-(f_\mathrm{S}/f_{\dot{e}})\dot{S} = cH(t) \quad (29)$$

Note

$$f_{\sigma0} = f_\sigma(e_0, \dot{e}_0, \sigma_0, \dot{\sigma}_0, S_0), \text{ etc.} \quad (30)$$

that is, the coefficients a and b of the delta function $\delta(t)$ and its derivative $\delta'(t)$ are constants because they are evaluated only at the time ($t = 0$) when the delta function and its derivative have nonzero values. However, the coefficients R_e and c need not be constants. In creep experiments for which steady state creep eventually results, $\ddot{e}(\infty) = 0$ (ignoring the possibility of accelerating creep for the moment), (26) requires that either R_e goes to zero at large strains (or times) or there is a continuous (constant) structure change (as given by (29)).

If we assume that R_e and c are constants, then (26) is the usual creep equation obtained when first-order kinetics are assumed to hold (except for the source terms on the right-hand side). If we assume $R_e = r_e + k\dot{e}$, then we obtain the second-order kinetics equation of *Johnston and Gilman* [1959], *Li* [1963], *Akulov* [1964], and *Ajaja and Ardell* [1979].

Obviously, much more complicated variations of R_e, with respect to the state variables, are possible. For example, it can be a function of the strain, strain rate, and structure. Note, the stress, stress rate, and temperature are all held constant, or, at least, assumed constant, during the test. We note that assuming R_e is constant does not necessarily infer that the partial derivatives of f with respect to strain and strain rate are constants. If both f_e and $f_{\dot{e}}$ have the same functional variation with the state variables, their ratio could be constant, while each of them could vary greatly during the test.

If we assume that first-order kinetics hold, then $R_e = r_e$ = a temporal constant, and letting $c = \dot{e}_s r_e$ (a temporal constant), the solution to (26) is

$$e(t) = \{e_e + \dot{e}_s t + e_\mathrm{T}[1 - \exp(-r_e t)]\} H(t) \quad (31)$$

where

$$a = e_e \qquad b = \dot{e}_s + r_e(e_\mathrm{T} + e_e) \quad (32)$$

This is the McVetty-Garofalo equation (3), which was obtained theoretically by *Webster et al.* [1969] by assuming first-order kinetics. If the elastic strain e_e is small in relation to the transient strain e_T, we have

$$\dot{e}_i = \dot{e}_s + r_e e_\mathrm{T} \simeq b = -\sigma_0 f_{\sigma0}/f_{\dot{e}0} \quad (33)$$

If this expression is to be independent of stress and temperature, as has been found experimentally (at least, approximately, over limited temperature and stress ranges), then the partial derivatives of the constitutive equation must have the same temperature variation (or be temperature independent) and have a stress dependence which just cancels the linear stress dependence expressed by σ_0. This may not be unreasonable because (1) the partial derivatives are evaluated at the initial values of the state variables and (2) the initial values of the stress rate and strain rate are both infinite. This could lead to a diminished dependence or variation of these partial derivatives on the stress and temperature.

If the ratio $\dot{e}_s/r_e$ is to be independent of stress and tem-

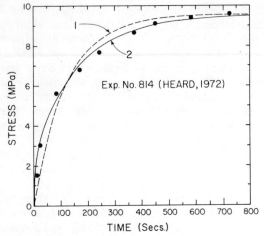

Fig. 2. Fit of a one-mechanism equation (curve 1, dashed: r_σ = 8.2/ks, σ_s = 9.60 MPa, rms error = 0.65 MPa) and a two-mechanism equation (curve 2, solid: r_1 = 5.90/ks, σ_1 = 7.15 MPa, r_2 = 84.7/ks, σ_2 = 2.43 MPa, rms error = 0.16 MPa) to Heard's data for a constant-strain-rate experiment on polycrystalline salt. (T = 300°C, confining pressure = 200 MPa and $\dot{e}$ = 0.154/ks; experiment 814, [*Heard*, 1972]).

perature (again, as has been found experimentally for limited temperature and stress ranges), we must have

$$-(f_S/f_{\dot{e}})\dot{S}/r_e^2 = cH(t)/r_e^2 = (\dot{e}_s/r_e)H(t) \qquad (34)$$

independent of temperature and stress.

Constant-Strain-Rate Experiment

In this case we have $\dot{e}(t) = \dot{e}_s H(t)$ and the condition $\sigma(\infty) = \sigma_s$ (i.e., we assume a steady state stress for the final condition). Then (25) gives, for $dT = 0$,

$$(d\dot{\sigma}/dt) + R_\sigma\dot{\sigma} = A\delta(t) + BH(t) \qquad (35)$$

where now

$$R_\sigma = R_\sigma(e, \dot{e}, \sigma, \dot{\sigma}, S, T) = f_\sigma/f_{\dot{\sigma}} \qquad (36)$$

$$A = -(f_{\dot{e}0}/f_{\dot{\sigma}0})\dot{e}_s \qquad B = C - (f_e/f_{\dot{\sigma}})\dot{e}_s = 0 \qquad (37)$$

and we have assumed that

$$-(f_S/f_{\dot{\sigma}})\dot{S} = CH(t) = (f_e/f_{\dot{\sigma}})\dot{e} \qquad (38)$$

The above assumption is necessary if we are to satisfy the condition $\sigma(\infty) = \sigma_s$ (or $\dot{\sigma}(\infty) = 0$) and have R_σ remain finite at $t = \infty$. This follows from (35), which must hold at all times. If we assume that first-order kinetics hold (i.e., $R_\sigma = r_\sigma$ = a constant), the solution of (35) is:

$$\sigma(t) = \sigma_s[1 - \exp(-r_\sigma t)]H(t) \qquad (39)$$

where

$$A = \sigma_s r_\sigma = -(f_{\dot{e}0}/f_{\dot{\sigma}0})\dot{e}_s \qquad (40)$$

If we are to have $\dot{e}_s/r_\sigma$ independent of temperature and pressure, then the ratio $f_{\dot{\sigma}0}/f_{\dot{e}0}$ must be independent of

temperature and inversely proportional to the steady state stress σ_s.

If the condition given by (38) does not hold, then the stress rate will not go to zero at large times, and the material will exhibit strain hardening. That is, we will have

$$\sigma(t) = \{\sigma_s[1 - \exp(-r_\sigma t)] + \dot{\sigma}_s t\}H(t) \qquad (41)$$

where

$$\dot{\sigma}_s r_\sigma = B = -(f_{S0}/f_{\dot{\sigma}0})\dot{S}_s - (f_{e0}/f_{\dot{\sigma}0})\dot{e}_s$$

$$\dot{\sigma}_s + r_\sigma\sigma_s = A = -(f_{\dot{e}0}/f_{\dot{\sigma}0})\dot{e}_s \qquad (42)$$

and we have assumed that

$$(f_S/f_{\dot{\sigma}})\dot{S}(t) = (f_{S0}/f_{\dot{\sigma}0})\dot{S}_s H(t)$$

$$(f_e/f_{\dot{\sigma}})\dot{e}(t) = (f_{e0}/f_{\dot{\sigma}0})\dot{e}_s H(t) \qquad (43)$$

A comparison of (39) with the measured stress obtained from a constant-strain-rate experiment shows that it gives reasonably good fits for high-temperature experiments $T > 0.7T_M$. Good fits of the creep-strain equation

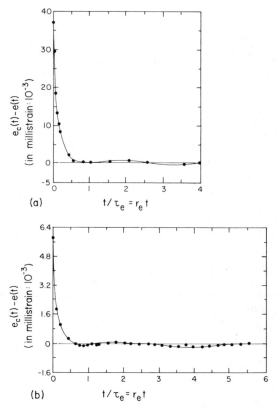

Fig. 3. (a) The difference between the calculated strain $e_c(t)$ and the measured strain $e(t)$, from a creep experiment on a Cu-15 at.% Al alloy. The calculated strain was obtained from the McVetty-Garofalo equation with r_e = 0.412/ks; experimental conditions: tensile stress, σ_s = 75.8 MPa and T = 483°C [*from Evans and Wilshire*, 1970]. (b) The same strain difference for a creep experiment on a Ni-20% Cr alloy. With r_e = 1.026/ks; experimental conditions: σ_s = 69.0 MPa and T = 750°C [*from Sidey and Wilshire*, 1969].

(31) to experimental data are also obtained when the temperature is near the melting temperature T_M (as noted earlier, see *Garofalo* [1963], *Evans & Wilshire* [1968], *Webster et al.* [1970], etc).

Constant-Strain (Relaxation) Experiment

In this case we have $e(t) = e_0 H(t)$, and making the same assumptions that were made for the constant-strain-rate case (namely, $R_0 = r_0 =$ constant and $B = 0$, see (38)), we obtain:

$$\sigma(t) = \sigma_0 \exp(-r_0 t) H(t) \qquad (44)$$

where

$$\sigma_0 = -e_0 (f_{\dot{e}0}/f_{\dot{\sigma}0}) \qquad (45)$$

Once again, if (38) does not hold, we have

$$\sigma(t) = \{\sigma_0 \exp(-r_\sigma t) + \sigma_1 [1 - \exp(-r_\sigma t)]\} H(t) \quad (46)$$

where

$$\sigma_1 r_\sigma = -(f_{S0}/f_{\dot{\sigma}0}) S_0 - (f_{e0}/f_{\dot{\sigma}0}) e_0 \qquad (47)$$

and we have assumed that $\dot{S}(t) = S_0 \delta(t)$ and σ_0 is given by (45). Comparing (47) and (42), we see that the threshold stress σ_1 of a relaxation experiment is related to the stress rate of a constant-strain-rate experiment as $\dot{\sigma}_s = k\sigma_1 r_\sigma$ for $\dot{e}_s = ke_0 r_\sigma$, provided $\dot{S}_s = kS_0 r_\sigma$, where k is some constant. If these relationships hold in experiments, it will provide a test of the assumptions that were made in arriving at the above results.

Comparison With Experimental Data

A comparison of the constant-strain-rate expression (39) with experiment is illustrated in Figure 2, where curve 1 is (39) fitted to one of *Heard*'s [1972] experiments on polycrystalline salt. While this curve gives a reasonably good fit (the rms error is 0.65 MPa), it is clear that the deviations between the fitted curve and the measured points are much larger than the errors in the data points themselves. This led us to consider the case of two mechanisms operating during the transient portion of the deformation curve. We had already found [*Parrish and Gangi*, 1977, 1981] that two mechanisms operate simultaneously during steady state creep. That is, we found a much better fit to *Heard*'s [1972] data for steady state creep in polycrystalline salt (as a function of temperature and differential stress) by using a two-mechanism equation.

Two-Mechanism Cases
Independent Mechanisms

If we assume there are two mechanisms operating simultaneously, there must be a constitutive equation for each mechanism. If the mechanisms are independent, then the two constitutive relationships are of the form

$$f_1(e_1, \dot{e}_1, \sigma_1, \dot{\sigma}_1, T_1, S_1) = 0 \qquad (48)$$

$$f_2(e_2, \dot{e}_2, \sigma_2, \dot{\sigma}_2, T_2, S_2) = 0 \qquad (49)$$

where e_1, $\dot{e}_1$, σ_1, etc. are the state variables that describe the conditions experienced by mechanism 1, while those with subscript 2 are those that relate to mechanism 2. Both constitutive equations can be written in differential form (as in (25)), and each differential form of the equation must be satisfied. When a creep experiment is being performed, we assume both mechanisms experience the same stress $\sigma_1 = \sigma_2$ and temperature $T_1 = T_2$. Each differential equation is solved to determine the strain for each mechanism, and the total (measured strain) is assumed to be the sum of the two strains:

$$e(t) = e_1(t) + e_2(t) \qquad (50)$$

For a relaxation (or constant-strain-rate) experiment, on the other hand, we assume both mechanisms experience the same strain $e_1 = e_2$ (or strain rate $\dot{e}_{s1} = \dot{e}_{s2}$) and temperature $T_1 = T_2$. Then each differential equation is solved independently to determine the stress for each mechanism, and the total measured stress is assumed to be the sum of the two stresses:

$$\sigma(t) = \sigma_1(t) + \sigma_2(t) \qquad (51)$$

Creep Experiment

If the above assumptions are valid, the strain variation for a creep experiment, for which two independent mechanisms operate, will be given by (compare (31))

$$e(t) = \{e_e + (\dot{e}_{s1} + \dot{e}_{s2})t + e_{T1}[1 - \exp(-r_1 t)]$$
$$+ e_{T2}[1 - \exp(-r_2 t)]\} H(t) \qquad (52)$$

where we assume each mechanism has a different rate constant ($r_1 \neq r_2$), different transient strain ($e_{T1} \neq e_{T2}$), and different steady state strain rates ($\dot{e}_{s1} \neq \dot{e}_{s2}$). This equation gives a better fit to the experimental data (as might be expected since two additional parameters, e_{T2} and r_2, are introduced). We see this equation is almost identical to the McVetty-Garofalo equation (see (3)), except for the term

$$-e_{T2} \exp(-r_2 t) H(t) = e(t) - \{e_e + \dot{e}_s t$$
$$+ e_{T1}[1 - \exp(-r_1 t)] + e_{T2}\} \qquad (53)$$

Evans and Wilshire [1970] and *Sidey and Wilshire* [1969] both plot the difference between the calculated strain (obtained by using the McVetty-Garofalo equation) $e_c(t)$ and the measured strain $e(t)$. Their results are reproduced in Figure 3. In both cases we see that the difference has the form of an exponentially decaying term, exactly as predicted by the two-mechanism equation (and (53)). This would also explain the big difference in the measured ratio of ($\dot{e}_i/\dot{e}_s \simeq 100$) compared with the calculated ratio ($\dot{e}_0/\dot{e}_s \simeq 4$) obtained by *Evans and Wilshire* [1970]. The measured initial strain rate ($\dot{e}_i$) is that associated with the faster of the two mechanisms, while the calculated initial strain rate ($\dot{e}_0$) is based on the sums of the steady state strain rates and the transient strains (see (33)) of both mechanisms. Two mechanisms operate simultaneously for

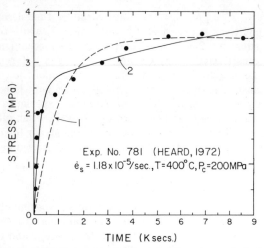

Fig. 4. Fit of a one-mechanism equation (curve 1, dashed: r_σ = 0.997/ks, σ_s = 3.50 MPa, rms error = 0.49 MPa) and a two-mechanism equation (curve 2, solid: r_1 = 4.52/ks, σ_1 = 2.59 MPa, r_2 = 0.088/ks, σ_2 = 1.98 MPa, rms error = 0.22 MPa) to Heard's data for constant-strain-rate experiment on polycrystalline salt. (T = 400°C, confining pressure = 200 MPa, and e = 0.0118/ks; experiment 781 [Heard, 1972]).

some rocks tested in creep also; this is clear from the data given by *Carter and Kirby* [1978] for salt, dunite (wet and dry), and clinopyroxenite (see their Figure 2).

Constant-Strain-Rate Experiment

Assuming two independent mechanisms operate during a constant-strain-rate experiment, we obtain, for the total stress (51) on the test specimen (compare (39)):

$$\sigma(t) = \{\sigma_{s1}[1 - \exp(-r_{\sigma 1}t)]$$
$$+ \sigma_{s2}[1 - \exp(-r_{\sigma 2}t)]\} H(t) \qquad (54)$$

This equation has been fitted to *Heard*'s [1972] salt data in Figure 2 (curve 2); clearly, a much better fit of the experimental data is obtained. The rms error decreases from 0.65 MPa to 0.16 MPa, more than a factor of four. However, even this does not appear to be the correct equation, because the estimated error of the stress during the experiment is less than 0.16 MPa. Nevertheless, it is clear that a better fit to the behavior of the salt during the test is obtained by assuming two independent mechanisms. Heard has kindly supplied us with his stress data, obtained for salt under various constant strain rates and temperatures [*Heard*, 1972]. In all cases, much better fits of the transient portion of his curves are obtained by the two-mechanism equation (54). In most cases, the rms errors decrease by large factors (greater than 2 to 4) and approach the expected experimental error more closely (R. Price, personal communication, 1979).

Relaxation Experiment

Similar assumptions can be made for a relaxation test, and if two independent mechanisms are operating, the re-sulting equation for the stress will be the sum of two terms similar to those given by (44) or (46).

Dependent Mechanisms

In trying to fit the two-mechanisms equation (54) to *Heard*'s [1972] polycrystalline-salt data, we found that we could not fit some of his data and that we could not improve significantly upon the poor fits obtained with the one-mechanism equation. In Figure 4, we show the one-mechanism fit (dashed curve) and the two-independent-mechanism fit (solid curve) to Heard's data for his experiment 781 ($\dot{e}_s$ = 1.18 × 10^{-5}/s, T = 400°C, confining pressure P_c = 200 MPa) on polycrystalline salt. While the rms error of the fit decreases by more than a factor of 2 between the two cases (0.49 MPa versus 0.22 MPa), it is clear that the two-mechanism equation is not a good fit to the data. For example, the data have an upward-concave trend (between about 300 to 1500 s, or between 2.0 and 2.5 MPa differential stress) which cannot be fitted by (54). The latter can give rise only to downward-concave curves. One way to achieve the upward-concave character for the curves is to allow second-order kinetics to hold (as *Johnston and Gilman* [1959] and *Ajaja and Ardell* [1979] did). However, this leads to nonlinear differential equations, so we looked for a simpler approach to obtain the desired results.

In the above, we have assumed that the mechanisms are independent and do not interact. Even if we had assumed second-order kinetics for each mechanism, we could still have assumed no interaction. Instead, we decided to investigate the case in which the mechanisms interact but each still satisfies first-order kinetics. We assume the constitutive equations for the two mechanisms now depend upon both mechanisms as

$$f_1(e_1, \dot{e}_1, e_2, \sigma_1, \dot{\sigma}_1, \sigma_2, T, S_1) = 0$$
$$f_2(e_2, \dot{e}_2, e_1, \sigma_2, \dot{\sigma}_2, \sigma_1, T, S_2) = 0 \qquad (55)$$

In the above we assume the constitutive equation depends not only on the strain and strain rate of one mechanism, but also upon the strain experienced by the other mechanism. We assume a similar dependence for the stress terms.

Now the differential forms of the constitutive equation constitute a set of coupled differential equations. In the following we will treat the cases of a relaxation experiment or a creep experiment simultaneously by letting $y(t)$ represent the dependent variable (i.e., the stress in a relaxation experiment) and $x(t)$ represent the specified variable (i.e., the strain in a relaxation experiment). The cases of (1) a constant-strain-rate experiment and (2) a constant-stress-rate experiment are also implicit in the following solution as well; in these cases, an additional time integration will be necessary to determine the measured stress or strain, respectively. Using these conditions, the differential form for the constitutive equations become

$$-\frac{d\dot{y}_1}{dt} = R_{11}\dot{y}_1 + R_{12}\dot{y}_2 + A_1\delta(t) + B_1\delta'(t) + C_1H(t)$$

$$ \tag{56}$$

$$-\frac{d\dot{y}_2}{dt} = R_{21}\dot{y}_1 + R_{22}\dot{y}_2 + A_2\delta(t) + B_2\delta'(t) + C_2H(t)$$

where

$$R_{11} = f_{1y_1}/f_1' \qquad R_{21} = f_{2y_1}/f_2'$$

$$R_{12} = f_{1y_2}/f_1' \qquad R_{22} = f_{2y_2}/f_2'$$

$$A_1 = x_0 f_{1x}/f_1' \qquad A_2 = x_0 f_{2x}/f_2'$$

$$B_1 = x_0 f_{1\dot{x}}/f_1' \qquad B_2 = x_0 f_{2\dot{x}}/f_2' \tag{57}$$

$$C_1 = S_{1_0} f_{1S_1}/f_1' \qquad C_2 = S_{2_0} f_{2S_2}/f_2'$$

$$f_1' = \partial f_1/\partial\dot{y}_1 \qquad f_2' = \partial f_2/\partial\dot{y}_2$$

$$f_{1y_1} = \partial f_1/\partial y_1 \qquad f_{1y_2} = \partial f_1/\partial y_2$$

and soon, and we have assumed that

$$x_1(t) = x_2(t) = x_0 H(t)$$
$$S_1(t) = S_{1_0}H(t) \qquad S_2(t) = S_{2_0}H(t) \tag{58}$$

We again assume first-order kinetics, that is, we let

$$R_{ij} = r_{ij} = \text{constants} \qquad (i, j = 1, 2) \tag{59}$$

and we let S_{1_0} and S_{2_0} be constants. We can solve the coupled differential equations by using standard Laplace-Transform techniques (see, for example, *Gardner and Barnes*, [1952]). We can express the total dependent variable as an inverse Laplace Transform:

$$y(t) = y_1(t) + y_2(t) = -L^{-1}\{[\eta_1(p) + \eta_2(p)]/D(p)\} \tag{60}$$

where

$$\eta_1(p) = (p + r_{22} - r_{21})(p^2 B_1 + p A_1 + C_1)$$

$$\eta_2(p) = (p + r_{11} - r_{12})(p^2 B_2 + p A_2 + C_2)$$

$$D(p) = p^2(p + p_+)(p - p_-)$$

where p is the Laplace Transform variable and

$$2p_\pm = -(r_{11} + r_{22}) \pm ((r_{11} - r_{22})^2 + 4r_{12}r_{21})^{1/2} \tag{61}$$

are singularities where the Laplace Transform is evaluated. The poles at the origin, corresponding to the p^2 term in the denominator, will give rise to the steady state part of the solution and the singular derivatives at $t = 0$. The poles $p_\pm$, on the other hand, give rise to exponential terms of the form

$$\exp(-p_+ t) \qquad \exp(-p_- t) \tag{62}$$

The complete solution to this problem is fairly complicated and will not be given here. Nevertheless, there are two interesting characteristics of the solution that can be determined from the above terms. First of all, we note that the coefficients r_{12} and r_{21} may be positive or negative and are independent of the coefficients r_{11} and r_{22}. We note that if the product of r_{12} and r_{21} is negative and if

$$|4r_{12}r_{21}| > (r_{11} - r_{22})^2 \tag{63}$$

the square root in (61) becomes imaginary, and we have a damped oscillatory solution. This oscillatory behavior is observed sometimes in experimental data (though rarely, if ever, commented upon) as can be seen from the data in Figure 3. There the oscillations are quite small and of the order of the experimental error, but they are visible. Similar oscillations can be seen in Garofalo's data for austenitic stainless steel (see, in particular, Figure 1 of *Li* [1963], where he plots the strain rate of Garofalo's 1960 data). The data points for polycrystalline salt in Figure 4 also show an oscillatory variation, at least relative to the two-mechanism curve, and this oscillatory character may give rise to the concave-upward shape of the data that was mentioned earlier. We are continuing work on this solution and its fit to experimental data. However, many more data points than those shown in Figure 4 are necessary to obtain reasonable constraints on all the parameters now available, namely $A_1, B_1, C_1, A_2, B_2, C_2, r_{11}, r_{12}, r_{21}$, and r_{22}.

The second interesting characteristic of the solution is the fact that it can give rise to accelerating or tertiary creep. For example, if r_{12} and r_{21} are both large in relation to r_{11} and r_{22}, and are of the same sign, the square root term of $p_\pm$ (in (61)) can exceed the sum of r_{11} and r_{22}. Thus one exponential term (see (62)) decreases with time, while the other increases with time. This exponential growth must eventually terminate in failure of the material. Both mechanisms contain the two exponential terms, and, therefore, their dependent parameters will both have the exponential growth.

It is instructive to look at the coupled differential equations in this case. We see that (because $r_{12} \gg r_{11}$ and $r_{21} \gg r_{22}$) the rate of change of the dependent parameter of one mechanism depends, primarily, on the amount of the other mechanism (cf. (56)). Thus as one increases in magnitude, the rate of the other increases, and this leads to the exponential growth of the dependent parameters.

An example of the interaction of two flow mechanisms that could give rise to accelerating creep would be that between microfractures and dislocations. It would be expected that the presence of microfractures would enhance the generation of dislocations (by way of generating stress concentrations and voids), and the presence of dislocations would enhance the generation of microfractures (by coalescence of the dislocations). Clearly, under easily achieved conditions, microcracks would be more effective in generating dislocations than the dislocations themselves; also, the presence of dislocations could easily be more effective in generating microcracks than the microcracks themselves.

It is interesting that the simple assumptions made above for interacting, multiple mechanisms can give rise to such a wide diversity in the flow behavior of materials. In the above we have assumed we have linear differential equations for the constitutive equations (i.e., first-order kinetics and constant coefficients); even this apparently highly restrictive assumption gives rise to diverse behavior. An advantage of using linear differential equations is that superposition can be used to treat the flow characteristics in experiments with variable independent parameters (e.g., cyclic relaxation experiments, etc.).

While we have only treated as many as two simultaneously acting mechanisms in the above, it should be clear that a similar analysis would be possible if more than two mechanisms give measurable contributions during an experiment. While there generally are more than two possible flow mechanisms for any material, it is usual that no more than two are of significant magnitude to have a measurable effect in any one experiment. Of course, the two dominant mechanisms can be different for experiments under different conditions. However, the past success of one-mechanism equations in characterizing experimental results indicates that, for a significant range of conditions and materials, only one mechanism is dominant (cf. *Amin et al* [1970]). As the accuracy, reproducibility, and reliability of the experiments increase, the effects of more than one mechanism operating will become apparent and will need to be taken into account (cf., e.g., Figures 3*a* and 3*b*, which shows the difference between the measured data and that expected for a single mechanism). Also, at least two mechanisms are useful (necessary?) if we are to obtain oscillatory flow behavior and accelerating flow.

Summary

A postulatory approach and macroscopic viewpoint has been taken to determine the constitutive equations for the one-dimensional flow of solids. If one flow mechanism is dominant, then first-order kinetics and a linear differential constitutive equation suffice to describe the flow behavior. One mechanism is generally dominant when the material is close to its melting temperature [*Amin et al.*, 1970]. At lower temperatures (and lower confining pressures for rocks), two independent mechanisms can describe the data by using first-order kinetics and linear differential equations, but now two constitutive equations are required, one for each mechanism. The total flow characteristic is assumed to be the sum of the flow characteristic of the individual mechanisms.

At other conditions of temperature, confining pressure, strain rate, etc., the flow behavior is described better by assuming two (or more) mechanisms interact while assuming first-order kinetics and linear differential constitutive equations. It is shown that interacting mechanisms can cause oscillatory behavior or accelerating behavior that leads to failure. The latter behavior might be considered a candidate for the generation of earthquakes both near the surface and at depth.

Only the case of one-dimensional flow has been considered; however, the concepts and methods used are directly applicable to flow (and anelasticity) in three dimensions. In this case, the stress and strain must both be treated as second-order tensors, and the constitutive equation itself

must also be a tensor relationship, one for each mechanism.

Acknowledgments. I thank Robert Kranz and Hugh Heard for their review of the manuscript; their suggestions materially improved its clarity and accuracy. This work was supported, in part, by the Battelle Memorial Institute, Office of Waste Isolation, Columbus, Ohio, under contract E512-00900.

References

Ajaja, O., and A. J. Ardell, A phenomenological theory of transient creep, *Philos. Mag. Part A, 39*, 75-90, 1979.

Akulov, N. S., On dislocation kinetics, *Acta Metall., 12*, 1195-1196, 1964a.

Akulov, N. S., The statistical theory of dislocations, *Philos. Mag., 9*, 767-779, 1964b.

Amin, K. E., A. K. Mukherjee, and J. E. Dorn, A universal law for high-temperature diffusion controlled transient creep, *J. Mech. Phys. Solids, 18*, 413-426, 1970.

Andrade, E. N. daC., On the viscous flow in metals, and allied phenomena, *Proc. Roy. Soc. London, Ser. A, 84*, 1-12, 1910.

Bailey, R. W., Note on the softening of strain-hardened metals and its relation to creep, *J. Inst. Metals, 35*, 27-43, 1926.

Carter, N. L., and S. H. Kirby, Transient creep and semibrittle behavior of crystalline rocks, *Pure Appl. Geophys., 116*, 807-839, 1978.

Conway, J. B., and M. J. Mulliken, Techniques for analyzing combined first- and second-stage creep, data, *Trans. Metall. Soc. AIME, 236*, 1629-1632, 1966.

Cottrell, A. H., and V. Aytekin, Andrade's creep law and the flow of zinc crystals, *Nature, 160*, 328-329, 1947.

Dorn, J. E., Some fundamental experiments on high temperature creep, *J. Mech. Phys. Solids, 3*, 85-116, 1954.

Evans, W. J., and B. Wilshire, Transient and steady-state creep of nickel, zinc, and iron, *Trans. Metall. Soc. AIME, 242*, 1303-1307, 1968.

Evans, W. J., and B. Wilshire, Transient and steady-state behavior of a copper-15 at. % aluminum alloy, *Met. Sci., 4*, 89-94, 1970.

Gardner, M. F., and J. L. Barnes, *Transients in Linear Systems*, vol. 1, *Lumped-Constant Systems*, pp. 136-139, John Wiley, New York, 1952.

Garofalo, F., Resistance to creep deformation and fracture in metals and alloys, *ASTM Spec. Tech. Publ. 283*, 82-98, 1960.

Garofalo, F., *Fundamentals of Creep and Creep Rupture*, Macmillan, New York, 1965.

Gittus, J. H., Strain, recovery and work-hardening during creep due to dislocations, *Philos. Mag., 23*, 1281-1296, 1971.

Heard, H. C., Steady-state flow in polycrystalline halite at pressures of 2 kilobars, in *Flow and Fracture of Rocks, Geophys. Monogr. Ser.*, vol. 16, edited by H. C. Heard et al., pp. 191-209, AGU, Washington, D. C., 1972.

Hollomon, J. H., The mechanical equation of state, *Met. Technol. N.Y.*, *13*, 1-9, 1946.

Ishida, Y., and D. McLean, Effect of nitrogen and manganese on recovery rate and friction stress during creep of iron, *J. Iron Steel Inst.*, *205*, 88-93, 1967.

Johnston, W. G., and J. J. Gilman, Dislocation velocities, dislocation densities, and plastic flow in lithium fluoride crystals, *J. Appl. Phys.*, *30*, 129-144, 1959.

Lacombe, M. J. de, Un mode de representation des courbes de fluage, *Rev. Metall. Paris*, *36*, 178-188, 1939.

Li, J. C. M., A dislocation mechanism of transient creep, *Acta Metall.*, *11*, 1269-1270, 1963.

McVetty, P. G., Working stresses for high-temperature service, *Mech. Eng.*, *56*, 149-154, 1934.

Mitra, S. K., and D. McLean, Work hardening and recovery in creep, *Proc. Roy. Soc. London, Ser. A*, *295*, 288-299, 1966.

Orowan, E., The creep of metals, *J. West Scotl. Iron Steel Inst.*, *54*, 45-96, 1947.

Parrish, D. K., and A. F. Gangi, A nonlinear, least squares fitting approach for determining activation energies for high-temperature creep, (abstr.), *Eos Trans.*, *AGU*, *58*, 514, 1977.

Parrish, D. K., and A. F. Gangi, A nonlinear, least squares technique for determining multiple mechanism, high-temperature-creep flow laws, this volume, 1981.

Sidey, D., and B. Wilshire, Mechanisms of creep and recovery in Nimonic 80A, *Met. Sci.*, *3*, 56-60, 1969.

Webster, G., A. Cox, and J. Dorn, A relationship between transient and steady-state creep at elevated temperatures, *Met. Sci.*, *3*, 221-225, 1969.

Zener, C., and J. H. Hollomon, Problems in non-elastic deformation of metals, *J. Appl. Phys.*, *17*, 69-82, 1946.

A Nonlinear Least Squares Technique for Determining Multiple-Mechanism, High-Temperature Creep Flow Laws

DAVID K. PARRISH[1]

Center for Tectonophysics, Departments of Geophysics and Geology
Texas A&M University, College Station, Texas 77843

ANTHONY F. GANGI

Department of Geophysics, Texas A&M University, College Station, Texas 77843

Current methods for determining flow laws for high-temperature creep in rocks, ceramics, and metals assume that a single mechanism dominates and therefore produces the observed strain rate. Our calculations show that this assumption may lead to errors in (1) the magnitudes of the critical flow law parameters and (2), consequently, conclusions about the atomic mechanisms causing the flow. An alternative approach is to assume that several independent mechanisms operate and to fit the data derived from creep experiments ($\dot{\epsilon}$, T, σ) to a flow law of the form $\dot{\epsilon} = \sum_{i=1}^{I} A_i \sigma^{n_i} \exp(-Q_i/RT)$, where $\dot{\epsilon}$ is the total steady state strain rate resulting from I independent flow mechanisms operating at the temperature T and stress σ, and A_i, Q_i, and n_i are the structure constants, apparent activation energies, and stress exponents for each mechanism. Since the fitting equation above is nonlinear, an iterative, nonlinear least squares technique is used to determine the best estimates of A_i, Q_i, and n_i. Nonlinear least squares fits to published experimental data for Yule marble demonstrate the equivalence of the nonlinear and multilinear fitting techniques when only one mechanism is dominant. Nonlinear least squares fits to published data for polycrystalline halite provide further examples which suggest that two or more flow mechanisms independently contribute to the total strain rate.

Introduction

Identification of creep mechanisms is necessary for the prediction of creep in solids subjected to stress at high temperatures for long periods. Abundant experimental data indicate that steady state strain rates in metals, ceramics, and rocks are controlled by thermally activated mechanisms (see, for example, review papers by *Garofalo* [1965] and *Sherby and Burke* [1967] for metals, *Evans and Langdon* [1976] for ceramics, and *Carter* [1976] for rocks). The steady state strain rate relationship, which is valid for an individual creep mechanism, is given by [*Evans and Langdon*, 1976]

$$\dot{\epsilon}_i = H_i (D\mu_i b_i/RT)(b_i/G_i)^{m_i}(\sigma/\mu_i)^{n_i} \qquad (1)$$

where $\dot{\epsilon}_i$ is the strain rate, H_i, m_i, and n_i are dimensionless parameters, R is the gas law constant, T is temperature, b_i is the magnitude of a Burger's vector, G_i is the grain size,

and μ_i is the shear modulus. D_i is a diffusion coefficient given by

$$D_i = D_{i0} \exp(-q_i/RT) \qquad (2)$$

where D_{i0} is a material parameter and q_i is the activation energy for the particular creep mechanism. Commonly, the structural parameters, H_i, μ_i, b_i, and G_i are assumed constant, so that (1) is written as

$$\dot{\epsilon}_i = A_i \sigma^{n_i} \exp(-Q_i/RT) \qquad (3)$$

where

$$A_i = (H_i b_i/RT)(b_i/G_i)^{m_i} \mu_i^{n_i-1}$$

and Q_i is the apparent activation energy for creep. The activation energy, q_i, and the apparent activation energy for creep, Q_i, are slightly different. From (2), q_i is given by

$$q_i = -R \frac{\partial \ln D_i}{\partial (1/T)}$$

_____
[1] Now at RE/SPEC, Inc., Rapid City, South Dakota 57709.

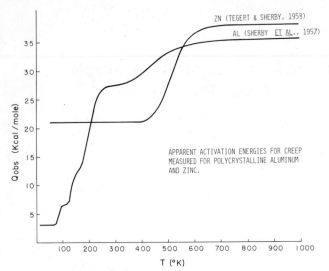

ZN (TEGERT & SHERBY, 1958)

AL (SHERBY ET AL., 1957)

APPARENT ACTIVATION ENERGIES FOR CREEP
MEASURED FOR POLYCRYSTALLINE ALUMINUM
AND ZINC.

Fig. 1. Observed apparent activation energy for creep (Q_{obs}) in creep experiments on zinc and aluminum.

and from (3), Q_i is given by

$$Q_i = -R \frac{\partial \ln \dot{\epsilon}_i}{\partial (1/T)}$$

The apparent activation energy for creep is a function of q_i as well as the temperature, internal structure, and elastic moduli of the deforming solid. *Evans and Langdon* [1976, p. 377] show that under constant stress conditions and assuming that the H_i are independent of temperature, q_i and Q_i are related through the expression

$$Q_i = q_i - RT \left[1 + \frac{(n_i - 1) T}{\mu_i} \left(\frac{\partial \mu_i}{\partial T} \right)_\sigma \right]$$

The relevant diffusion coefficients D_i for ceramics and rocks are usually not known, so determination of the apparent activation energy for creep is a very important step toward identifying the creep mechanism.

An expression such as (3) has been used extensively to identify the parameters A_i, Q_i, and n_i for a number of materials under specific experimental conditions at which a single creep mechanism controls the total steady state strain rate. Creep, however, is a complex process, during which several mechanisms participate at all times to a greater or lesser extent. The different mechanisms may operate independently, in which case each mechanism contributes to the total rate; or the mechanisms may be interdependent, in which case the slowest mechanism will dominate the total creep rate.

If we consider those cases where independent, thermally activated creep mechanisms are operating, the observed strain rate $\dot{\epsilon}_{obs}$ is the sum of the strain rates due to each mechanism:

$$\dot{\epsilon}_{obs} = \sum_{i=1}^{I} \dot{\epsilon}_i = \sum_{i=1}^{I} A_i \sigma^{n_i} \exp (-Q_i/RT) \quad (4)$$

where A_i, Q_i, and n_i are the structure parameter, the apparent activation energy, and the stress exponent, respectively, for the ith creep mechanism.

Depending on the exact values of A_i, Q_i, and n_i, each term in (4) may dominate all others over some range of stress and temperature values. If the strain rates can be measured over that range of stresses and temperatures where only one creep mechanism dominates, then (3) can be used to determine the parameters A, Q, and n for a particular mechanism. The apparent activation energy Q can be determined from a temperature differential creep test, the stress exponent n can be determined from a stress differential creep test, and all three parameters can be determined by multilinear least squares fitting of (3) to steady state creep data.

When the experimentally applied flow stresses and temperatures are such that the measured strain rate is not identical to the strain rate of a single mechanism, the observed parameters A, Q, and n determined using the above procedures do not correspond to the parameters for any of the independent mechanisms.

In the following sections the procedures for determining creep mechanism parameters from experiments in which a single mechanism is dominant are examined to demonstrate the need for a nonlinear least squares procedure for analyzing steady state creep data.

The results of applying the least squares procedure to published data for Yule marble [*Heard*, 1963; *Heard and Raleigh*, 1972] support the previous conclusion that a single creep mechanism dominates the flow in the temperature and stress ranges, $400 < T < 800°C$ and $15 < \sigma < 150$ MPa, respectively. However, a two-mechanism flow law seems more appropriate for polycrystalline halite [*Heard*, 1972] over the temperature and stress ranges, $100 < T < 400°C$ and $1.5 < \sigma < 15$ MPa. The behavior at the low temperatures and low strain rates predicts lower stresses when the halite flow law is extrapolated to important geological conditions.

Calculation of Apparent Activation Energy

Temperature differential creep tests have been used extensively to calculate apparent activation energies for a number of materials. During temperature differential tests the stress is held constant while the temperature is changed abruptly from T_1 to T_2, causing a change in strain rate from $\dot{\epsilon}_1$ to $\dot{\epsilon}_2$. If the crystal structure remains constant immediately after the temperature change, an observed apparent activation energy can be calculated from (3):

$$Q_{obs} = \frac{-Rd (\ln \dot{\epsilon})}{d (1/T)} \cong \frac{RT_1 T_2}{T_2 - T_1} \ln (\dot{\epsilon}_2 / \dot{\epsilon}_1) \quad (5)$$

Apparent activation energies for zinc [*Tegert and Sherby*, 1958] and aluminum [*Sherby et al.*, 1957] have been calculated in this way (Figure 1). Independent thermally activated mechanisms are identified by the distinct plateaus in $Q_{obs} - T$ plots. Zinc, for instance, apparently has two independent mechanisms corresponding to two

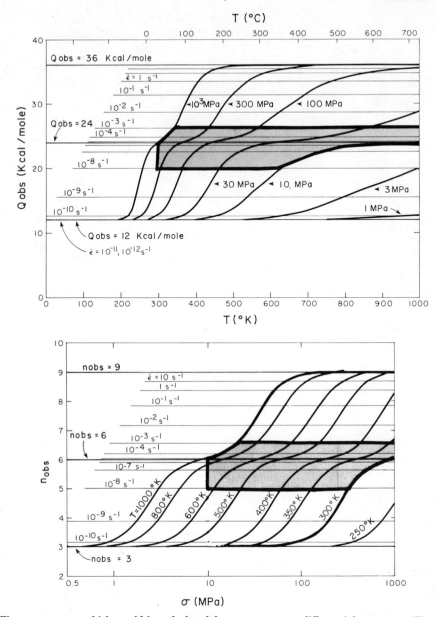

Fig. 2. Flow parameters which would be calculated from temperature differential creep test (Figure 2a) and stress differential creep tests (Figure 2b) on a solid which obeys a multiple mechanism flow: $\sum_1^3 A_i \exp(-Q_i/RT)\sigma^{n_i}$, where $A_i = (1, 30, 0.3)$, $Q_i = (12, 24, 36)$, and $n_i = (3, 6, 9)$. (a) Observed activation energy Q_{obs} as a function of temperature for various values of stress. (b) Observed stress exponent n_{obs} as a function of stress for various values of temperature.

temperature ranges where the observed activation energy is constant (Figure 1). Aluminum, on the other hand, has at least two distinct plateaus and possibly as many as five, as indicated by inflections in the observed activation energy function. These inflections may be indistinct plateaus corresponding to temperature and stress conditions at which there is no clear dominance by any one term in (4).

Poorly defined plateaus such as those evident in the activation energy function for aluminum (Figure 1) are expected if (4) is the flow equation for the material and if

three or more independent mechanisms contribute to the total strain rate. To illustrate the effect of three independent mechanisms, the observed activation energy has been calculated using (4) and the following flow parameters:

$$A_1 = 1\,(\text{MPa})^{-3}\,\text{s}^{-1} \qquad Q_1 = 12\,\text{kcal/mol}\,(50\,kJ) \qquad n_1 = 3$$

$$A_2 = 30\,(\text{MPa})^{-6}\,\text{s}^{-1} \qquad Q_2 = 24\,\text{kcal/mol}\,(100\,kJ) \qquad n_2 = 6$$

$$A_3 = 0.3\,(\text{MPa})^{-9}\,\text{s}^{-1} \qquad Q_3 = 36\,\text{kcal/mol}\,(150\,kJ) \qquad n_3 = 9$$

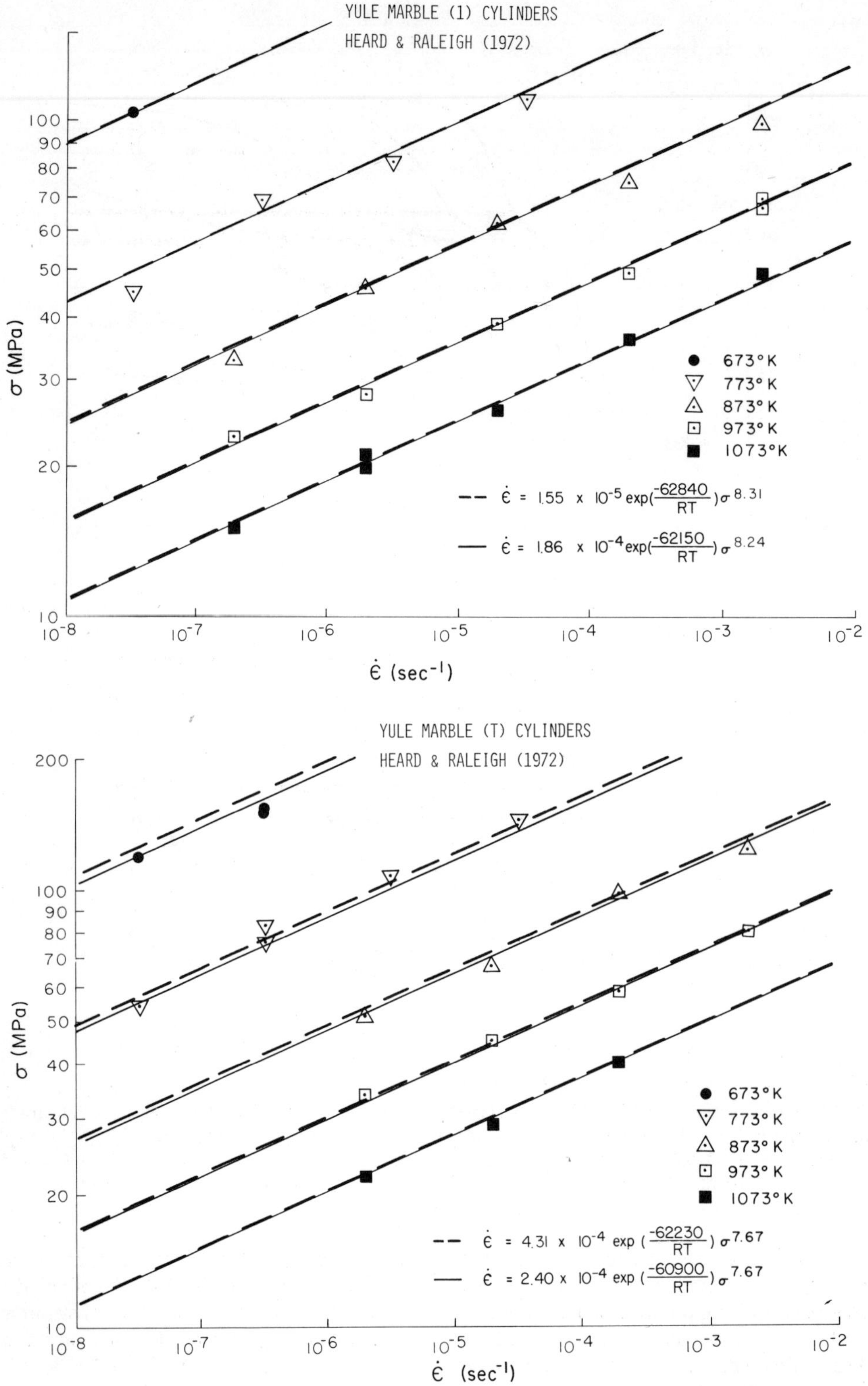

Fig. 3. Experimental data for steady state creep of Yule marble [*Heard and Raleigh,* 1972]. Solid lines show the best multilinear fit obtained by *Heard and Raleigh* [1972]. Dashed lines show the best nonlinear least squares fit. (*a*) The *l* cylinders. (*b*) The *T* cylinders.

The observed apparent activation energy at any stress level is given by

$$Q_{obs} = \sum Q_i \dot{\epsilon}_i / \sum \dot{\epsilon}_i \qquad (6a)$$

$$Q_{obs} = \sum Q_i A_i \sigma^{n_i} \exp(-Q_i/RT) / \sum A_i \sigma^{n_i} \exp(-Q_i/RT) \qquad (6b)$$

A family of observed activation energy functions obtained by evaluating (6) at increasing values of stress (σ = 1.0 to 1000 MPa) illustrates the variation of Q_{obs} with temperature and stress (Figure 2a). Clearly, there are temperature and stress conditions where one activation energy is defined by a distinct plateau. For example, the observed activation energy Q_{obs} approaches Q_3 at temperatures greater than 750°K and stresses greater than 300 MPa. The reason is that above 750°K the third term in the summations in (6) is so large that the other two terms are insignificant. As the flow stress (and, consequently, the strain rate) increases above 300 MPa, the third term becomes even larger, so the temperature at which Q_{obs} approaches Q_3 is lower (i.e., the edge of the plateau shifts to lower temperatures in Figure 2a).

At sufficiently low temperatures (less than 500°K) the observed activation energy approaches the lowest activation energy (Q_1) for all stresses less than 10 MPa. At these conditions the first term in the summations in (6) is so large that the second and third terms are insignificant. Increasing the stress reduces the temperature range of the plateau corresponding to the low activation energy because the contributions of the second and third terms of (6) increase when the stress is increased. When the stress is less than 1.0 MPa, the first term dominates (6) for all temperatures used in these calculations. Therefore the strain rate observed in steady state creep experiments conducted at these very low stresses would be dominated by only one mechanism at all temperatures (up to 1000°K). Naturally, only one activation energy would be determined as a result of such a set of experiments.

A plateau corresponding to Q_2 = 24 kcal (100 kJ) (Figure 2a) is well defined only at high temperatures (900°-1000°K) and 10 MPa. Only at these particular conditions does the second term in the summations in (6) become so large that the first and third terms are not significant. At stresses less than 10 MPa the first term is dominant, and at stresses greater than 10 MPa the only indication of an activation energy intermediate between Q_1 and Q_3 is a poorly defined plateau (inflection) similar to those in the observed activation energy functions for aluminum (Figure 1). Estimations of the apparent activation energy from these indistinct plateaus may be very inaccurate. Therefore, whenever possible, laboratory tests should be conducted at conditions where a single mechanism is dominant.

It is often impossible to attain the optimum conditions for steady state creep dominated by a single mechanism. The restricting parameter for many rock-forming minerals is the strain rate. Each stress and temperature state

effects a steady state strain rate for each mechanism. When contours of the observed strain rates are superposed on the observed activation energy functions (Figure 2a), it is clear that when the stress and temperatures are low enough to measure low activation energies, the strain rates are very low. In many cases the required strain rates are so low that the appropriate experiment would require years for completion.

The three-process model shown in Figure 2a presumes that experiments at all stresses and temperatures necessary to define distinct plateaus are available. In reality, because of laboratory apparatus limitations the entire range of experimental conditions may not include the conditions where a single mechanism is dominant enough to define a distinct plateau. Suppose, for instance, that laboratory apparatus were designed for the stress values, temperatures, and strain rates within the shaded region of Figure 2a. Even though the experimental data would contain information about all the activation energies, the exact value of Q_2 could not be extracted.

Calculation of Stress Exponents

Stress exponents (n_i in (4)) are commonly determined by conducting stress differential creep tests. The temperature is held constant while a small, abrupt change is made from an initial stress σ_1 to a final stress σ_2, causing a change in strain rate from $\dot{\epsilon}_1$ to $\dot{\epsilon}_2$. The crystal structure remains constant to first order after the stress change; therefore an observed stress exponent can be calculated from (3):

$$n_{obs} = \frac{d \ln \dot{\epsilon}}{d \ln \sigma} \cong \frac{\ln(\dot{\epsilon}_2/\dot{\epsilon}_1)}{\ln(\sigma_2/\sigma_1)} \qquad (7)$$

This equation will give the stress exponent of one creep mechanism if a single creep mechanism is so dominant that (3) is a valid approximation to (4). As in the case of the observed activation energy, an unequivocal determina-

TABLE 1. Revisions to Differential Stresses After Recorrecting for Aluminum Jacket Strength

Experiment	Strain Rate, s⁻¹	Temperature		Differential Stress at 10% Strain, MPa	
		°C	°K	Heard [1972]	Heard† (1980)
783	1.15×10^{-3}	300	573	11.0	12.7
858	1.15×10^{-4}	300	573	8.8	8.1
782	1.18×10^{-5}	300	573	4.9	6.2
814	1.54×10^{-4}	300	573	7.4	9.5
787	1.18×10^{-6}	300	573	3.1	3.9
806	1.15×10^{-7}	300	573	2.2	3.0
840	1.15×10^{-8}	300	573	(1.6) *	(2.4)

* Extrapolated to 10% strain.
† Personal communication.

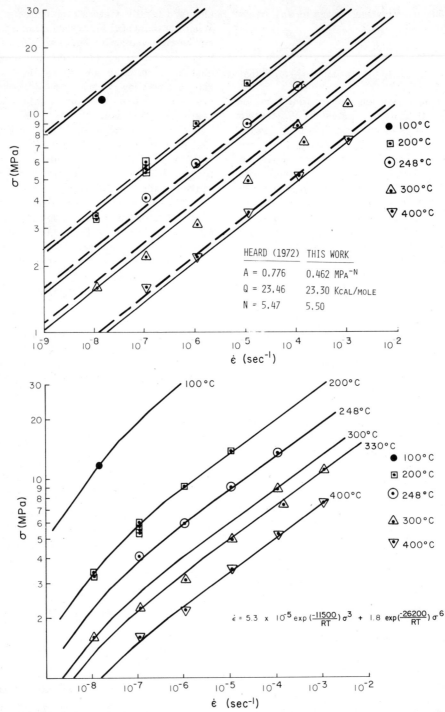

Fig. 4. Experimental data for polycrystalline halite [*Heard*, 1972]. Solid lines show the best multilinear fit obtained by *Heard* [1972]. Dashed lines show the best nonlinear least squares fit. (*a*) One-mechanism flow law is the fitting function. (*b*) Two-mechanism flow law is the fitting function.

tion of n depends on the material properties (A_i, Q_i, and n_i) and the experimental conditions (σ, T, and $\dot{\epsilon}$). The observed stress exponent n_{obs} is calculated using (4) as

$$n_{obs} = \sum n_i \dot{\epsilon}_i \bigg/ \sum \dot{\epsilon}_i \qquad (8a)$$

$$n_{obs} = \sum n_i A_i \sigma^{n_i} \exp\,(-Q_i/RT) \bigg/ \sum A_i\, \sigma^{n_i} \exp\,(-Q_i/RT) \quad (8b)$$

Equations (8) have exactly the same form as (6). Consequently, it is not surprising that a family of curves showing the variation of n_{obs} with stress at constant tem-

TABLE 2. Multiple-Mechanism Flow Law Parameters Determined by Nonlinear Least Squares Fit of Polycrystalline Halite Data

One-Mechanism Flow Law $\dot{\epsilon} = A\sigma^n \exp(-Q/RT)$	Two-Mechanism Flow Law $\dot{\epsilon} = \sum_{i=1}^{2} A_i \sigma^{n_i} \exp(-Q_i/RT)$	
	Mechanism 1	Mechanism 2
$A = 0.462$	$A_1 = 1.83$	$A_2 = 5.3 \times 10^{-5}$
$\ln A = 0.722 \pm 0.66$	$\ln A_1 = 0.606 \pm 0.94$	$\ln A_2 = -9.85 \pm 2.9$
$Q = 23.30 \pm 0.43$ kcal/mol	$Q_1 = 26.20 \pm 2.4$ kcal/mol	$Q_2 = 12.0 \pm 8.5$ kcal/mol
$n = 5.50 \pm 0.17$	$n = 6.00 \pm 0.5$	$n_2 = 3.0 \pm 1.9$
rms fractional error = 0.42	rms fractional error = 0.35	

From [Heard, 1972]. Units of stress are MPa; units of strain rate are s^{-1}. The uncertainty in each parameter is ±1 standard deviation.

peratures (Figure 2b) looks very similar to the family of curves describing the variation of Q_{obs} with temperature (Figure 2a). The same values of A_i, Q_i, and n_i were used in the calculations for Figures 2a and 2b.

As with the observed activation energy function (at constant stress), the observed stress exponent function (at constant temperature) exhibits distinct plateaus over stress ranges at which one mechanism dominates the flow. In our example (Figure 2b) a plateau defining the lowest stress exponent ($n_1 = 3$) is well defined for all calculated stresses when the temperature is 200°K, but when the temperature is 400°K, a distinct plateau for n_1 is well defined only at stresses below 10 MPa. The observed stress exponent n_{obs} approaches $n_3 = 9$ only at the highest temperatures and stresses used in the calculation. A distinct plateau defining the intermediate stress exponent ($n_2 = 6$) is not well defined at any combination of temperature and stress. Only the poorly defined plateaus (inflections) suggest that a third mechanism (A_2, Q_2, n_2) contributes to the total strain rate. As with the observed activation energy, if the experimental stress, temperature, and strain rate conditions necessary to define a constant value (plateau) for n_{obs} cannot be attained, estimates of the stress exponent may be inaccurate.

The similarity between the families of curves describing the temperature variation of the observed activation energy and the stress variation of the observed stress exponent reflects the similarity between (6) and (8). In both (6) and (8) the observed parameter is a weighted sum of either the apparent activation energies (equations (6)) or the stress exponents (equations (8)). The 'weighting parameters' in each case are the strain rates for each independent creep mechanism ($\dot{\epsilon}_i$ in (6a) and (8a)). As the strain rate for each (ith) mechanism increases with increasing temperature and stress (equations (6b) and (8b)), the weight given the ith observed strain rate (equation (4)), activation energy (equation (6a)) or stress exponent (equation (8a)) increases to a limiting value of $A_i \sigma^{n_i}$.

Linear Least Squares Method

If a single creep mechanism is dominant in a suite of steady state creep experiments, a linear least squares minimization procedure can be used to determine the pa-

rameters A, Q, and n in (3). The procedure is to linearize (3) to the form

$$\ln \dot{\epsilon} = \ln A - (Q/RT) - n \ln \sigma \qquad (9)$$

Standard multilinear least squares minimization procedures are applied to determine the parameters in A, Q, and n which provide the best fit of (9) to experimental values of $\dot{\epsilon}$, T, and σ.

Multilinear least squares fitting was used to obtain the steady state flow equation for Yule marble [Heard and Raleigh, 1972]. The linear isothermal lines on a log-log plot of steady state stress σ versus strain rate $\dot{\epsilon}$ obtained at constant temperature (Figure 3a) demonstrate that (3) is an adequate description of the steady state creep data for Yule marble. Moreover, the linear relation between $\ln \dot{\epsilon}$ and $\ln \sigma$ suggests that a single creep mechanism dominates the steady state creep of Yule marble over the ranges of temperature, stress, and strain rate measured. Heard and Raleigh [1972] note that the rate-controlling process in their experiments may be 'dislocation climb to form polygonal subgrain boundaries along with intragranular and intergranular recrystallization.'

Multilinear least squares fitting is appropriate for creep data from experiments in which a single mechanism dominates the total strain rate. However, when no single creep mechanism dominates the flow, a linear relationship between $\ln \dot{\epsilon}$ and $\ln \sigma$ does not exist, and the data points would not align along linear isotherms, as the Yule marble data do (Figures 3a and 3b).

Nonlinear Least Squares Technique

To analyze those steady state creep experiments in which multiple independent mechanisms are contributing to the steady state flow, a nonlinear least squares technique for fitting (4) to the strain rates, temperatures, and stresses must be used.

The approach is to determine the parameters A_i, Q_i, and n_i, which minimize the mean squared fractional error δ^2 between the experimentally measured strain rate $\dot{\epsilon}_k$ and the theoretical rate $\dot{\epsilon}_0(\sigma, T)$:

$$\delta^2 = \frac{1}{K} \sum_{k=1}^{K} \left(\frac{\dot{\epsilon}_k - \dot{\epsilon}_0(\sigma, T)}{\dot{\epsilon}_k} \right)^2 \qquad (10)$$

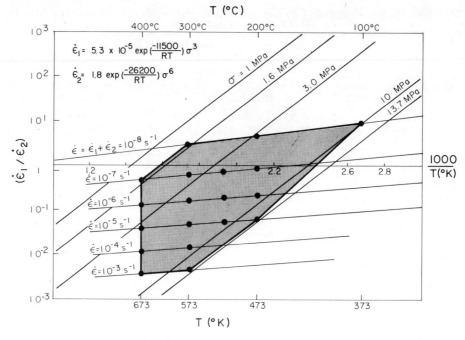

Fig. 5. Ratio of the independent strain rates in a two-mechanism flow law for polycrystalline halite. Steeply sloping lines are lines of constant stress difference. Gently sloping lines are lines of constant total strain rate ($\dot{\epsilon} = \dot{\epsilon}_1 + \dot{\epsilon}_2$). Shaded area encompasses the experimental data [*Heard*, 1972] that were used to determine the two-mechanism flow law. Ratios greater than unity indicate that mechanism 2 dominates mechanism 1. Ratios less than unity indicate that mechanism 1 dominates mechanism 2.

The fitting function is nonlinear, so the system of equations derived from the minimization of (10) will be nonlinear.

A derivation of the procedure and the algorithm used to determine the parameters A_i, Q_i, and n_i is outlined in Appendix A. In brief, the algorithm fits the data interatively to a linear approximation of (4) until the best fit of the constants A_i, Q_i, and n_i is determined.

Application of Nonlinear Least Squares Technique

Yule marble. Using the nonlinear least squares technique eliminates the assumption that a single mechanism dominates; however, if only one process is present, the nonlinear technique converges to approximately the same value as those determined using (9). To determine whether this is true, we applied the nonlinear least squares procedure to the Yule marble data of *Heard and Raleigh* [1972] (Figures 3a and 3b). The resulting parameters A, Q, and n are the same (within statistical error) as those obtained by the linear least squares procedure they used. The parameters differ slightly because different error parameters are minimized in the two procedures. The multilinear least squares procedure [*Heard and Raleigh*, 1972] minimizes the squared error in the logarithm of strain rate, whereas the nonlinear procedure (Appendix A) minimizes the squared fractional error in the strain rate. The equivalent results obtained by linear and nonlinear analyses of steady state strain rate data for Yule

marble support *Heard and Raleigh*'s [1972] conclusion that a single creep mechanism dominates the flow of Yule marble at the temperatures, stresses, and strain rates of their experiments.

Polycrystalline halite. Application of the nonlinear fitting technique to steady state flow data for polycrystalline halite, obtained by *Heard* [1972], indicates that over the temperature and stress ranges of his experiments, more than one mechanism contributes to the total strain rate. However, H. C. Heard (personal communication, 1980) has recorrected some of the data published earlier [*Heard*, 1972]. These recorrected data are listed in Table 1. The nonlinear least squares technique was applied to the recorrected data, and it indicates that the recorrected data are much more variable than the data published earlier [*Heard*, 1972]. Consequently, only the flow law parameters for one mechanism are obtained by the best fit. Moreover, because of the increased variability the fit obtained by the nonlinear technique is different than the fit obtained by the linear technique. Analyses of both the *Heard* [1972] data and the recorrected data (H. C. Heard, personal communication, 1980) is presented here. Although the parameters obtained from the *Heard* [1972] data cannot be accepted as realistic, they provide an example of the application of the nonlinear technique to published data. Analysis of the H. C. Heard (personal communication, 1980) data illustrates the effect that variability in data may have on the fit.

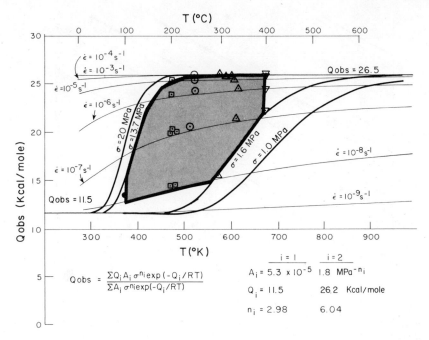

Fig. 6.   Observed activation energy Q_{obs} calculated for polycrystalline halite using a two-mechanism flow law.

Nonlinear least squares fitting of (4) to polycrystalline-halite data [*Heard*, 1972] was used to determine the best fit parameters for both a one-mechanism and a two-mechanism strain rate law. The one-mechanism parameters (Figure 4a) are the same (within statistical error) as those obtained by *Heard* [1972] using the linear least squares procedure. The apparent activation energy and the stress exponent for the one-mechanism law are approximately 23.5 and 5.5 kcal/mol, respectively. However, the one-mechanism model does not accommodate all the data well. The 100°C data point does not lie on the 100° isotherm in a logarithmic plot of strain versus stress (Figure 4a), and in general, the high-temperature (400°C) data appear to fit the one-mechanism law better than the low-temperature data do. The 300°C data do not fit the 300°C isotherm. These are the data that have been recorrected by H. C. Heard (personal communication, 1980).

A two-mechanism law ($I = 2$ in (4)) provides a better fit to the halite data (solid lines in Figure 4b), as indicated by the reduction of the root mean square (rms) fractional error (δ of (10)) from 0.42 to 0.35 (see Table 2). Because of the similar activation energies (26.2 versus 23.3 kcal/mol), stress exponents (6.0 versus 5.5), and structure constants ($A = 0.462$ versus 1.83) it appears that one of the mechanisms detected by the two-mechanism model is the same as that determined from a single-mechanism model. The modification of the higher activation energy values from 23.5 to 26.2 kcal/mol and change in the stress exponent n from 5.5 to 6.0 might be expected if a single-mechanism function were imposed on a set of data resulting from two flow mechanisms. If *Heard*'s [1972] data are

correct, the better fit obtained using a two-mechanism function would imply that a second mechanism with a lower activation energy (12 kcal/mol) and a weaker stress dependence ($n = 3$) contributes to the strain rate.

Although the rms fractional error for the overall fit is lower for the two-mechanism flow law than the one-mechanism flow law, the uncertainties for each parameter (± 1 standard deviation) determined for the two-mechanism flow law are larger than those for the one-mechanism flow law. No analytic form exists for calculating the uncertainties in each parameter determined by a nonlinear least squares fitting procedure. The uncertainties shown in Table 2 were calculated according to a method suggested by *Bevington* [1969] (see Appendix B). The increased error bounds for the two-mechanism fit are apparently due to the fact that six parameters were determined using the same number of data points as were used to determine only three parameters in the one-mechanism flow law. Although the two-mechanism flow law fits all the data better than the one-mechanism flow law, the uncertainty in each parameter of the two-mechanism flow law is greater than the uncertainty in the one-mechanism flow law. Although the error bounds on each parameter in the two-mechanism flow law are very large, they do not overlap. That is, even though the uncertainties in each parameter are large, the parameters of two independently operating flow mechanisms are statistically distinct.

The contribution from each mechanism can be calculated using (4) and the two-mechanism flow law. The ratio of the two strain rates may then be plotted as a function of temperature and stress (Figure 5) to determine at

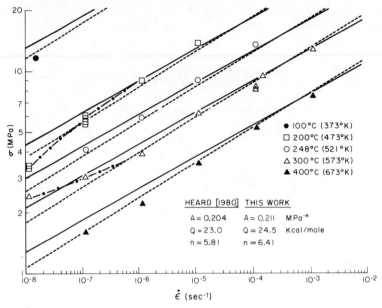

Fig. 7. Experimental data for polycrystalline halite recorrected for aluminum jacket strength (H. C. Heard, personal communication, 1980). Solid lines show the best fit using the nonlinear least squares fitting technique. Dashed lines show the fit obtained by H. C. Heard (Personal communication, 1980) using the linear fitting technique.

which conditions each dominates the flow and to what degrees. The shaded area in Figure 5 encompasses the actual data used to obtain the flow law. Steeply sloping lines are constant stress contours. Gently sloping lines are constant total stain rate contours ($\dot{\epsilon}_1 + \dot{\epsilon}_2$). In four experiments (one at 100°C, one at 300°C, and two at 200°C) the mechanism with the low activation energy and low stress exponent dominates, but slightly ($\dot{\epsilon}_1/\dot{\epsilon}_2 < 10$ compared with $\dot{\epsilon}_2/\dot{\epsilon}_1 > 100$ at higher temperatures and strain rates).

The temperature and stress conditions at which each mechanism dominates are made clear by calculating the apparent activation energy for creep using (6) and the two-mechanism flow law parameters (Figure 6). Observed strain rate contours are superimposed on Figure 6 (as in Figure 2a). The two plateaus at 26 and 12 kcal/mol correspond to complete dominance by one mechanism. The low activation energy mechanism never overwhelms the other mechanism at the physical conditions encompassed by the experimental data.

H. C. Heard (personal communication, 1980) has discovered that an incorrect aluminum jacket strength was used to correct all the 300°C tests for the effect of the jacket on the strength of the NaCl. The recorrected data (Figure 7) show no suggestion that two mechanisms contribute to the flow of polycrystalline halite. According to H. C. Heard (personal communication, 1980) '...almost all of the data points are translated up, that is, to give a more consistent fit with the other data.' The fit by H. C. Heard (personal communication, 1980) of (9) to the new data is slightly better (Table 3).

Using the nonlinear least squares technique to fit (3) to the recorrected data, a poorer fit is obtained (Figure 7). The difference between the fits obtained by the linear technique and the nonlinear technique results from the

fact that the techniques weigh each data point differently. The linear fit to (9) gives less weight to the data points at low strain rates than the nonlinear least squares technique which weighs all data points equally. When the point at 300°C, 10^{-8} s^{-1}, 22 MPa is given a weight equal to all the other 300° data, the trend in the 300°C data appears to be opposite to the trend in the 200°C data (see the dash-dot lines in Figure 7). These two opposing trends in the data are treated by our nonlinear technique as data variability. Because data variability is treated as statistical uncertainty, a poorer fit is obtained by the nonlinear technique for the recorrected data.

Conclusions

The usual assumption that a single dominant flow mechanism controls the steady state strain rate in laboratory creep experiments may lead to errors in the values of the critical parameters in flow laws, and it may cause other contributing mechanisms to be overlooked. A nonlinear least squares fit of a multiple-mechanism flow law to steady state creep data determines not only the parameters in a single-mechanism flow law but also the parame-

TABLE 3. Flow Law Parameters Determined for
Polycrystalline Halite

	Heard [1972]	Recorrected Data H. C. Heard (personal communication, 1980)
log A	−0.254 ± 0.40	−1.17 ± 0.36
Q	23.46 ± 0.93 kcal/mol	23.00 ± 0.78 kcal/mol
n	5.47 ± 0.21	5.81 ± 0.19

Recorrected data is *Heard* [1972] data in which the 300°C test data have been recorrected for aluminum jackets. Uncertainty in each parameter is 1 standard deviation.

ters for a multiple-mechanism flow law which may be more appropriate for the data. The technique has been tested using synthetic data to determine that it converges to the correct values, and it has been applied to experimental data for Yule marble and polycrystalline halite. The results support the previous conclusion that a single mechanism dominates the flow over the range of conditions used in previous Yule marble experiments. A two-mechanism flow law seems more appropriate for the polycrystalline-halite data published by *Heard* [1972]. However, these data have been recorrected for the aluminum jacket strength by H. C. Heard (personal communication, 1980). Because the recorrected data exhibit more variability, a single-mechanism flow law fits the data best.

Appendix A: Nonlinear Least Squares Fitting of Multimechanism Creep Laws

If we assume that the strain rate in steady state creep is due to multiple mechanism, it can be expressed as (see equation (4))

$$\dot{\epsilon}_0(\sigma, T; A_i, n_i, Q_i) = \sum_{i=1}^{I} A_i \sigma^{n_i} \exp{(-Q_i/RT)} \qquad (A1)$$

where all the terms have been defined previously.

To determine the flow law parameters A_i, n_i, and Q_i in a least squares sense, we minimize the mean square fractional error:

$$\sigma^2(A_i, n_i, Q_i) = \frac{1}{K} \sum_{k=1}^{K} \left\{ \frac{\dot{\epsilon}_k - \dot{\epsilon}_0(T_k, \sigma_k; A_i, n_i, Q_i)}{\dot{\epsilon}_k} \right\}^2 \qquad (A2)$$

with respect to the unknown flow law parameters. In the above expression the $\dot{\epsilon}_k$ are the measured strain rates at the temperatures T_k and stresses σ_k, while $\dot{\epsilon}_0(T_k, \sigma_k; A_i, n_i, Q_i)$ is the theoretical expression for the total (observed) strain rate.

The mean square fractional error δ^2 will be minimized with respect to the flow law parameters when the partial derivatives of δ^2 are all zero; that is,

$$\frac{K}{2}\frac{\partial \delta^2}{\partial A_i} = 0 = \sum_{k=1}^{K} \left(\frac{\dot{\epsilon}_k - \dot{\epsilon}_0(T_k, \sigma_k)}{\dot{\epsilon}_k^2} \right) \frac{\partial \dot{\epsilon}}{\partial A_i}\Big|_{A_i, n_i, Q_i}$$

$$\frac{K}{2}\frac{\partial \delta^2}{\partial n_i} = 0 = \sum_{k=1}^{K} \left(\frac{\dot{\epsilon}_k - \dot{\epsilon}_0(T_k, \sigma_k)}{\dot{\epsilon}_k^2} \right) \frac{\partial \dot{\epsilon}}{\partial n_i}\Big|_{A_i, n_i, Q_i} \quad (A3)$$

$$\frac{K}{2}\frac{\partial \delta^2}{\partial Q_i} = 0 = \sum_{k=1}^{K} \left(\frac{\dot{\epsilon}_k - \dot{\epsilon}_0(T_k, \sigma_k)}{\dot{\epsilon}_k^2} \right) \frac{\partial \dot{\epsilon}}{\partial Q_1}\Big|_{A_i, n_i, Q_i}$$

$i = 1, 2, \cdots, I$. This is a set of nonlinear equations which is difficult, if not impossible, to solve directly for the correct (optimum) parameter values A_i, n_i, and Q_i.

However, this set of equations can be linearized if estimates of the parameter values are known. These estimates can be obtained directly from the experimental data with a little computation; however, accurate estimates are not necessary because the iteration procedure used to determine the parameter values (and which is described below) converges to the correct values even when bad initial guesses are made.

The equations are linearized by expanding the theoretical expression (equation (A1)) about the guesses of the parameters (call them, say $A_i^{(0)}$, $n_i^{(0)}$, $Q_i^{(0)}$):

$$\dot{\epsilon}(T_k, \sigma_k; A_i, n_i, Q_i) \approx \dot{\epsilon}(T_k, \sigma_k; A_i^{(0)}, n_i^{(0)}, Q_i^{(0)})$$

$$+ \sum_i \left\{ dA_i \frac{\partial \dot{\epsilon}}{\partial A_i}\Big|_0 + dn_i \frac{\partial \dot{\epsilon}}{\partial n_i}\Big|_0 + dQ_i \frac{\partial \dot{\epsilon}}{\partial Q_i}\Big|_0 \right\} \quad (A4)$$

where the partial derivatives are evaluated at the guesses. Evaluating the partial derivatives in (A3) at the guesses for the parameter values, a set of linear equations is obtained. This set of equations can be written in matrix form as

$$
\begin{bmatrix}
\sum_k \dfrac{\Delta\dot{\epsilon}}{\dot{\epsilon}_k} \dfrac{\partial \dot{\epsilon}}{\partial A_1}\Big|_0 \\[2ex]
\sum_k \dfrac{\Delta\dot{\epsilon}}{\dot{\epsilon}_k} \dfrac{\partial \dot{\epsilon}}{\partial n_1}\Big|_0 \\[2ex]
\sum_k \dfrac{\Delta\dot{\epsilon}}{\dot{\epsilon}_k} \dfrac{\partial \dot{\epsilon}}{\partial Q_1}\Big|_0
\end{bmatrix}
$$

$$
=
\begin{bmatrix}
\sum_k \dfrac{1}{\dot{\epsilon}_k^2} \dfrac{\partial \dot{\epsilon}}{\partial A_1}\Big|_0 \dfrac{\partial \dot{\epsilon}}{\partial A_1}\Big|_0 & \sum_k \dfrac{1}{\dot{\epsilon}_k^2} \dfrac{\partial \dot{\epsilon}}{\partial A_1}\Big|_0 \dfrac{\partial \dot{\epsilon}}{\partial n_1}\Big|_0 & \cdots \\[2ex]
\sum_k \dfrac{1}{\dot{\epsilon}_k^2} \dfrac{\partial \dot{\epsilon}}{\partial n_1}\Big|_0 \dfrac{\partial \dot{\epsilon}}{\partial A_1}\Big|_0 & \sum_k \dfrac{1}{\dot{\epsilon}_k^2} \dfrac{\partial \dot{\epsilon}}{\partial n_1}\Big|_0 \dfrac{\partial \dot{\epsilon}}{\partial n_1}\Big|_0 & \cdots \\[2ex]
\sum_k \dfrac{1}{\dot{\epsilon}_k^2} \dfrac{\partial \dot{\epsilon}}{\partial Q_1}\Big|_0 \dfrac{\partial \dot{\epsilon}}{\partial A_1}\Big|_0 & \sum_k \dfrac{1}{\dot{\epsilon}_k^2} \dfrac{\partial \dot{\epsilon}}{\partial Q_1}\Big|_0 \dfrac{\partial \dot{\epsilon}}{\partial n_1}\Big|_0 & \cdots
\end{bmatrix}
\begin{bmatrix}
dA_1 \\[2ex] dn_1 \\[2ex] dQ_1
\end{bmatrix}
$$

or

$$y^{(0)} = M^{(0)} \, dp^{(0)}$$

where $\Delta\dot{\epsilon} = \dot{\epsilon}_k - \dot{\epsilon}_0(T_k, \sigma_k; A^{(0)}, n^{(0)}, Q^{(0)})$. This matrix equation is solved to give parameter increments which can be added to the initial guesses to obtain a better guess for the parameters:

$$M^{(0)-1} y^{(0)} = dp^{(0)}$$

$$p^{(1)} = p^{(0)} + dp^{(0)}$$

that is,

$$
\begin{bmatrix}
A_1^{(1)} \\ n_1^{(1)} \\ Q_1^{(1)} \\ A_2^{(1)} \\ n_2^{(1)} \\ Q_2^{(1)} \\ \vdots
\end{bmatrix}
=
\begin{bmatrix}
A_1^{(0)} \\ n_1^{(0)} \\ Q_1^{(0)} \\ A_2^{(0)} \\ n_2^{(0)} \\ Q_2^{(0)} \\ \vdots
\end{bmatrix}
+
\begin{bmatrix}
dA_1 \\ dn_1 \\ dQ_1 \\ dA_2 \\ dn_2 \\ {}^\circ Q_2 \\ \vdots
\end{bmatrix}
$$

This procedure is continued by recomputing the matrix M (sometimes called the curvature matrix) at the new values of the parameters (denoted by the parameter vector $p^{(1)}$) to give

$$p^{(2)} = p^{(1)} + dp^{(1)} + M^{(1)^{-1}} y^{(1)}$$

The iteration procedure is stopped when the change in the parameter vector is smaller than some predefined value. The criterion we used was to stop the iteration when (say, at the jth iteration)

$$\sum_i \left[\left(\frac{dA_i^{(j)}}{A_i^{(j)}} \right)^2 + \left(\frac{dn_i^{(j)}}{n_i^{(j)}} \right)^2 + \left(\frac{\partial Q_i^{(j)}}{Q_i^{(j)}} \right)^2 \right] \leq 10^{-8}$$

This guaranteed that the values of the parameters are within 1 part in 10^4 of their optimum value. A weaker criterion could have been used, but it was found that once the parameters were within 1% of their optimum values, only two or three iterations were needed to satisfy the above criterion.

Some problems were encountered in using this technique. These generally occurred only when there were large errors in the measured values and if the initial guesses of the parameters were not very good. The mean square fractional error δ^2 was also printed out during the iteration procedure, and it would vary greatly when the parameter values were not near the optimum ones. However, when the optimum values were approached, the mean square error would decrease and reach a stable value, which was usually constant to three or four significant figures for the last couple of iterations. A search for good initial guesses of the parameters can be made by varying the parameters one at a time and determining where the mean square error is small. Convergence was also improved by treating $\ln A_i$ rather than A_i as a flow law parameter.

Appendix B: Estimation of Errors

No analytic form exists for the uncertainties in each of the parameters determined by a nonlinear least squares fitting procedure because an iterative search rather than analytical solution is used. However, *Bevington* [1969, p. 242] shows that the expression for the uncertainty of each parameter for a linear least squares fit is a reasonable approximation to the uncertainty of each parameter for a nonlinear least squares fit. The correct analytical expression for the standard deviation of each parameter resulting from a linear least squares fit is related to the curvature matrix M used in the solution procedure:

$$\sigma_{p_j}^2 = E_{jj}$$

The terms E_{jj} are the diagonal terms of the inverted curvature matrix M. The variance of each data point, σ_i^2, is usually incorporated in M. However, the σ_i are not known, so they have been treated as constant: $\sigma_i = \sigma$. The data variance σ can be approximated by the sample variance s^2; thus

$$\sigma_i \simeq \sigma \simeq s^2 = \frac{1}{K-n} \sum \left(\frac{\dot{\epsilon}_k - \dot{\epsilon}(T_k, \sigma_k)}{\dot{\epsilon}} \right)^2.$$

The sample variance can then be used to evaluate the variance of each parameter p_j, as

$$\sigma_{p_j}^2 = s^2 E_{jj}$$

where E_{jj} is evaluated for $\sigma_i = \sigma = 1$.

Acknowledgments. This work was supported, in part, by the Battelle Memorial Institute, Office of Waste Isolation, Columbus, Ohio, under contract E512-00900.

References

Bevington, P. R., *Data Reduction and Error Analysis for the Physical Sciences*, 336 pp., McGraw-Hill, New York, 1969.

Carter, N. L., Steady-state flow of rocks, *Rev. Geophys. Space Phys.*, *14*, 301–360, 1976.

Evans, A. F., and T. G. Langdon, Structural ceramics, *Prog. Mater. Sci.*, *21*, 171–441, 1976.

Garofalo, F., *Fundamentals of Creep and Creep Rupture in Metals*, 258 pp., McMillan, New York, 1965.

Heard, H. C., Effect of large changes in strain rate in the experimental deformation of Yule marble, *J. Geol.*, *71*, 162–195, 1963.

Heard, H. C., Steady state flow in polycrystalline halite at pressures of 2 kilobars, in *Flow and Fracture of Rocks*, *Geophys. Monogr. Ser.*, vol. 16, edited by H. C. Heard, et al., pp. 191–209, AGU, Washington, D. C., 1972.

Heard, H. C., and C. B. Raleigh, Steady-state flow in marble at 500° to 800°C, *Geol. Soc. Am. Bull.*, *82*, 935–946, 1972.

Sherby, O. D., and P. M. Burke, Mechanical behavior of crystalline solids at elevated temperature, *Prog. Met. Phys.*, *13*, 325–390, 1967.

Sherby, O. D., J. L. Lytton and J. E. Dorn, Activation energies for creep of high-purity aluminum, *Acta Metal.*, *5*, 219, 1957.

Tegart, W. J. M., and O. .D. Sherby, Activation energies for high temperature creep of polycrystalline zinc, *Philos. Mag.*, *3*, 1287, 1958.

Wavefronts in Transversely Isotropic Media

H. Odé

Shell Research and Development Bellaire Research Center, Bellaire, Texas 77401

A simple estimate of the amplitude of the wave front in a transversely isotropic medium is made as follows. It is assumed that the shot causes a uniform radial displacement. In any direction the amplitudes of the two types of 'compressional' waves, which together will yield this unit radial displacement, can be computed. For each of these waves the velocity and shape of the wave front can be computed by well-known methods. A brief discussion of these methods, dating back as far as 1911, is included in the paper.

Introduction

Finely layered sequences of isotropic homogeneous media can be considered as transversely isotropic media. These are characterized by five, instead of two elastic moduli. Such media display some curious properties in connection with energy propagation. In seismic exploration it is almost always assumed that the medium is isotropic, and in consequence it is taken for granted that the direction of energy propagation is perpendicular to the wave front. This is generally not true for anisotropic media.

It is probable that many sediments behave as transversely isotropic media not only because platelets and grains are arranged in semiordered patterns, but also because of fine layering. *Rudzki* [1911] and later *Postma* [1955] have drawn attention to this fact. Since then, a considerable amount of work was done [*Kraut*, 1963; *Berryman*, 1979] and efforts were made to compare the properties of transversely isotropic media to actual wave propagation in the earth [*Levin*, 1979, 1980]. In this paper, which is based on work the author did for the Shell Development Company in 1975, we shall not be concerned with the magnitudes of the five elastic constants, but rather with the consequences of the anisotropy on the propagation of simple plane waves. With the advent of more sophisticated velocity determinations and amplitude recording in seismic exploration it is necessary to investigate the implications of these techniques that arise from deviations from the isotropic behavior of rocks.

In a transversely isotropic medium, energy is radiated away from the source in a nonuniform manner. In particular, several consecutive pulses can be observed at points away from the source, carrying amounts of energy that depend on direction. The variation of the amplitude with respect to direction is the main subject of this report. All computations are made for direct arrivals from a source considered to be so small in dimensions with respect to the travel distance as to be a point. We also assume that the same source in an isotropic medium would radiate equally in all directions.

The Transversely Isotropic Medium

In view of the fact that fine layering of isotropic materials has the effect of giving the medium transverse isotropy, we consider here only transverse isotropy. It is possible that a regular arrangement of almost spherical grains produces orthotropic or cubic anisotropy, but such a case will not be considered here.

The equations of motion for a transversely isotropic medium are (see, for example, *Postma* [1955] or *White* [1965])

$$\rho \frac{\partial^2 u}{\partial t^2} = A \frac{\partial^2 u}{\partial x^2} + N \frac{\partial^2 u}{\partial y^2} + L \frac{\partial^2 u}{\partial z^2}$$
$$+ (A - N) \frac{\partial^2 v}{\partial x \partial y} + (F + L) \frac{\partial^2 w}{\partial x \partial z}$$

$$\rho \frac{\partial^2 v}{\partial t^2} = N \frac{\partial^2 v}{\partial x^2} + A \frac{\partial^2 v}{\partial y^2} + L \frac{\partial^2 v}{\partial z^2}$$
$$+ (A - N) \frac{\partial^2 u}{\partial x \partial y} + (F + L) \frac{\partial^2 w}{\partial y \partial z} \quad (1)$$

$$\rho \frac{\partial^2 w}{\partial t^2} = L \frac{\partial^2 w}{\partial x^2} + L \frac{\partial^2 w}{\partial y^2} + C \frac{\partial^2 w}{\partial z^2}$$
$$+ (F + L) \frac{\partial^2 u}{\partial x \partial z} + (F + L) \frac{\partial^2 v}{\partial y \partial z}$$

in which ρ is the density of the medium, u, v, w are the displacement components, the z axis is the axis of symmetry, and A, N, L, F, and C are the five elastic moduli. These equations follow simply by introducing the pertinent symmetry conditions into the strain energy function, which for transversely isotropic materials is [see *Love*, 1944]

$$2\Psi = A\left(e_{xx}^2 + e_{yy}^2\right) + Ce_{zz}^2 + 2F\left(e_{yy} + e_{xx}\right)e_{zz}$$
$$+ 2\left(A - 2N\right)e_{xx}e_{yy} + 4L\left(e_{yz}^2 + e_{zx}^2\right) + 4Ne_{xy}^2 \qquad (2)$$

Then

$$\sigma_x = \frac{\partial\Psi}{\partial e_{xx}} = Ae_{xx} + \left(A - 2N\right)e_{yy} + Fe_{zz}$$

$$\sigma_y = \frac{\partial\Psi}{\partial e_{yy}} = \left(A - 2N\right)e_{xx} + Ae_{yy} + Fe_{zz} \qquad (3)$$

$$\tau_{xy} = \frac{\partial\Psi}{\partial e_{xy}} = 2Ne_{xy} \qquad \tau_{yz} = 2Le_{yz} \qquad \tau_{zx} = 2Le_{zx}$$

Because Ψ must be positive definite, certain conditions must be satisfied by the moduli A, C, L, F, and N. A theorem of algebra states that the necessary and sufficient condition that a quadratic form such as the right-hand side of (2) is positive definite, is that all indicated determinants of the matrix

$$\begin{vmatrix} A & A-2N & F & 0 & 0 & 0 \\ A-2N & A & F & 0 & 0 & 0 \\ F & F & C & 0 & 0 & 0 \\ 0 & 0 & 0 & L & 0 & 0 \\ 0 & 0 & 0 & 0 & L & 0 \\ 0 & 0 & 0 & 0 & 0 & N \end{vmatrix}$$

must be positive. This gives the five independent conditions:

$$A > 0 \qquad A > N \qquad (A-N)C > F^2 \qquad L > 0 \qquad N > 0 \qquad (4)$$

From these conditions it also follows that $C > 0$; however, F could be negative.

Plane Waves

We assume that a plane wave disturbance of the type

$$u = u_0 f\ (l + my + nz - ct)$$
$$v = v_0 f\ (lx + my + nz - ct) \qquad . \quad (5)$$
$$w = w_0 f\ (lx + my + nz - ct)$$

satisfies (1). The directional cosines include l, m, and n, and c is the phase velocity in that direction. Upon substitution into (1) we obtain the equation

$$\begin{vmatrix} (Al^2 + Nm^2 + Ln^2 - \rho c^2) & (A-N)lm & (F+L)ln \\ (A-N)lm & (Nl^2 + Am^2 + Ln^2 - \rho c^2) & (F+L)mn \\ (F+L)ln & (F+L)mn & (Ll^2 + Lm^2 + Cn^2 - \rho c^2) \end{vmatrix} = 0 \qquad (6)$$

This equation is satisfied by

$$N(m^2 + l^2) + Ln^2 = \rho c^2$$

$$[L(l^2 + m^2) + Cn^2 - \rho c^2][A(l^2 + m^2) + Ln^2 - \rho c^2]$$
$$= (F + L)^2 n^2 (m^2 + l^2) \qquad (7)$$

As could have been expected, the phase velocity c is independent of rotation around the z axis because l and m only occur in the combination $l^2 + m^2$. Thus for simplicity we set in the following $m = 0$. The equations in (7) reduce to the isotropic case for $A = C$, $L = N$, and $A - N = F + L$ and become

$$N = \rho c^2 \qquad (\rho c^2 - N)(\rho c^2 - A) = 0$$

showing that there are then only 2 different types of waves instead 3 types for (7).

Horizontally Polarized Wave

The wave type corresponding to the first of the equations in (7) has a phase velocity

$$c = \left(\frac{Nl^2 + Ln^2}{\rho}\right)^{1/2} \qquad (8)$$

In general it is a plane wave for which $u_0 = w_0 = 0$ and $v_0 \neq 0$ so that the direction of oscillation is perpendicular to the plane $y = 0$ (consequence of choosing $m = 0$). It might be called the 'horizontally polarized wave,' usually denoted by the symbol Sh (Figure 1). From (8) it follows that the phase velocity in the x direction is

$$c_x = \left(\frac{N}{\rho}\right)^{1/2} = V_h \qquad (9a)$$

and in the z direction it is

$$c_z = \left(\frac{L}{\rho}\right)^{1/2} = V_z \qquad (9b)$$

It is convenient in this discussion to consider the surface obtained by the end points of a radius vector equal to the magnitude of the phase velocity and drawn in the appropriate direction (phase velocity surface or indicatrix). For the Sh wave, this surface is obtained by setting $c = r$; $l = x/r$; $n = z/r$ in (8), so that

$$r^4 = V_h^2 x^2 + V_z^2 z^2 \qquad (10)$$

a fourth degree surface. Let us now consider the totality of plane waves of the type Sh in any direction. At any time the envelope of the plane waves forms a surface (in two dimensions a curve) that is called the 'wave front' and that is denoted here by $E = 0$. The construction is indicated in Figure 2 for the cases $A = 16$, $C = 9$, $L = F = 1$, and $N = 4$.

This surface shows how far a disturbance has progressed from a source S after a lapse of time. The meaning of Figure 2 is as follows: a planar element of the wave at S propagates with a phase velocity c in a direction ϕ (i.e., toward the point x_0, z_0 on the indicatrix). However,

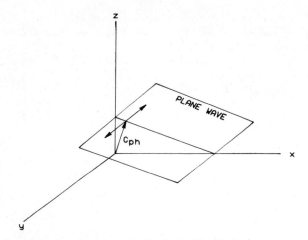

Fig. 1. Geometry of a horizontally polarized wave.

its energy streams in a direction determined by the wave surface $E = 0$, i.e., in the direction of the point x, z, where the plane wave is tangent to the envelope $E = 0$. This direction is called the ray. Thus, rays are not perpendicular to the wave fronts. Hence, energy propagates in a particular wave front by moving forward with it and shifting sideways simultaneously.

For the Sh wave the wave front E is particularly simple. Let $z_0 = kx_0$, where k is a direction coefficient related to n by $n^2 = k^2/(1 + k^2)$, considered here as a parameter. Then the equation of E in parametric form is obtained as follows: the tangent to E through the point x_0, z_0 of the indicatrix is given by

$$z - z_0 = -\frac{1}{k}(x - x_0)$$

Hence the parametric equation of $E = 0$ can be written as

$$x = x_0 - k\frac{dx_0}{dk} - k^2\frac{dz_0}{dk}$$

$$z = z_0 + \frac{dx_0}{dk} + k\frac{dz_0}{dk} \qquad (11)$$

Now, from (10), which is the equation of the indicatrix, we have

$$(1 + k^2)^2 x_0^4 = V_h^2 x_0^2 + V_z^2 k^2 x_0^2$$

or

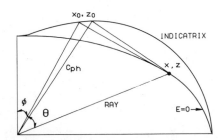

Fig. 2. Indicatrix and wave front for the horizontally polarized wave: $A = 16$, $C = 9$, $L = F = 1$, and $N = 4$.

so that, using $z_0 = kx_0$,

$$x = x_0 - k^2 x_0 - \frac{k(1 + k^2)}{2x_0}\frac{df}{dk} = \frac{V_h^2}{(V_h^2 + k^2 V_z^2)^{1/2}}$$

$$z = 2kx_0 + \frac{1 + k^2}{2x_0}\frac{df}{dk} = \frac{kV_z^2}{(V_h^2 + k^2 V_z^2)^{1/2}}$$

hence the wave front $E = 0$ is the ellipse

$$\frac{x^2}{V_h^2} + \frac{z^2}{V_z^2} = 1$$

To appreciate better what this means, we have shown in Figure 3 two elliptical wave fronts at times t_1 and t_2, which are caused by a propagating disturbance and thus are similar in shape.

The phase velocity is given by $\Delta s/(t_2 - t_1)$, where Δs is the perpendicular distance between the two plane waves tangent to both wave fronts; however, the ray velocity is $\Delta s/(t_2 - t_1)\cos\theta^*$ and is in a direction making an angle θ^* with the phase velocity direction. This difference in direction between phase velocity and ray velocity is a most significant difference between wave propagation in isotropic and anisotropic media. In the former case the indicatrix and the wave front are spheres (and identical); in the lat-

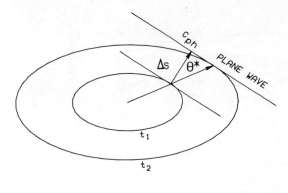

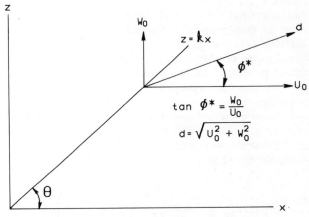

Fig. 4. Geometry of the compressional waves.

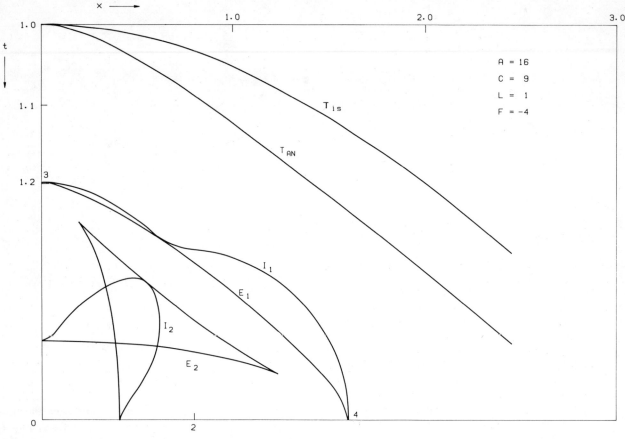

Fig. 5.   Indicatrix, wave front, and travel time curve for $A = 16$, $C = 9$, $L = 1$, and $F = -4$.

ter case, phase and ray velocity are equal only in the direction of the principal axes of the ellipse $E = 0$.

Indicatrix and Wave Front for the Other Two Types of Waves

The analysis for the wave fronts corresponding to the second equation of (7) is somewhat more involved. The equations in (6) show that for these waves we have for $m = 0$, $u_0 \neq 0$, $w_0 \neq 0$, and $v_0 = 0$. This means that in a direction given by $z = kx$ two waves propagate for which u_0 and w_0 are not zero. Each of these waves is a combination of shear motion and compressional motion; thus we see that the fundamental distinction in shear and compressional waves as solution of the characteristic equation does not carry over in the case of anisotropic media. Then the two fundamental waves are combinations of shear and compressional motion.

The compressional part of one of these composite waves is given (see Figure 4) by the component in the k direction of the displacement vector and the shear part by its normal component; i.e.,

$$S_{\mathrm{com}} = d \cos (\theta - \phi^*)$$

$$S_{\mathrm{shear}} = d \sin (\theta - \phi^*)$$

The phase velocities of these waves are given as function of direction by

$$\rho c^2 = \frac{(A + L) + (C - A)n^2}{2} \pm \frac{1}{2} \{(A - L)^2 + [2(A + L)$$

$$\cdot (C - A) + 4(2AL - AC - L^2 + (F + L)^2)]n^2\}^{1/2}$$

$$\div + [(C - A)^2 - 4(AL - AC - L^2$$

$$+ LC + (F + L)^2)]n^4$$

$$m = 0 \qquad l^2 = 1 - n^2 \qquad (12)$$

The phase velocity in the direction of the x axis ($n = 0$) is

$$\rho c_x{}^2 = \frac{A + L}{2} \pm \frac{A - L}{2}$$

or

$$c_{x1}{}^2 = \frac{A}{\rho} \qquad c_{x2}{}^2 = \frac{L}{\rho}$$

Along the z axis ($n = 1$) we have

$$\rho c_z{}^2 = \frac{L + C}{2} \pm \frac{C - L}{2}$$

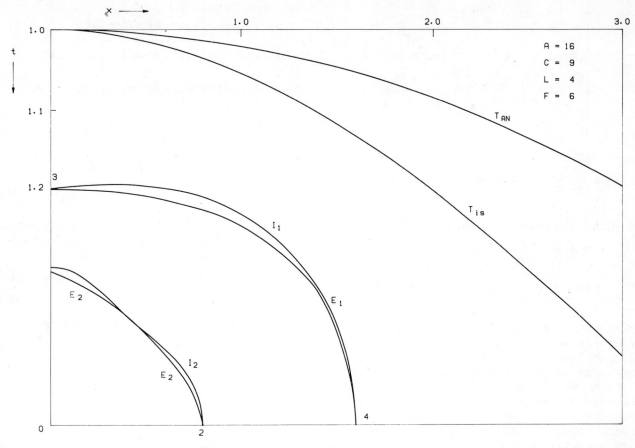

Fig. 6. Idem for $A = 16$, $C = 9$, $L = 4$, and $F = 6$.

or

$$c_{z1}^2 = \frac{C}{\rho} \qquad c_{z2}^2 = \frac{L}{\rho}$$

To facilitate convenient discussion, we shall speak of the surfaces corresponding to these waves as 'sheet one' and 'sheet two.'

Postma designed a geometrical method to construct wave fronts, which we shall not use here. He found that one of the wave fronts sometimes was very peculiar in shape, in that it possessed two cusps. It is this particular

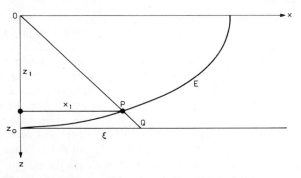

Fig. 7. Geometry used in computation of travel time curves.

behavior of the wave front that shows us that several pulses of energy will arrive at some distance from the source.

First, we will investigate the geometry of sheet one and sheet two. Using $z_{0i} = kx_{0i}$, the parametric equations of the wave front (11) can be rewritten as

$$x_i = (1 - k^2) x_{0i} - k (1 + k^2) \frac{dx_{0i}}{dk}$$

$$z_i = 2kx_{0i} + (1 + k^2) \frac{dx_{0i}}{dk} \tag{13}$$

where the subscript i indicates the sheet ($i = 1$ or 2). From this it follows that

$$\frac{dx_i}{dx} = -k \left[2x_{0i} + 4k \frac{dx_{0i}}{dx} + (1 + k^2) \frac{dx_{0i}^2}{dk^2} \right]$$

$$\frac{dz_i}{dk} = \left[2x_{0i} + 4k \frac{dx_{0i}}{dk} + (1 + k^2) \frac{dx_{0i}^2}{dk^2} \right] \tag{14}$$

A cusp in an analytic curve is necessarily determined by the simultaneous condition that $dx_i/dk = dz_i/dk = 0$ (it is not a sufficient condition). Thus for the existence of a cusp one must have at least

$$2x_{0i} + 4k \frac{dx_{0i}}{dk} + (1 + k^2) \frac{dx_{0i}^2}{dk^2} = \frac{d^2}{dk^2} [(1 + k^2) x_{0i}] = 0 \tag{15}$$

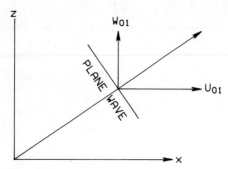

Fig. 8.   Condition of orthogonality for plane wave.

in which x_{0i} is a root of the equation

$$x_{0i}^4 - \frac{(L + A/\rho) + (L + C/\rho)k^2}{(1 + k^2)^2} x_{0i}^2$$

$$+ \frac{(A + Lk^2/\rho)(L + Ck^2/\rho) - (F + L/\rho)^2k^2}{(1 + k^2)^4} = 0 \qquad (16)$$

which is the equation of the indicatrix surface corresponding to the second of the equations in (7). For simplicity we set in the following $\rho = 1$. It means that instead of A, etc., we consider coefficients $A' = A/\rho$, etc.

Before discussing the geometry of the wave fronts $E_1 = 0$ and $E_2 = 0$ we might consider two special cases; namely, those in which the expression under the root sign in (12) is a perfect square. A simple analysis shows that this happens only in two cases:

$$(F + L) = 0$$

$$(F + L)^2 = (A - L)(C - L)$$

For $(F + L) = 0$ we find for the phase velocities

$$(\rho c^2)_1 = Al^2 + Ln^2$$

$$(\rho c^2)_2 = Ll^2 + Cn^2 \qquad (17)$$

Because these indicatrix surfaces are similar in every respect to those for the Sh wave, the corresponding wave fronts are ellipses; that is, curves without cusps.

For $(F + L)^2 = (A - L)(C - L)$ we find similarly

$$(\rho c^2)_1 = Al^2 + Cn^2$$

$$(\rho c^2)_2 = L$$

Therefore one wave front is an ellipse, and the other is a circle.

To discuss the geometry of the general case we set

$$R = [(L + A)^2 + (L + C)k^2]^2$$

$$- 4[(A + Lk^2)(L + Ck^2) - (F + L)^2k^2]$$

$$= [(A - L) - (C - L)k^2]^2 + 4(F + L)^2k^2 \qquad (18)$$

so that from (16) we derive

$$x_{01} = \frac{1}{1 + k^2} \sqrt{\frac{1}{2}[(L + A) + (L + C)k^2 + \sqrt{R}]}$$

$$x_{02} = \frac{1}{1 + k^2} \sqrt{\frac{1}{2}[(L + A) + (L + C)k^2 - \sqrt{R}]} \qquad (19)$$

Thus a necessary condition for the existence of a cusp can be written by using (15):

$$\frac{d^2}{dk^2}[(L + A) + (L + C)k^2 \pm \sqrt{R}]^{1/2} = 0 \qquad (20)$$

Now consider the wave front $E_1 = 0$ (plus sign in (19)). The function

$$\Phi = + \sqrt{(L + A) + (L + C)k^2 + (R)^{1/2}}$$

is increasing in value with k increasing in the range $0 \le k < \infty$. The rate of increase of Φ will be faster for $F \ne -L$ than for $F = -L$. But for $F = -L$ we have

$$\Phi = + \sqrt{(L + A) + (L + C)k^2 + (A - L) - (C - L)k^2} \qquad (21)$$

so that for the existence of a cusp one must have

$$\frac{d^2}{dk^2}(A + Lk^2)^{1/2} = 0 \qquad (22)$$

or

$$\frac{AL}{(A + Lk^2)^{3/2}} = 0$$

There is no real and finite value of k that satisfies (22), and hence no cusp can occur in the range of values k can assume. For $F \ne -L$, the function Φ will increase faster, and hence it certainly cannot have cusps. Thus the wave front E_1 has no cusps, a result first obtained already by Rudzki.

The situation is quite different for the wave front $E_2 = 0$ (negative sign in (19)). In that case there might be cusps; however, this is not necessarily true, as we have seen in the discussion of the two cases where the root of (12) can be taken exactly. We shall not investigate this matter further here but merely note that only the second sheet can have cusps.

Examples

To show the nature of the wave fronts, we have made a number of computations for various values of the moduli A, C, L, F, and N. The results of these computations are shown in Figures 5 and 6. Each figure was obtained by computing for a sequence of values of k, corresponding to regular increases of the angle ϕ, the points (x_{01}, z_{01}) and (x_{02}, z_{02}) of both indicatrices (equation (19)). The wave fronts E_1 and E_2 are computed from (13), in which dx_{01}/dk is eliminated by means of (19). For the first and second sheet, one needs only to specify A, C, L, and F, but N may be left arbitrary, provided it satisfies inequality (4).

Figures 5 and 6 show the following: (1) both wave fronts for the indicated anisotropic case (E_1 and E_2); (2) both indicatrices (I_1 and I_2); and (3) the travel time curve for a reflector at unit time depth for the isotropic (T_{is}) and the anisotropic (T_{AN}) case (sheet 1).

Here the computation of the travel time curves needs some explanation; the method indicated by Figure 7 is most convenient. Let OP be the ray and E_1 the wave front corresponding to unit time for the anisotropic medium, and E_{is} is the wave front corresponding to an isotropic medium. By assumption, the time OP is unity. Thus

$$t_{OQ} = t_{OP} \frac{z_0}{z_1} = \frac{z_0}{z_1}$$

in which z_1 follows from (13). In our examples we take $z_0 = 3$ (vertical phase velocity is $C/\rho = 3$). ξ, half the source to receiver distance then follows from

$$\xi = x_1 \cdot \frac{z_0}{z_1} \tag{23}$$

The times for the isotropic wave front E_{is} (which is a circle) with which our times must be compared, follow most easily from the proportionality of the times for isotropic and anisotropic wave propagation along OQ:

$$t_{OQ} \text{ (anisotropic)} = \frac{OQ}{V_{\text{Ray}}} \qquad t_{OQ} \text{ (isotropic)} = \frac{OQ}{3}$$

hence

$$t_{is} = \frac{V_{\text{Ray}} \cdot t_{OQ} \text{ (anisotropic)}}{3} = \frac{V_{\text{Ray}}}{3} \cdot \frac{z_0}{z_1} = \frac{\sqrt{x_1{}^2 + z_1{}^2}}{z_1}$$

because $V_{\text{Ray}} = \sqrt{x_1{}^2 + z_1{}^2}$.

Concerning the results we note the following facts:
1. Figure 5 shows a case of well developed cusps in one of the wave fronts; the other is without cusps. Moreover, it is clear that the ray velocity for quite an extended range of ϕ is smaller than the isotropic one, although finally the horizontal ray velocity will exceed the vertical one.
2. Negative values of F are permitted, provided they satisfy inequality 4. When $(F + L)^2$ has the same value as $(-F + L)^2$ (for instance, $F = -4$ and $L = 1$ and $F = +2$ and $L = 1$), identical figures are obtained.
3. For very large values of $(F + L)$ the cusps can be located outside the quadrant we consider.

Energy Propagation From an Idealized Source

From (6) we find for the ratio of the amplitude components for the first and second sheet, ($\rho = 1$)

$$\gamma_i = \frac{u_{0i}}{w_{0i}} = \frac{(F + L) \, ln}{c_i{}^2 - Al^2 - Ln^2} = \frac{c_i{}^2 - Ll^2 - Cn^2}{(F + L) \, ln} \tag{24}$$

in which the subscript i of c is either '1' or '2', depending on the sheet considered. Hence, for any direction of the phase velocity (direction of propagation of the plane wave) one can compute the ratios γ_1 and γ_2. These are two different waves, both with displacement in the x-z planes and none in the y plane. For each of these waves the oscillation is neither purely a shear motion nor a compressional motion, because in general the ratio $w_0/u_0 = \tan \phi$ is not equal to the value of $k = \tan \theta$, the direction of the phase propagation (see Figure 4). By combining two plane waves in the (l, $m = 0$, n) direction of appropriate amplitude we can construct a 'composite' wave with a displacement parallel to the specified direction. The third wave, discussed as the Sh wave, may be omitted for this purpose because its oscillation has no component perpendicular to the front. It is assumed here that such a wave front can be generated as the result of an explosion at the source (considered so small that it is a point). Such disturbance immediately falls apart into two separate waves; one represented by an envelope of the first kind (sheet one), and the other by an envelope of the second kind (sheet two). In each of these two waves, energy is transported in various directions in densities that can be computed, as we shall do. In this method of generating the 'composite' wave the initial condition of geometrical nature is that the initial displacement is perpendicular to the spherical surface of the small charge cavity. Instead one could also require that initially a constant pressure acts on the shot hole walls. We have preferred our method mainly for the reason that it is much more simple to use in making estimates of energy propagation in various directions. To keep the computations simple, we consider a cylindrical rather than a spherical cavity.

Assuming that the 'composite' plane wave at the source carries a normalized amount of energy in any direction, the partition of this energy over both plane waves and the direction in which the energy in these plane waves is transported away from the source (ray direction) can be computed as follows. Let the sum of both oscillations at the source be

$$\begin{aligned} u &= \alpha u_{01} + \beta u_{02} \\ w &= \alpha w_{01} + \beta w_{02} \end{aligned} \tag{25}$$

The condition that the displacement u, w is perpendicular to the cylindrical cavity surface (see Figure 8) is (for $k = \tan \theta$)

$$\alpha w_{01} + \beta w_{02} = k (\alpha u_{01} + \beta u_{02})$$

By using (24) we get,

$$\frac{\alpha w_{01}}{\beta w_{02}} = \frac{k \gamma_2 - 1}{1 - k \gamma_1} = \frac{1}{\lambda} \tag{26}$$

For any direction θ ($\tan \theta = k$) γ_1 and γ_2 are known (c_1 and c_2 can be computed) and hence λ is known. It is convenient to assume that originally the amplitudes w_{01} and w_{02} are

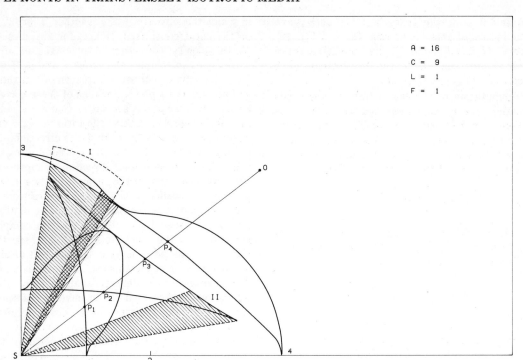

Fig. 9. Arrival of various wave fronts.

unity and to compute the amplitudes of both plane waves by means of (24), then find the total amplitude from (25) and finally to normalize back to unity, i.e.,

$$w_{01} = w_{02} = 1$$

$$u_{01} = \gamma_1 \qquad u_{02} = \gamma_2$$

so that

$$u = \alpha\gamma_1 + \beta\gamma_2$$

$$w = \alpha + \beta$$

But from (26) we have for $w_{01} = w_{02} = 1$

$$\frac{\alpha}{\beta} = \frac{1}{\lambda}$$

so that

$$u = \alpha(\gamma_1 + \lambda\gamma_2)$$

$$w = \alpha(1 + \lambda)$$

We now normalize the compound wave to unity:

$$1 = \sqrt{u^2 + w^2} = \alpha\sqrt{(\gamma_1 + \lambda\gamma_2)^2 + (1 + \lambda)^2}$$

Hence we find for α

$$\frac{1}{\alpha} = \sqrt{(\gamma_1 + \lambda\gamma_2)^2 + (1 + \lambda)^2} \qquad (27)$$

Thus the amplitudes of the two waves u_{01}, w_{01} and u_{02}, w_{02}, which when compounded give a unit amplitude wave in the direction $k = \tan\theta$, are

$$u_{01} = \alpha\gamma_1 \qquad u_{02} = \lambda\alpha\gamma_2$$
$$w_{01} = \alpha \qquad w_{02} = \lambda\alpha \qquad (28)$$

We show now that one must have

$$\gamma_1\gamma_2 + 1 = 0 \qquad (29)$$

Namely, the total energy of the composite wave front being unity, this energy is partitioned in the ratio

$$\frac{E_1}{E_2} = \frac{u_{01}^2 + w_{01}^2}{u_{02}^2 + w_{02}^2}$$

over the wave fronts (1) and (2), so that one must have

$$u_{01}^2 + w_{01}^2 + u_{02}^2 + w_{02}^2 = (u_{01} + u_{02})^2 + (w_{01} + w_{02})^2$$

or

$$u_{01} u_{02} + w_{01} w_{02} = 0$$

which is the same as (29), when we use (28). The same result also follows immediately from (24). Then

$$\gamma_1\gamma_2 = \frac{u_{01}}{w_{01}} \frac{u_{02}}{w_{02}} = \frac{c_1^2 - Ll^2 - Cn^2}{c_2^2 - Al^2 - Ln^2}$$

Because (from (12)) ($\rho = 1$)

$$c_1^2 + c_2^2 = (A + L)l^2 + (L + C)n^2$$

Equation (32) follows immediately.

Of some interest here is the special case $F = -L$. The phase velocities are given by (17). For the first kind of these waves it follows from the third equation in (5), $m =$

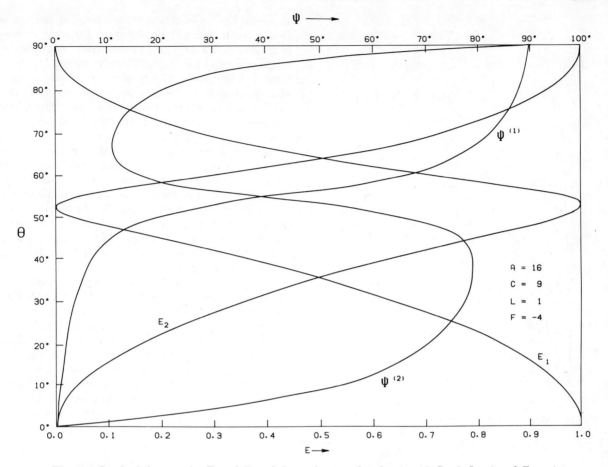

Fig. 10.   Graph of the energies E_1 and E_2 and the angles ψ_1 and ψ_2 for $A = 16$, $C = 9$, $L = 1$, and $F = -4$.

0, that $w_0 = 0$, and for the second kind it follows from the first equation in (5) that $u_0 = 0$. Hence these waves correspond to plane waves that oscillate in a horizontal or in a vertical direction only. In other words, $\gamma_1 = \infty$ and $\gamma_2 = 0$. The condition that the amplitude of the compound wave is perpendicular to the initial wave front now is (see Figure 8)

$$w_{02} = k u_{01}$$

hence

$$u_{01} = \frac{1}{\sqrt{1 + k^2}} \qquad w_{01} = 0$$

$$u_{02} = 0 \qquad w_{02} = \frac{k}{\sqrt{1 + k^2}}$$

Taking the energy proportional to the square of the amplitude, we have $E_1/E_2 = 1/k^2$ independent of the elastic moduli.

After having obtained the energy partition ratio, we can compute for a given direction θ of propagation of the plane wave the two corresponding points each on its appropriate wave front, in the direction of which the partitioned energies are transported. With angle θ there cor-

respond thus two angles $\psi^{(1)}$ and $\psi^{(2)}$, for sheet one and sheet two, respectively, given by

$$\tan \psi^{(1)} = \frac{z_1}{x_1} \qquad \tan \psi^{(2)} = \frac{z_2}{x_2}$$

Let it now be assumed that between two angles θ_0 and $(\theta_0 + \Delta\theta°)$ a unit amount of energy is radiated away. A part $E_1 = (u_{01}^2 + w_{01}^2)$ streams out between the directions $\psi^{(1)}$ and $\psi^{(1)} + \Delta\psi^{(1)}$, and a part E_2 between the directions $\psi^{(2)}$ and $\psi^{(2)} + \Delta\psi^{(2)}$. For the purpose of easy computation we have taken $\Delta\theta°$ equal to $1°$.

Depending on the angles $\Delta\psi^{(1)}$ and $\Delta\psi^{(2)}$, given by

$$\Delta\psi^{(1)} = \psi^{(1)} \ (\theta_0 + 1°) - \psi^{(1)} \ (\theta_0)$$

$$\Delta\psi^{(2)} = \psi^{(2)} \ (\theta_0 + 1°) - \psi^{(2)} \ (\theta_0)$$

this energy is concentrated (when $\Delta\psi < 1°$) or spread out (when $\Delta\psi > 1°$). We may now investigate the energy (or amplitude) an observer would measure at some location in the medium away from the shot hole. Because the position of the observer is measured by the angle ψ, we must find the two corresponding angles θ of the original plane waves, which propagate their energy in the direction ψ.

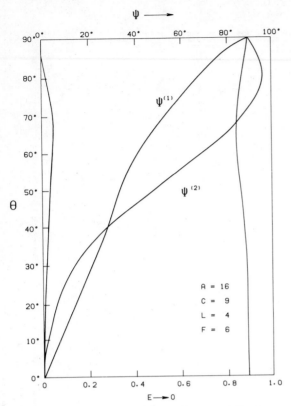

Fig. 11. Idem for $A = 16$, $C = 9$, $L = 4$, and $F = 6$.

However, the times at which the energy of the two wave fronts reach him will be different. Figure 9 shows the propagation of energy in a typical case. The original composite wave front at S distributes its energy in two directions. Let the phase velocity of the composite wave front at S progress in a direction within the vertically dashed cone. Then the energy for the first wave front (sheet) streams in a direction within the cone I and for the second sheet in a direction within the cone II.

Figure 9 shows that in this particular case the energy density over the wave fronts I and II becomes less dense because it gets spread out because of the increase of the apical angle. If one assumes that unit amounts of energy are radiated away in any direction, this energy will, because the shift in direction is different for any direction, become either more or less dense. This is expressed in the fact that the apex of both cones I and II are larger than the apex of the densely dashed cone.

Let us now assume that an observer is located somewhere along the line SO of Figure 9. This line cuts the two wave fronts in four places: P_1, P_2, P_3, and P_4. The energy corresponding to the arrival at P_1 is derived from the second wave, which has left the source S at a small angle. Likewise, the energy corresponding to P_3 is also derived from the second wave, which has left the source at an angle close to 45°; similarly, the energy corresponding to P_2 is derived from the second wave front, leaving the source at a large angle. Only the energy corresponding to P_4 is derived from the first wave, which has left the source

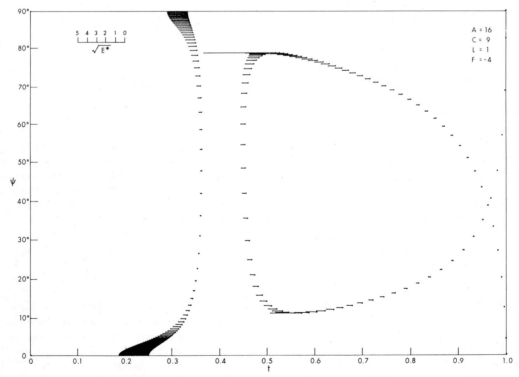

Fig. 12. Schematic view of energy pulses at various angular positions for $A = 16$, $C = 9$, $L = 1$, and $F = -4$ at unit distance from the source.

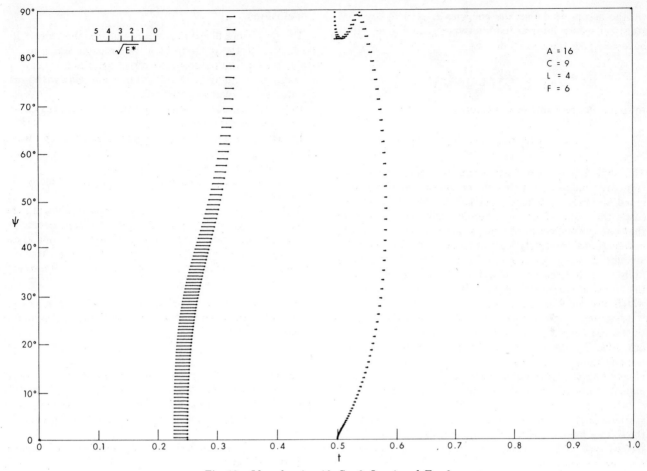

Fig. 13. Idem for $A = 16$, $C = 9$, $L = 4$, and $F = 6$.

at an intermediate angle. Each of these arrivals reaches the observer at a different time, because for each the ray velocity is different. Thus he will see four different arrivals each at a different time. It is possible that the observer has a position in which he only will receive two pulses (close to either the x and z axis) or three pulses (direction of double point of second sheet). At the cusps the values of $|\Delta\psi|$ are small with respect to $\Delta\theta°$, hence the energy density is large. This follows immediately from the fact that x_i and z_i are stationary (see (14)) for the parameter $k(\theta_0)$ at a cusp. Thus an observer located in a direction corresponding to that of the cusps will observe particularly energetic second pulses. Their times of arrival are computed from the ray velocities, which are merely the lengths of the radius vectors SP_1, SP_2, SP_3, and SP_4. Thus the energy of the first type of wave (SP_4) arrives first and that of the second wave radiated at low angles (SP_1) arrives last.

The energy itself in any direction is computed from (28). When a direction θ is specified ($z = kx$), l and n are known ($k = \tan \theta = n/l$), so that γ_1 and γ_2 can be computed from (24) and α from (27) and λ from (26). Assuming initially uniform energy radiation in any direction, one can compute for any θ the energies E_1 and E_2 and the angles $\psi^{(1)}$ and $\psi^{(2)}$ of the corresponding ray direction.

In Figures 10–11, the results are shown for two cases. There are two horizontal scales: one for the E, the other for the angle ψ. The vertical scale is on θ. Note that when there is a cusp in the second wave front, the curve for $\psi^{(2)}$ has for some range of values of ψ more than one intercept with a vertical line (it is shaped like the letter S). Because the first wave front has no cusps, $\psi^{(1)}$ is uniformly increasing. The sum $E_1 + E_2$ is precisely unity so that both energy curves are each other's mirror image.

In Figures 12 and 13 we have plotted the relative amplitudes $\sqrt{E_1}{}^*$ and $\sqrt{E_2}{}^*$ (see equation (30)) for the various pulses the observer sees at his location. An estimate of these amplitudes is made from a computation of $\Delta\psi$ corresponding to varying ϕ in finite increments of 1°. Such an estimate will improve with choosing smaller increments. Somewhat arbitrarily (neglecting three-dimensional effects) the energy densities are defined:

$$E_1{}^* = \frac{E_1}{(\Delta\psi^{(1)})} \qquad E_2{}^* = \frac{E_2}{(\Delta\psi^{(2)})} \qquad (30)$$

On the assumption that the energies E_1 and E_2 are constant for values intermediate between ϕ_0 and $\phi_0 + 1°$, we have assigned the energy density

$$\frac{E_1 \, (\phi_0 + 1)}{|\psi^{(1)} \, (\phi_0 + 1) - \psi^{(1)} \, (\phi_0)|}$$

to the 'average' value of ψ, namely

$$\bar{\psi} = \frac{1}{2} \, [\psi^{(1)} \, (\phi_0) + \psi^{(1)} \, (\phi_0 + 1)]$$

This should give, away from the cusps in the second wave front, reasonable estimates of the energy densities, provided we have sufficiently small intervals in ϕ. At the cusps the value of $|\Delta\psi|$ can become arbitrarily small, so that one can expect at the cusps large amplitudes which, however, are extremely local in nature, because the total amount of energy passing in these directions must remain finite. For that reason it would have perhaps been better to plot the amount of energy passing through an angle $\Delta\psi$, which also could have been done, but was not.

Assuming that the observer is at a unit distance from the source, the times at which he observes pulses of energy are given by the inverses of the ray velocity. They are computed for the entire sequence of ϕ. For each ϕ, two corresponding ψ's are computed also and the values of $\Delta\psi^{(1)}$ and $\Delta\psi^{(2)}$ are determined together with the angle $\bar{\psi}$.

In Figures 12 and 13 we have plotted a horizontal bar starting at the point t_1, ½$[\psi^{(1)} \, (\phi_0 + 1) - \psi^{(1)} \, (\phi_0)]$, toward the left, of length $\sqrt{E_1^*}$; $\sqrt{E_2^*}$ makes a similar plot for the second type of wave front. The time scale is horizontal, and the ψ scale is vertical.

This type of plot serves the purpose of seeing at a glance the relative magnitude of the pulses and their arrival times for various values of ψ. The isotropic case would show only a single arrival of constant amplitude at angle ψ.

Some final remarks concerning the Figures 12-13 should be made. In most cases the first wave propagates most of the energy in a horizontal and vertical direction and rarely at an intermediate angle. The presence of cusps is indicated by the looped character of the second curve. The second wave front, which may arrive rather soon after passage of the first, transports most energy at the cusps.

Conclusions

The preceding discussion serves only as the merest introduction to the problem of wave propagation in anisotropic media. The purpose is to direct attention to some of its aspects that may be important in connection with present-day data processing methods.

In many respects an anisotropic medium behaves differently from an isotropic one:

1. The rays are no longer perpendicular to the wave fronts.

2. There will be several energy pulses.

3. T-x curves for a reflection off a horizontal plane are not hyperbolic.

A number of examples have been computed showing the relative energy content, relative arrival times, geometry of wave fronts, etc. At the present time we do not know whether the values of the elastic moduli used in these examples are significant for finely layered sedimentary rocks.

Acknowledgment. The author thanks the management of the Shell Development Company for permission to publish this work.

References

Berryman, J. S., Long-wave elastic anisotropy in transversely isotropic media, *Geophysics*, *44*, 896-917, 1979.

Kraut, E. A., Advances in the theory of anisotropic elastic wave propagation, *Rev. Geophys. Space Phys.*, *1*, 401-448, 1963.

Levin, F. K., Seismic velocities in transversely isotropic media, *Geophysics*, *44*, 918-936, 1979.

Levin, F. K., Seismic velocities in transversely isotropic media II, *Geophysics*, *45*, 3-17, 1980.

Love, M. A., *A Treatise on the Mathematical Theory of Elasticity*, 2nd ed., p. 643, Dover Publications, New York, 1944.

Postma, S. W., Wave propagation in a stratified medium, *Geophysics*, *20*, 780-806, 1955.

Rudzki, M. P., Parametrische Darstellung der elastischer Wellen in anisotropen Medien, *Bull. Acad. Cracovie*, *2*, 503, 1911.

White, J. E., *Seismic Waves*, McGraw Hill, New York, 1965.

A Discussion of the Approximation of Subsurface (Burial) Stress Conditions in Laboratory Experiments

R. A. NELSON

Amoco Production Co., Geological Research, Tulsa, Oklahoma 74102

Because the state of stress due to burial in the subsurface is distinctly anisotropic ($\sigma_v \neq \sigma_h$), the direct use of standard laboratory testing procedures in burial simulations is difficult if not misleading. Estimates of subsurface reservoir quality (porosity and permeability) can be made by either (1) performing new more sophisticated triaxial stress tests tracking either hydraulic fracture-derived stress gradients or uniaxial strain test-derived stress ratios for the rock, or (2) manipulating isotropic stress test data and anisotropic subsurface stress states to utilize existing, though somewhat inappropriate, permeability measurements at confining pressure. Reservoir permeability estimates can be made with reasonable accuracy from hydrostatic confining pressure tests by equating confining pressure and mean stress in the subsurface. Stress-strain curves derived from triaxial compression tests at constant confining pressure do not define the ductility of rocks at equilibrium in the subsurface. Indeed, at equilibrium, ductility is ill defined. These curves do, however, define the potential response of subsurface rocks in reaction to additional tectonic loads. The appropriate curves to use in modeling this response are dependent on the direction of the newly applied load with respect to the in situ anisotropic overburden stress components ($\sigma_v \neq \sigma_h$).

Introduction

Over the years, rock mechanics laboratory experiments have focused on the effect of individual environmental variables on rock behavior and deformation (confining pressure, temperature, strain rate, pore pressure, fluid chemistry, etc.). While attempts have been made to combine these variables [*Handin et al.*, 1963], the direct simulation of depth of burial has not been addressed in detail.

Depth of burial simulations are generally made by either hydrostatic confining pressure tests or by triaxial compression tests at constant confining pressure. Basic to the direct use of these tests is the assumption that the vertical and horizontal components of the subsurface state of stress are at least initially equal ($\sigma_v = \sigma_h$). However, all theory and in situ stress measurements indicate that this is not the case ($\sigma_v \neq \sigma_h$; see the discussion of 'subsurface stresses' in this report).

The faithful representation of existing subsurface stress states in depth simulations is not necessarily of paramount importance in general rock mechanics research. As *Handin et al.* [1963] (p. 753) point out

'One approach would simulate the natural environment of deeply buried rocks as realistically as possible in every test and this would have the advantage that a wealth of empirical data could be collected relatively rapidly. However, because of the complex interrelations among the significant variables, the favored method was to study the effects of these variables separately in order to acquire at least a qualitative understanding of how each contributed to the deformational behaviors of the rocks.'

Such faithful representation of subsurface stress states is, however, very important in the direct application of rock mechanics data in the mining and petroleum industries. The purpose of this paper is, therefore, to demonstrate how standard laboratory rock mechanics data can be manipulated to simulate more accurately the depth of burial, subsurface stress conditions, and rock response or behavior. While other variables such as temperature and pore fluid chemistry vary with depth, this paper will address the effect of stress state only.

Discussion

Types of Tests Generally Used to Simulate Depth of Burial

Changes in the mechanical behavior of rocks with increasing depth of burial or overburden are most frequently approximated in the laboratory with hydrostatic confining pressure tests or with triaxial compression tests at constant confining pressure. In the hydrostatic con-

TABLE 1. Total and Effective Burial Stress Ratios as Derived From Hydraulic Fracture Experiments

Reference	S_H/S_V, Average	Range	$\sigma_h/\sigma_v{}^*$, Average
Bredehoeft et al. [1976]			
Mesaverde			
Cameron 702	0.75	0.58–0.91	0.50
Cameron 704	0.77	0.75–0.78	0.54
Shell 23X-2	0.59	0.49–0.68	0.18
McGarr and Gay [1978]			
Average for all 16 with S_1 dipping 70° or more	0.56	0.24–0.82	0.12
Freidman and Heard [1974]	...		Maximum = 0.83
Texas Gulf Coast			
Howard and Fast [1970]			
Average for Texas, Louisiana, and Gulf Coast areas	0.70	...	0.40
Average for Oklahoma, Kansas, North Texas, and New Mexico areas	0.60	...	0.20

*Assuming $P_p = 1/2 S_v$ when not stated in the reference.

fining pressure test, a jacketed rock sample is subjected to equal and increasing external loads. These loads are applied by various geometries of fluid pressure and solid piston displacement. The tests are often run at elevated temperatures and pore pressures. Repeated measurements are frequently made of some petrophysical parameter of interest (pore volume, permeability, sonic velocity, compressibility, etc.) and plotted as a function of hydrostatic stress. These tests are the oldest and most frequently performed tests used to simulate reservoir rock properties with depth [*Fatt*, 1952; *McLatchie et al.*, 1958; *Dobrynin*,

1962; *Knutson and Bohor*, 1962; *Gray et al.*, 1963; *Nelson*, 1976; *Nelson and Handin*, 1977].

In the triaxial compression test, a fluid confining pressure is applied to the outside of a jacketed rock sample in a manner similar to the hydrostatic confining pressure test. From some initial equal external pressure condition (equal to confining pressure) an additional external load is generated by a pair of opposing steel pistons. In general, the fluid confining pressure is held constant throughout the test, while the piston load is gradually increased until either the strength of the material, piston travel, or the load capabilities of the apparatus are exceeded. Pore pressure and temperature can be varied as well as the displacement rate of the pistons.

In these tests the loads and displacements are measured and related to the deformation of the sample, usually by a stress-strain diagram. In recent years these tests have become more sophisticated with the addition of simultaneous petrophysical measurements. In these, changes in pore volume, axial permeability, and sonic velocity are determined incrementally during the test and related to the loading conditions. Triaxial compression tests are most often used to simulate both rock behavior during deformation and rock properties at depth.

Subsurface Stresses

The state of stress in the subsurface has often been modeled in the laboratory as hydrostatic (isotropic stress tensor). The in situ state of stress in the subsurface is, however, distinctly nonhydrostatic. This is shown either by hydraulic fracture data [*Cleary*, 1959; *Howard and Fast*, 1970] or by a simple elastic solution for rock subjected to the weight of overlying material [*Price*, 1959, 1966, 1974; *Teeuw*, 1971; *Hubbert and Willis*, 1972]. These common techniques will be discussed in the following paragraphs.

In hydraulic fracture treatments the vertical effective

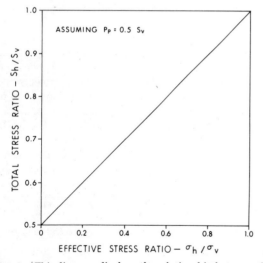

COMPARISON OF
TOTAL TO EFFECTIVE STRESS RATIOS

Fig. 1. This diagram displays the relationship between the total stress ratios (S_h/S_v) and effective stress ratios (σ_h/σ_v) assuming a 'normal' pore pressure ratio of 0.5 psi/ft (1.13×10^{-3} MPa/m).

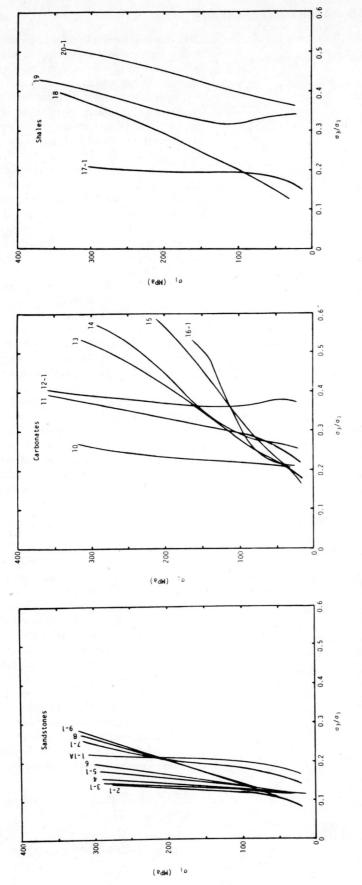

Fig. 2. These diagrams display uniaxial strain test data from *Rigert* [1980]. The data are plotted as stress ratio (σ_3/σ_1) as a function of axial stress (σ_1). Curves are shown for shales, carbonates, and sandstones. Within the sandstone data, porosity generally increases with sample number from 0.5 to 18.7%. For later discussions, the Nugget Sandstone sample shown in Figures 8–10 is curve number 4 on this diagram. For a further discussion of the samples shown, see *Rigert* [1980].

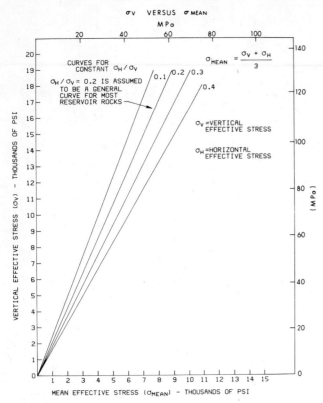

Fig. 3. This diagram displays mean effective stress in the subsurface as a function of vertical effective stress and effective stress ratio.

stress (σ_v) is assumed to be equal to the weight of the overburden minus the pore pressure, and the horizontal effective stress (σ_h) is calculated by subtracting reservoir pore pressure from the wellbore initial shut in pressure. Because the formation pressure or pore pressure (P_p) is often not accurately known at the time of treatment, results are usually presented as total stress components ($S_v = \sigma_v + P_p$, $S_h = \sigma_h + P_p$). Some published results are shown in Table 1. A correlation of total stress ratios (S_h/S_v) to effective stress ratios (σ_h/σ_v) assuming a 'normal' pore pressure gradient (0.5 psi/ft or 1.13 MPa/m) is given in Figure 1. As shown by the compiled data of Howard and Fast in Figure 1, most sedimentary rocks have effective stress ratios (σ_h/σ_v) between 0.2 and 0.4.

Elastic analysis showing the anisotropic nature of the state of stress due to burial is generally shown with the use of Poisson's ratio. As is shown in *Gray et al.* [1963], and *Hubbert and Willis* [1972] the horizontal stress component due to burial (σ_h) can be related to the vertical stress component (σ_v) by

$$\sigma_h = \left(\frac{\gamma}{1-\gamma}\right)\sigma_v \qquad (1)$$

where γ is Poisson's ratio, the ratio of transverse to longitudinal strain in compression tests. As used in *Gray et al.*

[1963], a Poisson's ratio of 0.25 yields a σ_h equal to 0.33 σ_v. Such a figure would mean that at any depth, the horizontal component of overburden stress would be one third of the vertical component.

If hydraulic fracture data for particular rocks in specific areas are not available, estimations of burial stress ratios can be made from specific laboratory experiments. Such experimentation is in fact necessary because a direct calculation of burial stress ratios from Poisson's ratio via (1) is not appropriate. The reason for this lies, I believe, in the measurement of Poisson's ratio in the laboratory.

Static Poisson's ratio is determined by shortening a sample in response to an applied axial load and allowing it to bulge or deflect laterally into the constant pressure confining media. The ratio is calculated by dividing the fractional change in transverse diameter of the sample by its fractional longitudinal shortening. In the subsurface, however, I assume that rocks are confined by other rigid rock and are not able to deform or strain substantially in the horizontal plane in response to overburden load. Therefore, rocks in the subsurface would tend to transmit stress and strain differently than in most laboratory experiments. Lateral displacements in response to burial alone must be small, if not zero. Poisson's ratio, which is calculated by allowing unconstrained lateral displacement under laboratory conditions, is therefore an inappropriate parameter to use in the calculation of the horizontal components of burial stress. To approximate the components of stress resulting from burial, compression tests could be performed with the transverse displacement of the sample held at zero. Such conditions are met in uniaxial strain tests [*Brace and Riley*, 1972]. These tests give idealized end member stress ratios that may approximate many burial stress conditions.

Data on uniaxial strain tests have been reported in *Price* [1966], *Brace and Riley* [1972], *Schock et al.* [1973], *Butters* [1974], and *Rigert* [1980]. A compilation of *Rigert's* [1980] data, which was initiated and funded by Amoco Production Company, is shown in Figure 2. Given the uniaxial strain assumption, these data give the maximum and minimum principal stress components and stress ratios likely for burial stresses alone in selected sandstones, limestones, and shales. In lieu of hydraulic fracture data in a formation of interest, these data could be used to approximate burial stress components in the subsurface. It must be emphasized that such data should be considered as the lower limiting values of burial stress ratios for the various rocks. Any residual tectonic load or lateral displacements would alter these values. A complete discussion of the data shown in Figure 2 will be left to Rigert (manuscript in preparation, 1980).

Depth of Burial Simulations Using Standards or Existing Laboratory Data

In the petroleum industry there are two major rock mechanics problems for which more accurate depth of burial

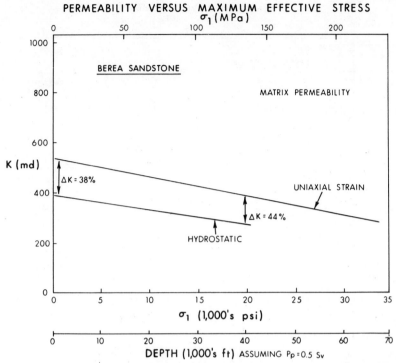

Fig. 4.   This diagram shows permeability (K) as a function of effective maximum principal stress (σ_1). Curves are shown for the Berea sandstone tested in a hydrostatic confining pressure test and in a uniaxial strain test. Equivalent depths are shown assuming maximum stress equals vertical stress equals 0.5 psi/ft (1.13×10^{-3} MPa/m).

simulations are needed. One involves the prediction of pore volume and fluid flow capabilities in subsurface reservoirs, while the other involves prediction of the relative deformational response (brittle versus ductile) of a particular suite of subsurface rocks. Predictions of each of these must be made somewhat differently, depending on the type of results desired.

Estimates of reservoir quality (porosity and permeability) with depth of burial and reservoir depletion. Standard laboratory techniques for estimating reservoir properties under subsurface loads are, as mentioned before, generally made under isotropic (hydrostatic) stress states. As also mentioned before, subsurface stress states are distinctly anisotropic ($\sigma_1 > \sigma_2 \geq \sigma_3$). The two alternative methods available for making more accurate depth simulations of porosity and permeability are (1) new measurements in more sophisticated triaxial stress tests tracking either hydraulic fracture-derived stress gradients for a particular area or uniaxial strain test-derived stress gradients for the rock, and (2) manipulation and incorporation of both isotropic stress test data and anisotropic subsurface stress data to recalculate appropriate depths on permeability/pressure curves. Of these two, the second is more practical.

Certainly new triaxial loading tests should be performed using logical stress conditions. However, the necessary equipment to do such testing with pore volume and per-

meability measuring capabilities is quite expensive and does not exist in many testing facilities. In addition, it is quite difficult to discard 25–30 years of accrued confining pressure/permeability data. It is imperative, therefore, to develop not only new testing procedures but also new techniques by which correlative points can be determined between the anisotropic stress state (depth) and the isotropic stress test data (confining pressure test). This section will discuss a possible technique of correlating such data.

Gray et al. [1963], *Wilhelmi and Somerton* [1967], and *Mordecai and Morris* [1971] have shown that there is a significant difference in matrix permeability measured under hydrostatic and triaxial loading conditions. Permeability reductions under triaxial loading conditions can be obtained by selecting appropriate equivalent pressure points on the hydrostatic loading curve for that same rock. For a consolidated rock, the hydrostatic pressure point that corresponds most closely to a particular triaxial state of stress is the calculated mean stress [$\sigma_{mean} = (\sigma_1 + 2\sigma_3)/3$].

In a nontectonic area, appropriate subsurface stress conditions can be predicted by using the measurements in Table 1. Assuming these stress ratios, Figure 3 depicts subsurface mean stress as a function of vertical effective stress (depth). This mean stress value can then be used as the corresponding pressure point on a matrix permeability

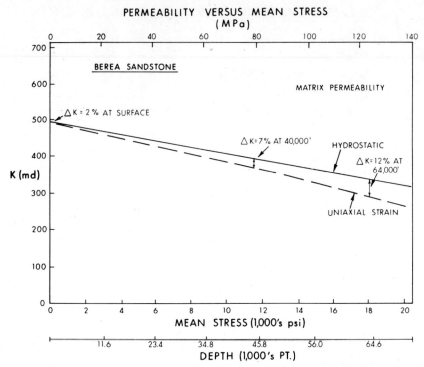

Fig. 5.   This diagram shows permeability (K) as a function of effective mean stress. Curves are shown for the same uniaxial strain and hydrostatic confining pressure tests shown in Figure 4.

versus confining pressure plot. Figures 4–7 show the results of this approach from experiments on Berea Sandstone.

These tests address the effect of stress state alone in burial simulations. The testing consisted of (1) a hydrostatic confining pressure test to 20,000 psi (138 MPa), (2) a uniaxial strain test to an axial stress of about 35,000 psi (241 MPa), and (3) a series of four triaxial compression tests at confining pressures of 2,000, 4,000, 8,000, and 10,000 psi (14, 28, 55, 69 MPa). Permeability (K) parallel to the piston load (σ_1) was measured incrementally in all tests.

Figure 4 displays a portion of these permeability data as a function of the maximum effective principal stress component (σ_1). Maximum effective principal stress is equated with vertical effective reservoir stress and, therefore, depth of burial by most workers ($\sigma_1 = \sigma_v \cong$ depth). By using this approach, Figure 4 shows a 38% to 44% higher permeability in the subsurface stress condition (uniaxial strain) than in the hydrostatic confining pressure test. Such discrepancies could adversely affect a petroleum play based on the direct use of confining pressure data alone.

Figure 5 presents this same permeability data as a function of mean stress. These curves show the correlation of permeabilities at correlative mean stress points between the hydrostatic and uniaxial loading conditions. At a simulated depth of 40,000 ft (12,192 m), calculated from one half the maximum principal stress component from the uniaxial strain curve, the discrepancy between the two curves is only 7% in comparison with a value at the same depth on Figure 4 of 44%. Differences at surface conditions are reduced from 38% to 2%. At a depth of about 65,000 ft (19,812 m) on Figure 5, the two curves are only 12% apart.

Plotting in terms of mean stress, therefore, gives us fairly consistent permeability values in the Berea Sandstone measured under drastically different loading conditions. This situation appears to be true for many brittle rocks. In rocks that deform in a ductile manner over a wide range of stress conditions, this somewhat fortuitous relationship generally does not hold. The conclusion is, then, that for rocks similar to the Berea Sandstone, we can equate a predicted or measured mean stress value at a depth in the subsurface with a confining pressure point on a permeability versus hydrostatic confining pressure curve and expect a reasonable representation of the reservoir permeability.

Estimates of permeability reduction with reservoir depletion can be made by assuming that pore or formation pressure reduction increases the effective stress values by an amount shown by the law of effective stress. Rock mechanics researchers usually assume the relationships $\sigma_v = S_v - P_p$ and $\sigma_h = S_h - P_p$ [Handin et al., 1963].

Figures 6 and 7 show the relative position of the triaxial compression test data in comparison with the hydrostatic and uniaxial strain curves. These tests are not correlative in depth with the previously discussed data. We see on Figure 6 that the triaxial compression tests plot inter-

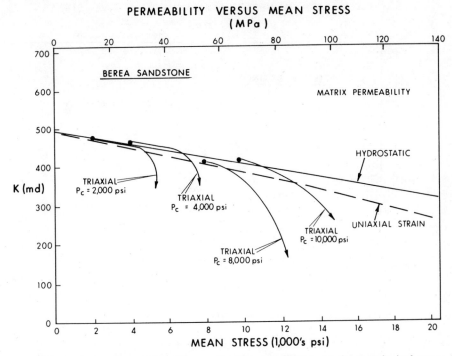

PERMEABILITY VERSUS MEAN STRESS
(MPa)

Fig. 6. This diagram shows permeability (*K*) as a function of effective maximum principal stress (σ_1). This plot shows the relative position of the curves for the triaxial compression tests at constant confining pressure (*Pc*) with respect to the uniaxial strain and hydrostatic confining pressure curves from Figure 4.

mediate to the others until ductile flow is initiated. This is indicated by the strong inflection points and nonlinear behavior of these curves. In Figure 7 the triaxial compression curves are, because of their initial hydrostatic stress state, initiated at the hydrostatic curve and reduced drastically in permeability due to large increases in stress difference. Because these curves represent constant confining pressure and increasing piston load, they can have little or no bearing on burial stresses. These curves can only represent permeability reductions in response to some additional tectonic deformation and not depth of burial alone.

Data manipulation using fracture system permeability must be made in a slightly different manner. These techniques will not be discussed in this paper.

Speculations on the estimation of relative deformational response with depth. In both structural geology and reservoir engineering it is often important to know the relative mechanical properties or deformational response of rocks due to locally changing stress states. The structural geologist may indeed be interested in determining which portions of the stratigraphic section will undergo compaction or dilation during tectonism; while the reservoir engineer is more interested in determining relative rock response during pore pressure depletion in a reservoir. I will briefly address the first of these two questions.

In other than tectonically active areas, subsurface rocks are at equilibrium. Rock materials in triaxial compression tests are, however, constantly deforming and are not in equilibrium. In such tests, deformation is continually taking place in response to the advancing pistons. It is, therefore, impossible to define directly the ductility of a rock at rest in the subsurface from such test data. This fact may seem obvious, but it is surprising how many people misinterpret plots of ductility versus depth such as those found in *Handin et al.* [1963] (pp. 748 and 749). While carefully not stated as such in the text, many readers imply from these diagrams that a rock at depth has some inherent ductility. This implication is, as shown below, incorrect.

Brittle and ductile response and the transition between the two are defined in a rock by the geometry of a series of stress-strain curves taken from triaxial compression test data. Because these terms are defined by the amount of permanent deformation or the degree of linearity or nonlinearity of these curves, they are ill-defined in a subsurface environment where no displacements are currently taking place. It is, therefore, improper to state that a particular sandstone is ductile at a 10,000 ft (3,048 m) depth of burial because its stress-strain diagram at 5,000 psi (34 MPa) confining pressure in a triaxial compression test shows ductile response above a certain axial stress. With no subsurface displacements, the rock plots at one point on this curve and is neither brittle nor ductile. What can be said is that with the addition of an external load to create disequilibrium, the rock has a potential for ductile response if the loads are sufficiently high.

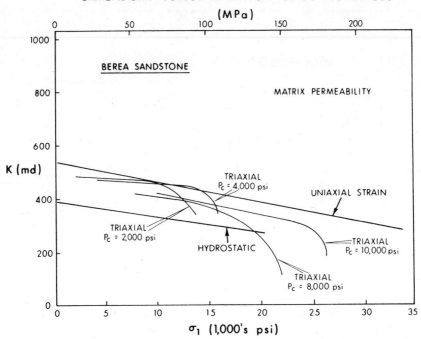

Fig. 7. This diagram shows permeability (K) as a function of effective mean stress. This plot shows the relative position of the curves for the triaxial compression tests at constant confining pressure (Pc) with respect to the uniaxial strain and hydrostatic confining pressure curves from Figure 5.

The problem in applying such laboratory experiments to subsurface conditions is further complicated by considering the interaction of burial and tectonic stress states. As discussed in the previous paragraph, the additional nonhydrostatic load necessary to cause disequilibrium and deformation can act in any direction. A variation in the maximum stress direction of the two stress states causes us to use different paths on the suite of stress-strain curves. The following is a speculation on one possible way of describing these differences. This approach ignores the effect of the intermediate effective principal stress (σ_2).

Figures 8 and 9 present data from triaxial compression and uniaxial strain tests performed on Nugget Sandstone core samples at Texas A&M University. If we assume an effective stress ratio (σ_h/σ_v) for this rock of 0.17 and a vertical effective reservoir stress of 30,000 psi (207 MPa), we calculate a horizontal effective reservoir stress of about 5,000 psi (34 MPa). Assuming that burial stresses are developed with depth along the uniaxial strain curve, point A on Figure 8 represents the equilibrium reservoir condition. This assumption appears reasonable owing to the good correlation between the principal stress ratios derived from hydraulic fracture measurements and uniaxial strain tests (see the discussion of 'subsurface stresses' in this report). This uniaxial vertical strain situation must, however, change during tectonic deformation (from which triaxial strains can be interpreted from deformed geometries). Deformation paths must, therefore, be simulated

by a combination of uniaxial strain and triaxial compression tests in the laboratory. Possible deformation paths from subsurface point A in response to either unidirectional vertical or horizontal tectonic loads are shown below (Figures 8 and 9, respectively).

The addition of a vertical tectonic load should move this rock from point A up along the nearly coincident triaxial compression and uniaxial strain curves until the constant confining pressure curve (equal to the horizontal effective reservoir stress (5,000 psi, 34 MPa)) departs from linearity. At this point the rock would deform along this triaxial compression test curve as shown by B, Figure 8.

The addition of a horizontal tectonic load, on the other hand, should at first drive this same reservoir equilibrium condition (point A, Figure 9) down to zero stress difference, possibly along the uniaxial strain curve, as the horizontal tectonic load approaches the vertical reservoir stress. From there additional horizontal load would cause this rock to deform along the triaxial compression curve at a constant confining pressure equal to the vertical effective reservoir stress (30,000 psi, 207 MPa) as shown by C, Figure 9.

This Nugget Sandstone experiences a brittle to ductile transition at a confining pressure between 30,000 and 45,000 psi (207 and 310 MPa). Samples were deformed by macroscopic fracturing at confining pressures lower than 30,000 psi (207 MPa), and by homogenous flow at 45,000 psi or higher (310 MPa), as shown on Figures 8 and 9. The

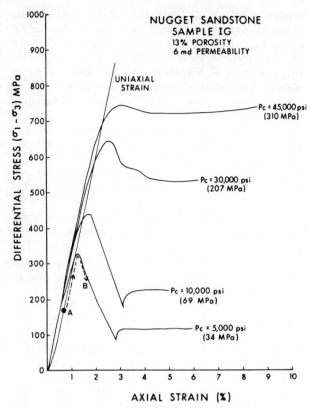

Fig. 8. This diagram is a plot of differential stress $(\sigma_1 - \sigma_3)$ versus axial strain. The curves represent a uniaxial strain test and a series of triaxial compression tests at constant confining pressure (Pc) all run on companion samples of Nugget Sandstone. Point A represents an assumed initial subsurface stress condition. Line A–B indicates a possible deformation path for this rock in response to the addition of a vertical tectonic load.

addition of either vertical or horizontal tectonic loads would cause this rock to deform in a brittle manner from its initial subsurface condition (point A). However, because the vertical effective reservoir stress is near the brittle to ductile transition, a moderate increase in reservoir depth or depletion in pore pressure might cause this rock to respond in a brittle manner in response to a vertical tectonic load, while in a ductile manner in response to a horizontal tectonic load.

The Nugget Sandstone data presented in Figures 8 and 9 show a close correspondence between the uniaxial strain test curve and the initial linear portions of the triaxial compression test curves. This correspondence is undoubtedly due to the elastic nature of this rock prior to failure. Other rock materials that do not exhibit an extended initial linear elastic response in triaxial compression tests will display a much poorer coincidence of curves than rocks like the Nugget Sandstone. In such rocks it may be necessary to test the material by first loading along the uniaxial strain curve to the predicted reservoir conditions and then to simulate an additional tectonic load; confining pressure should be held fixed at this horizontal reservoir stress con-

dition and a triaxial compression test should be performed from this point on. The difference between this stress-strain curve and that of the traditional triaxial compression test, which starts at an initial hydrostatic state of stress equal to the horizontal reservoir stress, should be a function of the ductility of the rock. This disparity in curves would, I believe, highlight a problem in strain path dependency in these rocks.

Whether or not these speculations on the simulation of relative deformational response at depth are correct, it is clear that the direct use of standard triaxial compression test data in such simulations is not straightforward. These tests give us good basic data on rock response. However, the application of these data in particular geological problems will require more specific analytical and testing procedures than are available at the present time.

Conclusions

On the basis of the material discussed in this report, several major conclusions can be drawn:

1. Standard laboratory testing procedures cannot be directly applied to depth of burial simulations. The application of these techniques requires either new testing pro-

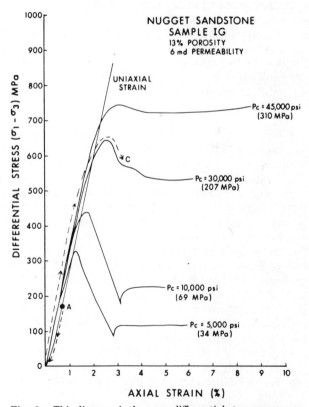

Fig. 9. This diagram is the same differential stress versus axial strain plot for Nugget Sandstone shown in Figure 8. Point A represents an assumed initial subsurface stress condition. Line A–C indicates a possible deformation path for this rock in response to the addition of a horizontal tectonic load.

cedures or manipulation of existing data in a form commensurate with the type of result desired.

2. Estimations of reservoir porosity and permeability with depth of burial should be made by tracking true subsurface stress states. These burial stress components can in part be derived either from hydraulic fracture gradients or, if necessary, from uniaxial strain test idealizations.

3. In lieu of permeability measurements under loading conditions tracking true subsurface stress states, correlative permeability values can be derived from hydrostatic confining pressure tests by equating hydrostatic confining pressure and mean stress in the subsurface.

4. Triaxial compression tests at constant confining pressure do not define the ductility of rocks in the subsurface. They do, however, indicate the potential response (brittle versus ductile) of the material if an additional tectonic load is applied. The appropriate curve to be used is dependent on the direction of the applied load with respect to the anisotropic overburden stress state ($\sigma_v > \sigma_h$).

Appendix

List of Symbols

σ_1 maximum principal effective stress component;
σ_2 intermediate effective principal stress component;
σ_3 minimum principal effective stress component;
σ_v vertical effective stress component;
σ_h horizontal effective stress component;
P_p pore or formation pressure;
γ Poisson's ratio;
S_v total vertical stress ($\sigma_v + P_p$); and
S_h total horizontal stress ($\sigma_h + P_p$).

Acknowledgments. I wish to thank Amoco Production Company for granting clearance of the material included in this report for publication. In addition, I wish to thank J. S. Bradley and S. Serra of Amoco Research and J. M. Logan and Mel Friedman of Texas A&M University and Hans Swolfs of the U.S. Geological Survey for their helpful suggestions and review of this manuscript. All of the experimental work reported was conducted for Amoco by the Center for Tectonophysics of Texas A&M University, College Station, Texas. The Berea sandstone tests were performed by Henrietta Gatto-Bauer and John M. Logan, both of Texas A&M University. The Nugget Sandstone tests were performed by James A. Rigert of Notre Dame University and John M. Logan of Texas A&M University. I would also like at this time to thank John Handin for his guidance while I attended Texas A&M and for his pioneering work in experimental rock deformation, which is the very foundation of much of what has been done since.

References

Brace, W. F., and D. K. Riley, Static uniaxial deformation of 15 rocks to 30 kb, *Int. J. Rock Mech. Mining Sci., 9,* 271-288, 1972.

Bredehoeft, J. D., R. G. Wolff, W. S. Keys, and E. Shuter, Hydraulic fracturing to determine the regional in situ stress field, Piceance Basin, Colorado, *Geol. Soc. Am. Bull., 87,* 250-258, 1976.

Butters, S. W., R. R. Nielson, A. H. Jones, and S. J. Green, Material properties of grouts and of tuffs from selected drill holes, *Rep. DNA3383F,* Terra Tek, Inc., Salt Lake City, Utah, 1974.

Cleary, J. M., Hydraulic fracture theory, Part III—Elastic properties of sandstones, *Circ. Illinois State Geol. Surv., 182,* 44, 1959.

Dobrynin, V. M., Effect of overburden pressure on some properties of sandstone, *J. Pet. Technol., 12,* 360-366, 1962.

Fatt, I., and D. H. Davies, Reduction in permeability with confining pressure, *Trans. Am. Inst. Mech. Eng., 195,* 329-330, 1952.

Friedman, M., and H. C. Heard, Principal stress ratios in Cretaceous limestones from Texas Gulf Coast, *Am. Assoc. Pet. Geol. Bull., 58,* 71-78, 1974.

Gray, D. M., I. Fatt, and G. Berganini, The effect of stress on permeability of sandstone cores, *Soc. Pet. Eng. J., 2,* 203, 1963.

Handin, J., R. V. Hager, Jr., M. Friedman, and J. N. Feather, Experimental deformation of sedimentary rocks under confining pressure: Pore pressure tests, *Am. Assoc. Pet. Geol. Bull., 47,* 717-755, 1963.

Howard, G. C., and C. R. Fast, *Hydraulic Fracturing, Monogr Ser.,* vol. 2, Society of Petroleum Engineers, Dallas, Tex., 1970.

Hubbert, M. K., and D. G. Willis, Mechanics of hydraulic fracturing, *Mem. Am. Assoc. Pet. Geol., 18,* 239-257, 1972.

Knutson, C. F., and B. F. Bohor, Reservoir rock behavior under moderate confining pressure, paper presented at 5th Symposium on Rock Mechanics, Am. Inst. of Mech. Eng., Minneapolis, Minn., May, 1962.

McGarr, A., and N. C. Gay, State of stress in the earth's crust, in *Annu. Rev. Earth Planet. Sci., 6,* 405-436, 1978.

McLatchie, A. S., R. A. Hemstock, and J. W. Young, The effective compressibility of reservoir rock and its effect on permeability, *Trans. Am. Inst. Mech. Eng., 213,* 386-388, 1958.

Mordecai, M., and L. N. Morris, An investigation into changes of permeability occurring in sandstones when failed under triaxial stress conditions, paper presented at 12th Symposium on Rock Mechanics, Am. Inst. of Mech. Eng., Rolla, Mo., 1971.

Nelson, R. A., An experimental study of fracture permeability in porous rock, paper presented at 17th Symposium on Rock Mechanics, Am. Inst. of Mech. Eng., Snowbird, Utah, Aug. 25-27, 1976.

Nelson, R. A., and J. W. Handin, An experimental study of fracture permeability in porous rock, *Bull. Am. Assoc. Pet. Geol., 61,* 227-236, 1977.

Price, N. J., Mechanics of jointing in rocks, *Geol. Mag., 46,* 149-167, 1959.

Price, N. J., *Fault and Joint Development in Brittle and Semi-Brittle Rock*, Pergamon, New York, 1966.

Price, N. J., The development of stress systems and fracture patterns in undeformed sediments, paper presented at the Third Congress on Rock Mechanics, Int. Soc. Rock Mech., Denver Colo., Sept. 1-7, 1974.

Rigert, J. A., Uniaxial and controlled-lateral strain tests on selected sedimentary rocks, Ph.D. dissertation, Texas A&M Univ., College Station, 1980.

Schock, R. N., H. C. Heard, and D. R. Stephens, Stress-strain behavior of a granodiorite and two graywackes on compression to 20 kilobars, *J. Geophys. Res.*, *78*, 5922-5941, 1973.

Teeuw, D., Predictions of formation compaction from laboratory compressibility data, *Soc. Pet. Eng. J.*, *11*, 263-271, 1971.

Wilhelmi, B., and W. H. Somerton, Simultaneous measurement of pore and elastic properties of rock under triaxial stress conditions, *Soc. Pet. Eng. J.*, *7*, 283-294, 1967.

Reflections on the Value of Laboratory Tests on Rocks

P. E. GRETENER

Department of Geology & Geophysics, University of Calgary, Alberta T2N 1N4, Canada

There is no doubt that laboratory tests and experiments under controlled conditions have contributed greatly to our understanding of geological processes. Such understanding is of importance both from a scientific and practical point of view. Apparent contradictions to field observations have usually been the result of overzealous interpretations of laboratory results without sufficient consideration for the limitations imposed by the simplified conditions under which such tests are performed.

John Handin, his collaborators, and his students have provided the geological fraternity with numerous laboratory tests on rocks over the past three decades. To make such tests requires special skill and highly sophisticated equipment. These is little point for all of us to repeat valid tests made by experimentalists whom one explicitly trusts just to verify this very point. However, it is usually impossible for the experimentalist to squeeze all the ramifications out of his tests and integrate them with other tests and, most important, reconcile the laboratory results with field observations. There is room for the synthesizers, the clients of the laboratory testers, and the following remarks are made in this capacity.

Extreme views exist on the value of laboratory tests for the science of geology. A small minority will deny them any value, claiming that the complex processes in nature can never be duplicated in the laboratory. At the other end of the scale an increasing number of geologists are willing to accept the last digit the computer spits out as significant. Clearly, neither point of view is rational and productive in terms of recognizing the laboratory experiment as an integral part of modern geological research.

Let me first examine the view that a laboratory experiment can never simulate natural conditions. This view rests on the misunderstanding that experimentalists are attempting to reproduce in the laboratory specific geological structures as found in nature. Certainly this was true to some extent in the early days of geological modeling, but it has long since been abandoned as a hopeless task and an incorrect approach. With the recognition of the fact that any specific geological structure is always the result of both fundamental processes and incidental conditions [*Dahlstrom*, 1970], any such approach is doomed. Only the fundamental processes can be tested, and since the final result depends so much on uncontrollable incidental conditions, the modeling of specific structures amounts to the futile task of reproducing both the 'signal and the noise.' In reality, modern experiments are under-

taken because one precisely recognizes the fact that nature is a poor experimenter. Not only the incidental conditions tend to obscure the final result, but, at the same time, nature has a way of always changing several parameters at once, which tends to obliterate the effect of any individual contribution. The laboratory experiment that investigates the sole effect of confining pressure [*Handin and Hager*, 1957] is deliberately unrealistic. Generally, in nature, confining pressure is increased by additional burial: sedimentary or tectonic. In reality this cannot happen without increasing the temperature. Thus, in nature, both confining pressure and temperature increase (or decrease) together. But the question arises: which contributes what? After all, whereas burial produces a reasonably fixed increase in confining pressure (also not universally true), the normal geothermal gradient in a sedimentary sequence can vary by a factor of almost 3. Both hot and cool burial are possible. For any reasonable forecast of mechanical behavior with depth, the relative contributions of both effects must be known and separately investigated [*Handin and Hager*, 1957, 1958]. The modern laboratory experiments deliberately deviate from natural boundary conditions, and the objection that such tests are carried out under unrealistic conditions is just no argument. The tests are not supposed to be realistic. In fact 'realistic tests' that are designed to simulate specific structures are frowned upon because they usually achieve this goal only under uncontrolled conditions and, therefore, add little to our understanding of geological processes.

At the other end of the scale we find an increasing number of geologists who are suitably impressed by the sophisticated nature of most electronically controlled modern laboratory equipment. They accept all numerical values obtained as gospel truth. This, in fact, may be the more dangerous of the extreme attitudes, and it is often the underlying cause for apparent contradictions between laboratory results and field observations. In the following a number, though not all, of the problems are discussed that put limitations on the direct application of laboratory test results to large-scale rock bodies.

1. *The problem of the avoided flaw.* Somebody once said (the name escapes me) that rock mechanics is the science of the flaws: the pores, the micro- and macrodiscontinuities, or simply the imperfections in general. Modern research has fully confirmed this view. The laboratory sample seldom exceeds a few cubic centimeters in size. As such it does contain microflaws, and their effects are well understood today. In fact it is one of the major achievements of modern laboratory testing to have illuminated this subject matter. However, it is equally obvious that such laboratory samples are free of macroflaws by definition.

Before pursuing this subject any further it becomes necessary to define the sphere of interest of the writer. As a geophysicist and geologist, long associated with the oil industry, I can say that broadly speaking my attention is focused on the top 10 km of the earth's skin. Rocks now or previously at a greater depth are generally not of interest to the oil explorationist since they have lost their porosity and the hydrocarbons have been 'overcooked.' The same is also true for my scientific interest in the structural deformation of the outer parts of thrust belts where the structural deformation is generally confined to the uppermost 10 km. It is well known that the importance of macroflaws will diminish with depth. However, at 10 km the effective overburden stress is only about 1.5 kbar (150 MPa) and the temperature usually less than 400°C. In this realm, macro-discontinuities are undoubtedly of great importance for the transmission of seismic energy, the mechanical strength, and the flow of subsurface fluids, to name just a few. The case of the Denver earthquakes [*Evans*, 1966] and the subsequent Rangely experiments of the USGS [*Raleigh et al.*, 1972] demonstrate this quite clearly.

In tests on laboratory samples the avoidance of macroflaws must be accounted for when translating the results to large rock masses. Engineers have long recognized this. In fact for mechanical strength they apply the size factor to account for the observation that strength diminishes drastically as sample size increases. Equally, the long and ongoing debate about the tensile strength of rocks is a futile one. Even small laboratory samples usually have a negligible tensile strength of about 1/20 of the compressive strength. Larger rock masses, because of the ever present macroflaws, must have zero tensile strength, a view long advocated by King Hubbert.

This contemplation in no way diminishes the validity of sample testing, but it makes it clear that the resulting numerical values cannot be applied to large rock masses without certain correction factors.

2. *The problem of the sample bias.* Personally I am much interested in the sedimentary sequence, a stack of layers of strongly varying mechanical properties. The total response to a given boundary condition can only be assessed when the properties of all individual layers are sufficiently well known. Generally, carbonates and sandstones can be sampled much easier than the intervening shales. As a result our knowledge is strongly biased toward these 'beams' in the sedimentary sequence, with a dire lack of good data on such elusive rocks as shales and coals. Yet it becomes more and more obvious that it is the behavior of these weak and impermeable rocks that governs the overall response of the sequence in terms of mechanical deformation as well as fluid flow.

3. *The problem of heterogeneity and anisotropy.* Large rock masses are invariably heterogeneous, and often also anisotropic. The effect of such quantities is obvious when one thinks of the Hubbert sandbox [*Hubbert*, 1951]. It ingeniously demonstrates the formation of normal and thrust faults in principle. However, due to the isotropic and homogeneous nature of the sand, the thrust faults lack the stepping so typical for their appearance in the anisotropic sedimentary sequence and first shown by *Rich* [1934]. The same of course is true for the *Hafner* [1951] model, which also is based on a homogeneous and isotropic material. Inhomogeneity and anisotropy will severely modify any results obtained in a laboratory test under conditions of isotropy and homogeneity. The laboratory experiment can, however, only become misleading when its limitations are deliberately neglected.

4. *The problem of the boundary conditions.* Many papers have been written on this subject. The scale in space has already been discussed under 1. But it is equally impossible to model certain geological processes at their natural speed. Many attempts have been made to assess the effect of strain rates. High to medium strain rates are applied, followed by an extrapolation on a logarithmic scale that makes the conclusions more palatable. However, it is generally recognized that the time factor introduces another element of uncertainty into the evaluation of laboratory tests.

Triaxial testing employs a fixed confining pressure. Engineers, in their computation of virgin subsurface stresses, often assume lateral confinement of the sequence. Is the laboratory test appropriate? In view of the fact that rocks in nature do fail by faulting, which contradicts rigid lateral confinement, one can say that the triaxial test is in order. In reality the term 'triaxial testing' is a misnomer, insofar as both the intermediate and the least principle compressive stresses are equal. This, of course, means that the radial orientation of the fracture plane is controlled by an incidental flaw rather than the stress ellipsoid, which in this case is of rotational symmetry. In terms of actual strength it is assumed (and has been shown) that the value of the intermediate compressive stress is of little significance.

5. *The problem of sample preparation and extraction.* For the surface samples the degree of freshness (lack of weathering) is critical. Samples may also become

nonrepresentative because of improper sampling procedures. Samples obtained from quarries may contain additional flaws introduced by the process of blasting. An extreme case are the sidewall cores extracted by brute force and well known to be only poorly representative of the undisturbed rock. Equally, samples from deep cores may have developed micro- or macroflaws, owing to stress relief. A case in question is the discing of diamond cores from holes drilled into the highly stressed rocks in advance of deep mine addits.

Despite all these limitations, laboratory tests have provided valuable insights into rock behavior. In the following a few selected examples are chosen. Clearly, the biased interest of the reviewer is apparent in the choice.

From a practical as well as scientific point of view, one of the most important mechanical properties of rocks is the case of brittle versus ductile behavior. Brittle deformation leaves behind a fractured permeable and porous rock. Ductile deformation results in a dense rock.

Commercial concentrations of oil and gas, as well as base metals, require the circulation of subsurface fluids. In fact, many lead-zinc, copper, and like deposits are comparable to tar sands, migrated in a liquid condition and now fixed as solid deposits. Oil and gas differ only insofar as they have retained their fluid nature. For the accumulation of natural resources the migration of subsurface fluids is a necessity, which in turn focuses our attention on the retention of porosity and permeability. In the case of the still fluid deposits of oil and gas, preservation of minimum porosity and permeability, to this day, is a must in order to permit commercial recovery.

Many of the major oil/gas reservoirs, such as the Asmari/Qum limestone in Iran, the Reforma limestone in Mexico, and the Paleozoic carbonate reservoirs of the southern Canadian Rockies are fractured. Brittle rock behavior is the key to their commercial functioning.

In the particular case of the southern Canadian Rockies the reservoirs are secondary dolomite, such rock being far more brittle than the original limestones [Handin et al., 1963]. Tight folding of the frontal end (leading edge) of the Paleozoic thrust slivers produced the extension fractures that run parallel to the axes of these elongate fields with length-to-width ratios of about 10 to 1. The fractures provide the necessary axial permeability, which is the key to commercial production. Nondolomitized sections are usually noncommercial. By deduction one concludes that dolomitization preceded the Laramide deformation.

In addition, one must point out the important fact that fractures oriented in a near-vertical position to the bedding planes tend to homogenize the fluid flow in any reservoir. Due to the intercalations of shaly, less permeable, streaks, the natural permeability parallel to bedding is often 1 or more orders of magnitude higher than the one perpendicular to bedding. It is the fractures that alleviate this deficiency and permit a more effective overall drainage of the reservoirs.

In the light of such considerations the paper by *Handin et al.* [1963] must be considered a milestone. In their Figure 27 (p. 749) they relate ductility for different rock types to confining pressure (or depth). Later, the depth scale had to be modified in light of the concept of effective stress and abnormal pore pressures. Abnormally high pore pressures can extend the realm of brittle failure far beyond what was originally envisaged, as pointed out by *Secor* [1965]. The enormous significance of brittle versus ductile deformation for both the scientific and exploration geologists needs no further elaboration.

It is without question that the laboratory testing has contributed much to focus our attention on this concept and all its practical and theoretical ramifications.

Already mentioned above as closely related is the concept of effective stress. Initially developed for unconsolidated materials by Terzaghi in 1923, it was firmly introduced into geological thinking by *Hubbert and Rubey* [1959]. (In fairness it should be mentioned that *Goguel* [1952] clearly formulated the concept for rocks at a much earlier date.) The adoption of this concept once more constitutes the recognition that rocks are not just assemblages of minerals (as still taught in most universities) but are porous materials full of flaws. The openings are usually filled with brine under a certain pressure, referred to as pore pressure. The concept states that any total outside stress S applied to such a system is supported jointly by the matrix (effective) stress σ and the pore pressure p:

$$S = \sigma + p \qquad \sigma = S - p$$

Numerous laboratory tests over the last two decades were concerned with this concept. Today, we can say with confidence that the concept as stated above is valid for the case of shear failure. The introduction of this concept into geological thinking has, in my mind, proved more fruitful than the more glamorous and increasingly popular concept of plate tectonics. Effective stress, or rather the lack of it, explains such features as solid intrusions, both magmatic and sedimentary; it opens vistas for the possible mechanical understanding of large overthrusts; it deepens the realm of brittle fracturing; and it led to the idea that movements on earthquake-producing faults may eventually be controlled by man. The basic understanding has been produced by numerous laboratory experiments that have been carried out in the past two decades. The early paper by *Handin et al.* [1963] has already been mentioned. *Brace*'s [1968] paper on dilatancy hardening must be rated as another classic. The rare (expensive) field test, such as the one at Rangely [*Raleigh et al.*, 1972], has fully vindicated *Evans'* [1966] explanation of the Denver earthquakes, and at the same time has firmly established the validity of the concept of effective stress as a working tool.

In fact this latter episode deserves a bit of elaboration. *Hubbert and Rubey* [1960] have shown, in a most elegant manner, how an increase in pore pressure can result in failure by shifting the Mohr circle toward the failure envelope. When incorporating into their graph the concept of preexisting discontinuities [Gretener, 1972], it easily follows that movement on such planes can be reactivated in exactly the same manner. Thus the 'Hubbert and Rubey Theory,' supported by numerous laboratory tests, predicted what Evans [1966] suggested had happened in nature, and the USGS experiments at Rangely [Raleigh et al., 1972] fully confirmed one of the most remarkable chains of progress in the history of geology. The moral of the story is twofold:

1. Good theory, good experimentation, and good field observations never clash. Apparent disagreements always were prompted by the incomplete appreciation of the limitations of one of the approaches, as for instance shown by Brace [1968] for the case of dilatancy hardening.

2. In situ experiments are the ultimate test and the dream of any geological experimentalist. Costs are usually prohibitive and can only be justified when commercial interests are at stake (the Rangely experiments are the rare exception). This is the compelling reason for scientists and practitioners to collaborate. Often industrial tests (well logging, well fracing, experimental addits, mine pillar surveillance, etc.) can be made far more meaningful by simply recording a few more details than are necessary for the practical aspects of the job, just a little more human effort with virtually no additional costs. It is absolutely foolish for the scientific community to stick to 'pure science' and not ride the coattails of industry. The benefits of cooperation are immense for both parties.

From a practical point of view we still stand at the beginning. The recognition that abnormally high pore pressures may be the rule rather than the exception indicates that the effective overburden stress may be far more variable than initially believed. What is the effect of high pore pressure and consequently low overburden stress on such properties as porosity and permeability? Amos Nur and his students are busily engaged in evaluating this question [Sprunt and Nur, 1977; Walls and Nur, 1980]. Again, the results will be of equal importance to the academic and the explorationist in geology.

Some 30 years ago, a young John Handin made the decision to follow in the footsteps of his teacher, the late Dave Griggs, and devote himself to the laboratory testing of rocks. Was it a wise decision? In my humble opinion as a practitioner and teacher of geology and geophysics the answer can only be an unqualified 'yes.' One of the satisfactions in teaching is the feeling of progress, to be able to explain a few observations that were classified as mysteries during one's own attendance at university. Today we are in this fortunate position, and the laboratory testing of rocks, since its infancy in my student days, has played no small part in this. Thank you, John!

References

Brace, W. F., The mechanical effects of pore pressure on fracturing of rocks, Geol. Surv. Pap., Geol. Surv. Can., 68-52, 113-124, 1968.

Dahlstrom, C. D. A., Structural geology in the eastern margin of the Canadian Rocky Mountains, Bull. Can. Pet. Geol., 18(3), 332-406, 1970.

Evans, D. M., The Denver area earthquakes and the Rocky Mountain Arsenal disposal well, Mt. Geol., 3(1), 23-36, 1966.

Goguel, J., Traité de Tectonique, p. 383, Masson et Cie, Paris, 1952.

Gretener, P. E., Thoughts on overthrust faulting in a layered sequence, Bull. Can. Pet. Geol., 20(3), 583-607, 1972.

Hafner, W., Stress distribution and faulting, Geol. Soc. Am. Bull., 62(4), 373-398, 1951.

Handin, J., and R. V. Hager, Jr., Experimental deformation of sedimentary rocks under confining pressure: Tests at room temperature on dry samples, Am. Assoc. Pet. Geol. Bull., 41(1), 1-50, 1957.

Handin, J., and R. V. Hager, Jr., Experimental deformation of sedimentary rocks under confining pressure: Tests at high temperature on dry samples, Am. Assoc. Pet. Geol. Bull., 42(12), 2892-2934, 1958.

Handin, J., R. V. Hager, Jr., M. Friedman, and J. N. Feather, Experimental deformation of sedimentary rocks under confining pressure: Pore pressure tests, Am. Assoc. Pet. Geol. Bull., 47(5), 717-755, 1963.

Hubbert, M. K., Mechanical basis for certain familiar geologic structures, Geol. Soc. Am. Bull., 62(4), 355-372, 1951.

Hubbert, M. K., and W. W. Rubey, Role of fluid pressure in mechanics of overthrust faulting, I, Mechanics of fluid-filled porous solids and its application to overthrust faulting, Geol. Soc. Am. Bull., 70(2), 115-166, 1959.

Hubbert, M. K., and W. W. Rubey, Role of fluid pressure in mechanics of overthrust faulting—A reply, Geol. Soc. Am. Bull., 71(5), 611-628, 1960.

Raleigh, C. B., J. H. Healy, and J. D. Bredehoeft, Faulting and crustal stress at Rangely, Colorado, Flow and Fracture of Rocks, Geophys. Monogr. Ser., vol. 16, edited by Heard et al., pp. 275-284, AGU, Washington, D. C., 1972.

Rich, J. L., Mechanics of low-angle overthrust faulting as illustrated by Cumberland thrust block, Virginia, Kentucky, and Tennessee, Am. Assoc. Pet. Geol. Bull., 18(12), 1584-1596, 1934.

Secor, D. T., Role of fluid pressure in jointing, Am. J. Sci., 263, 633-646, 1965.

Sprunt, E. S., and A. Nur, Destruction of porosity through pressure solution, Geophysics, 42(4), 726-741, 1977.

Walls, J., and A. Nur, The effects of pore pressure and confining pressure on permeability of sandstones (abstr.), Geophysics, 45(3), 1980.